Econophysics and Sociophysics

Edited by
*Bikas K. Chakrabarti, Anirban Chakraborti,
and Arnab Chatterjee*

Related Titles

Ashenfelter, O., Levine, P.B., Zimmermann, D.J.
**Statistics and Econometrics:
Methods and Applications**
320 pages
2006
Softcover
ISBN 0-470-00945-4

Koop, G.
Analysis of Economic Data
224 pages
2004
Softcover
ISBN 0-470-02468-2

DeGroot, M. H.

Optimal Statistical Decisions
WCL Edition
489 pages
2004
Softcover
ISBN 0-471-68029-X

Econophysics and Sociophysics

Trends and Perspectives

Edited by
Bikas K. Chakrabarti, Anirban Chakraborti, and Arnab Chatterjee

WILEY-VCH Verlag GmbH & Co. KGaA

The Editors

Bikas K. Chakrabarti
Theoretical Condensed Matter Physics Division/
Centre for Applied Mathematics and
Computational Science
Saha Institute of Nuclear Physics
1/AF Bidhannagar, Kolkata 700 064
India
bikask.chakrabarti@saha.ac.in

Anirban Chakraborti
Department of Physics
Banaras Hindu University
Varanasi 221 005
India
achakraborti@yahoo.com

Arnab Chatterjee
Theoretical Condensed Matter Physics Division/
Centre for Applied Mathematics and
Computational Science
Saha Institute of Nuclear Physics
1/AF Bidhannagar, Kolkata 700 064
India
arnab.chatterjee@saha.ac.in

Cover
Simulation of pedestrian streams in the city center of Dresden, Germany (Copyright by Landeshauptstadt Dresden, Städtisches Vermessungsamt, Dresden, Germany).

All books published by Wiley-VCH are carefully produced. Nevertheless, authors, editors, and publisher do not warrant the information contained in these books, including this book, to be free of errors. Readers are advised to keep in mind that statements, data, illustrations, procedural details or other items may inadvertently be inaccurate.

Library of Congress Card No.:
applied for

British Library Cataloguing-in-Publication Data
A catalogue record for this book is available from the British Library.

Bibliographic information published by the Deutsche Nationalbibliothek
The Deutsche Nationalbibliothek lists this publication in the Deutsche Nationalbibliografie; detailed bibliographic data is available in the Internet at <http://dnb.ddb.de>.

Printing Strauss GmbH, Mörlenbach

Binding Littges & Dopf Buchbinderei GmbH, Heppenheim

© 2006 WILEY-VCH Verlag GmbH & Co. KGaA, Weinheim

All rights reserved (including those of translation into other languages). No part of this book may be reproduced in any form – by photoprinting, microfilm, or any other means – nor transmitted or translated into a machine language without written permission from the publishers. Registered names, trademarks, etc. used in this book, even when not specifically marked as such, are not to be considered unprotected by law.

Printed in the Federal Republic of Germany
Printed on acid-free paper

ISBN-13: 978-3-527-40670-8
ISBN-10: 3-527-40670-0

Contents

Preface *XIX*

List of Contributors *XXIII*

1	**A Thermodynamic Formulation of Economics** *1*	
	Juergen Mimkes	
1.1	Introduction *1*	
1.2	Differential Forms *2*	
1.2.1	Exact Differential Forms *2*	
1.2.2	Not Exact Differential Forms *3*	
1.2.3	The Integrating Factor *4*	
1.2.4	The First and Second Law of Differential Forms *5*	
1.2.5	Not Exact Differential Forms in Thermodynamics and Economics *5*	
1.3	The First Law of Economics *6*	
1.3.1	The First Law: Capital Balance of Production *6*	
1.3.2	Work (W) *7*	
1.3.3	Surplus (ΔQ) *7*	
1.3.4	Capital (E) *8*	
1.4	The Second Law of Economics *8*	
1.4.1	The Second Law: Existence of a System Function (S) *8*	
1.4.2	The Integrating Factor (T) *8*	
1.4.3	Entropy and Production Function (S) *9*	
1.4.4	Pressure and Personal Freedom *9*	
1.4.5	The Exact Differential ($dS(T,V)$) *9*	
1.4.6	The Maxwell Relation *10*	
1.4.7	Lagrange Function *10*	
1.5	Statistics *11*	
1.5.1	Combinations *11*	
1.5.2	Normal Distribution *11*	

1.5.3	Polynomial Distribution *11*
1.5.4	Lagrange Function in Stochastic Systems *12*
1.5.5	Boltzmann Distribution *13*
1.6	Entropy in Economics *16*
1.6.1	Entropy as a Production Function *16*
1.6.2	Entropy of Commodity Distribution *17*
1.6.3	Entropy of Capital Distribution *19*
1.6.4	Entropy of Production *20*
1.6.5	Summary of Entropy *21*
1.7	Mechanism of Production and Trade *21*
1.7.1	The Carnot Process *21*
1.7.2	The Origin of Growth and Wealth *23*
1.7.3	World Trade Mechanism *25*
1.7.4	Returns *26*
1.8	Dynamics of Production: Economic Growth *27*
1.8.1	Two Interdependent Systems: Industry and Households *27*
1.8.2	Linear and Exponential Growth ($0 < p < 0.5$) *28*
1.8.3	Trailing Economies: USA and Japan ($0.5 < p < 1$) *29*
1.8.4	Converging Economies, West and East Germany ($p > 1$) *29*
1.9	Conclusion *29*
	References *33*

2	**Zero-intelligence Models of Limit-order Markets** **35**
	Robin Stinchcombe
2.1	Introduction *35*
2.2	Possible Zero-intelligence Models *39*
2.3	Data Analysis and Empirical Facts Regarding Statics *41*
2.4	Dynamics: Processes, Rates, and Relationships *45*
2.5	Resulting Model *49*
2.6	Results from the Model *50*
2.7	Analytic Studies: Introduction and Mean-field Approach *51*
2.8	Random-walk Analyses *54*
2.9	Independent Interval Approximation *59*
2.10	Concluding Discussion *60*
	References *62*

3	**Understanding and Managing the Future Evolution of a Competitive Multi-agent Population** **65**
	David M.D. Smith and Neil F. Johnson
3.1	Introduction *65*
3.2	A Game of Two Dice *67*
3.3	Formal Description of the System's Evolution *77*

3.4	Binary Agent Resource System	*81*
3.5	Natural Evolution: No System Management	*83*
3.6	Evolution Management via Perturbations to Population's Composition	*87*
3.7	Reducing the Future–Cast Formalism	*91*
3.8	Concluding Remarks and Discussion	*95*
	References	*97*

4 Growth of Firms and Networks *99*

Yoshi Fujiwara, Hideaki Aoyama, and Wataru Souma

4.1	Introduction	*99*
4.2	Growth of Firms	*101*
4.2.1	Dataset of European Firms	*101*
4.2.2	Pareto–Zipf's Law for Distribution	*103*
4.2.3	Gibrat's Law for Growth	*103*
4.2.4	Detailed Balance	*105*
4.3	Pareto–Zipf and Gibrat under Detailed Balance	*107*
4.3.1	Kinematics	*108*
4.3.2	Growth of Firms and Universality of Zipf's Law	*109*
4.4	Small and Mid-sized Firms	*113*
4.4.1	Data for Small and Mid-sized Firms	*113*
4.4.2	Growth and Fluctuations	*114*
4.5	Network of Firms	*117*
4.5.1	Shareholding Networks	*117*
4.5.1.1	Degree Distribution	*117*
4.5.1.2	Correlation Between Degree and a Firm's Growth	*119*
4.5.1.3	Simple Model of Shareholding Network	*120*
4.5.2	Business Networks	*122*
4.5.2.1	Bankruptcy of Firms	*122*
4.5.2.2	Distribution of Debt	*123*
4.5.2.3	Lifetime of Bankrupted Firms	*124*
4.5.2.4	Chain of Bankruptcy and Business Network	*125*
4.6	Conclusion	*127*
	References	*128*

5 A Review of Empirical Studies and Models of Income Distributions in Society *131*

Peter Richmond, Stefan Hutzler, Ricardo Coelho, and Przemek Repetowicz

5.1	Introduction	*131*
5.2	Pareto and Early Models of Wealth Distribution	*132*
5.2.1	Pareto's Law	*132*

5.2.2	Pareto's View of Society *135*	
5.2.3	Gibrat and Rules of Proportionate Growth *137*	
5.2.4	The Stochastic Model of Champernowne *138*	
5.2.5	Mandelbrot's Weighted Mixtures and Maximum Choice *139*	
5.3	Current Studies *140*	
5.3.1	Generalized Lotka–Volterra Model *141*	
5.3.2	Family Network Model *143*	
5.3.3	Collision Models *145*	
5.4	A Case Study of UK Income Data *148*	
5.5	Conclusions *157*	
	References *158*	

6 Models of Wealth Distributions – A Perspective *161*
Abhijit Kar Gupta

6.1	Introduction *161*
6.2	Pure Gambling *164*
6.3	Uniform Saving Propensity *166*
6.4	Distributed Saving Propensity *169*
6.4.1	Power Law from Mean-field Analysis *171*
6.4.2	Power Law from Reduced Situation *172*
6.5	Understanding by Means of the Transition Matrix *173*
6.5.1	Distributions from the Generic Situation *177*
6.6	Role of Selective Interaction *180*
6.7	Measure of Inequality *183*
6.8	Distribution by Maximizing Inequality *185*
6.9	Confusions and Conclusions *187*
	References *189*

7 The Contribution of Money-transfer Models to Economics *191*
Yougui Wang, Ning Xi, and Ning Ding

7.1	Introduction *191*
7.2	Understanding Monetary Circulation *194*
7.2.1	Velocity of Money Circulation and its Determinants *194*
7.2.2	Holding Time versus Velocity of Money *196*
7.2.2.1	From Holding Time to Velocity of Money *196*
7.2.2.2	Calculation of Average Holding Time *198*
7.2.3	Keynesian Multiplier versus Velocity of Money *199*
7.3	Inspecting Money Creation and its Impacts *201*
7.3.1	Money Creation and Monetary Aggregate *202*
7.3.1.1	A Simplified Multiplier Model *202*
7.3.1.2	A Modified Money-transfer Model *203*
7.3.2	Money Creation and Velocity of Money *205*

7.3.2.1	Velocity of Narrow Money	205
7.3.2.2	Velocity of Broad Money	208
7.4	Refining Economic Mobility	210
7.4.1	Concept and Index of Measurement	211
7.4.2	Mobility in a Money-transfer Model	212
7.4.3	Modification in the Measurement Index	214
7.5	Summary	216
	References	216
8	**Fluctuations in Foreign Exchange markets**	**219**
	Yukihiro Aiba and Naomichi Hatano	
8.1	Introduction	219
8.2	Modeling Financial Fluctuations with Concepts of Statistical Physics	220
8.2.1	Sznajd Model	220
8.2.2	Sato and Takayasu's Dealer Model	222
8.3	Triangular Arbitrage as an Interaction among Foreign Exchange Rates	225
8.4	A Macroscopic Model of a Triangular Arbitrage Transaction	228
8.4.1	Basic Time Evolution	230
8.4.2	Estimation of Parameters	231
8.5	A Microscopic Model of Triangular Arbitrage Transaction	236
8.5.1	Microscopic Model of Triangular Arbitrage: Two Interacting ST Models	237
8.5.2	The Microscopic Parameters and the Macroscopic Spring Constant	240
8.6	Summary	246
	References	246
9	**Econophysics of Stock and Foreign Currency Exchange Markets**	**249**
	Marcel Ausloos	
9.1	A Few Robust Techniques	251
9.1.1	Detrended Fluctuation Analysis Technique	251
9.1.2	Zipf Analysis Technique	254
9.1.3	Other Techniques for Searching for Correlations in Financial Indices	255
9.2	Statistical, Phenomenological and "Microscopic" Models	258
9.2.1	ARCH, GARCH, EGARCH, IGARCH, FIGARCH Models	259
9.2.2	Distribution of Returns	260
9.2.3	Crashes	265
9.2.4	Crash Models	267

9.2.5	The Maslov Model *268*
9.2.6	The Sandpile Model *268*
9.2.7	Percolation Models *269*
9.2.8	The Cont–Bouchaud model *270*
9.2.9	Crash Precursor Patterns *272*
9.3	The Lux–Marchesi Model *274*
9.3.1	The Spin Models *275*
	References *276*

10 **A Thermodynamic Formulation of Social Science** *279*
Juergen Mimkes

10.1	Introduction *279*
10.2	Probability *280*
10.2.1	Normal Distribution *280*
10.2.2	Constraints *281*
10.2.3	Probability with Constraints (Lagrange Principle) *282*
10.3	Elements of Societies *283*
10.3.1	Agents *284*
10.3.2	Groups *284*
10.3.3	Interactions *286*
10.3.4	Classes *287*
10.3.5	States: Collective vs Individual *289*
10.4	Homogenious Societies *291*
10.4.1	The Three States of Homogeneous Societies *291*
10.4.1.1	Atomic Systems: H_2O *291*
10.4.1.2	Social Systems: Guided Tours *292*
10.4.1.3	Economic Systems: Companies *292*
10.4.1.4	Political Systems: Countries *293*
10.4.2	Change of State, Crisis, Revolution *294*
10.4.3	Hierarchy, Democracy and Fertility *294*
10.5	Heterogeneous Societies *296*
10.5.1	The Six States of Binary Societies *296*
10.5.2	Partnership *298*
10.5.3	Integration *299*
10.5.4	Segregation *300*
10.6	Dynamics of Societies *301*
10.6.1	Hierarchy and Opinion Formation *301*
10.6.2	Simulation of Segregation *304*
10.6.2.1	Phase Diagrams *304*
10.6.2.2	Intermarriage *305*
10.6.3	Simulation of Aggression *307*
10.7	Conclusion *308*
	References *309*

11	**Computer Simulation of Language Competition by Physicists** *311*	
	Christian Schulze and Dietrich Stauffer	
11.1	Introduction *311*	
11.2	Differential Equations *312*	
11.3	Microscopic Models *320*	
11.3.1	Few Languages *320*	
11.3.2	Many Languages *321*	
11.3.2.1	Colonization *321*	
11.3.2.2	Bit-string Model *323*	
11.4	Conclusion *329*	
11.5	Appendix *331*	
11.5.1	Viviane Colonization Model *331*	
11.5.2	Our Bit-string Model *331*	
	References *337*	

12	**Social Opinion Dynamics** *339*	
	Gérard Weisbuch	
12.1	Introduction *339*	
12.2	Binary Opinions *341*	
12.2.1	Full Mixing *342*	
12.2.2	Lattices as Surrogate Social Nets *343*	
12.2.3	Cellular Automata *344*	
12.2.3.1	Growth *344*	
12.2.4	INCA *346*	
12.2.5	Probabilistic Dynamics *348*	
12.2.6	Group Processes *349*	
12.3	Continuous Opinion Dynamics *349*	
12.3.1	The Basic Case: Complete Mixing and one Fixed Threshold *350*	
12.3.2	Social Networks *352*	
12.3.3	Extremism *354*	
12.4	Diffusion of Culture *357*	
12.4.1	Binary Traits *357*	
12.4.2	Results *358*	
12.4.3	Axelrod Model of Cultural Diffusion *359*	
12.5	Conclusions *360*	
12.5.1	Range and Limits of Opinion Dynamics Models *360*	
12.5.2	How to Convince *360*	
12.5.3	How to Make Business *361*	
12.5.4	Final Conclusions *364*	
	References *364*	

13 Opinion Dynamics, Minority Spreading and Heterogeneous Beliefs 367
Serge Galam

13.1 The Interplay of Rational Choices and Beliefs 367
13.2 Rumors and Collective Opinions in a Perfect World 370
13.3 Arguing by Groups of Size Three 372
13.4 Arguing by Groups of Size Four 372
13.5 Contradictory Public Opinions in Similar Areas 375
13.6 Segregation, Democratic Extremism and Coexistence 378
13.7 Arguing in Groups of Various Sizes 381
13.8 The Model is Capable of Predictions 388
13.9 Sociophysics is a Promising Field 390
References 391

14 Global Terrorism versus Social Permeability to Underground Activities 393
Serge Galam

14.1 Terrorism and Social Permeability 394
14.2 A Short Introduction to Percolation 395
14.3 Modeling a Complex Problem as Physicists do 396
14.4 The World Social Grid 398
14.5 Passive Supporters and Open Spaces to Terrorists 400
14.6 The Geometry of Terrorism is Volatile 404
14.7 From the Model to Some Real Facts of Terrorism 406
14.8 When Regional Terrorism Turns Global 409
14.9 The Situation Seems Hopeless 412
14.10 Reversing the Strategy from Military to Political 413
14.11 Conclusion and Some Hints for the Future 415
References 416

15 How a "Hit" is Born: The Emergence of Popularity from the Dynamics of Collective Choice 417
Sitabhra Sinha and Raj Kumar Pan

15.1 Introduction 417
15.2 Empirical Popularity Distributions 419
15.2.1 Examples 421
15.2.1.1 City Size 421
15.2.1.2 Company Size 422
15.2.1.3 Scientists and Scientific Papers 422
15.2.1.4 Newspaper and Magazines 424
15.2.1.5 Movies 424
15.2.1.6 Websites and Blogs 428

15.2.1.7	File Downloads	*430*
15.2.1.8	Groups	*431*
15.2.1.9	Elections	*431*
15.2.1.10	Books	*434*
15.2.1.11	Language	*435*
15.2.2	Time-evolution of Popularity	*436*
15.2.3	Discussion	*437*
15.3	Models of Popularity Distribution	*438*
15.3.1	A Model for Bimodal Distribution of Collective Choice	*440*
15.4	Conclusions	*444*
	References	*446*

16 Crowd Dynamics *449*
Anders Johansson and Dirk Helbing

16.1	Pedestrian Modeling: A Survey	*449*
16.1.1	State-of-the-art of Pedestrian Modeling	*450*
16.1.1.1	Social-force Model	*450*
16.1.1.2	Cellular Automata Models	*451*
16.1.1.3	Fluid-dynamic Models	*451*
16.1.1.4	Queueing Models	*451*
16.1.1.5	Calibration and Validation	*451*
16.2	Self-organization	*452*
16.2.1	Lane Formation	*453*
16.2.2	Strip Formation	*454*
16.2.3	Turbulent and Stop-and-go Waves	*455*
16.3	Other Collective Crowd Phenomena	*456*
16.3.1	Herding	*456*
16.3.2	Synchronization	*456*
16.3.3	Traffic Organization in Ants	*457*
16.3.4	Pedestrian Trail Formation	*457*
16.4	Bottlenecks	*458*
16.4.1	Uni-directional Bottleneck Flows	*458*
16.4.1.1	Analytical Treatment of Evacuation Through an Exit	*459*
16.4.1.2	Intermittent Flows and Faster-is-slower Effect	*461*
16.4.1.3	Quantifying the Obstruction Effect	*461*
16.4.2	Bi-directional Bottleneck Flows	*463*
16.5	Optimization	*463*
16.5.1	Pedestrian Flow Optimization with a Genetic Algorithm	*463*
16.5.1.1	Boolean Grid Representation	*464*
16.5.1.2	Results	*466*
16.5.2	Optimization of Parameter Values	*467*
16.6	Summary and Selected Applications	*470*
	References	*471*

17	**Complexities of Social Networks: A Physicist's Perspective** *473*	
	Parongama Sen	
17.1	Introduction *473*	
17.2	The Beginning: Milgram's Experiments *474*	
17.3	Topological Properties of Networks *474*	
17.4	Some Prototypes of Small-world Networks *477*	
17.4.1	Watts and Strogatz (WS) Network *478*	
17.4.2	Networks with Small-world and Scale-free Properties *478*	
17.4.3	Euclidean and Time-dependent Networks *479*	
17.5	Social Networks: Classification and Examples *479*	
17.6	Distinctive Features of Social Networks *481*	
17.7	Community Structure in Social Networks *482*	
17.7.1	Detecting Communities: Basic Methods *483*	
17.7.1.1	Agglomerative and Divisive Methods *483*	
17.7.1.2	A Measure of the Community Structure Identification *483*	
17.7.2	Some Novel Community Detection Algorithms *485*	
17.7.2.1	Optimization Methods *486*	
17.7.2.2	Spectral Methods *487*	
17.7.2.3	Methods Based on Dissimilarity *488*	
17.7.2.4	Another Local Method *489*	
17.7.3	Community-detection Methods Based on Physics *489*	
17.7.3.1	Network as an Electric Circuit *489*	
17.7.3.2	Application of Potts and Ising Models *490*	
17.7.4	Overlap of Communities and a Network at a Higher Level *491*	
17.7.4.1	Preferential Attachment of Communities *493*	
17.8	Models of Social Networks *493*	
17.8.1	Static Models *493*	
17.8.2	Dynamical Models *496*	
17.9	Is it Really a Small World? Searching: Post Milgram *498*	
17.9.1	Searching in Small-world Networks *498*	
17.9.2	Searching in Scale-free Graphs *499*	
17.9.3	Search in a Social Network *499*	
17.9.4	Experimental Studies of Searching *500*	
17.10	Endnote *501*	
17.11	Appendix: The Indian Railways Network *502*	
	References *502*	
18	**Emergence of Memory in Networks of Nonlinear Units: From Neurons to Plant Cells** *507*	
	Jun-ichi Inoue	
18.1	Introduction *507*	
18.2	Neural Networks *508*	

18.2.1	The Model System	*509*
18.2.2	Equations of States	*510*
18.2.2.1	$p = 1$ Case	*510*
18.2.2.2	Entropy of the System	*512*
18.2.2.3	Internal Energy Density	*513*
18.2.2.4	Compressibility	*513*
18.2.2.5	Overlap at the Ground State for $\mu = 0$	*514*
18.2.3	Replica Symmetric Calculations for the Case of Extensive Patterns	*514*
18.2.4	Evaluation of the Saddle Point	*517*
18.2.5	Phase Diagrams	*518*
18.2.5.1	Para-spin-glass Phase Boundary	*519*
18.2.5.2	Critical Chemical Potential μ_c at $T = 0$	*520*
18.2.5.3	Saddle-point Equations for $\mu < \mu_c$ at $T = 0$	*520*
18.2.6	Entropy of the System	*521*
18.2.6.1	High-temperature Limit	*522*
18.2.6.2	At the Ground State	*522*
18.2.7	Internal Energy	*523*
18.2.8	The Compressibility	*524*
18.3	Summary: Neural Networks	*525*
18.4	Plant Intelligence: Brief Introduction	*525*
18.5	The I–V Characteristics of Cell Membranes	*526*
18.6	A Solvable Plant-intelligence Model and its Replica Analysis	*527*
18.6.1	Replica Symmetric Solution	*527*
18.6.2	Phase Diagrams	*528*
18.6.2.1	Saddle-point Equations	*529*
18.6.2.2	$T = 0$ Noiseless Limit	*529*
18.6.2.3	Spin-glass Para-phase Boundary	*529*
18.6.3	Phase Diagrams for $T \neq 0$	*530*
18.6.4	Negative λ case	*530*
18.7	Summary and Discussion	*531*
	References	*533*
19	**Self-organization Principles in Supply Networks and Production Systems** *535*	
	Dirk Helbing, Thomas Seidel, Stefan Lämmer, and Karsten Peters	
19.1	Introduction	*535*
19.2	Complex Dynamics and Chaos	*537*
19.3	The Slower-is-faster Effect	*539*
19.3.1	Observations in Traffic Systems	*541*
19.3.1.1	Panicking Pedestrians	*541*
19.3.1.2	Freeway Traffic	*541*

19.3.1.3	Intersecting Vehicle and Pedestrian Streams 543
19.3.2	Relevance to Production and Logistics 545
19.3.2.1	Semi-conductor Chip Manufacturing 545
19.3.2.2	Container Terminals 545
19.3.2.3	Packaging and Other Industries 547
19.4	Adaptive Control 550
19.4.1	Traffic Equations for Production Systems 550
19.4.2	Re-routing Strategies and Machine Utilization 552
19.4.3	Self-organized Scheduling 554
19.5	Summary and Outlook 557
	References 558

20 Can we Recognize an Innovation?: Perspective from an Evolving Network Model *561*

Sanjay Jain and Sandeep Krishna

20.1	Introduction 561
20.2	A Framework for Modeling Innovation: Graph Theory and Dynamical Systems 563
20.3	Definition of the Model System 564
20.4	Time Evolution of the System 566
20.5	Innovation 567
20.6	Six Categories of Innovation 572
20.6.1	A Short-lived Innovation: Uncaring and Unviable Winners 572
20.6.2	Birth of an Organization: Cooperation Begets Stability 573
20.6.3	Expansion of the Organization at its Periphery: Incremental Innovations 575
20.6.4	Growth of the Core of the Organization: Parasites Become Symbionts 575
20.6.5	Core-shift 1: Takeover by a New Competitor 576
20.6.6	Core-shift 2: Takeover by a Dormant Innovation 578
20.7	Recognizing Innovations: A Structural Classification 578
20.8	Some Possible General Lessons 581
20.9	Discussion 583
20.10	Appendix A: Definitions and Proofs 584
20.10.1	Derivation of Eq. (20.1) 584
20.10.2	The Attractor of Eq. (20.1) 585
20.10.3	The Attractor of Eq. (20.1) when there are no Cycles 585
20.10.4	Graph-theoretic Properties of ACSs 585
20.10.5	Dominant ACS of a Graph 586
20.10.6	Time Scales for Appearance and Growth of the Dominant ACS 587
20.11	Appendix B: Graph-theoretic Classification of Innovations 587
	References 590

Color Plates *593*

Subject Index *607*

Author Index *613*

Preface

In a proverbial Indian story (Buddhist Udana 68–69), a few blind people touched different parts of an elephant: the trunk, tusk, leg, tail, etc., and interpreted them as different animate/inanimate objects following their own perceptions, ideas or experiences. We, the scientists: physicists, biologists, economists or sociologists, all tend to do the same. In all its manifestations, inanimate, biological or sociological, nature does perhaps employ the same elegant code, like the genetic code of the elephant, but suppressed partially and expressed differently for various parts of its body. We perceive them differently, depending on our training and background. Nature hardly cares whether our views are physical, biological, or sociological. The complexity studies aim to capture these universal codes, manifested differently in different parts of the same body of natural phenomena.

This grand unification search is at a very inspiring stage today and this book reports on a part of these interdisciplinary studies, developed over the last ten to fifteen years and classified mainly under the headings *econophysics* or *sociophysics*. It was not the success of the studies that motivated us to collect the authentic reviews on intriguing developments in this volume; but it was rather the promise and novelty of this research which has been our guide in selecting them.

The contents of this book may be divided into two parts. The first nine chapters can be broadly categorized as econophysics and the rest as sociophysics, although there are obvious overlaps between the two.

In the first chapter, J. Mimkes shows how exact differentials can be formed out of inexact ones, and then identifies and exploits the correspondences between such functions in thermodynamics and in economics. In the next chapter, R. Stinchcombe shows how limit-order financial markets can be faithfully modeled as nonequilibrium collective systems of "particles" (orders) depositing, evaporating, or annihilating, at rates determined by the price and market condition. After establishing a general "complex adaptive" framework, starting from simple games and well-known limiting cases, like minority games, D. M. D. Smith and N. F. Johnson show how "general managers" could be designed for the evolution of competitive multi-agent populations. In the following chapter, Y. Fujiwara et al. analyzed exhaustively the data for firm sizes and their growths in Europe and Japan, establishing the power-law regimes and the conditions for detailed balance in their growth dynamics. In the next chapter P. Richmond et al. briefly review the wealth/income distributions in various societies, and describe some of the successful statistical physics models, and the asset exchange model with random savings, in particular, to capture such intriguing distribution forms. In the next chapter, A. Kar Gupta concentrates on one class of such (random asset exchange) models, studying them using a transition-matrix approach, and identifies some correspondence in formalism with one-dimensional diffusion and aggregation of particles. Y. Wang et al. then discuss how such asset exchange models can be used to figure out the monetary circulation process and to improve the measurement of economic mobility. The mechanical modeling of the triangular arbitrage advantages in the foreign exchange market is described next by Y. Aiba and N. Hatano. M. Ausloos in the next chapter, describes the general features of the fluctuations in the stock and foreign exchange markets, emphasizing measuring techniques and subsequent statistical and microscopic-like models; including also the specificity of crash patterns.

In the tenth chapter, J. Mimkes extends the thermodynamical correspondence of free energy minimization to the corresponding optimization of "happiness" in society. C. Schulze and D. Stauffer next discuss the intriguing problem of growth and decay (due to competition and/or regional/global dominance) of languages and computer simulation models for such dynamics. In the following chapter, G. Weisbuch reviews the evolutionary dynamics of collective social opinions using cellular automata and percolation models. S. Galam, in the next chapter, describes how spread and decay of conflicting public opinion can be modeled using statistical physics. Next, he argues how social percolation of an extreme opinion (say, of terrorism) occurs, and identifies the global spread/percolation of such terrorism with the event of the September 11, 2001 attack on the USA. S. Sinha and R. K. Pan, in the next chapter, identify some robust features (e.g., log-normal form and bimodality) in the distribution and growth of popularity in several social phenomena, such as movies, elections, blogs, languages, etc. and describes how some agent-based models can capture these features. In chapter sixteen, A. Johansson and D. Helbing describe the unique features of dynamics of dense crowds under constraints, and review the various cellular automata and flow-like continuity equation models used to describe them. P. Sen, in the next chapter, describes the distinctive static and dynamic properties of social networks including those of the railway networks and citation networks. The emergence of "collective memory" in many such social phenomena, including games, are very characteristic and J.-I. Inoue, in chapter eighteen, describes the celebrated

Hopfield model for associative memory and its extension to (nonfrustrating) networks of plant cells for the emergence of "intelligence" in them. D. Helbing et al. describe in the next chapter, how some fluid/traffic-like flow models can be adopted for optimized production in various manufacturing industries. In the last chapter, S. Jain and S. Krishna describe how one can identify the effects of various innovations in the context of evolving network models.

We sincerely hope that these wonderful and up-to-date reviews in such a wide landscape of emerging sciences of econophysics and sociophysics will benefit the readers with an exciting feast of relevant ideas and information. We are indeed thankful to our esteemed contributors for their efforts and outstanding co-operation. We are also grateful to Wiley-VCH for their encouragement and constant support in this project.

Bikas K. Chakrabarti, Anirban Chakraborti and Arnab Chatterjee

Kolkata and Varanasi, May 2006

List of Contributors

Yukihiro Aiba Ch. 8
Institute of Industrial Science
University of Tokyo
Komaba 4-6-1, Meguro
Tokyo 153-8505
Japan
aiba@iis.u-tokyo.ac.jp

Hideaki Aoyama Ch. 4
Department of Physics
Graduate School of Science
Kyoto University, Yoshida
Kyoto 606-8501
Japan
aoyama@phys.h.kyoto-u.ac.jp

Marcel Ausloos Ch. 9
SUPRATECS
B5 Universite de Liege
Sart Tilman
4000 Liege
Belgium
Marcel.Ausloos@ulg.ac.be

Ricardo Coelho Ch. 5
School of Physics
University of Dublin
Trinity College
Dublin 2
Ireland
coelhorj@tcd.ie

Ning Ding Ch. 7
Department of Systems Science
School of Management
Beijing Normal University
Beijing 100875
P. R. China

Yoshi Fujiwara Ch. 4
ATR Network Informatics
Laboratory
Seika-chou Hikari-dai 2-2-2
Souraku-gun
Kyoto 619-0288
Japan
yfujiwar@atr.jp

Serge Galam Ch. 13, 14
Centre de Recherche en
Epistémologie Appliquée (CREA)
Ecole Polytechnique,
1, rue Descartes
75005 Paris
France
galam@ccr.jussieu.fr
serge.galam@polytechnique.edu

Naomichi Hatano Ch. 8
Institute of Industrial Science
University of Tokyo
Komaba 4-6-1, Meguro
Tokyo 153-8505
Japan
hatano@iis.u-tokyo.ac.jp

Econophysics and Sociophysics: Trends and Perspectives.
Bikas K. Chakrabarti, Anirban Chakraborti, Arnab Chatterjee (Eds.)
Copyright © 2006 WILEY-VCH Verlag GmbH & Co. KGaA, Weinheim
ISBN: 3-527-40670-0

List of Contributors

Dirk Helbing   Ch. 16, 19
Institute for Transport & Economics
TU Dresden
Andreas-Schubert-Str. 23
01062 Dresden
Germany
helbing@trafficforum.org

Stefan Hutzler   Ch. 5
School of Physics
University of Dublin
Trinity College
Dublin 2
Ireland
shutzler@maths.tcd.ie

Jun-ichi Inoue   Ch. 18
Complex Systems Engineering
Graduate School of Information
Science & Technology
Hokkaido University
N14-W9, Kita-ku
Sapporo 060-0814
Japan
j_inoue@complex.eng.hokudai.ac.jp

Sanjay Jain   Ch. 20
Department of Physics and
Astrophysics
University of Delhi
Delhi 110 007
India

Santa Fe Institute
1399 Hyde Park Road
Santa Fe, NM 87501
USA

Jawaharlal Nehru Centre for
Advanced Scientific Research
Jakkur, Bangalore 560 064
India
jain@physics.du.ac.in

Anders Johansson   Ch. 16
Institute for Transport & Economics
TU Dresden
Andreas-Schubert-Str. 23
01062 Dresden
Germany
johansson@vwi.tu-dresden.de

Neil F. Johnson   Ch. 3
Clarendon Laboratory
Oxford University
Parks Road
Oxford OX1 3PU
United Kingdom
n.johnson@physics.ox.ac.uk

Abhijit Kar Gupta   Ch. 6
Physics Department
Panskura Banamali College
Panskura, East Midnapore
Pin: 721 152, West Bengal
India
abhijit_kargupta@rediffmail.com

Sandeep Krishna   Ch. 20
Niels Bohr Institute
Blegdamsvej 17
Copenhagen 2100
Denmark
sandeep@nbi.dk

Stefan Lämmer   Ch. 19
Institute for Transport & Economics
TU Dresden
Andreas-Schubert-Str. 23
01062 Dresden
Germany

Juergen Mimkes   Ch. 1, 10
Physics Department
Paderborn University
Warburgerstr. 100
33100 Paderborn
Germany
mimkes@zitmail.uni-paderborn.de

Raj Kumar Pan   Ch. 15
The Institute of Mathematical
Sciences
CIT Campus, Taramani
Chennai 600 113
India
rajkp@imsc.res.in

Karsten Peters Ch. 19
Institute for Transport & Economics
TU Dresden
Andreas-Schubert-Str. 23
01062 Dresden
Germany
peters@vwi.tu-dresden.de

Przemek Repetowicz Ch. 5
School of Physics
University of Dublin
Trinity College
Dublin 2
Ireland
repetowp@tcd.ie

Peter Richmond Ch. 5
School of Physics
University of Dublin
Trinity College
Dublin 2
Ireland
richmond@tcd.ie

Christian Schulze Ch. 11
Institute for Theoretical Physics
Cologne University
50923 Köln
Germany

Thomas Seidel Ch. 19
Institute for Transport & Economics
TU Dresden
Andreas-Schubert-Str. 23
01062 Dresden
Germany

Parongama Sen Ch. 17
Department of Physics
University of Calcutta
92 A. P. C. Road
Kolkata 700 009
India
psphy@caluniv.ac.in

Sitabhra Sinha Ch. 15
The Institute of Mathematical
Sciences
CIT Campus, Taramani
Chennai 600 113
India
sitabhra@imsc.res.in

David M. D. Smith Ch. 3
Clarendon Laboratory
Oxford University
Parks Road
Oxford OX1 3PU
United Kingdom
d.smith3@physics.ox.ac.uk

Dietrich Stauffer Ch. 11
Institute for Theoretical Physics
Cologne University
50923 Köln
Germany
stauffer@thp.uni-koeln.de

Robin Stinchcombe Ch. 2
Rudolf Peierls Centre for
Theoretical Physics
Oxford University
1 Keble Road
Oxford OX1 3NP
United Kingdom
r.stinchcombe1@physics.ox.ac.uk

Wataru Souma Ch. 4
ATR Network Informatics
Laboratory
Seika-chou Hikari-dai 2-2-2
Souraku-gun
Kyoto 619-0288
Japan
souma@atr.jp

Yougui Wang Ch. 7
Department of Systems Science
School of Management
Beijing Normal University
Beijing 100875
P. R. China
ygwang@bnu.edu.cn

Gérard Weisbuch Ch. 12
Laboratoire de Physique
Statistique ENS
24 rue Lhomond 75005 Paris
France
weisbuch@lps.ens.fr

Ning Xi Ch. 7
Department of Systems Science
School of Management
Beijing Normal University
Beijing 100875
P. R. China

1
A Thermodynamic Formulation of Economics
Juergen Mimkes

The thermodynamic formulation of economics is based on the laws of calculus. Differential forms in two dimensions are generally not exact forms (δQ), the integral from (A) to (B) is not always the same as the integral from (B) to (A). It is possible to invest little in one way and gain a lot on the way back, and to do this periodically. This is the mechanism of energy production in heat pumps, of economic production in companies and of growth in economies. Not exact forms may be turned into exact forms (dS) by an integrating factor T, $dS = \delta Q/T$. The new function (S) is called entropy and is related to the probability (P) as $S = \ln P$. In economics the function (S) is called production function. The factor (T) is a market index or the standard of living, GNP/capita, of countries. The dynamics of economic growth is based on the Carnot process, which is driven by external resources. Economic growth and capital generation – like heat pumps and electric generators – depend on natural resources like oil. GNP and oil consumption run parallel for all countries. Markets and motors, economic and thermodynamics processes are all based on the same laws of calculus and statistics.

1.1
Introduction

In the last ten years new interdisciplinary approaches to economics and social science have been developed by natural scientists. The problems of economic growth, distribution of wealth, and unemployment require a new understanding of markets and society. The dynamics of social systems has been introduced by W. Weidlich (1972) [17] and H. E. Stanley (1992) [15] has coined the term econophysics. A thermodynamic approach to socio-economics has been favored by D. K. Foley (1994) [4], J. Mimkes (1995) [10] and Drăgulescu and V. M. Yakovenko (2001) [3]. Financial markets have been discussed by M. Levy et al. (2000) [8], S. Solomon and Richmond (2001) [14], Y. Aruka (2001) [1] and many others. Many conferences have been held to enhance the communication between natural and socio-economic sciences with topics like

Econophysics and Sociophysics: Trends and Perspectives.
Bikas K. Chakrabarti, Anirban Chakraborti, Arnab Chatterjee (Eds.)
Copyright © 2006 WILEY-VCH Verlag GmbH & Co. KGaA, Weinheim
ISBN: 3-527-40670-0

econophysics, complexity in economics and socio-economic agent systems. In the first chapter, the mechanism of economic production is discussed on the basis of calculus and statistics. The two mathematical fields will be applied to economics in a similar way to thermodynamics, this is the thermodynamic formulation of economics.

1.2 Differential Forms

1.2.1 Exact Differential Forms

The total differential of a function $f(x,y)$ is given by (see, e.g., W. Kaplan [6])

$$df = (\partial f/\partial x)\, dx + (\partial f/\partial y)\, dy \tag{1.1}$$

The second (mixed) derivative of the function $f(x,y)$ is symmetric in x and y,

$$\frac{\partial^2 f}{\partial x\, \partial y} = \frac{\partial^2 f}{\partial y\, \partial x} \tag{1.2}$$

In the same way every differential form

$$df = a(x,y)\, dx + b(x,y)\, dy \tag{1.3}$$

is called total or exact, if the second derivatives

$$\partial a(x,y)/\partial y = \partial b(x,y)/\partial x \tag{1.4}$$

are equal. Exact differential forms are marked by the "d" in df. The function $f(x,y)$ exists and may be determined by a line integral,

$$\int_A^B df = \int_A^B \left(\frac{\partial f}{\partial x}\, dx + \frac{\partial f}{\partial y}\, dy \right) = f(B) - f(A) \tag{1.5}$$

The closed integral of an exact differential form is zero: The closed integral may be split into two integrals from A to B on path (1) and back from B to A on path (2). Reversing the limits of the second integral changes the sign of the second integral. Since both integrals depend on the limits A and B only, the closed integral of an exact differential is zero:

$$\oint df = \int_A^B df_{(1)} + \int_B^A df_{(2)} = \int_A^B df_{(2)} - \int_A^B df_{(2)} = 0 \tag{1.6}$$

Example:

$$f(x,y) = x^3 y^5$$
$$df = (3x^2 y^5)\, dx + (5x^3 y^4)\, dy$$
$$\frac{\partial^2 f}{\partial x \partial y} = 3 \cdot 5 \cdot x^2 y^4 = \frac{\partial^2 f}{\partial y \partial x} \tag{1.7}$$
$$\oint df = 0$$

1.2.2
Not Exact Differential Forms

In one dimension all differential forms are exact. A two-dimensional differential form δg

$$\delta g = a(x,y)\, dx + b(x,y)\, dy \tag{1.8}$$

is not always an exact differential form. The second derivatives are generally not equal,

$$\partial a(x,y)/\partial y \neq \partial b(x,y)/\partial x \tag{1.9}$$

These differential forms are called not exact and are marked by the "δ" in δg. A function $g(x,y)$ does not exists in general and may not be determined by a line integral, because the line integral of not exact differential forms depends on the integral limits A and B and on the path of integration. Any different path of integration will lead to a new function $g(x,y)$. A closed integral from A to B along path (1) and back from B to A along path (2) will not be zero,

$$\oint \delta g = \int_A^B \delta g_{(1)} - \int_A^B \delta g_{(2)} \neq 0 \tag{1.10}$$

Example: We may construct a non exact differential form by dividing df (1.7) by y:

$$\delta g = df(x,y)/y = (3x^2 y^4)\, dx + (5x^3 y^3)\, dy \tag{1.11}$$
$$12x^2 y^3 = \partial a(x,y)/\partial y \neq \partial b(x,y)/\partial x = 15x^2 y^3$$

In Fig. 1.1. the closed integral of δg is calculated for path (1) from point $A = (1;1)$ along the line $(y=1)$ to $B = (2;1)$ and then along the line $(x=2)$ to point $C = (2;2)$.

$$\int_{1;1}^{2;2} \delta g_1 = \int_1^2 (3x^2 (y=1)^4)\, dx + \int_1^2 5(x=2)^3 y^3\, dy$$
$$= (2^3 - 1^3) + 5 \cdot 2^3 (2^4 - 1^4)/4 = 157$$

The second integral of δg is calculated for path (2) from point $A = (1;1)$ along the line $(x = 1)$ to point $D = (1;2)$ and then along the line $(y = 2)$ to point $C = (2;2)$:

$$\int_{1;1}^{2;2} \delta g_2 = \int_1^2 (3x^2(y=2)^4 \, dx + \int_1^2 5(x=1)^3 y^3 \, dy$$
$$= 2^4(2^3 - 1^3) + 5 \cdot 1^3(2^4 - 1^4)/4 = 130,75$$

The closed line integral along path (1) from $A = (1;1)$ via $B = (2;1)$ to $C = (2;2)$ and back along path (2) from C via $D = (1;2)$ to $A = (1;1)$ – see Fig. 1.1 – is

$$\oint \delta g = \int_A^B \delta g_{(1)} + \int_B^A \delta g_{(2)} = 157 - 130,75 = 26,25 \neq 0$$

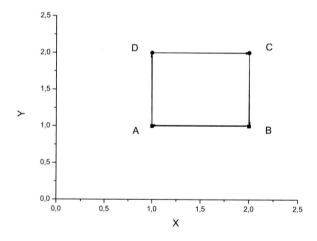

Fig. 1.1 The closed line integral along path (1) from $A = (1;1)$ via $B = (1;2)$ to $C = (2;2)$ and back along path (2) from C via $D = (2;1)$ to A is not zero!

1.2.3
The Integrating Factor

A two-dimensional differential form δg may be made exact by an integrating factor $1/T$:

$$df = \delta g/T = [a(x,y) \, dx + b(x,y) \, dy]/T \qquad (1.12)$$

(Sometimes the integrating factor is also called λ, but with respect to thermodynamics the factor $1/T$ will be used here.) In two dimensions this factor always exists.

Example: The non total differential form δg (1.11) has been obtained by dividing the exact form df (1.7) by "y". Accordingly, $y = 1/T$ will be the integrating factor of δg.

1.2.4
The First and Second Law of Differential Forms

The results above may be stated in two laws for differential forms:
The first law of differential forms:

Two-dimensional differential forms δg will generally not be exact.

The second law of differential forms:

A not exact differential form may be made exact by an integrating factor $1/T$.

1.2.5
Not Exact Differential Forms in Thermodynamics and Economics

Why are differential forms and calculus of not exact forms so important for thermodynamic and economics?

1. Thermodynamics. Heat is a function of at least two variables, temperature and pressure. According to the first law of differential forms, heat (δQ) will be a non exact differential form. The value of the integral from (A) to (B) will not be the same as from (B) to (A). This makes the first law of differential forms so important in the application to periodic machines: it is possible to invest little heat in one way and gain a lot of heat on the way back and to do this periodically.

Heat pumps. *A heat pump or generator may periodically invest 1 kWh of energy in one way and gain 5 kWh on the way back!*

The second law makes it possible to calculate a complicated technical process of not exact differential forms by a simple function. This makes thermodynamics so important for theory and application (see, e.g. R. Fowler and Guggenheim (1960) [5]).

2. Economics. Economic growth is a function of at least two variables, labour and capital. According to the first law of differential forms profit (δQ) will not be an exact differential form. The value of the integral from (A) to (B) will not be the same as from (B) to (A). This makes the first law of differential forms so attractive in the application to periodic production: it is possible to invest little capital in one way and gain a lot of capital on the way back and to do this periodically:

Banks. A bank may periodically invest 4% of interest in savings and collect 10% interest from investors.

Companies. A company may periodically pay as little as possible to workers and collect periodically as much as possible from customers.

The second law makes it possible to calculate a complicated technical process of not exact differential forms by a simple function. This makes the thermodynamic formulation of economics so important for theory and application in economics and business.

1.3
The First Law of Economics

1.3.1
The First Law: Capital Balance of Production

Economic production in farms, automobile plants, medical offices or banks is achieved by hard work, but the output will be different for each type of production. The output depends on each specific production process, or in mathematical terms, on the path of integration. Production may be modeled by calculus of not exact differential forms,

$$-\oint \delta W = \oint \delta Q \tag{1.13}$$

The surplus (Q) is a result of work input $(-W)$. The cyclic process of economic production Eq. (1.13) may be split into two parts, the integral from A to B and back from B to A,

$$-\oint \delta W = \oint \delta Q = \int_A^B \delta Q_{(1)} - \int_A^B \delta Q_{(2)} = Y - C = \Delta Q \tag{1.14}$$

The economic process consists of output (Y) and input (C), the difference is the surplus ΔQ. (The letter S for surplus in standard economics has been replaced by ΔQ.) Income (Y) and costs or consumption (C) are both part of the same production cycle, and depend on the specific production and consumption processes. Surplus and economic growth cannot be calculated in advance (*ex ante*), unless the whole process is entirely known.

Equation (1.14) may be written in differential forms. If two not exact differentials $(-\delta W)$ and (δQ) are equal along the same path of integration, they may differ only by a total differential form (dE), which will always vanish for closed integrals,

$$\delta Q = dE - \delta W \tag{1.15}$$

This is the first law of economics in differential form. It is a capital balance of production. The surplus (δQ) will increase the capital (dE) and requires the input of work $(-\delta W)$. The main feature of this first law of economics is the fact that the capital balance of production cannot be expressed by definite functions. The capital balance can only be given by not exact differential forms! Eq. (1.13) will be discussed in more detail below.

1.3.2
Work (W)

Work (W) is the effort and know-how which we invest in our job. The function (W) does not exist as a general function, it always depends on the path of integration, the production process. Accordingly, (W) cannot be calculated *"ex ante"*. Work is not equivalent to labor, which defines the number and kind of people in the production process. Work is equivalent to the production process. The dimension of (W) is capital, the same as capital (E) and surplus (Q). In thermodynamics the function (W) refers to work of machines. In economics the function (W) may refer to people as well as to machines! The thermodynamic formulation of economics reveals a problem of modern production namely that people and machines work according to the same laws, Eq. (1.13). If people do not work efficiently, they may be replaced by machines: in construction labor will be taken over by cranes and motors, in offices work may be done by computers.

1.3.3
Surplus (ΔQ)

The surplus (ΔQ) is the result of work (W), Eq. (1.13), and again cannot be calculated *"ex ante"*, as (δQ) is a not exact differential form. The integral depends on the path of integration, the surplus depends on the production process. The thermodynamic formulation of economics makes it possible to compare economic production to work in thermodynamics.

Heat pumps. *A heat pump is close to the energy reservoir of a river or garden. A heat pump or generator may periodically invest 1 kWh of energy one way and gain 5 kWh on the way back! The heat output (Q) is larger than the work input (W). Where does the heat come from?*

In each cycle the heat (ΔQ) is pumped from the environment, the garden or river, which will be cooled down when the heat pump is operating. In gardens or rivers the energy loss will be filled up from the reservoir of the environment.

Banks. *A bank is close to the capital reservoir of savers. A bank may periodically invest 4% of interest in savings and collect 10% interest from investors. The output (Q) is larger than the input (W)! Where does the surplus come from?*

For banks the surplus capital for each cycle is taken from the growth of the saving community. It is not only the first member of an economic chain which exploits nature and the environment, but all other members of an economic chain do the same in each production cycle. This mechanism of economic growth will be discussed in more detail in the Carnot process, below.

1.3.4
Capital (E)

The capital (E) is the basis of economic production (W). The farm is the capital of the farmer, the production plant is the capital of a company, the investment the capital of investors.

Without labour (W) capital cannot grow. Only by an input of work may the capital increase. Of course, capital may also decrease by mismanagement or failures. But every economic system has to produce positive surplus in order to survive. After each production cycle the surplus (ΔQ) has to be in a reasonable relationship to the invested capital (E). The relation

$$r = - \oint \delta W / E = \oint \delta Q / E = \Delta Q / E \tag{1.16}$$

is called the efficiency of the production cycle (δW). The ratio "r" is called the interest rate and is given in percent. The efficiency or interest rate measures the success of a production cycle (δW) and determines, whether people or machines will be employed in a specific production process.

1.4
The Second Law of Economics

1.4.1
The Second Law: Existence of a System Function (S)

The not exact differential form δQ may be changed into an exact differential form dS by an integrating factor T. This is called the second law:

$$dS = \frac{1}{T} \delta Q \tag{1.17}$$

Equation (1.17), is a law for the existence of a system function (S), which is called entropy in physics and information science. Economists usually call this function the production or utility function. $1/T$ is the integrating factor.

1.4.2
The Integrating Factor (T)

$1/T$ is the integrating factor of the capital balance (1.13). T is proportional to the mean capital (E) of N agents of the specific economic system,

$$E = cNT \tag{1.18}$$

c is a proportional factor. T may be regarded as an "economic temperature". In a market of N commodities, T is proportional to the mean price level. In a society of N households T is proportional to the mean capital per household, or standard of living. In countries, T is proportional to the GDP per capita. T is introduced by the second law as the main variable in all economic functions.

1.4.3
Entropy and Production Function (S)

Inserting Eq. (1.17) into (1.13) we find

$$-\oint \delta W = \oint T dS \tag{1.19}$$

The entropy or production function (S) is closely related to the work function (W). But in contrast to (W) the function (S) is independent of the production process, it has the dimension of a (production) number and may be calculated "*ex ante*". The functions (W) and (S) represent mechanism and calculation in all economic processes:

The work function (W). The work function (W) is defined by the production process and may be different for each process. This makes it possible to invest little in one part of the process and gain much in another part of the production process in order to obtain a surplus.

The production function (S). The entropy or production function (S) depends on the system and makes it possible to calculate the economic process "*ex ante*".

1.4.4
Pressure and Personal Freedom

In Eq. (1.17) the non exact form δQ has been expressed by dS: $\delta Q = TdS$. In the same way the non exact form of production (δW) may be expressed by the exact differential form dV,

$$\delta W = -p\, dV \tag{1.20}$$

The parameter p may be called the pressure, V may be regarded as space or personal freedom, which may be reduced due to the external economics or social pressure.

1.4.5
The Exact Differential ($dS(T, V)$)

According to Eq. (1.17) and (1.20) the entropy (dS) may be written as an exact differential form of T and V:

$$dS(T, V) = \frac{\partial S}{\partial T} dT + \frac{\partial S}{\partial V} dV = \frac{1}{T}(dE(T, V) + p(T, V)\, dV)$$

$$dS(T,V) = \frac{1}{T}\frac{\partial E}{\partial T}dT + \frac{1}{T}\left(\frac{\partial E}{\partial V} + p\right)dV \qquad (1.21)$$

The exact differential dS may be integrated independent from the path of integration. The production function $S(T,V)$ depends on capital $E(T,V)$ and economic pressure $p(T,V)$. However, other variables, like $S(T,p)$ are also possible, they require additional calculations and will be discussed at a later point.

1.4.6
The Maxwell Relation

The functions E and p cannot be chosen arbitrarily, as the mixed differentials of the exact form dS in Eq. (1.21) have to be equal. This leads to (*exercise!*)

$$\frac{\partial p}{\partial T} = \frac{1}{T}\left(\frac{\partial E}{\partial V} + p\right) = \frac{\partial S}{\partial V} \qquad (1.22)$$

These "Maxwell relations" are general conditions for all model functions $E(T,V)$, $p(T,V)$ and $S(T,V)$.

Equation (1.21) leads to the existence of a function $F(T,V)$,

$$F(T,V) = E(T,V) - TS(T,V) \qquad (1.23)$$

(see exercise). F may be called the "effective costs" function, which will be a minimum for stable economic systems. The function F corresponds to the "Helmholtz free energy" function of thermodynamics.

Exercise: *The total differential $dF = d(E - TS) = dE - TdS - SdT$ may be transformed by Eqs (1.15), (1.17) and (1.20) into*

$$dF = -p\,dV - T\,dS$$

The mixed second derivative of dF is given by the Maxwell relation Eq. (1.22).

1.4.7
Lagrange Function

Dividing the function $F(T,V)$ in Eq. (1.23) by $(-T)$ we obtain

$$L(T,V) = S(T,V) - (1/T)E(T,V) \to \text{maximum!} \qquad (1.24)$$

L is the Lagrange function which maximizes the production function (S) under constraints of costs (E) with a Lagrange multiplier $\lambda = (1/T)$. This is the result of the first and second law. The discussion of the Lagrange function will be continued in Section 1.5.4.

1.5 Statistics

1.5.1 Combinations

The distribution of decisions is given by the mathematics of combinations. For two possible decisions – left/right or yes/no – we find the probability P

$$P(N_L; N_R) = \frac{N_0!}{N_L! N_R!} \cdot \frac{1}{2^N} \qquad (1.25)$$

N_L is the number of decisions for the left side and N_R the number of decisions for the right side, $N_L + N_R = N_0$. ($N!$ stands for the product $4! = 1 \cdot 2 \cdot 3 \cdot 4$ and $0! = 1$).

1.5.2 Normal Distribution

For large numbers N_0 the probability function $P(N_L, N_R)$ in Eq. (1.25) leads to a normal distribution,

$$P(N) = \frac{1}{\sqrt{2\pi}\sigma} \cdot e^{-\frac{(N-\bar{N})^2}{2\sigma^2}} \qquad (1.26)$$

with $0 \leq N \leq N_0$, $\bar{N} = N_0/2$ and $2\sigma = \sqrt{N}$. The normal distribution is one of the most important probability functions in the natural, social and economic sciences.

1.5.3 Polynomial Distribution

In many cases we have more than two decisions, e.g., we can choose the color of a car to be black, white, red, blue, etc. For $N_0 = N_1 + \ldots + N_K$ and K possible equal decisions we obtain

$$P(N_1; \ldots; N_K) = \frac{N_0!}{N_1! \cdot \ldots \cdot N_K!} \cdot \frac{1}{K^N} \qquad (1.27)$$

If the probability of the decisions is not equal, we have to introduce the probability q_k of the decision k. The sum of all q_k will be equal to one, $\Sigma\, q_k = 1$:

$$P(N_1; \ldots; N_K) = \frac{N_0!}{N_1! \cdot \ldots \cdot N_K!} \cdot q_1^{N_1} \cdot \ldots \cdot q_K^{N_K} \qquad (1.28)$$

If we have N_0 cars with K different colors and each color has the probability q_k, then $P(N_1; \ldots; N_K)$ is the probability to find, in a street with N cars, N_1 cars of color 1, N_2 cars of color 2 and N_k cars of color k.

1.5.4
Lagrange Function in Stochastic Systems

What is the most probable distribution of apples, pears and bananas under given prices (E)? Or, what is the most probable distribution of N goods in K different price categories?

The probability (P) will always tend to be a maximum. The most probable distribution of N commodities with constraints of price E may be calculated by the Lagrange function (L),

$$L(N_k) = \ln P(N_k) - (1/T)\Sigma N_k E_k \rightarrow \text{maximum!} \tag{1.29}$$

$P(N_k)$ is the probability according to Eq. (1.28), N_k is the number of goods, E_k the price in price class (k), ($1/T$) is the Lagrange multiplier. Equation (1.29) is the Lagrange function Eq. (1.24) of systems that follow the laws of probability.

Example: The Munich beer garden.

1. A Munich beer garden has N_1 permanent and N_2 temporary employees. The wages are $E_1 = 15€$ per hour for the permanent and $E_2 = 7,5€$ per hour for the temporary staff. The Lagrange function calculates the optimal output per hour under the constraints of wages E,

$$E = N_1 E_1 + N_2 E_2 = N(x_1 E_1 + x_2 E_2) \tag{1.30}$$

where $x_k = N_k/N$ is the relative number and N the total number of staff. The entropy for two types of employees is given by

$$S = N \ln N - N_1 \ln N_1 - N_2 \ln N_2 = -N(x_1 \ln x_1 + x_2 \ln x_2) \tag{1.31}$$

The Lagrange function is maximized,

$$L = N \ln N - N_1 \ln N_1 - N_2 \ln N_2 - (1/T)(N_1 E_1 + N_2 E_2) = \text{max!} \tag{1.32}$$

At a maximum the derivatives with respect to N_1 and N_2 will be zero. The relative numbers of permanent and temporary staff x_1 and x_2 follow a Boltzmann distribution. With the given values of E_1 and E_2 we obtain

$$x_1 = \exp(-E_1/T) = 0.38$$

$$x_2 = \exp(-E_2/T) = 0.62$$

$$T = 15.8$$

$$S/N = 0.664$$

$$E/N = 10.35$$

The relative numbers of permanent and temporary staff x_1 and x_2, the Lagrange parameter T, the mean output per person S/N and the mean wages per person may be calculated from the wages E_1 and E_2 without any further assumptions.

2. In standard economics the Cobb Douglas production function U

$$U = N_1^\alpha N_2^{1-\alpha} = N x_1^\alpha x_2^{1-\alpha} \qquad (1.33)$$

is often used in the Lagrange function. In addition to wages E_i, the employees x_i are rated by an additional elasticity parameter α. For an arbitrary value $\alpha = 0.7$ we obtain

$$x_1 = \alpha/[\alpha/E_1 + (1-\alpha)/E_2]/E_1 = 0.538$$

$$x_2 = \alpha/[\alpha/E_1 + (1-\alpha)/E_2]/E_2 = 0.462$$

$$T = (E_1/\alpha)(N_2/N_1)^{1-\alpha} = 22.44$$

$$S/N = x_1^\alpha x_2^{1-\alpha} = 0.5141$$

$$E/N = (E_1 x_1 + E_2 x_2) = 11.54$$

For all values of α the mean output S/N is lower and the mean wage costs E/N are higher compared to the entropy $S = \ln P$. The Cobb Douglas function is obviously not the optimal production function. The entropy and Cobb Douglas function look very similar and differ by a factor of about 1.4, this is shown in Figs 1.2 and 1.3.

1.5.5
Boltzmann Distribution

In Fig. 1.4 we have $N = 10$ buyers looking for automobiles. There are now $K = 5$ different car models on the market with the attractiveness $q_k = 1$. The constraint for each model is the price E_k. As a result we find N_k buyers for each car model k.

What is the most probable distribution?

The problem is solved by the Lagrange function (1.29). The probability $P(N_k)$ is given by (1.28):

$$L(N_k) = \ln\{N!/(\Pi N_k!)\Pi q_k^{N_k} - (1/T)\Sigma_k(N_k E_k)\} = \text{maximum!} \qquad (1.34)$$

For large numbers N the faculty may be replaced by the Stirling formula

$$N! = N \ln N - N \qquad (1.35)$$

where N may be replaced by $N = (\Sigma N_k)$. The Lagrange function is now given by

$$L(N_k) = \Big\{(\Sigma N_k)\ln(\Sigma N_k) - \Sigma(N_k \ln N_k) + \Sigma(N_k \ln q_k)$$

$$- (1/T)\Sigma N_k E_k\Big\} \to \text{maximum!} \qquad (1.36)$$

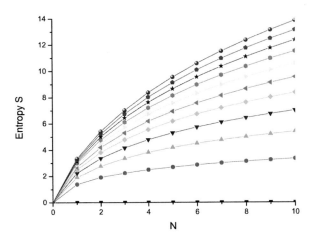

Fig. 1.2 The entropy $S(N_2) = (N_1 + N_1)\ln(N_1 + N_1) - N_1 \ln N_1 - N_2 \ln N_2$ plotted versus N_2 in the range from 0 to 10. The parameter is N_1 in the range from 0 to 10.

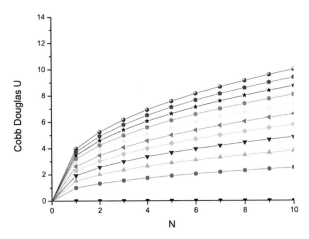

Fig. 1.3 The Cobb Douglas function $U(N_2) = N_1^\alpha N_2^{1-\alpha}$ plotted versus N_2 in the range from 0 to 10. The parameter is N_1 in the range from 0 to 10. In this range the Cobb Douglas Function is smaller than entropy by a factor of about 1.4 for all values of α. The closest match between the functions is obtained for $\alpha = 0.4$.

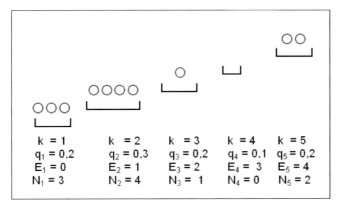

Fig. 1.4 $N = 10$ identical buyers are looking for automobiles. There are now $K = 5$ different car models on the market with the attractiveness $q_k = 1$. The constraint for each model is the price E_k. As a result we find N_k buyers for each car model.

At equilibrium the Lagrange function may be differentiated with respect to N_k,

$$\partial L/\partial N_k = \{\ln(\Sigma N_k) - \ln N_k\} + \ln q_k - E_k/T = 0 \tag{1.37}$$

This leads to the distribution of N_k different objects as a function of price E_k,

$$N_k/N = q_k \exp(-E_k/T) \tag{1.38}$$

The Boltzmann distribution is the most probable distribution of N elements in K categories under constraints (E).

Figure 1.5 shows the distribution of cars sold in the German automobile market in 1998. According to the German tax laws there are four classes of cars, 1.5 liters. 1.8 liters. 2.4 liters and above. The diamond points in Fig. 1.5 are the data of cars sold in Germany in 1998 as given by the automobile industry. The distribution does not yield a Boltzmann distribution, nearly six million units are missing in the lowest category at 1.5 liters or 20 000 DEM. However, the number of seven million used cars is reported for Germany in 1998 by the German Automobile Agency in Flensburg. If this number is added to the new cars of 1998 in the lowest price category, a Boltzmann distribution is obtained. Obviously, the complete automobile market is determined by new and used cars!

Figure 1.6 shows traffic violators given by the German Traffic department in Flensburg, in 2000. The number of repeated violators is shown as a function of the fine (in number of points proportional to a fine in €). The number of violators decreases with a growing fine according to the Boltzmann distribution, Eq. (1.38). The data in Fig. 1.6 exactly follow the calculations. Decisions on

16 | *1 A Thermodynamic Formulation of Economics*

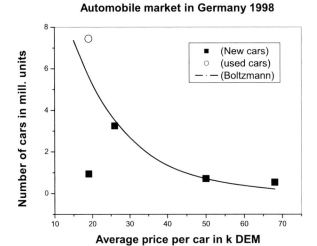

Fig. 1.5 Production of new cars in Germany 1998. According to the German tax classes four types of cars are reported by the industry, 1.5 liters, 1.8 liters, 2.4 liters and above. A Boltzmann distribution, Eq. (1.38), is obtained only if the number of seven million used cars is added to the lowest price category. The complete automobile market is only given by new and used cars!

buying cars or violating traffic rules depends exponentially on the price (E) in relation to the standard of living (T).

1.6
Entropy in Economics

1.6.1
Entropy as a Production Function

In stochastic systems the production function (S) is given by

$$S(N_k) = \ln P(N_k). \tag{1.39}$$

The term entropy (S) is used in mathematics, physics and information science, and has been first introduced to economics by N. Georgescu-Roegen (1974) [7] and more recently by D. K. Foley and J. Mimkes (1994) [4]. In stochastic systems, entropy replaces the Cobb Douglas function of standard economics as a production function. There are several reasons for this replacement.

1. Entropy is a natural system function without additional parameters. The Cobb Douglas function has an arbitrary "elasticity parameter" α.

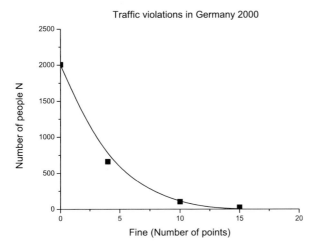

Fig. 1.6 The Boltzmann distribution of repeated traffic violating agents as a function of fine (in points) in Germany 2000. Decisions on violations depend exponentially on the fine (E).

2. The Cobb Douglas function has been found by fitting data, there is no theoretical foundation for this function. The similarity between the functions in Figs 1.2 and 1.3 suggests that entropy would fit the data as well.

3. Figure 1.2 and 1.3 indicate that entropy leads to higher values of production and lower values of costs than does the Cobb Douglas function.

4. In the Lagrange principle the value of different groups of labor are characterized by their wages. They do not need an additional characterization by a parameter α. The Lagrange function with entropy is a sufficient characterization of labor groups.

5. Entropy has a very great significance in production and trade, entropy characterizes the change in the distribution of commodities and money during the process of production and trade. This will be discussed in the following section.

1.6.2 Entropy of Commodity Distribution

We will now discuss the significance of entropy in economics in more detail.

Example: A farmer sells ten apples. Before the transaction the apples are unevenly distributed, the customers have none and the farmer has all the apples, Fig. 1.7.

Fig. 1.7 Before selling the ten apples are unevenly distributed, the customers have none and the farmer has all the apples.

The probability for the distribution of ten apples to one out of five persons is given by Eq. (1.28):

$$P_1 = 10!/(0!0!0!0!10!)/5^{10} = 5^{-10}$$

$$S_1 = -10\ln 5 = -16.094$$

The farmer sells two apples to each customer and keeps two for himself, Fig. 1.8. Probability and entropy are now

$$P_2 = 10!/(2!2!2!2!2!)/5^{10} = 0.0116$$

$$S_2 = \ln(0.0116) = -4.557$$

The entropy of these distributions is negative, since the probability is always $P \leq 1$.

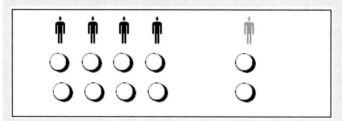

Fig. 1.8 After selling, the apples are evenly distributed, the farmer and each customer has two apples.

In the process of selling, the entropy of apple distribution has changed by

$$\Delta S = S_2 - S_1 = -4.557 + 16.094 = 11.437$$

Selling (distribution) of commodities is equivalent to an increase in entropy. At the end of the sale the distribution of apples has reached equilibrium, all have the same number of apples and the probability and entropy are at a maximum,

$$S_2 = \ln P_2 = \text{maximum}!$$

The trade of commodities is generally finished when equilibrium has been reached.

1.6.3 Entropy of Capital Distribution

Example: The farmer sells his ten apples for 1 € each. Before the transaction the farmer and each of his four customers have two 1 € coins in their pockets. The coins are evenly distributed, Fig. 1.9.

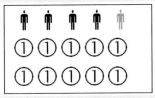

Fig. 1.9 Before selling the ten 1 € coins are evenly distributed: all five persons have two 1 € coins each in their pocket.

The probability for a distribution of two 1 € coins for each of five person is given by

$$P_1 = 10!/(2!2!2!2!2!)/5^{10} = 0.0116$$

$$S_1 = \ln(0.0116) = -4.557$$

The farmer sells two apples to each customer and keeps two for himself, Fig. 1.10. He collects 2 € from each customer. The farmer has now 10 € in his pocket, the customers have no more 1 € coins. The probability and entropy are now

$$P_2 = 10!/(0!0!0!0!10!)/5^{10} = 5^{-10}$$

$$S_2 = -10 \ln 5 = -16.094$$

Fig. 1.10 After selling the apples the 1 € coins are unevenly distributed: the farmer has all coins and the customers have none.

In the process of selling, the entropy of the 1 € coins distribution has changed by

$$\Delta S = S_2 - S_1 = -16.094 + 4.557 = -11.437$$

The negative entropy difference indicates that the coins have been collected. The example shows a new aspect of entropy (S). A positive change of entropy in a system of elements (commodities, capital) is equivalent to distributing, while a negative change of entropy corresponds to collecting elements (commodities, capital). In trading the entropies of commodities and capital change in the opposite direction, the absolute entropy difference of commodities and capital is the same.

1.6.4
Entropy of Production

In the production line of an automobile plant, workers have to assemble parts of a car according to the construction plan. In Fig. 1.11 there are still N different parts, which may be assembled in $P = N!$ possible ways. Assembling means ordering N parts together in one and only one way according to the construction plan.

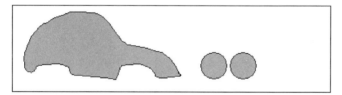

Fig. 1.11 Before assembling N parts of a product there are still $P = N!$ possibilities. Before production the car is still in disorder.

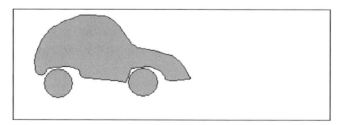

Fig. 1.12 Assembling means ordering N parts in one and only one way according to the construction plan. Production is ordering.

Assembling and ordering according to a plan means entropy reduction, $\Delta S < 0$. Work and production are always accompanied by entropy reduction. This applies to manual work as well as mental work. A puzzle

$$d+i+c+n+o+o+p+r+t+u = \text{production}$$

may be rearranged into a meaningful word. Mental work is ordering many ideas into one meaningful master plan or theory.

1.6.5
Summary of Entropy

Entropy may have many different aspects, but the main result for entropy may be stated as follows.

$$\Delta S = S_2 - S_1 > 0 \tag{1.40}$$

corresponds to distributing elements like commodities or money and creating disorder.

$$\Delta S = S_2 - S_1 < 0 \tag{1.41}$$

corresponds to collecting elements like commodities or money and creating order.

1.7
Mechanism of Production and Trade

1.7.1
The Carnot Process

The mechanism of production and trade is based on the Carnot process. Equation (1.19) may be integrated along the closed path with $T = $ constant and $S = $ constant in the T-S diagram, Fig. 1.13

$$-\oint \delta W = \oint \delta Q = \oint T\, dS = \int_1^2 T_1\, dS + \int_3^4 T_2\, dS = Y - C = \Delta Q \tag{1.42}$$

Automobile production is a typical economic process that can be modeled by the Carnot process in Fig. 1.13. The cycle of production starts and ends at point (1).

Example: Carnot cycle of automobile production.
(1) → (2): Automobile production starts at point (1). Workers with a low standard of living (T_1) produce the automobile according to the production plan (ΔS). The total production costs are given by material (E) and labor: $C = E + T_1 \Delta S$. The costs could be reduced by building the cars

according to the same production plan (ΔS) and same material (E) in a place with low standard of living (T_1).

(2) → (3): Transport of cars from production plant (T_1) to market (T_2).

(3) → (4): The automobiles are sold to customers at a market with a high standard of living (T_2). The sales price is given by the material and the market: $Y = E + T_2 \Delta S$.

(4) → (1): The cycle is closed by recycling the automobile.

The surplus is $\Delta Q = Y - C = \Delta T \Delta S$ and corresponds to the enclosed area in Fig. 1.13.

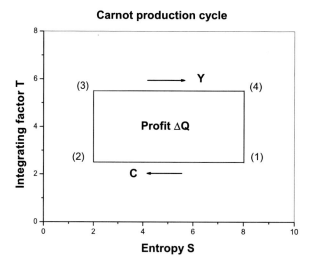

Fig. 1.13 In the Carnot production cycle the flow of goods starts and ends at point (1), (see text).

Example: Carnot cycle of apple farming.

(1) → (2): In the fall a farmer collects apples from his trees and stores them in his cellar. The work of collecting apples from the trees leads to a reduction of entropy of apple distribution, ($\Delta S < 0$). The production costs are $C = T_1 \Delta S$. T_1 is the price level of apples in the fall.

(2) → (3): The apples are stored in the cellar without changing the distribution, S = constant.

(3) → (4): In spring the apples are distributed ($\Delta S > 0$) from the farm to the market at the higher price level (T_2) of apples, the total amount of income from the apples is $Y = T_2 \Delta S$.

(4) → (1): The apples are sold and there is no change in the distribution of apples until the fall, $S = $ constant. The cycle starts again. The surplus of the apple production cycle is $\Delta Q = Y - C = \Delta T \Delta S$ and corresponds to the enclosed area in Fig. 1.2. The capital flow starts from point (1) and ends at point (1), but capital flow is opposite to the flow of apples, the work (W) and the profit (Q) in Eq. (1.42) have the opposite sign.

(1) → (4): The apple farmer goes to the market. The money in his pocket does not change, ($\Delta S = 0$).

(4) → (3): At the market the apple farmer collects ($\Delta S < 0$) money from the customers.

(3) → (2): The apple farmer returns home without spending the money in his pocket, ($\Delta S = 0$).

(2) → (1): At home the apple farmer distributes ($\Delta S < 0$) part of the money to the apple pickers.

Income (Y), costs (C) and profits (ΔQ) of labor are determined by

$$Y = E + T_2(S_4 - S_3) \tag{1.43}$$

$$C = E + T_1(S_2 - S_1) \tag{1.44}$$

$$\Delta Q = Y - C = \Delta T \Delta S \tag{1.45}$$

$\Delta Q = Y - C$ is the profit given by the enclosed area of Fig. 13. The materials (E) are the same in production and consumption and do not enter the calculations in a closed cycle. The Carnot process is the basis of all economic processes and will now be discussed in more detail. In economics every company, bank, and every person is a Carnot-like machine. In thermodynamics every motor and energy generator is a Carnot-like machine. In biology every living cell is a Carnot-like machine. The Carnot process is the common mechanism in economics, thermodynamics and biology.

1.7.2
The Origin of Growth and Wealth

What is the mechanism of economic interaction? If a baker sells bread to his customers, where does the wealth come from? From his work? From his customers?

If buyers and sellers just exchange values, there is no change in wealth and nobody will become richer. Since people make profit and do get rich by economic interactions, agents must take it from somewhere. If one agent takes it from the other agent, there will be no economic transactions, nobody wants to go to a market where he gets robbed.

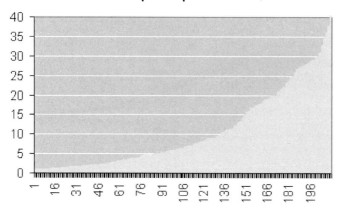

Fig. 1.14 shows the world GDP per capita in US$ for all countries in 2004. South Asian countries have the lowest and North America the highest GDP per capita.

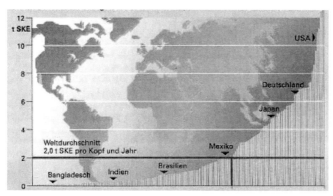

Fig. 1.15 shows the energy consumption in tons of coal in 122 countries in 1991. South Asian countries have the lowest and North America the highest energy consumption. (UN Statistical Yearbook 1991). The shape of the functions in Figs 1.14 and 1.15 are very similar and correspond to nearly the same countries.

The answer is the Carnot process. A heat pump extracts heat from a cold river and heats up a warm house. A bank may extract capital from poor savers and give it to rich investors. Economic interaction of two partners is only possible by exploitation of a third party due to clever manipulation (work). The most common objects of exploitation are natural resources, the environment and common property like water, air, coal and oil. A motor runs on oil, industrial production also runs on oil. Figure 1.14 and 1.15 show the world distribution of wealth (GDP per capita) and the world energy consumption per capita. Both run nearly parallel in all countries. This shows indeed that all wealth comes from exploitation. And this is not only true for the first person in the economic chain (like miners and farmers), but for everybody in the economic chain who wants to make a profit.

1.7.3
World Trade Mechanism

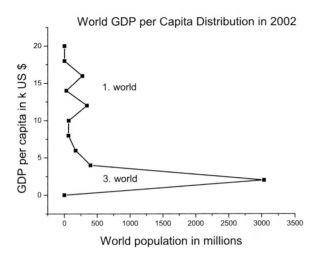

Fig. 1.16 shows the distribution of wealth in the world (CIA World Factbook, USA, 2004). The number of people in different income classes is given by the gross domestic product (GDP) per person. The distribution is clearly divided into two parts. The majority of people (about three billion) in the "third world" live below or close to 2 000 US$ per person. The minority of about one billion people in the "first world" have an income between 12 000 and 15 000 US$ per capita. (The gap at 14 000 US$ per capita is an artifact and due to US–EU currency fluctuations.) About two billion people, the "second world" live in between the two extremes.

Machines like motors or heat pumps always require or create two different temperatures (T). Inside the motor it is hot and therefore it needs water or air cooling outside. The heat pump works with a cold river and a warm house, the refrigerator has a cold inside and a warm outside. This difference in temperatures is necessary to make the Carnot process work. The areas of different temperatures have to be separated. If the door of the refrigerator stays open, it will not work, the efficiency is $r = 0$.

In all economic systems, work (W) according to the Carnot process always creates two different price or income levels (T). Buying and selling must create two price levels, otherwise there is no reason to do business. But also within a country, production (W) will create poor workers and rich capitalists.

The two-level system is also observed in the world distribution of wealth. Figure 1.16 shows the GDP distribution of the world and the corresponding number of people. The wealth of nations is clearly divided into two parts. In the "third world" more than three billion people live below or close to a GDP of 2 000 US$ per capita (1995). And in the "first world" about one billion people live between 12 000 and 16 000 US$ per capita. (The small dip at 14 000 US$ per capita is more or less artificial and due to fluctuations of the US$ and EU currencies.)

The Carnot process will stabilize the two different standards of living in the world population. The lower standard will grow with time, but also the difference will grow in order to enhance the efficiency. But the distribution of countries does not have to stay that way for ever. Some countries like China and India will emerge from the bottom and come closer to the top after a decade or two. Some richer countries may stay in the same position or even drop their standard of living. This will be discussed in Section 1.8 on the economic growth of interdependent systems.

1.7.4
Returns

Profits $\Delta Q = \Delta T \Delta S$ rise with the difference in price and costs ΔT of the product. The ideal efficiency of production in Eq. (1.16) may now be given by

$$r = \frac{Y-C}{C} = \frac{T_2 - T_1}{T_1} \tag{1.46}$$

Example: Dutch Importation of furniture from Indonesia

The GDP per capita in Holland and Indonesia are

$T_{\text{Holland}} = 12\,000$ US$ per capita and

$T_{\text{Indonesia}} = 3\,000$ US$ per capita.

Importing local commodities like furniture from Indonesia to Holland leads to an (ideal) efficiency

$r = (12\,000 - 3\,000)/3\,000 = 3$ or 300%.

Returns are ideally independent of the type of commodity. The efficiency is determined only by the difference in the standard of living ΔT.

The difference in complexity ΔS does not appear in efficiency calculations. For this reason the entropy function (S) has little importance in macro economics. However, in micro economics, the entropy (S) is linked to the probability (P) of the system. This is important for stock markets. A high entropy difference ΔS indicates a high probability of continuity or security of a share. A company with simple products may quickly be replaced. A company with complex products will last longer. Production (ΔW) creates a certain area $\Delta T \Delta S$, Fig. 1.13. The area with a large ΔT and a small ΔS will be highly efficient but less secure. A large ΔS and a small ΔT indicates that this company creates complex products with less efficiency and high security. The area $\Delta T \Delta S$ is determined by the work invested in the product. The shape of the area indicates whether a share is speculative and (perhaps) profitable or secure and less profitable. The optimum for a portfolio of stocks may be a nearly square area $\Delta T \Delta S$ with medium efficiency and medium security.

1.8
Dynamics of Production: Economic Growth

1.8.1
Two Interdependent Systems: Industry and Households

The dynamics of economic systems is again based on the Carnot process. So far all equations have been static, as in the thermodynamic formulation of economics the first and second laws do not contain time. But the length of a Carnot cycle is a natural time scale; a day, a month or a year. Economic growth may be handled like a starting motor. Both the inside and outside of the motor will get warmer, depending on how the heat is distributed. In economic systems the profit of each cycle has to be divided between the two sides of the production cycle, Y and C. If the lower level (C) gets the share "p" and the higher level (Y) the share $(1-p)$ of the profit (ΔQ), we obtain:

$$dY_1 = p \Delta Q \, dt \tag{1.47}$$

$$dY_2 = (1-p) \Delta Q \, dt \tag{1.48}$$

$$\Delta Q = Y_2 - Y_1 \tag{1.49}$$

The solution of this set of differential equations is:

$$Y_1(t) = Y_0 + p[Y_{20} - Y_{10}][\exp(\alpha t) - 1] \tag{1.50}$$

$$Y_2(t) = Y_{20} + (1-p)[Y_{20} - Y_{10}][\exp(\alpha t) - 1] \tag{1.51}$$

with

$$\alpha = (1 - 2p) \tag{1.52}$$

According to Eqs (1.47)–(1.52) a rising standard of living (Y) in two interdependent economic systems is determined by the share of the profit "p" of the group at the lower level (Y_1). The results are shown in Figs 1.17–1.22.

1. $p = 0$; Fig. 1.17. If all profit goes to the richer party (Y_2), the standard of living of group (2) will grow exponentially, the standard of living of the first party stays constant, (Y_{10}).

2. $p = 0.25$; Fig. 1.17. at 25% of the profit for the poorer party (Y_1) and 75% for the rich party (Y_2) both parties will grow exponentially. Examples are Japan and Germany after World War II, both economies were depending on the US and were growing exponentially, this is indicated in Fig. 1.18.

3. $p = 0.50$; Fig. 1.17. An even split between the two parties leads to a linear growth of both parties. The efficiency of the interaction is reduced with time.

4. $p = 0.75$; Fig. 1.19. The growth of both parties is leveling off not much above the initial standard of living. An example is the present US–Japanese economic relationship; both economies are close to each other without much economic growth, as shown in Fig. 1.20.

5. $p = 1.00$; Fig. 1.19. If all profit goes to the poor side, the standard of living of the poor party soon reaches the constant standard of living of the rich party.

6. $p = 1.25$; Fig. 1.21. If more than 100% of the profit goes to the poor party, (Y_2) will decrease, and (Y_1) will catch up with (Y_2). This has been observed in the relationship of West and East Germany after reunion in 1990, Fig. 1.22.

The data in Figs 1.17–1.22 can only indicate the results of Eqs (1.47)–(1.52), as all countries also have other (less important) interactions with other countries. The results may also be applied to other binary interactive economic systems like industry and households, or in trade. For industries and households the distribution of profit p is determined by the interacting agents of unions and industry, whereas in trade we have buyer and seller. This is now discussed in more detail.

1.8.2
Linear and Exponential Growth ($0 < p < 0.5$)

Figure 1.17 shows the problem of unions and industry in more detail. Unions tend to ask for high raises in payments, industry urges to invest the profits. Indeed, the fair deal, a split of profits 50:50 between workers and industry (dashed line) in Fig. 1.17 is not the best deal and will only result in linear growth. Workers and industry are much better off in a deal where 90% of the profits are reinvested (solid lines). Low increase in wages will lead to exponential growth for industry and later for workers as well. But workers (as well as their managers) will have to be more patient with pay raises, as in Germany or Japan after World War II, Fig. 1.18.

1.8.3
Trailing Economies: USA and Japan ($0.5 < p < 1$)

The opposite picture is shown in Fig. 1.19. A high factor p leads to decreasing efficiency, (Y_1) is trailing a decreasing (Y_2). After Japan and Germany have acquired many production plants, the factor p has grown and the efficiency of the exports has started to decrease. In Fig. 1.20 the economic level (Y_1) of Japan is now trailing the slowly decreasing level (Y_2) of the USA.

1.8.4
Converging Economies, West and East Germany ($p > 1$)

If the poor side (Y_1) profits very much, $p = pN_1/(N_1 + N_2) > 1$, both parties will converge, as shown in Fig. 1.21. This happened during the reunification of West and East Germany, Fig. 1.22. The standard of living in East Germany grew by 100% within six years, as the standard of living in West Germany was declining. The economic levels (Y_1) and (Y_2) in East and West Germany have nearly converged and differ now, after more than 15 years, by only 20%.

1.9
Conclusion

In the thermodynamic formulation of economics, the laws of markets and societies have been derived from calculus and statistical laws. No further assumptions have been used to derive the laws of economics. The calculated functions are supported by data, which all seem to agree very well. Many arguments indicate that the thermodynamic formulation of economics is a very general approach, which fits the data very well and explains economics and economic growth on the basis of natural science.

However, economic interactions are not only governed by statistical laws. Many economic interactions are restricted by traditional customs, civil laws or by agreements between trading partners. These additional laws will influence the interactions of free economic agents. How will they affect the results above? This will be discussed in more detail in the chapter on "The Thermodynamic Formulation of Social Science".

30 | *1 A Thermodynamic Formulation of Economics*

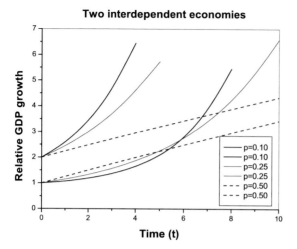

Fig. 1.17 The development of the standard of living of two interdependent economic systems starting at $Y_1 = 1$ and $Y_2 = 2$. The profit for the poor side varies from $p = 0.10$ to $p = 0.40$. After some time the standard of living of workers (Y_1) will grow with a lower pay raise p!!

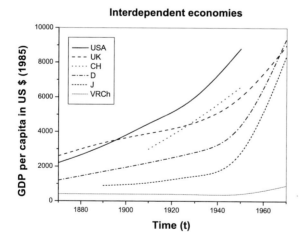

Fig. 1.18 Economic growth of US, UK, Switzerland, Japan, Germany and China between 1870 and 1990. The victorious allies USA and UK have grown exponentially. Japan and Germany only started to grow exponentially after World War II by international trade at low wages. China was excluded and did not take part in economic growth at that stage.

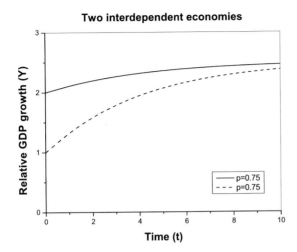

Fig. 1.19 The development of the standard of living of two interdependent economic systems starting at $Y_1 = 1$ and $Y_2 = 2$. At high values of profit for the poor side, $p = 0.75$, economic growth is declining with time.

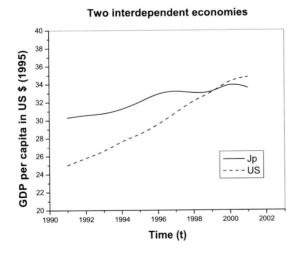

Fig. 1.20 The development of the standard of living (GDP/person) of the USA and Japan in quarters between 1980 and 2000. The interdependent economic systems are declining with time.

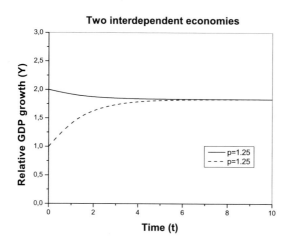

Fig. 1.21 The development of the standard of living of two interdependent economic systems starting at $Y_1 = 1$ and $Y_2 = 2$. At very high values of profit for the poor side, $p > 1$ both economies will converge below $Y_2 = 2$.

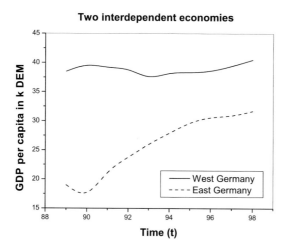

Fig. 1.22 Real standard of living in West and Germany between 1989 and 1998 due to productivity and capital transfer. In 1998 East Germany reached about 80% of the living standard in West Germany.

References

1 ARUKA, Y. (Ed.), *Evolutionary Controversies in Economics*, Springer, Tokyo, **2001**

2 *CIA World Factbook*, USA, **2004**

3 DRĂGULESCU, A. A., YAKOVENKO, V. M., *Eur. Phys. J. B* 20 (**2001**)

4 FOLEY, D. K., *Journal of Economic Theory* 62, No. 2, (**1994**)

5 FOWLER, R., GUGGENHEIM, E. A., *Statistical Thermodynamics*, Cambridge University Press, **1960**

6 KAPLAN, W., *Advanced Calculus*, Addison-Wesley, **2003**

7 GEORGESCU-ROEGEN, N., *The Entropy Law and the Economic Process*, Cambridge, Mass. Harvard Univ. Press, **1974**

8 LEVY, M., LEVY, H., SOLOMON, S., *Microscopic Simulation of Financial Markets*, Academic Press, New York, **2000**

9 MIMKES, J., *J. Thermal Analysis* 60 (**2000**), p. 1055–1069

10 MIMKES, J., *J. Thermal Analysis* 43 (**1995**), p. 521–537

11 MIMKES, J., WILLIS, G., in *Econophysics of Wealth Distributions*, CHATTERJEE, A., YARLAGADDA, S., CHAKRABARTI, B. K., (Eds.), Springer-Verlag Italia, Milan, **2005**

12 MIMKES, J., ARUKA, Y., in *Econophysics of Wealth Distributions*, CHATTERJEE, A., YARLAGADDA, S., CHAKRABARTI, B. K., (Ed.), Springer-Verlag Italia, Milan, **2005**

13 SILVA, A. C, YAKOVENKO, V. M., *Europhysics Letters* 69 (**2005**), p. 304–310

14 SOLOMON, S, RICHMOND, P., in *Economics with Heterogeneous Interacting Agents*, KIRMAN, A., ZIMMERMANN, J.-B. (Ed.), Springer-Verlag, Heidelberg **2001**

15 STANLEY, H. E., AMARAL, L. A. N., CANNING, D., GOPIKRISHNAN, P., LEEA Y. AND LIUA, Y., *Physica A* 269 (**1999**), p. 156–169

16 WEIDLICH, W., *Sociodynamics*, Amsterdam: Harwood Acad. Publ., **2000**

17 WEIDLICH, W., *The use of statistical models in sociology*, Collective Phenomena 1 (**1972**), p. 51

18 WILLIS, G., MIMKES, J., Cond-mat/0406694.

2
Zero-intelligence Models of Limit-order Markets
Robin Stinchcombe

Financial markets are similar to common macroscopic physical systems in being collective many body assemblies with out-of-equilibrium stochastic dynamics. Perhaps the simplest of all financial markets is the limit-order market, where an order book records placement and removal of orders to buy or sell, and their settlement. Regarding the orders as "particles" depositing, evaporating or annihilating, at prescribed rates, on a price axis provides both a physical analogy and the starting point for a class of market models. These are the "zero-intelligence" models of limit-order markets (ones where individual agents' choices and strategies are not allowed for beyond a general stochastic rule related to the market condition). The construction and analysis of such models and their properties and limitations will be discussed here. It will be outlined how, in the last few years, detailed temporal limit-order market data became available electronically which made it possible to infer the stochastic dynamic processes which operate in the market and how their rates are connected between themselves and to the market condition. The interactive feedback involved here is the origin of the most interesting collective properties of the system, and determines the nature of its stability and fluctuating behavior, while the stochastic nature of the processes makes all predictions statistical. Analytic analysis of such models has made some progress using techniques which will be outlined. A concluding discussion concerns the possible universality of the particular model arrived at, and also how the zero-intelligence description contains agent-based aspects. An explanation of the empirical way they are incorporated (through the phenomenological rate relations inferred from the market data) needs a complementary many-player approach, such as a generalized minority game description.

2.1
Introduction

A market is where traders arrange transactions. Markets have existed since the earliest times that humans formed social groups, and they at first involved

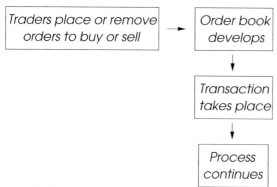

Fig. 2.1 Schematic representation of limit-order market.

barter. As money systems emerged, the traders began to sell and buy, in increasing amounts and in more and more regulated conditions [1]. Stock markets eventually emerged in which particular stock is floated and traded, with sometimes very hectic trading, for example, when the stock is in great demand.

Among many different types of financial market which have emerged up to now [1–3], the limit-order market is one for patient traders. A particular feature is the order book, in which traders may place orders to buy ("bids") or sell ("asks") (for a particular stock) at prices they regard as realistic or achievable, being prepared to wait until that price may be realized. The state of the order book evolves until transactions eventually occur, and the process continues, as shown schematically in Fig. 2.1. A more detailed description is as follows: as more and more orders are placed (or removed) or shifted in price the "spread", i.e., the gap between the limiting orders to buy (i.e., the highest or "best" bid) or sell (the lowest ask) changes, and may decrease to such a value that new ("market") orders are attracted at one or other of the two limiting prices, and transactions immediately take place until no more matches exist.

Figure 2.2 shows schematically a part of this process corresponding to order placement (but, for simplicity, not removal or any other such process, e.g., shift) changing the state of the order book but not yet leading to transactions (i.e., no market orders involved in the evolution covered by the figure). The figure introduces the price axis, which is fundamental in modeling such markets, and is intended to suggest that, in modeling, (a) the movement of the gap edges is a crucial feature to follow, and (b) that the detailed working of the market is tied in with the specific order processes (placement, etc.) and how their probabilities depend on price, etc., and (c) that such markets differ from other financial markets because of the separation of gradual evolution and fast transactions.

2.1 Introduction

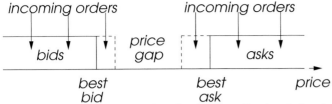

Fig. 2.2 Schematic representation of order disposition in a limit-order market, and of the contribution of order placement to its evolution between transactions.

A virtue of studying limit-order markets is the existence of the order book, which allows (i) a detailed quantitative assessment of the market condition to be made at any time, and (ii) (from its update) the actual processes to be identified. So, at the very least, quantitative empirical studies are, in principle, possible. We will return to this point, and its exploitation, subsequently, but first we consider ways one might approach such systems (see also [2], [3]).

As hinted at above, one way is to try to capture the working of the market just in terms of the price movement of the limiting orders (best bid, best ask), i.e., of the gap edges. In the most basic view (Fig. 2.3) these each undergo some form of random walk. An unconstrained unbiased free random walk is clearly an inadequate representation: the difference in price between the lowest sell order and the highest buy order is never negative (as two independent random walks would give) so at least the walks should be taken to be mutually reflecting; moreover, the actual walks are biased (as represented in

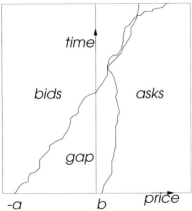

Fig. 2.3 Schematic representation of time-development of disposition on price axis of orders to buy (bids) and sell (asks) and the price gap between them (spread). The lower and upper limits of the gap (the best bid and best ask) are here labelled by $-a$ and b respectively. Both are time dependent, as illustrated.

the figure) in line with overall price movements such as inflation. The biased, and reflecting behavior should follow from a simple walk picture, provided it captures the relevant processes and their rates. Improved descriptions would need to build-in mechanisms for trends and correlations. Such approaches are illustrated in Section 2.8

It is expected that the condition of the market will affect the action of traders. This must be one source of correlations. Indeed it must introduce some feedback effects which could strongly modify the evolution of the market, which might vary from stabilization to instabilities and even crashes. For example, one could expect that some general price movement might trigger some noticeable collective reaction (like herding for example, [4]).

Such collective effects have analogies in many other contexts, ranging from social behavior and vehicular traffic to physical systems. All such examples involve (i) many constituents, and (ii) interactions. The latter can be direct interactions between individual constituents, or interactions mediated via other individuals or via groups of them. Another form of interaction is that between each individual and a common quantity, often called a mean field. An example of the latter is the noise level at a drinks party, to which the individuals present have to react by speaking more loudly to be heard, which introduces a destabilizing feedback. The obvious analogy suggests that another example might be some measure, evident to traders, of the average state of a financial market at each time.

The collective effects seen in macroscopic systems such as crowds (stampedes, riots, etc.) and traffic (crawls, jams, etc.) are familiar. Those in physical systems may not be so well known outside of physics, but they are ubiquitous, and many techniques have been developed for them, some of which are now being applied to nonphysical examples including those mentioned above.

Microscopic assemblies, of atoms etc., were the systems for which mean-field theory was first developed: the mean-field viewpoint can explain equilibrium phase transitions such as condensation (e.g., gas to liquid), or magnetism, etc., which are extreme examples of collective behavior [5]. It developed from phenomenological descriptions based on empirical observations and relationships.

A more detailed understanding of physical collective systems often involves their representation in terms of simplified models. Those (the minimal models) containing only the barest essentials are often the springboard for analytic solutions ([5,6]).

In common with crowds, traffic, etc., markets are stochastic nonequilibrium systems, so one needs all the procedures to be found in such sources as [7] and, beyond that, the models and approaches which have proved effective for the nonequilibrium collective systems in physics (see, for example, [8]).

This chapter will mainly adopt such approaches in the description of limit-order markets, because of their parallels with the physical cooperative systems, and with the other nonphysical macroscopic systems to which these approaches are being successfully applied.

We stress here that such an approach does not consider possible effects which could arise from the difference of traders from automatons, such as their ability to choose wildly individual strategies. So our approach is of the zero-intelligence type (cf. [9]).

We will return to this very important point in the concluding section (2.10) of the chapter.

To set up a minimal collective model, rather than to follow approaches such as analysis of time series (see, e.g., [2, 3]), involves consideration and identification of internal mechanisms. Simulation or analytical investigation of the resulting model could lead to detailed quantitative descriptions and predictions.

The chapter is laid out from here as follows. Section 2.2 gives some relevant background, including work on a range of possible models, especially those of [10–13]. Section 2.3 goes on to empirical data, collected from a particular order book, which can help to decide the most appropriate type of model [14], beginning with certain "static" properties. Section 2.4 contains a similar analysis of the data concerning dynamic properties and processes which distinguish between the possible models. Section 2.5 gives the resulting model while Section 2.6 lists some of its predictions [14] (see also [15, 16]). Section 2.7 introduces analytical approaches, especially for the model given in Section 2.5 and [17] for a forerunner model [12]. Random-walk descriptions generalized in the light of the identified processes are given in Section 2.8 [18, 19]. Further approximate analytical work [20] on the model given in Section 2.5 is briefly mentioned in 2.9. Section 2.10 is a concluding discussion.

In a substantial part of this chapter we follow closely the original accounts given in [14] and in [18], referring also to subsequent work in [15, 16, 20].

2.2
Possible Zero-intelligence Models

In physics, the usual sequence of procedures is: observing, and collecting empirical data, followed by inferring basic constituents, interrelationships and processes, modeling them, and then making predictions on the basis of the models, by calculation or theoretical analysis.

Ordinary financial markets are well suited to these methods because the copious instantaneous information available from market data allows, in principle, a far greater understanding of mechanisms as (well as inferences about microstructure [21, 22]).

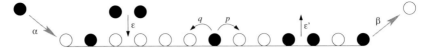

Fig. 2.4 Various particle exclusion processes on a chain: boundary injection, asymmetric hopping, pair creation/annihilation (or dimer deposition/evaporation), boundary ejection. Filled circles denote particles, open circles are vacancies.

Various sorts of minimal model have been proposed for modeling limit-order markets. In such models the evolution is regarded as due to stochastic order placement, and the analogy is made with particle systems where order size and price become particle mass and spatial position [10–15, 18–20, 23] (see also [21, 24]).

Much is known [8] about the behavior of (exclusion) particle models (not more than one particle at each site) having the types of state and processes exemplified in Fig. 2.4. These can model a vast range of systems from physical and chemical processes to traffic (of many things, from molecular motors to vehicles). That range is extended through equivalences to the configurations and processes shown in Fig. 2.5.

The equivalence involves treating clusters of adjacent particles in the first picture as columns in the second one. The latter picture can cover states and possible processes of a financial market (compare with Fig. 2.6), and so, through the equivalence, can the first.

This figure shows effects not just of deposition/placement, but also of removal. Market orders are not shown, but they would be asks placed at (or below) the best bid to effect immediate transactions there, and similarly with bid market orders. If the market order is greater than or equal to the accumulated order at the gap edge, the edge will move.

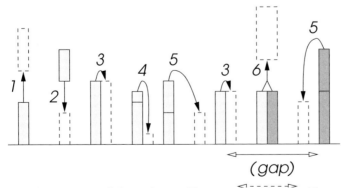

Fig. 2.5 Processes: Column picture with evaporation (1), deposition (2), asymmetric hopping (3), chipping (4), fragmentation (5), and pair annihilation (6), of two species (distinguished by shading).

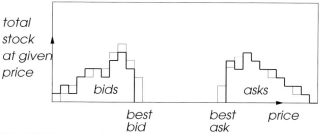

Fig. 2.6 Successive instantaneous states of limit-order market/book given as accumulated bids and asks (for a particular stock) as a function of price.

The main question concerns which processes to incorporate in the model.

In this regard the collective particle models proposed for limit-order markets fall into two families, one in which orders diffuse on the price axis [10, 11, 19] and the other where orders can be placed and executed [12, 21], and also cancelled [14, 15, 23].

Analysis of available market data makes it possible to distinguish between these families,

We will show, following [14],that there is clear evidence from such empirical studies that supports the second class of minimal model, and in particular shows that order diffusion on the price axis is unimportant. Instead, the data points to a specific nonequilibrium model with the processes of deposition, transaction and cancellation of orders, with time dependent probabilities.

2.3
Data Analysis and Empirical Facts Regarding Statics

This section provides the evidence for the conclusion just stated.

The data analyzed was obtained from the Island Electronic Communication Network (ECN) (www.island.com), part of the NASDAQ. This displays at each time step (tick) the 15 best orders of each type (bid or ask) and each stock (though leaving out the very small ones), and hence shows most of the activity of the market. Four stocks (Cisco (CSCO), Dell (DELL), Microsoft (MSFT) and Worldcom (WCOM)) were analyzed. Though orders are identified by a serial number, orders are not conserved because of partial filling and this makes a certain amount of reconstruction necessary, the errors of which are difficult to quantify. Nevertheless, the analysis goes beyond previous studies, which were mainly concerned with static properties of markets, and provides a dynamic analysis that is able to identify fundamental properties of such markets. In this way, by analyzing the available data from an appropriate point of view, we are able to arrive at the necessary ingredients for a minimal market model.

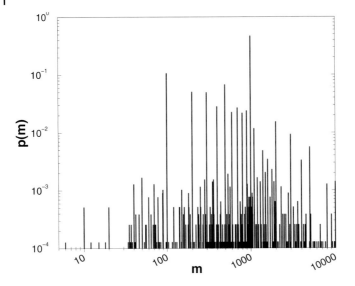

Fig. 2.7 Clustering seen in order size probability distribution $P(m)$ of bid size.

For each limit-order k its price, size (discrete), and lifetime (which is finite because of transaction, removal, or time out) are denoted by p, m, and τ respectively. A binary variable $\zeta_k \in \{B, A\} \equiv \{+1, -1\}$ can represent its type (bid or ask). This corresponds to a two-species particle model.

The well known clustering of orders in size is seen in Fig. 2.7 for the size distribution $P(m)$ for DELL. As with much of the other data displayed, this is for a particular time and day during 2000–2001 (detailed in the original paper [14]), but it is not noticeably time or stock dependent.

Figure 2.8 shows the lifetime histograms for all cancelled and for all fully-filled orders (bids and asks) from all four stocks. They can be fitted by power law decay with exponents -2.1 ± 0.1 and -1.5 ± 0.1, respectively. Time-out effects are evident in the peaks.

The shape of the price distributions, Fig. 2.9, is strongly time-dependent but also shows typical clustering, at about 16 tick separation; the tick is $ 1/256$. Hereafter we shall denote the position of gap edges on the price axis by $a(t), b(t)$, so that the gap is

$$g(t) = a(t) - b(t). \tag{2.1}$$

Δx denotes the price measured from the appropriate gap edge.

Figure 2.10 and 2.11 show coarse-grained versions of the order distribution on the price axis for two different stocks, on the same day, for bids ($\Delta x < 0$) and asks ($\Delta x > 0$). The convexity and the strong peaking at $\Delta x = 0$ are consistent with overdiffusive price behavior.

2.3 Data Analysis and Empirical Facts Regarding Statics | 43

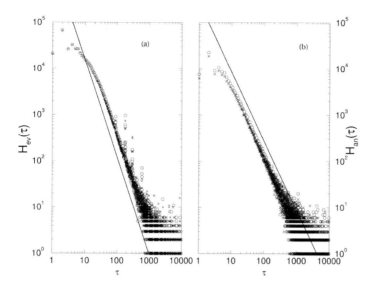

Fig. 2.8 Histograms of the lifetimes of bid (x) and ask orders (circles) for all stocks and all days. (a) histogram for cancelled orders. (b) histogram for fully filled orders. (The straight lines are power law fits, with exponents -2.1 and -1.5).

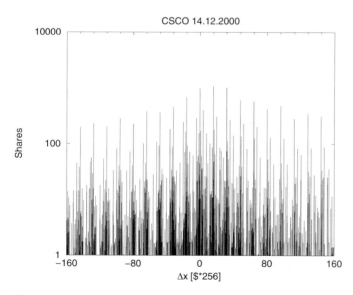

Fig. 2.9 Histogram of aggregated orders for CSCO, showing an order separation of about 16 ticks.

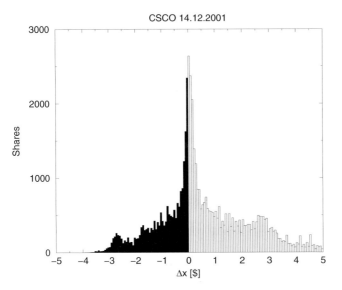

Fig. 2.10 Histogram of aggregated orders of CSCO. The left and right distributions are the buy and sell side, respectively.

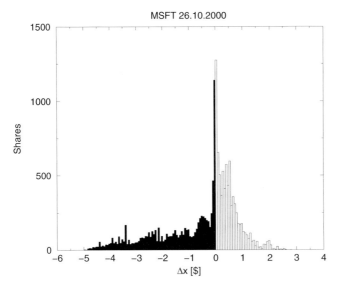

Fig. 2.11 As in Fig. 2.10 except that in this case the orders are for MSFT.

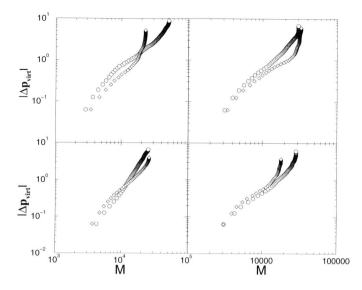

Fig. 2.12 Virtual impact of four days of CSCO. Circles and diamonds correspond to ask and bid sides, respectively.

The asymmetry of the distributions (bid and ask combined) and their time dependence, imply that the virtual impact function is non zero and time dependent. This function is an integrated measure of the effect of large orders on the price. It is shown in Fig. 2.12 for CSCO on four different days.

Still more important are the facts learnt from data analysis of dynamic properties. These identify the processes which must be incorporated in an appropriate model and their rates, and also what they may depend on. This part of the empirical study is given in the next section.

2.4
Dynamics: Processes, Rates, and Relationships

In the spirit of the particle models the limit orders can be regarded as two species of particle, A and B (asks and bids) so then placing or withdrawing an order corresponds to particle deposition or evaporation, while a transaction is the annihilation of an AB particle pair (immediately following deposition of a particle at one of the other type). The respective deposition, evaporation or annihilation rates, δ, η, α for each particle type (subscripts A or B) depend on their spatial position (price), and while they fluctuate they are correlated as is evident from the bid rates and their cross-correlations shown, for a particular stock and time window (of length 1500 s), in Fig. 2.13.

Algebraically time(-separation)-dependent rate autocorrelation functions (for DELL on 09.04.2001 are exhibited in Fig. 2.14 (b). as well as cross-

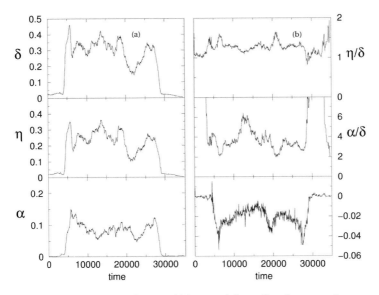

Fig. 2.13 (a) From top to bottom: bid rates of deposition δ, evaporation η and annihilation α in events/second, time averaged over a window of 1500 seconds of a day of DELL. (b) Ratios δ/η and δ/α, and $\delta - \eta - \alpha$ correlations.

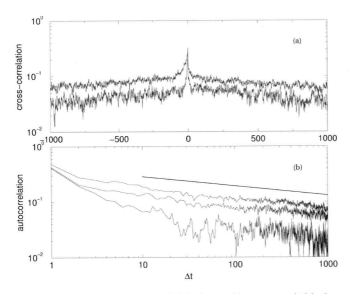

Fig. 2.14 (a) Cross-correlations $\langle \delta(t)\eta(t+\Delta t)\rangle$ (top) and $\langle \delta(t)\alpha(t+\Delta t)\rangle$. (b) Autocorrelation of the bid rates δ (top), η (middle) and α (bottom); the continuous line has a -0.17 exponent.

Figure 2.15 shows (again for DELL at 09.04.2001), for each type of event, the total number occurring in a small time interval proportional to $1/x$ (i.e., an accumulation of rates down to x). The accumulation gives its Poissonian tails. It can be seen from the figure that (as would be expected) deposition events dominate and, in contrast with some previous assumptions about dominant mechanisms, evaporation events are the next most frequent. This is supported by results [32] from other order markets.

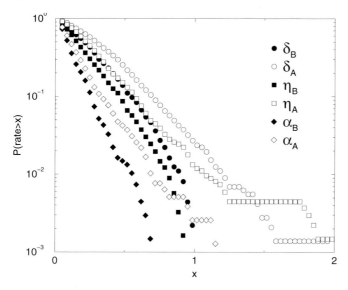

Fig. 2.15 Cumulative probability of event rates δ (circles), η (squares) and α (diamonds) of bid (filled symbols) and ask (plain symbols) sides.

We have not so far mentioned order diffusion, which is a key ingredient in one class of models [10, 11]. This corresponds to an individual order jumping in price. Figure 2.16 provides a measure of this, and shows that the degree of diffusion is extremely small: the vast majority of orders diffuse less than one effective tick during their life.

Another significant finding (from Fig. 2.13) is the nearly constant ratio of rates of evaporation and deposition.

All the above dynamic data concern temporal properties. We next turn to spatial (i.e., price-dependent) properties, and to spatio-temporal ones.

Figure 2.17 shows the dependence of event frequency on price, relative to gap edges. As could be expected, the activity is concentrated near the edges (cf. the order price distributions Figs 2.10 and 2.11).

Spatio-temporal dynamic properties are far harder to investigate, but there is empirical evidence that processes do depend on the time-dependent spatial characteristics of the market, as might be expected from the "feedback" idea mentioned earlier. In particular a dependence on the gap, g, (which is

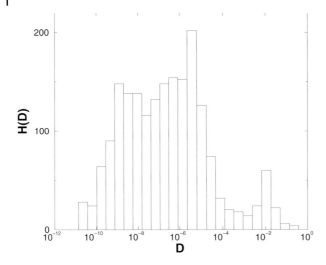

Fig. 2.16 Histogram $H(D)$ of individual order diffusion coefficients.

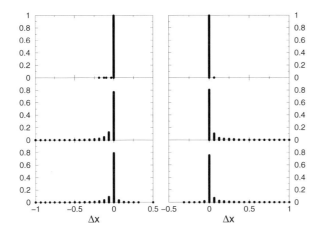

Fig. 2.17 Frequency of event versus the price to the relevant edge. Left (right) column refers to bid (ask) side. The events in each column, are, from the top, annihilation, evaporation, deposition.

indeed a quantity evident to traders – see the remark in Section 2.2) can be seen (Fig. 2.18) in the spatial width of deposition. Two slightly different measures of that are given, both showing that this is roughly constant or linear in g at relatively small or large g respectively. The dotted reference line in the lower graph is $y = 0.18\, x + C$. The linearity can be readily interpreted for orders placed in the gap: for orders placed elsewhere the dependence could have arisen from the small window set by the order book containing only the 15 closest-to-best orders. However, this seems not to be the case, since

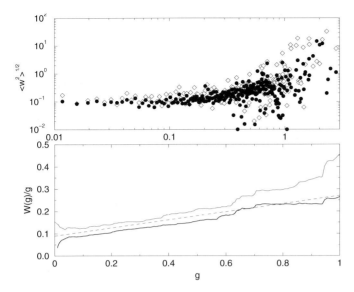

Fig. 2.18 Upper graph: spatial width $w(g)$ of deposition versus gap averaged over all stocks and all days (bids). Lower graph: normalized integral $W(g)/g$ of $w(g)$ (bids: thick curve, asks: thin curve).

(cf. Fig. 2.17) the activity occurs in a much smaller window than the one which those orders cover.

2.5 Resulting Model

The empirical evidence presented in the preceding section points to a collective model [14] with two species of particle (bids and asks) containing the processes of deposition, evaporation and annihilation. Diffusion appears to be negligible. Moreover, the evidence suggests that the associated probabilities should depend on price relative to the extremum and on the market condition, through the gap size. The price dependence allows annihilation to be included in the deposition process. For simplicity we take all orders to be of the same size, all events to be independent and time independent, and allow at most one order to be deposited in each time step.

The events are then:

(i) with probability δ_A an A particle (ask) is deposited at site x drawn at random from a probability density function $P_A(x,t)$; whenever x lies in the bulk of the B distribution, the annihilation is regarded as occurring at the relevant gap edge. Similarly for the B particle deposition.

(ii) particle A (or B) can evaporate with probability η_A (or η_B) at site x drawn from the appropriate $P(x)$.

We take each $P(x, t)$ to be centred on the relevant gap edge and to be Gaussian, with gap-dependent variance of the form

$$\sigma(t) = Kg(t) + C. \tag{2.2}$$

A set δ, η, K, C for each species comprize the parameters of the model. It can be convenient [15, 20] to work with dimensionless variables (scaled by characteristic measures of time, price interval, and share number).

Results from this simplest version of the model are outlined in the next section. In Section 2.5 various generalizations including distinct (and non-Gaussian) P's for the different processes will be considered.

How widely applicable the model is, and the possible need to extend beyond the zero-intelligence representation, will be considered in the conclusion.

2.6
Results from the Model

Monte Carlo simulations were carried out for a version of the model in which the parameters δ, η, K, C are taken to be species-independent (the same for bids and asks). The model gives:

(i) power law tails of returns,

(ii) volatility clustering,

(iii) rich market evolution,

(iv) asymptotically diffusive Hurst behavior.

The volatility clustering occurs in various parameter ranges and is connected with the power law decay of the gap autocorrelation function (compare [12]).

The market evolution given by the model (simulation) (Fig. 2.19) is strongly dependent on the parameter K measuring the feedback (via the gap) from the market condition. It can mimic behavior ranging from gradual inflation to crashes.

The Hurst exponent (defined in Section 2.5) increases with time from $1/4$ (underdiffusive) to $1/2$ (diffusive), as shown in the simulation results of Fig. 2.20

Possible generalizations of the model will be mentioned in the next and concluding sections, while analytic work on it is outlined in the next section.

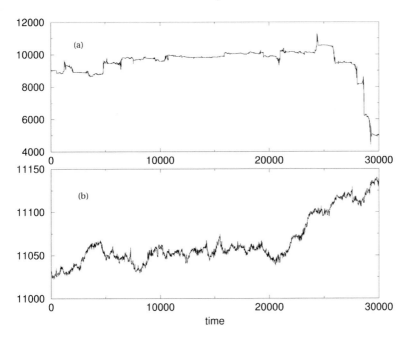

Fig. 2.19 Two time series of the market model with parameters $\alpha = 1/2$, $\delta = 1/100$, $C = 3$, and $K = 2$ (a) and $K = 1$ (b), showing the dramatic influence of K on the system's evolution.

2.7
Analytic Studies: Introduction and Mean-field Approach

We provide here an outline of analytic work on the three types of minimal model that have been proposed.

After early work (e.g., of [21, 22, 24]) on "toy" market models, theoretical work on the collective-particle models for limit-order markets have ranged from mean-field treatments, including the mean-field solution [17] of the Maslov model [12], through random walk and other approaches capturing the stochasticity of the diffusion models [10, 11, 19, 25] and possibly trends in generalized descriptions [18], to more advanced analytic work on the model [14–16] arrived at in the previous sections ([20]).

We discuss these approaches in the order just given.

Slanina [17] applied a mean-field approximation to the model of Maslov [12], assuming uniform order-density in the bulk bid and ask regions. The step-to-step evolution of the model was captured by multiplication of matrices, and a stationary distribution of price changes was found which has a power-law tail with exponent -2. The price autocorrelation function was also calculated.

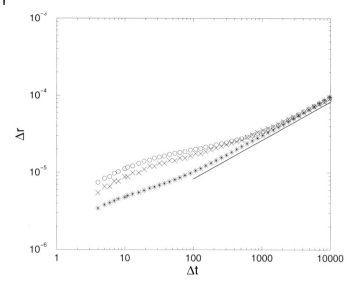

Fig. 2.20 Results from the market model: typical $\langle r_{\Delta t}\rangle$ versus Δt, showing that the Hurst exponent H is 1/4 for small times and tends to 1/2 for large times (○: at most one particle is deposited during t and $t+1$ $\delta = 0.5$, $\eta = 0.01$, $K = 1.5$, $C = 10$; x: simultaneous deposition of an average of δ particles of both types per unit of time (same parameters); ⋆: simultaneous deposition $\delta = 2$, $\eta = 0.04$, $K = 1.5$, $C = 10$).

In the same spirit, one can apply as follows, a mean-field approximation to the deposition, evaporation and annihilation model of [14] arrived at in Section 2.5.

We take the model in the generalized form involving separate probability density functions $P(x,t)$ for each bid or ask process, i.e., the functions $P_{X\nu_X}(x,t)$ with $X = A, B$ and $\nu_X = \delta_X, \eta_X$. Here we do not assume these

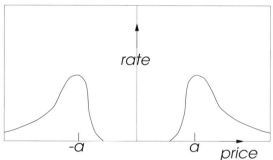

Fig. 2.21 Idealized symmetrized version of instantaneous order placement rates, for bids and asks as a function of price.

2.7 Analytic Studies: Introduction and Mean-field Approach

to be of Gaussian form. The special case having symmetry between bids and asks (illustrated in Fig. 2.21) is easier for analysis, but we do not use that here. For the general case, we can write down the following mean-field equations for the time evolution of each order-density profile ($\rho_X(x,t)$). For asks with $x \geq a(t)$ (where $a(t)$ is the relevant gap edge for asks), the equation is

$$\partial \rho_A(x,t)/\partial t = \delta_A P_{A\delta_A}(x,t)[1 - \rho_A(x,t)] - \rho_A(x,t)\eta_A P_{A\eta_A}(x,t)$$
$$-\delta(x - a(t)) \int_{-\infty}^{a(t)} \delta_A P_{A\delta_A}(x,t)\,dx \quad (2.3)$$

and similarly for bids. The mean-field equation treats ρ_A and P_{Av_A} as uncorrelated, even though in actual markets P_{Av_A} depends on the market condition, specified by ρ_A, ρ_B. It is in the same spirit as mean-field approximations for simple traffic models, such as the asymmetric exclusion process [8, 26], which give reasonable accounts of steady state traffic density profiles in space. With the adiabatic assumption that the time dependence of $\rho_X(x,t)$ is slow compared to that of $P_{Xv_X}(x,t)$ we obtain a quasi-static solution for $x > a$

$$\rho_A(x) \sim \delta_A P_{A\delta_A}(x,t) / [\delta_A P_{A\delta_A}(x,t) + \eta_A P_{A\eta_A}(x,t)] \quad (2.4)$$

Figure 2.13 suggests that the right-hand side is approximately time independent, so this result is not unreasonable.

From data used to obtain Fig. 2.17, $P_{A\delta_A}$ actually falls off with increasing positive $\Delta x \equiv x - a(t)$ faster than $P_{A\eta_A}$, which makes the ask density profile $\rho_A(x)$ a rapidly decreasing function of Δx. Similarly for the bids. So this mean-field argument gives a qualitative account (at least) of the schematic shape of

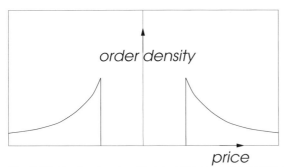

Fig. 2.22 Idealized symmetrized version of order-density profiles.

the order price distributions (density profiles) Fig. 2.10 and Fig. 2.11 inside the bid and ask regions. A schematic form is shown for the symmetric case in Fig. 2.22. Cut-offs at the best bid and ask appear from a self-consistency relation for these edges, which corresponds to a deterministic equation. The time-dependent solutions of Eq. (2.3) can also be found near the gap edges, and the approach can, in principle, be extended to allow for feedback from the market condition, for example, as represented by linear relationships of the widths of the $P_{X v_X}(x, t)$ to the gap. However, the mean-field approximation has removed the stochasticity, which is crucial for time-dependences, so it is better to carry out such investigations of the dynamics of the model allowing for the random walk of the gap edges, or by some generalization of the approaches applied (e.g., in [25]) to the diffusion-type models of [10] and [11].

2.8
Random-walk Analyses

Random-walk procedures can be applied in the following way [18] to versions of the model given in Section 2.5 and generalizations of it.

The main generalization is to an exclusion model in which rates change simultaneously. Such time-varying rates (involving periods where rates stay constant) are the key to understanding the observed overdiffusive behavior of financial markets. Overdiffusive behavior is where the logarithm of the price, p_t at time t, behaves like $p_t - p_0 \sim t^H$ with $H > 1/2$. H is the Hurst exponent.

Elements of the essential process, the placing of orders on the price axis, are as follows. Orders placed in the bulk are distinguished from those placed in the gap: the recipe is to place new ask orders, with probability δ_a at price $p_t = a(t) - \Delta$ ($\Delta > 0$) drawn from the distribution $P_a(\Delta)$ or similarly with probability δ'_a at price $a(t) + \Delta'$ ($\Delta' > 0$) drawn from the distribution $P'_a(\Delta')$; and similarly with bids. The exclusion condition is that no order can be placed at any price (site) already occupied by an order (so there is at most one order (particle) per site). Particles are cancelled with fixed probability η; with probability $\alpha_a(t)$ a market order annihilates the best ask at $a(t)$, and with probability $\alpha_b(t)$ the best bid at $b(t)$. In the following, all distributions (P's) are taken to be the same Gaussian function (including gap feedback of the market condition in its width). To mimic the temporal variation of rates seen in real markets various alternative protocols are considered:

(i) all rates are independently redrawn at random with probability p,

(ii) a variable $\sigma(t) = \pm 1$, associated with the direction of trend, changes sign with probability p, where the rates are of the cross-correlated form are:

$$\alpha_a(t) = \alpha_0 + \alpha_1 \sigma(t) \tag{2.5}$$

$$\alpha_b(t) = \alpha_0 - \alpha_1 \sigma(t) \tag{2.6}$$

$$\delta_a(t) = \delta_0 - \delta_1 \sigma(t) \tag{2.7}$$

$$\delta_b(t) = \delta_0 + \delta_1 \sigma(t) \tag{2.8}$$

(iii) as in (ii) except $\alpha_0 = \alpha_1$, $\delta_0 = \delta_1$ (so market orders and order deposition occur at opposite sides).

(iv) as in (iii) except $\sigma(t)$ only changes when the gap is zero ([19]).

Several gap properties can be calculated when cancellation is neglected and $P(\Delta) = \delta_{\Delta,1}$. Then $a(t), b(t)$ follow biased random walks and so does the gap, but with reflecting boundary conditions at zero (Fig. 2.23).

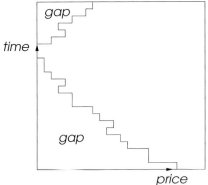

Fig. 2.23 Schematic representation of time-development of the bid-ask spread, i.e., the gap between limiting (best) orders to buy (best bid) and sell (best ask).

Gap closing is then a first passage problem [27] and the trend duration can be calculated analytically and compared to numerical results (see Fig. 2.24, for protocol (iii)).

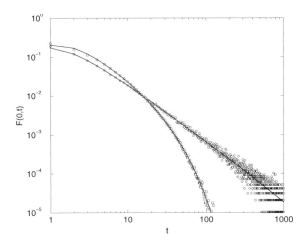

Fig. 2.24 Trend duration pdf for protocol (iii) ($\delta = 0.5$, $\alpha = 0.20$ (squares) and $\alpha = 0.48$ (circles)), 10^7 iterations. Continuous lines are theoretical predictions.

For all the protocols the price is overdiffusive ($H > 1/2$) at intermediate t, crossing over (at $t \sim 1/p$ for rules (i)–(iii)) to diffusive ($H = 1/2$) for large t. For protocol (i), neglecting the diffusion of the mid price gives

$$\langle (p_{t+\tau} - p_t)^2 \rangle = \left\langle \left(\sum_{t'=1}^{\tau} r_{t+t'} \right)^2 \right\rangle \sim \tau + \frac{2(1-p)}{p^2} \left[p\tau - 1 + (1-p)^{\tau} \right] \quad (2.9)$$

This result is shown as the dotted line in Fig. 2.25 and adequately represents the computed behavior except at very short times where underdiffusive behavior can occur.

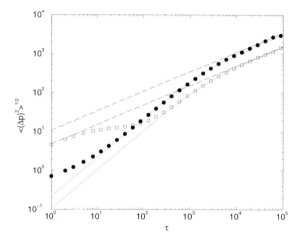

Fig. 2.25 Hurst plot of the price increments for protocol (i). Market order rates drawn uniformly between 0 and 0.1, deposition rate between 0 and 1 (circles), 0 and 0.5 (squares). $p = 0.001$, 10^6 iterations, no order cancellation. The dashed line represents normal random walk ($H = 1/2$) and dotted lines are obtained from the analysis.

Protocol (ii) yields similar behavior, though the overdiffusive region is pushed to increasingly large times as p increases (Fig. 2.26).

For protocol (iii) the price undergoes a persistent random walk (successive steps preferring to be in the same direction). The equivalence to an Ising chain allows an analytic solution at small p and large times where

$$\langle [p(t+\tau) - p(t)]^2 \rangle \simeq \alpha^2 \left[\tau + \frac{\tau(1-p)}{p} \tanh[\tau p/(1-p)] \right] \quad \tau \gg 1 \quad (2.10)$$

So H crosses over from 1 to 1/2 (Fig. 2.27). Protocol (iv) gives a similar behavior.

Analytic and numerical results for the price increment autocorrelation function have also been obtained. Results are shown in Fig. 2.28 for protocol (iii), as a function of p for times differing by $\tau = 1$.

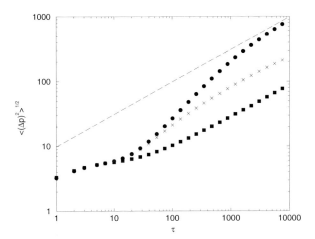

Fig. 2.26 Hurst plot for protocol (ii), showing characteristic under-diffusive prices at short times. ($\delta = 0.5$, $\alpha_0 = 0.22$, $\alpha_1 = 0.02$, $\eta = 0$, $p = 0.001$, 0.01 and 0.1, 10^6 iterations).

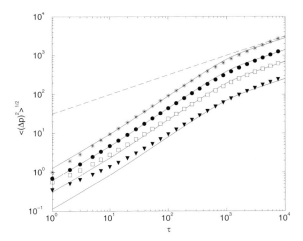

Fig. 2.27 Hurst plot for protocol (iii) for $\alpha = 0.9$, $\delta = 1$ (stars), $\alpha = 0.44$, $\delta = 0.5$ (circles), $\alpha = 0.22$, $\delta = 0.25$ (squares), and $\alpha = 0.08$, $\delta = 0.1$ (triangles) ($p = 0.001$, $\eta = 0$ and 10^6 iterations in all data sets).

Price increment distributions given by the protocols (i)–(iii) have exponential tails. It can be seen in Fig. 2.29 that the average order-density profiles computed, with gap feedback, have power-law tails (as do simulation results for the model in [14, 16]), and the functional shape of the average densities is the same as that of the price (cf. [23]).

In summary, this type of approach with time-varying rates can provide analytical results which capture typical features of the price movements of real

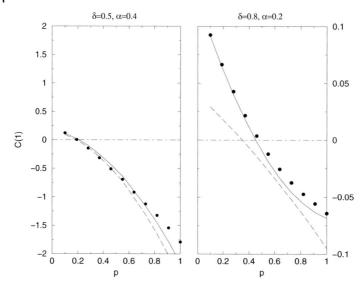

Fig. 2.28 Price increment autocorrelation function (left panel: $\delta = 0.5$, $\alpha = 0.4$; right panel: $\delta = 0.8$, $\delta = 0.4$. $\eta = 0$, average over 50 runs, 10^5 iterations); continuous and dashed lines are theoretical predictions.

markets.

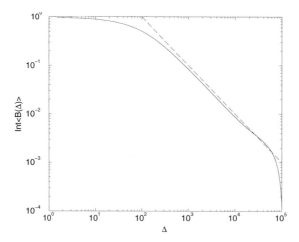

Fig. 2.29 Integrated average bid density $\langle B \rangle (\Delta)$ versus Δ, the position relative to the best bid, for $\eta = 0.01$ and gap feedback. The dashed line has exponent -1.

2.9
Independent Interval Approximation

It is possible to provide analyses of features that go beyond those just described by applying more sophisticated methods to the whole state of the market models. The work of [20] perhaps goes the furthest, so far, in that direction. It analyzes an exclusion version of the model of [14–16] starting from the full master equation for this nonequilibrium particle system and then uses the Independent Interval Approximation (IIA). This goes beyond mean-field theory in including correlations within, but not between, clusters (which can be the columns of accumulated orders at a given price or the intervals seperating them). The IIA can sucessfully capture many spatial and temporal features, including cluster size distributions, of nonequilibrium particle systems with a variety of rates and behavior ([28–31]). The application in [20] is more difficult than usual, because of the moving gap edges and the nature of the feedback. Nevertheless, considerable progress has been made. Among many other results obtained, density profiles (i.e., order density as a function of price) near the gap edges were found using the IIA as well as simulation. Figure 2.30 shows in schematic form a calculated profile for various values of a granularity parameter which characterizes the chunkiness of stored orders (and whch affects volatility). The analytic result agrees well with Monte Carlo simulation.

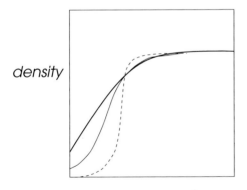

Fig. 2.30 Schematic representation of order-density profiles obtained from the independent interval analysis [20] of the model of [14–16] for various values of a granularity parameter.

2.10
Concluding Discussion

Evidence has been presented which strongly suggests the representation of limit-order markets in terms of a nonequilibrium particle model in which two species of particle play the role of orders. A particular set of processes (on a price axis) represent the placing, removal and transaction of orders. These processes have time-varying rates as a result of the effect, on the order processes, of the market condition. This feedback effect is expressed through the time-dependent spread, which governs the price dependence of the processes. Many properties of the model have been explored through simulation, including power law tails of returns, volatility clustering, and Hurst exponent behavior. The time series show very rich varieties of market evolution, set largely by the strength of the feedback, which strongly affects the stability. Various theoretical approaches which have been applied to the model have been outlined. Among these, random-walk descriptions capture the essence of the Hurst behavior and demonstrate the important consequences of the time-varying rates. The other approaches outlined range from a simple mean-field theory (whose generalization with added noise could be of value) to a more powerful approach using the independent interval approximation.

The evidence for the model was all taken from a specific limit-order market, so the question arises of whether it is special in some significant way. One possibly important thing which distinguishes markets is whether order removal is free. It is, in the Island ECN we investigated empirically, which raises the question of whether the large degree of order evaporation seen is due to this and not shared by other markets. However, work on other limit-order markets without free order removal find a similar proportion of evaporating orders [32]. So order removal cost aspects do not appear to affect the generality of the model.

Similarly one may ask whether the observed lack of diffusion is special to the market investigated. It does not appear to be so, from the available evidence.

A related question is whether no diffusion gives the behavior of a different class from that with small diffusion, i.e., whether, in the language of critical phenomena, diffusion is a relevant variable. There is here an argument that can be considered based on work on particle models with analogies to the proposed minimal market models. A simple example is the model in Fig. 2.4 but without the boundary processes. For the special case where the rates satisfy

$$p + q = \varepsilon + \varepsilon' \tag{2.11}$$

but are otherwise general, this can be exactly solved by first using a general mapping to quantum spin models [8, 33, 34] and then a transformation [35] to fermion variables [36]. One can use the results to see how the behavior of the model is affected by changing the rates, or even setting one to zero. Two distinct classes of asymptotic dynamics are found: exponential decay, from gapped spectrum, and power law decay, when there is no spectral gap. This is not decided by absence of diffusion (hopping) but by whether or not ε or $\varepsilon\prime$ vanishes.

The general idea of relevance of variables takes a much stronger form in the case of critical phenomena (near a continuous phase transition). For this, and other obvious reasons, it is important whether or not the financial markets are in any sense critical. Critical dynamics is normally found at large scales, of merely time in many nonequilibrium systems (e.g., having power-law asymptotic decay) or also when other scales are long (typically length scales in the case of physical phenomena). In the case of markets, the long-time power-law behavior appears to be critical, as do the large-price-difference distributions. This idea is reinforced by recent work [37] based on a phenomenological model of a multi-asset financial market that includes feedback on correlations which determine the optimal portfolio. Results from this model suggest that markets seem to self-organise to a near critical state. In such situations, i.e., in critical regions, both equilibrium [5,6] and nonequilibrium behavior (steady state and dynamic) [8] is governed by universality classes: all the members of a given class share precisely the same "universal" quantities, such as exponents quantifying steady-state power-law tails, and dynamic critical exponents (cf. the Hurst exponent). It is not parameters to which decide which universality class a critical system belongs, but more robust things (like symmetries in certain physical examples) usually related to one or two special relevant variables (which control the extent of symmetry breaking, etc.). The example of the soluble free fermion exclusion model given above, exhibits this. That gives minimal models a much more significant role in representing (exactly for critical behavior) the properties of wide classes of real sytems. So whether or not financial markets are near critical is of the greatest importance for their modeling.

The other main point we discuss here concerns the "zero-intelligence" [9] approach we have followed. The name relates to the apparent neglect of trader/agent choice. In any real market, traders are free to decide for themselves what strategies they use in placing or removing their orders. They do not all react in the same way to the market conditions, let alone all do the same thing all the time. So, approaches emphasising agent choice, for example, accommodated in the framework of a many-player game theory, such as the minority game [38,39], are currently being developed. In some sense what is being accomplished with such approaches is complementary to the results given in this chapter. Possibly what is needed is a combination of both approaches. Looked at from the perspectives we have used, the agents' choice (perception and strategies) must have partly given rise to the processes identified and to the form of the price dependence of all the $P(x,t)$'s appearing in the model, while the agents' reaction to the market condition is what introduced the dependence of the $P(x,t)$'s on the gap. To account for such forms, e.g., by incorporating an agent-based approach, such as the minority game, is a major challenge.

Acknowledgments

I am most grateful to Bikas Chakrabarti, Arnab Chatterjee and Anirban Chakraborti for the invitation to write this chapter. I wish to thank Richard Mash for economics guidance, and particularly Damien Challet for his extreme generosity in the collaborations on which much of the chapter is based. This work was supported by EPSRC under the Oxford Condensed Matter Theory Grants GR/R83712/01 and GR/M04426/01.

References

1 See, e.g., VARIAN, H. R., *Intermediate Microeconomics*, 6th Ed., Norton, New York, **2002**

2 BOUCHAUD, J. P., POTTERS, M., *Theory of Financial Risks*, Cambridge University Press, Cambridge, **2000**

3 MANTEGNA, R. N., STANLEY, H. E., *Introduction to Econophysics*, Cambridge University Press, Cambridge, **2000**

4 LUX, T., MARCHESI, M., *Nature* 397 (**1999**), pp. 498–500

5 YEOMANS, J. M., *Statistical Mechanics of Phase Transitions*, Oxford University Press, Oxford, **1992**

6 See, e.g., STINCHCOMBE, R. B., Phase Transitions, in *Order and Chaos in Nonlinear Physical Systems*, (Ed. S. Lundqvist, N. H. March, M. P. Tosi), Plenum, New York, **1988**, and references therein

7 VAN KAMPEN, N. G., *Stochastic Processes in Physics and Chemistry*, Elsevier, Amsterdam, **1992**

8 See, e.g., STINCHCOMBE, R. B., *Advances in Physics* 50, (**2001**) p. 431, and references therein

9 GODE, D., SUNDER, S., *J. of Political Economy* 101 (**1993**), p. 119

10 BAK, P., SHUBIK, M., PACZUSKI, M., *Physica A* 246 (**1997**) p. 430; e-print cond-mat/9609144

11 ELIEZER, D., KOGAN, I. I., *Capital Markets Abstracts, Market Microstructure* 2 (**1999**) 3, eprint cond-mat/9808240; CHAN, D. L. C., ELIEZER, D., KOGAN, I. I., (**2001**), e-print cond-mat/0101474

12 MASLOV, S., *Physica A* 278 (**2000**), p. 571, e-print cond-mat/9910502

13 MASLOV, S., MILLS, M., *Physica A* 299 **2000**, p. 234; e-print cond-mat/0102518

14 CHALLET, D., STINCHCOMBE, R. B., *Physica A* 300 (**2001**), p. 285, e-print cond-mat/0106114

15 DANIELS, M. G., FARMER, J. D., GILLEMOT, L., IORI, G., SMITH, E., *Phys. Rev. Lett.* 90 (**2003**), p. 108102, e-print cond-mat/0112422

16 CHALLET, D., STINCHCOMBE, R. B., PHYSICA A 324 (**2003**), p. 141, e-print cond-mat/0211082

17 SLANINA, F., *Phy. Rev. E* 64 (**2001**), 056136, preprint cond-mat/0104547

18 CHALLET, D., STINCHCOMBE, R. B., *Quant. Fin.* 3 (**2003**) p. 165, e-print cond-mat/0208025

19 WILLMANN, R. D., SCHÜTZ, G. M., CHALLET, D., *Physica A* 316 (**2002**), p. 526, e-print cond-mat/0206446

20 SMITH, E., FARMER, J. D., GILLEMOT, L., KRISHNAMURTHY, S., *Quant. Fin.* 3 (**2003**), p. 481, e-print cond-mat/0210475

21 DOMOWITZ, I., WANG, J., *J. of Econ. Dynamics and Control* 18 (**1994**), p. 29

22 BOLLERSLEV, T., DOMOWITZ, I., WANG, J., *J. of Econ. Dynamics and Control* 21 (**1997**), p. 1471

23 BOUCHAUD, J.-P., MÉZARD, M., POTTERS, M., *Quant. Fin.* 2 (**2002**) p. 251, e-print cond-mat/0203511

24 COHEN, K. J. et al., *J. of Political Economy* 89 (**1981**), p. 287

25 TANG, L.-H., TIAN, G.-S., *Physica A* 264 (**1999**), p. 543, e-print cond-mat/981114

26 DERRIDA, B., DOMANY, E., MUKAMEL, D., *J. Stat. Phys.* 69 (**1992**), p. 667

27 WEISS, G. H., *Aspects and Applications of the Random Walk*, North-Holland, Amsterdam, **1994**

28 MAJUMDAR, S. N., KRISHNAMURTHY, S., BARMA, M., *J. Stat. Phys.* 99 (**2000**), p. 1

29 REIS, F., STINCHCOMBE, R. B., *Phy. Rev. E* 70 (**2004**), p. 036109; e-print cond-mat/0406252

30 REIS, F., STINCHCOMBE, R. B., *Phy. Rev. E* 71 (**2005**), p. 026110; e-print cond-mat/0411122

31 REIS, F., STINCHCOMBE, R. B., *Phy. Rev. E* 72 (**2005**), p. 031109; e-print cond-mat/0509422

32 COPPEJANS, M., DOMOWITZ, I., (**1999**), http://www.smeal.psu.edu/faculty/ihd1/Domowitz.html

33 BARMA, M., GRYNBERG, M. D., STINCHCOMBE, R. B., *Phys. Rev. Lett.* 70 (**1993**), p. 1033; STINCHCOMBE, R. B., GRYNBERG, M. D., BARMA, M., *Phys. Rev. E* 47 (**1993**), p. 4018

34 ALCARAZ, F. C., DROZ, M., HENKEL, M., RITTENBERG, V., *Ann. Phys.* 230 (**1994**), p. 250

35 JORDAN, P., WIGNER, E., *Z. Phys.* 47 (**1928**), p. 631

36 GRYNBERG, M. D., STINCHCOMBE, R. B., *Phys. Rev. Lett.* 74 (**1995**), p. 1242; *Phys. Rev. Lett.* 76 (**1996**), p. 851

37 RAFFAELL, G., MARSILI, M., e-print physics/0508159

38 CHALLET, D., ZHANG, Y.-C., *Physica A* 246 (**1997**), p. 407

39 CHALLET, D., http://www.unifr.ch/econophysics/minority

3
Understanding and Managing the Future Evolution of a Competitive Multi-agent Population

David M.D. Smith and Neil F. Johnson

3.1
Introduction

Complex Adaptive Systems (CAS) are of great interest to theoretical physicists because they comprize large numbers of interacting objects or "agents" which, unlike particles in traditional physics, change their behavior based on experience [1]. Such adaptation yields complicated feedback processes at the microscopic level, which in turn generate complicated global dynamics at the macroscopic level. CAS also arguably represent the "hard" problem in biology, engineering, computation and sociology [1]. Depending on the application domain, the agents in CAS may be taken as representing species, people, cells, computer hardware or software, and are typically quite numerous, e.g., 10^2–10^3 [1, 2].

There is also great practical interest in the problem of predicting and subsequently controlling a Complex Adaptive System. Consider the enormous task facing a Complex Adaptive Systems "manager" in charge of overseeing some complicated computational, biological, medical, sociological or even economic system. He would certainly like to be able to predict its future evolution with sufficient accuracy that he could foresee the system heading towards any "dangerous" areas. However, prediction is not enough. He also needs to be able to steer the system away from this dangerous regime. Furthermore, the CAS manager needs to be able to achieve this *without* detailed knowledge of the present state of its thousand different components, nor does he want to have to shut down the system completely. Instead he is seeking for some form of "soft" control. Even in purely deterministic systems with only a few degrees of freedom, it is well known that highly complex dynamics such as chaos can arise [3] making any control very difficult. The "butterfly effect" whereby small perturbations can have huge uncontrollable consequences, comes to mind. One would think that things would be considerably worse in a CAS, given the much larger number of interacting objects. As an additional complication, a CAS may also contain stochastic processes at the microscopic and/or macroscopic levels, thereby adding an inherently random element to the system's dynamical evolution. The Central Limit Theorem tells us that the combined effect of a large number of stochastic processes tends fairly rapidly to a Gaussian distribution. Hence, one would think that even with reasonably complete knowledge of the present and past states of the system, the evolution would be essentially diffusive and hence difficult to control without imposing substantial global constraints.

In this chapter, we address this question of dynamical control for a simplified yet highly nontrivial model of a CAS. We show that a surprising level of prediction and subsequent control can be achieved by introducing small perturbations to the agent heterogeneity, i.e., "population engineering". In particular, the system's global evolution can be managed and undesired future scenarios avoided. Despite the many degrees of freedom and inherent stochasticity both at the microscopic and macroscopic levels, this global control requires only minimal knowledge on the part of the "system manager". For the somewhat simpler case of Cellular Automata, Israeli and Goldenfeld [4] have recently obtained the remarkable result that computationally irreducible physical processes can become computationally reducible at a coarse-grained level of description. Based on our findings, we speculate that similar ideas may hold for a far wider class of system comprizing populations of decision-taking, adaptive agents.

It is widely believed (see for example, [5]) that Brian Arthur's so-called El Farol Bar Problem [6,7] provides a representative toy model of a CAS where objects, components or individuals compete for some limited global resource (e.g., space in an overcrowded area). To make this model more complete in terms of real-world complex systems, the effect of network interconnections has recently been incorporated [8–11]. The El Farol Bar Problem concerns the collective decision-making of a group of potential bar-goers (i.e., agents) who use limited global information to predict whether they should attend a potentially overcrowded bar on a given night each week. The Statistical Mechanics community has adopted a binary version of this problem, the so-called Minority Game (MG) (see [12]–[25]), as a new form of Ising model which is worthy of study in its own right because of its highly nontrivial dynamics. Here we consider a general version of such multi-agent games which (a) incorporates a finite time-horizon H over which agents remember their strategies" past successes, to reflect the fact that the more recent past should have more influence than the distant past, and (b) allows for fluctuations in agent numbers, since agents only participate if they possess a strategy with a sufficiently high success rate [15]. The formalism we employ is applicable to any CAS which can be mapped onto a population of N objects repeatedly taking actions in the form of some global "game".

The chapter has several parts. Initially (Section 3.2), we discuss a very simple, two-state "game" to introduce and familiarize the reader with the nature of the mathematics which is explored in further detail in the rest of the chapter. We then more formally establish a common framework for describing the spectrum of future paths of the complex adaptive system (Section 3.3). This framework is general to any complex system which can be mapped onto a general BAR (Binary Agent Resource) model in which the system's future evolution is governed by past history over an arbitrary but finite time window H (the *Time-Horizon*). In fact, this formalism can be applied to any CAS whose internal dynamics are governed by a Markov Process [15], providing the tools whereby we can monitor the future evolution both with and without the perturbations to the population's composition. In Section 3.4, we discuss the BAR model in more detail, further information is provided in [26]. We emphasize that such BAR systems are *not* limited to the well-known El Farol Bar Problem and Minority Games – instead these two examples are specific limiting cases. Initial investigations of a finite time-horizon version of the Minority Game were first presented in [15]. In Section 3.5, we consider the system's evolution in the absence of any such perturbations, hence representing the system's natural evolution. In Section 3.6, we revisit this evolution in the presence of control, where this control is limited to relatively minor perturbations at the level of the heterogeneity of the population. In Section 3.7 we revisit the toy model of Section 3.2 to provide a reduced form of formalism for generating averaged quantities of the future possibilities. In Section 3.8 we discuss concluding remarks and possible extensions.

3.2
A Game of Two Dice

In this section we examine a very simple toy model employed to generate a time series (analogue output) and introduce the Future–Cast formalism to describe the model's properties. This toy model comprizes two internal states, A and B, and two dice also denoted A and B. We make these dice generic in that we assign their faces values and these are not equal in likelihood. The rules of the model are very simple. When the system is in state A, dice A is rolled and similarly for dice B. The outcome, δ_t, of the relevant dice is used to increment a time (price) series, whose update can be written

$$S_{t+1} = S_t + \delta_t \tag{3.1}$$

The model employs a very simple rule to govern the transitions between its internal states. If the outcome δ_t is greater than zero (recall that we have re-assigned the values on the faces) the internal state at time $t+1$ is A and consequently dice A will be used at the next step regardless of the dice used

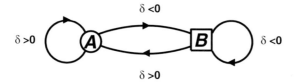

Fig. 3.1 The internal state transitions on the De Bruijn graph.

at this time step. Conversely, if $\delta_t < 0$, the internal state at $t+1$ will be B and dice B used for the next increment[1]. These transitions are shown in Fig. 3.1.

Let us prescribe our dice some values and observe the output of the system, namely rolling dice A could yield values $\{-5, +3, +8\}$ with probabilities $\{0.3, 0.5, 0.2\}$ and B yields values $\{-10, -8, +3\}$ with probabilities $\{0.1, 0.5, 0.4\}$. These are shown in Fig. 3.2. Let us consider that at some time t the system is in state A with the value of the time series being $S(t)$ and we wish to investigate the possible output over the next U time-steps ($S(t+U)$). Some examples of the system's change in output over the next 10 time-steps are shown in Fig. 3.3. The circles represent the system being in state A and the squares the system being in B.

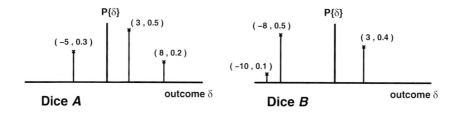

Fig. 3.2 The outcomes and associated probabilities of our two dice.

The stochasticity inherent in the system is evident in Fig. 3.3. Many possible paths could be realized even though they all originate from state A, and only a few of them are depicted. If one wanted to know more about the system's output after these 10 time-steps, one could consider many more runs and look at a histogram of the output for $U = 10$. This Monte Carlo [28] technique has been carried out in Fig. 3.4. It denotes the possible change in value of the systems analogue output after 10 time-steps and associated probability derived from 10^6 runs of the system all with the same initial starting state (A) at time t. However, this method is a numerical approximation. For accurate analysis

1) This state transition rule is not critical to the formalism. For example, the rule could be to use dice A if the last increment was odd and B is even, or indeed any property of the outcomes.

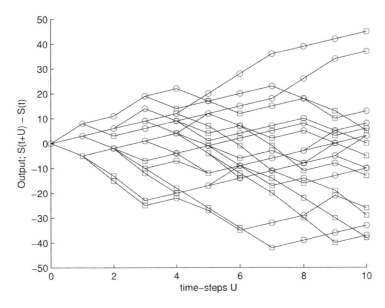

Fig. 3.3 The toy model's output over 10 time-steps for 20 realizations where the state at time t is **A** for all paths. The circles represent the system being in state **A** and the squares the system being in **B**.

of the system over longer time scales, or for more complex systems, it might prove both inaccurate and/or computationally intensive. For a more detailed analysis of our model, we must look at the internal state dynamics and how they map to the output. Let us consider all possible eventualities over 2 time-steps, again starting starting in state A at time t. All possible paths are denoted on Fig. 3.5. The resulting possible values of change in the output at time $t+2$ time-steps and their associated probabilities are also given explicitly. These values are exact in that they are calculated from the dice themselves. The *Future–Cast* framework which we now introduce will allow us to perform similar analysis over much longer periods. First, consider the possible values of $S(t+2) - S(t)$, which result in the system being in state A. The state transitions that could have occurred in order for these to arise are $A \to A \to A$ or $A \to B \to A$. The paths following the former can be considered a convolution (explores all possible paths and probabilities[2]) of the distribution of possible values of $S(t+1) - S(t)$ in state A with the distribution corresponding to the $A \to A$ transition. Likewise, the latter can be considered a convolution of the distribution of $S(t+1) - S(t)$ in state B with the distribution corresponding to a $B \to A$ transition. The resultant distribution of possibilities in state A at time

---

2) We define and use the discrete convolution operator $\otimes$ such that
$(f \otimes g)|_i = \sum_{j=-\infty}^{\infty} f(i-j)g(i)$.

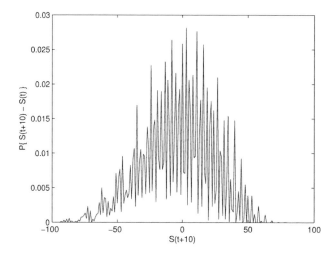

Fig. 3.4 The change in the system's output $S(t + U) - S(t)$ at $U = 10$, and associated probability as calculated for 10^6 time-series realizations.

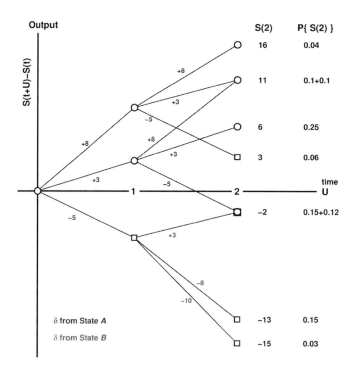

Fig. 3.5 All possible paths after two time-steps and associated probability. (For a color version please see color plate on page 593.)

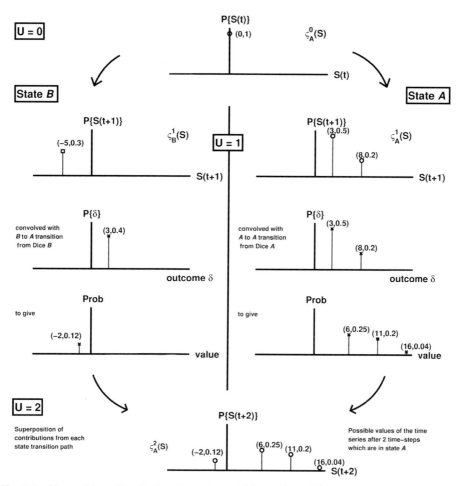

Fig. 3.6 All possible paths after two time-steps which could result in the system being in state **A**.

$t + 2$ is just the superposition of these convolutions as described in Fig. 3.6. We can set the initial value $S(t)$ to zero because the absolute position has no bearing on the evolution on the system. As such $S(t + U) - S(t) \equiv S(t + U)$.

So we note that if at some time $t + U$ there exists a distribution of values in our output which are in state A, then the convolution of this with the $B \to A$ transition distribution will result in a contribution at $t + U + 1$ to our distribution of S which will also be in A. This is evident in the right-hand side of Fig. 3.6. Let us define all these distributions. We denote all possible values in our output U time-steps beyond the present (time t) that are in state A as the function $\varsigma_A^U(S)$ and $\varsigma_B^U(S)$ is the function that describes the values of our output time series that are in state B (as shown in Fig. 3.6). We can similarly describe the transitions as prescribed by our dice. The possible values allowed and corresponding likelihoods for the $A \to A$ transition are denoted as $Y_{A \to A}(\delta)$ and similarly $Y_{A \to B}(\delta)$ for $A \to B$, $Y_{B \to A}(\delta)$ for $B \to A$ and $Y_{B \to B}(\delta)$ for the $B \to B$ state change. These are shown in Fig. 3.7.

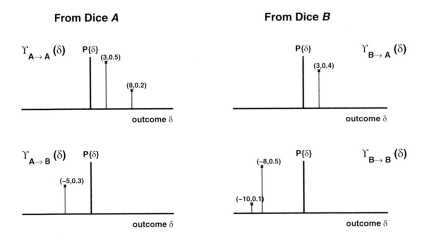

Fig. 3.7 The state transition distributions as prescribed by our dice.

We can now construct the Future–Cast. We can express the evolution of the output in their corresponding states as the superposition of the required convolutions:

$$\begin{aligned} \varsigma_A^{U+1}(S) &= Y_{A \to A}(\delta) \otimes \varsigma_A^U(S) + Y_{B \to A}(\delta) \otimes \varsigma_B^U(S) \\ \varsigma_B^{U+1}(S) &= Y_{A \to B}(\delta) \otimes \varsigma_A^U(S) + Y_{B \to B}(\delta) \otimes \varsigma_B^U(S) \end{aligned} \quad (3.2)$$

where $\otimes$ is the discrete convolution operator as defined in footnote 2. Recall that $S(0)$ has been set to zero such that we can consider the possible changes in output and the output itself to be identical. We can write this more concisely:

$$\underline{\varsigma}^{U+1} = \underline{\underline{Y}} \, \underline{\varsigma}^U \quad (3.3)$$

Where the element $Y_{1,1}$ contains the function $Y_{A \to A}(\delta)$ and the operator $\otimes$ and the element ς_1^{U+1} is the distribution $\varsigma_A^{U+1}(S)$. We note that this matrix of functions and operators is static, so only needs computing once. As such we can rewrite Eq 3.3 as

$$\underline{\varsigma}^U = \underline{\underline{Y}}^U \, \underline{\varsigma}^0 \quad (3.4)$$

such that $\underline{\varsigma}^0$ contains the state-wise information of our starting point (time t). This is the Future–Cast process. For a system starting in state A and with a start value of the time series of zero, the elements of $\underline{\varsigma}^0$ are as shown in

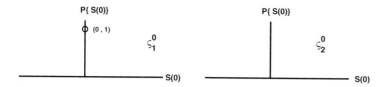

Fig. 3.8 The initial elements of our output distributions vector, $\underline{\varsigma}^0$ when starting the system in state **A** with an initial value of zero in our time series.

Fig. 3.8. Applying the Future–Cast process, we can look precisely at the systems potential output at any number of time-steps into the future. If we wish to consider the process over 10 steps again, we apply the following:

$$\underline{\varsigma}^{10} = \underline{\underline{Y}}^{10}\, \underline{\varsigma}^{0} \tag{3.5}$$

The resultant distribution of possible outputs (which we denote $\Pi^U(S)$) is then just the superposition of the contributions from each state. This distribution of the possible outputs at some time into the future is what we call the *Future–Cast*.

$$\Pi^{10} = \varsigma_1^{10} + \varsigma_2^{10} \tag{3.6}$$

This leads to distribution shown in Fig. 3.9.

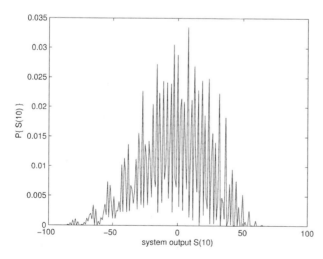

Fig. 3.9 The actual probability distribution $\Pi^{10}(S)$ of the output after 10 time-steps starting the system in state **A** with an initial value of zero in our time series.

Although the exact output as calculated using the Future–Cast process demonstrated in Fig. 3.9 compares well with the brute force numerical results of the Monte Carlo technique in Fig. 3.4, it allows us to perform some more interesting analysis without much more work, let alone computational exhaustion. Consider that we don't know the initial state of the system or that we want to know characteristic properties of the system. This might include wanting to know what the system does, on average, over one time-step increment. We could, for example, run the system for a very long time and investigate a histogram of the output movements over single time-steps as shown in Fig. 3.10. However, we can use the formalism to generate this result exactly.

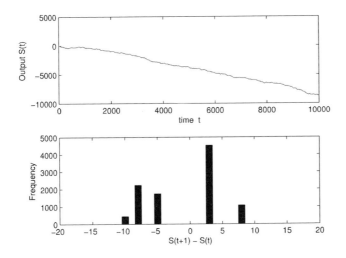

Fig. 3.10 The one step increments of our time-series as run over 10000 steps.

Imagine that the model has been run for a long time but we don't know which state it is in. Using the probabilities associated with the transitions between the states, we can infer the likelihood that the system's internal state is either A or B. Let the element Γ_1^t represent the probability that the system is in state A at some time t and Γ_2^t that the system is in state B. We can express these quantities at time $t+1$ by considering the probabilities of going between the states:

$$\begin{aligned} \Gamma_1^{t+1} &= T_{A \to A} \Gamma_1^t + T_{B \to A} \Gamma_2^t \\ \Gamma_2^{t+1} &= T_{A \to B} \Gamma_2^t + T_{B \to B} \Gamma_2^t \end{aligned} \quad (3.7)$$

Where $T_{A \to A}$ represents the probability that when the system is in state A it will be in state A at the next time-step. We can express this more concisely as:

$$\underline{\Gamma^{t+1}} = \underline{\underline{T}} \, \underline{\Gamma^t} \quad (3.8)$$

3.2 A Game of Two Dice

Such that $T_{1,1}$ is equivalent to $T_{A \to A}$. This is a Markov Chain. From the nature of our dice, we can trivially calculate the elements of $\underline{\underline{T}}$. The value for $T_{A \to A}$ is the sum over all elements in $Y_{A \to A}(\delta)$ or more precisely:

$$T_{A \to A} = \sum_{\delta=-\infty}^{\infty} Y_{A \to A}(\delta) \qquad (3.9)$$

The Markov Chain transition matrix $\underline{\underline{T}}$ for our system can thus be trivially written:

$$\underline{\underline{T}} = \begin{pmatrix} 0.7 & 0.4 \\ 0.3 & 0.6 \end{pmatrix} \qquad (3.10)$$

The static probabilities of the system being in either of its two states are given by the eigenvector solution to Eq. 3.11 with eigenvalue 1. This is equivalent to looking at the relative occurrence of the two states if the system were to be run over infinite time.

$$\underline{\Gamma} = \underline{\underline{T}}\,\underline{\Gamma} \qquad (3.11)$$

For our system, the static probability associated with being in state A is $\frac{4}{7}$ and obviously $\frac{3}{7}$ for state B. To look at the characteristic properties we are going to construct an initial vector similar to $\underline{\varsigma}^0$ in Eq. 3.2 for our Future–Cast formalism to act on but which is related to the static probabilities contained by the solution to Eq. 3.11. This vector is denoted $\underline{\kappa}$ and its form is described in Fig. 3.11. We employ $\underline{\kappa}$ in the Future–Cast over one time-step as in Eq. 3.12:

$$\underline{\varsigma}^1 = \underline{\underline{Y}}\,\underline{\kappa} \qquad (3.12)$$

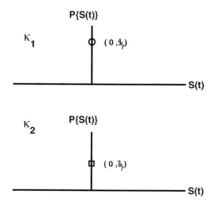

Fig. 3.11 Explicit depiction of the elements of vector $\underline{\kappa}$ which is used to analyze the characteristic behavior of the system.

The resulting distribution of possible outputs from superimposing the elements of $\underline{\varsigma}^1$ is the exact representation of the one time-step increments $(S(t+1) - \bar{S}(t))$ of the system if it were allowed to be run infinitely. We call this characteristic distribution Π_{char}^1. Applying the process a number of times will yield the exact distributions Π_{char}^U equivalent to looking at all values of $S(t+U) - S(t)$ which is a rolling window length U over an infinite time series, as in Fig. 3.12. This is also equivalent to running the system forward in time U time-steps from an unknown initial state, investigating all possible paths. The Markovian nature of the system means that this is not the same as the convolution of the one time-step characteristic Future–Cast convolved with itself U times.

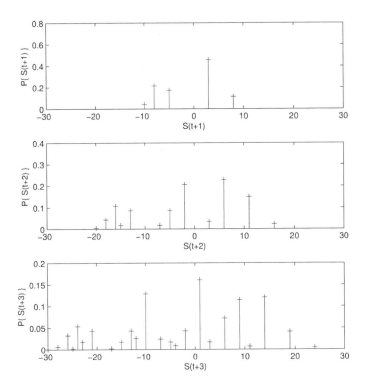

Fig. 3.12 The characteristic behavior of the system for $U = 1, 2, 3$ time-steps into the future from an unknown initial state. This is equivalent to looking at the relative frequency of the occurrence of changes in output values over 1, 2, and 3 time-step rolling windows

Clearly the characteristic Future–Cast over one time-step in Fig. 3.12 compares well with that of Fig. 3.10.

3.3
Formal Description of the System's Evolution

Here we provide a general formalism applicable to any Complex System which can be mapped onto a population of N species or "agents" who are repeatedly taking actions in some form of global "game". At each time-step each agent makes a (binary) decision $a_{\mu(t)}$ in response to the global information $\mu(t)$ which may reflect the history of past global outcomes. This global information is of the form of a bit-string of length m. For a general game, there exists some winning outcome $w(t)$ based on the aggregate action of the agents. Each agent holds a subset of all possible strategies – by assigning this subset randomly to each agent, we can mimic the effect of large-scale heterogeneity in the population. In other words, we have a simple way of generating a potentially diverse ecology of species, some of which may be similar but others quite different. One can hence investigate a typically diverse ecology whereby all possible species are represented, as opposed to special cases of ecologies which may themselves generate pathological behavior due to their lack of diversity.

The aggregate action of the population at each time-step t is represented by $D(t)$, which corresponds to the accumulated decisions of all the agents and hence the (analogue) output variable of the system at that time-step. The goal of the game, and hence the winning decision, could be to favor the minority group (MG), the majority group or indeed any function of the macroscopic or microscopic variables of the system. The individual agents do not themselves need to be conscious of the precise nature of the game, or even the algorithm for deciding how the winning decision is determined. Instead, they just know the global outcome, and hence whether their own strategies predicted the winning action[3]. The agents then reward the strategies in their possession if the strategy's predicted action would have been correct were that strategy implemented. The global history is then updated according to the winning decision. It can be expressed in decimal form as follows:

$$\mu(t) = \sum_{i=1}^{m} 2^{i-1}[w(t-i)+1] \qquad (3.13)$$

The system's dynamics are defined by the rules of the game. We will consider here the class of games whereby each agent uses his highest-scoring strategy at each time-step, and agents only participate if they possess a strategy with a sufficiently high success rate. (N.B. Both of these assumptions can be relaxed, thereby modifying the actual game being played). The following two scenarios might then arise during the system's evolution:

---

[3] The algorithm used by the "Game-master" to generate the winning decision could also incorporate a stochastic factor.

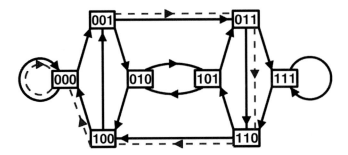

Fig. 3.13 A path of time-horizon length $H = 5$ (dashed line) superimposed on the de Bruin graph for $m = 3$. The 8 global outcome states represent the 8 possible bit-strings for the global information, and correspond to the global outcomes for the past $m = 3$ time-steps.

- An agent has two (or more) strategies which are tied in score and are above the confidence level, and the decisions from them differ.
- The number of agents choosing each of the two actions is equal, hence the winning decision is undecided.

We will consider these cases to be resolved with a fair "coin toss", thereby injecting stochasticity or "noise" into the system's dynamical evolution. In the first case, each agent will toss his own coin to break the tie, while in the second the Game-master tosses a single coin. To reflect the fact that evolving systems will typically be nonstationary, and hence the more distant past will presumably be perceived as less relevant to the agents, the strategies are rewarded as to whether they would have made correct predictions over the last H time-steps of the game's running. There is no limit on the size of H other than that it is finite and constant. The time-horizon represents a trajectory of length H on the de Bruijn graph in $\mu(t)$ (history) space [15] as shown in Fig. 3.13. The stochasticity in the game means that, for a given time-horizon H, and a given strategy allocation in the population, the output of the system is not always unique. We will denote the set of all possible outputs from the game at some number of time-steps beyond the time-horizon H, as the Future–Cast.

It is useful to work in a time-horizon space $\underline{\Gamma_t}$ of dimension 2^{m+H}. An element Γ_t corresponds to the last $m + H$ elements of the bit-string of global outcomes (or equivalently, the winning actions) produced by the game. This dimension is constant in time whereas for a nontime-horizon game it would grow linearly. For any given time-horizon state, Γ_t, there exists a unique score vector $\underline{G(t)}$ which is the set of scores $G_R(t)$ for all the strategies which an agent could possess. As such, for each particular time-horizon state, there exists a unique probability distribution of the aggregate action, $D(t)$. This distribution of possible actions when a specified state is reached will necessarily be the same each time that state is revisited. Thus, it is possible to construct a transition matrix (cf. Markov Chain [15]) $\underline{\underline{T}}$ of probabilities for the movements between these time-horizon states such that $\underline{P(\Gamma_t)}$ can be expressed as

$$\underline{P(\Gamma_t)} = \underline{\underline{T}}\,\underline{P(\Gamma_{t-1})} \tag{3.14}$$

where $\underline{P(\Gamma_t)}$ is a vector of dimension 2^{m+H} containing the probabilities of being in a given state Γ at time t.

The transition matrix of probabilities is constant in time and necessarily sparse. For each state, there are only two possible winning decisions. The number of nonzero elements in the matrix is thus $\leq 2^{(m+H+1)}$. We can use the transition matrix in an eigenvector–eigenvalue problem to obtain the stationary state solution of $\underline{P(\Gamma)} = \underline{\underline{T}}\,\underline{P(\Gamma)}$. This also allows calculation of some time-averaged macroscopic quantities of the game [15][4].

To generate the Future–Cast, we want to calculate the quantities in output space. To do this, we require:

- The probability distribution of $D(t)$ for a given time-horizon.
- The corresponding winning decisions, $w(t)$, for given $D(t)$.
- An algorithm generating output in terms of $D(t)$.

To implement the Future–Cast, we need to map from the transitions in the state space internal to the system to the macroscopic observables in the output space (often cumulative excess demand). We know that, in the transition matrix, the probabilities represent the summation over a distribution of possible aggregate actions which is binomial in the case where the agents are limited to two possible decisions. Using the output generating algorithm, we can construct an "adjacency" matrix $\underline{\underline{Y}}$ analogous to the transition matrix $\underline{\underline{T}}$, with the same dimensions. The elements of $\underline{\underline{Y}}$, contain probability distribution functions of change in output corresponding to the nonzero elements of the transition matrix together with the discrete convolution operator $\otimes$ whose form depends on that of the output generating algorithm.

The adjacency matrix of functions and operators can then be applied to a vector, $\underline{\varsigma^{U=0}(S)}$, containing information about the current state of the game and of the same dimension as $\underline{\Gamma_t}$. $\underline{\varsigma^{U=0}(S)}$ not only describes the time-horizon state positionally through its elements but also the current value in the output quantity S within that element. At $U = 0$, the state of the system is unique so

---

4) The steady state eigenvector solution is an exact expression equivalent to pre-multiplying the probability state vector $\underline{P(\Gamma_t)}$ by $\underline{\underline{T}}^\infty$. This effectively results in a probability state vector which is time-averaged over an infinite time interval.

there is only one nonzero element within $\underline{\varsigma}^{U=0}(S)$. This element corresponds to a probability distribution function of the current output value, its position within the vector corresponding to the current time-horizon state. The probability distribution function is necessarily of value unity at the current value or, for a Future–Cast expressed in terms of change in output from the current value, unity at the origin. The Future–Cast process for U time-steps beyond the present state can then be described by

$$\underline{\varsigma}^{U}(S) = \underline{\underline{Y}}^{U} \underline{\varsigma}^{0}(S) \tag{3.15}$$

The actual Future–Cast, $\Pi(S, U)$, is then computed by superimposing the elements of the output/time-horizon state vector:

$$\Pi^{U}(S) = \sum_{i=1}^{2^{(m+H)}} \varsigma_{i}^{U}(S). \tag{3.16}$$

Thus the Future–Cast, $\Pi^{U}(S)$, is a probability distribution of the outputs possible at U time-steps in the future.

As a result of the state dependence of the Markov Chain, Π is non-Gaussian. As with the steady-state solution of the state space transition matrix, we would like to find a "steady-state" equivalent for the output space[5] of the form

$$\Pi^{1}_{char}(S) = \langle \Pi^{1}(S) \rangle_{\infty} \tag{3.17}$$

where the one-time-step Future–Cast is time-averaged over an infinitely long period. Fortunately, we have the steady state solutions of $\underline{P(\Gamma)} = \underline{\underline{T}} \, \underline{P(\Gamma)}$ which are the (static) probabilities of being in a given time-horizon state at any time. By representing these probabilities as the appropriate functions, we can construct an "initial" vector, $\underline{\kappa}$, similar in form to $\underline{\varsigma}(S, 0)$ in Eq. 3.15 but equivalent to the eigenvector solution of the Markov Chain. We can then generate the solution of Eq. 3.17 for the *characteristic* Future–Cast, Π^{1}_{char}, for a given initial set of strategies. An element κ_i is again a probability distribution which is simply the point $(0, P_i(\Gamma))$, the static probability of being in the time-horizon state denoted by the elements position, i. We can then get back to the Future–Cast

$$\Pi^{1}_{char}(S) = \sum_{i=1}^{2^{(m+H)}} \varsigma_{i}^{1} \quad \text{where} \quad \underline{\varsigma}^{1} = \underline{\underline{Y}} \, \underline{\kappa}. \tag{3.18}$$

---

5) Note that we can use this framework to generate time-averaged quantities of *any* of the macroscopic quantities of the system (e.g., total number of agents playing) or volatility.

We can also generate characteristic Future–Casts for any number of time-steps, U, by pre-multiplying $\underline{\kappa}$ by $\underline{\underline{Y}}^U$

$$\Pi^U_{char}(S) = \sum_{i=1}^{2^{(m+H)}} \varsigma_i^U \quad \text{where} \quad \underline{\varsigma}^U = \underline{\underline{Y}}^U \underline{\kappa} \tag{3.19}$$

We note that Π^U_{char} is not equivalent to the convolution of Π^1_{char} with itself U times and as such is not necessarily Gaussian. The characteristic Future–Cast over U time-steps is simply the Future–Cast of length U from all the 2^{m+H} possible initial states where each contribution is given the appropriate weighting factor. This factor corresponds to the probability of being in that initial state. The characteristic Future–Cast can also be expressed as

$$\Pi^U_{char}(S) = \sum_{\Gamma=1}^{2^{(m+H)}} P(\Gamma) \, \Pi^U(S) \mid \Gamma \tag{3.20}$$

where $\Pi^U(S) \mid \Gamma$ is a normal Future–Cast from an initial time-horizon state Γ and $P(\Gamma)$ is the static probability of being in that state at a given time.

3.4
Binary Agent Resource System

The general binary framework of the BAR (Binary Agent Resource) system was discussed in Section 3.3. The global outcome of the "game" is represented as a binary digit which favors either those choosing option $+1$ or option -1 (or equivalently 1 or 0, A or B, etc.). The agents are randomly assigned s strategies at the beginning of the game. Each strategy comprizes an action $a^s_{\mu(t)}$ in response to each of the 2^m possible histories μ, thereby generating a total of 2^{2^m} strategies in the Full Strategy Space [6]. At each turn of the game, the agents employ their most successful strategy, being the one with the most virtual points. The agents are thus adaptive if $s > 1$.

We have already extended the BAR system by introducing the time-horizon H, which determines the number of past time-steps over which virtual points are collected for each strategy. We further extend the system by the introduction of a confidence level. The agents decide whether to participate or not depending on the success of their strategies. As such, the number of active

---

[6] We note that many features of the game can be reproduced using a Reduced Strategy Space of 2^{m+1} strategies, containing strategies which are either anti-correlated or uncorrelated with each other [12]. The framework established in the present paper is general to both the full and reduced strategy spaces, hence the full strategy space will be adopted here.

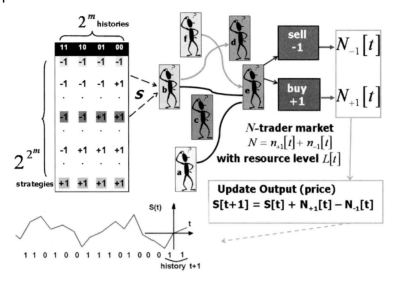

Fig. 3.14 Schematic diagram of the Binary Agent Resource (BAR) system. (For a color version please see color plate on page 594.)

agents $N(t)$ is less than or equal to N_{tot} at any given time-step. This results in a variable number of participants per time-step $V(t)$, and constitutes a "Grand Canonical" game. The threshold, τ, denotes the confidence level: each agent will only participate if he has a strategy with at least r points where

$$r = T(2\tau - 1). \tag{3.21}$$

Agents without an active strategy become temporarily inactive.

In keeping with typical biological, ecological, social or computational systems, the Game-master takes into account a finite global resource level when deciding the winning decision at each time-step. For simplicity, we will here consider the specific case[7] whereby the resource level $L(t) = \phi V(t)$ with $0 \leq \phi \leq 1$. We denote the number of agents choosing action $+1$ (or equivalently A) as $N_{+1}(t)$, and those that choose action -1 (or equivalently B) as $N_{-1}(t)$. If $L(t) - N_{+1}(t) > 0$ the winning action is $+1$ and vice versa. We define the winning decision 1 or 0 as follows:

$$w(t) = \text{step}[L(t) - N_{+1}(t)] \tag{3.22}$$

where we define $\text{step}[x]$ to be

$$\text{step}[x] = \begin{cases} 1 & \text{if } x > 0 \\ 0 & \text{if } x < 0 \\ \text{fair coin toss} & \text{if } x = 0 \end{cases} \tag{3.23}$$

---

[7] We note that ϕ itself could actually be a stochastic function of the known system parameters.

When $x = 0$, there is no definite winning option since $N_{+1}(t) = N_{-1}(t)$, hence the Game-master uses a random cointoss to decide between the two possible outcomes. We use a binary pay-off rule for rewarding strategy scores, although more complicated versions can, of course, be used. However, we note that nonbinary pay-offs (e.g., a proportional pay-off scheme) will decrease the probability of tied strategy scores, hence making the system more deterministic. Since we are interested in seeing the extent to which stochasticity can prevent control, we are instead interested in preserving the presence of such stochasticity. The reward function χ can be written

$$\chi[N_{+1}(t), L(t)] = \begin{cases} 1 & \text{for } w(t) = 1 \\ -1 & \text{for } w(t) = 0 \end{cases} \quad (3.24)$$

namely $+1$ for predicting the correct action and -1 for predicting the incorrect one. For a given strategy, R, the virtual points score is given by

$$G_R(t) = \sum_{i=t-T}^{t-1} a_R^{\mu(i)} \chi[N_{+1}(i), L(i)] \quad (3.25)$$

where $a_R^{\mu(t)}$ is the response of strategy, R, to the global information $\mu(t)$ summed over the rolling window of width H. The global output signal $D(t) = N_{+1}(t) - N_{-1}(t)$ is calculated at each iteration to generate an output time series.

3.5
Natural Evolution: No System Management

To realize all possible paths within a given game is necessarily computationally expensive. For a Future–Cast U timesteps beyond the current game state, there are necessarily 2^U winning decisions to be considered. Fortunately, not all winning decisions are realized by a specific game and the numerical generation of the Future–Cast can be made reasonably efficient.

Fortunately we can approach the Future–Cast analytically *without* having to keep track of the agents' individual microscopic properties. Instead we group the agents together via the population tensor of rank s given by $\underline{\underline{\Omega}}$, which we will refer to as the Quenched Disorder Matrix (QDM) [20]. This matrix is assumed to be constant over the time-scales of interest, and more typically is fixed at the beginning of the game. The entry $\Omega_{R2,R2,...}$ represents the number of agents holding the strategies $R1, R2, \ldots$ such that

$$\sum_{R,R',\ldots} \underline{\underline{\Omega}}_{R,R',\ldots} = N \quad (3.26)$$

For numerical analysis, it is useful to construct a symmetric version of this population tensor, $\underline{\underline{\Psi}}$. For the case $s = 2$, we will let $\underline{\underline{\Psi}} = \frac{1}{2}(\underline{\underline{\Omega}} + \underline{\underline{\Omega}}^{transpose})$ [17].

The output variable $D(t)$ can be written in terms of the decided agents $D_d(t)$ who act in a pre-determined way since they have a unique predicted action from their strategies, and the undecided agents $D_{ud}(t)$ who require an additional coin-toss in order to decide which action to take. Hence

$$D(t) = D_d(t) + D_{ud}(t) \tag{3.27}$$

We focus on $s = 2$ strategies per agent although the approach can be generalized. The element $\Psi_{R,R'}$ represents the number of agents holding both strategy R and R'. We can now write $D_d(t)$ as

$$D_d(t) = \sum_{R=1}^{Q} a_R^{\mu(t)} \mathcal{H}[G_R(t) - r] \sum_{R'=1}^{Q} (1 + \text{sgn}[G_R(t) - G_{R'}(t)]) \Psi_{R,R'} \tag{3.28}$$

where Q is the size of the strategy space, $\mathcal{H}$ is the Heaviside function and $\text{sgn}[x]$ is defined as

$$\text{sgn}[x] = \begin{cases} 1 & \text{if } x > 0 \\ -1 & \text{if } x < 0 \\ 0 & \text{if } x = 0 \end{cases} \tag{3.29}$$

The volume $V(t)$ of active agents can be expressed as

$$V(t) = \sum_{R,R'} \mathcal{H}[G_R(t) - r] \{ \text{sgn}[G_R(t) - G_{R'}(t)] + \tfrac{1}{2}\delta[G_R(t) - G_{R'}(t)] \} \Psi_{R,R'}$$

$$\tag{3.30}$$

where δ is the Dirac delta. The number of undecided agents $N_{ud}(t)$ is given by

$$N_{ud}(t) = \sum_{R,R'} \mathcal{H}[G_R(t) - r] \delta(G_R(t) - G_{R'}(t)) [1 - \delta(a_R^{\mu(t)} - a_{R'}^{\mu(t)})] \Psi_{R,R'} \tag{3.31}$$

We note that for $s = 2$, because each undecided agent's contribution to $D(t)$ is an integer, hence the demand of all the undecided agents $D_{ud}(t)$ can be written simply as

$$D_{ud}(t) \in 2 \, \text{Bin}\left(N_{ud}(t), \tfrac{1}{2}\right) - N_{ud}(t) \tag{3.32}$$

where $\text{Bin}(n, p)$ is a sample from a binomial distribution of n trials with probability of success p.

For any given time-horizon space-state Γ_t, the score vector $G(t)$ (i.e., the set of scores $G_R(t)$ for all the strategies in the QDM) is unique. Whenever this state is reached, the quantity $D_d(t)$ will necessarily always be the same, as

3.5 Natural Evolution: No System Management

will the distribution of $D_{ud}(t)$. We can now construct the transition matrix $\underline{\underline{T}}$ giving the probabilities for the movements between these time-horizon states. The element $\underline{\underline{T}}_{\Gamma_t|\Gamma_{t-1}}$ which corresponds to the transition from state Γ_{t-1} to Γ_t, is given for the (generalizable) $s = 2$ case by

$$\underline{\underline{T}}_{\Gamma_t|\Gamma_{t-1}} = \sum_{x=0}^{N_{ud}} \left\{ {}^{N_{ud}}C_x \left(\frac{1}{2}\right)^{N_{ud}} \delta\left[\text{sgn}\left(D_d + 2x - N_{ud}\right.\right.\right.$$
$$\left.\left. + V(1-2\phi)\right) + (2\mu_t\%2 - 1)\right]$$
$$+ {}^{N_{ud}}C_x \left(\frac{1}{2}\right)^{(N_{ud}+1)} \delta\left[\text{sgn}\left(D_d + 2x - N_{ud}\right.\right.$$
$$\left.\left.\left. + V(1-2\phi)\right) + 0\right]\right\} \quad (3.33)$$

where N_{ud}, D_d implies $N_{ud} \mid \Gamma_{t-1}$ and $D_d \mid \Gamma_{t-1}$, V implies $V(t-1)$, ϕ sets the resource level as described earlier and $\mu_t\%2$ is the required winning decision to get from state Γ_{t-1} to state Γ_t. We use the transition matrix in the eigenvector–eigenvalue problem to obtain the stationary state solution of $P(\Gamma) = \underline{\underline{T}}\, P(\Gamma)$. The probabilities in the transition matrix represent the summation over a distribution which is binomial in the $s = 2$ case. These distributions are all calculated from the QDM which is fixed from the outset. To transfer to output-space, we require an output generating algorithm. Here we use the equation

$$S(t+1) = S(t) + D(t) \quad (3.34)$$

hence the output value $S(t)$ represents the cumulative value of $D(t)$, while the increment $S(t+1) - S(t)$ is simply $D(t)$. Again, we use the discrete convolution operator $\otimes$ defined as

$$(f \otimes g)|_i = \sum_{j=-\infty}^{\infty} f(i-j) \times g(j) \quad (3.35)$$

The formalism could be extended for general output algorithms using differently defined convolution operators.

An element in the adjacency matrix for the $s = 2$ case can then be expressed as

$$\underline{\underline{Y}}_{\Gamma_t|\Gamma_{t-1}} = \left\{ \sum_{x=0}^{N_{ud}} \left((D_d + 2x - N_{ud}) \right. \right.$$

$$^{N_{ud}}C_x(\tfrac{1}{2})^{N_{ud}} \delta\left[\text{sgn}(D_d + 2x - N_{ud} + V(1-2\phi)) + (2\mu_t\%2 - 1) \right] +$$

$$\left. \left. ^{N_{ud}}C_x(\tfrac{1}{2})^{(N_{ud}+1)} \delta\left[\text{sgn}(D_d + 2x - N_{ud} + V(1-2\phi)) + 0 \right] \right) \right\} \otimes \quad (3.36)$$

where N_{ud}, D_d again implies $N_{ud} \mid \Gamma_{t-1}$ and $D_d \mid \Gamma_{t-1}$, V implies $V(t-1)$, and $\mu_t\%2$ is the winning decision necessary to move between the required states. The Future–Cast and characteristic Future–Casts ($\Pi^U(S)$, Π^U_{char}) U time-steps into the future can then be computed for a given initial quenched disorder matrix (QDM).

We now consider an example to illustrate the implementation. In particular, we provide the explicit solution of a Future–Cast in the regime of small m and H, given the randomly chosen quenched disorder matrix

$$\underline{\underline{\Omega}} = \begin{pmatrix} 0 & 0 & 1 & 0 & 0 & 1 & 0 & 1 & 0 & 0 & 1 & 0 & 0 & 0 & 0 & 0 \\ 0 & 0 & 0 & 1 & 0 & 0 & 1 & 0 & 2 & 0 & 0 & 1 & 0 & 1 & 0 & 0 \\ 0 & 1 & 1 & 0 & 1 & 0 & 1 & 0 & 0 & 0 & 0 & 2 & 1 & 0 & 0 & 1 \\ 0 & 1 & 0 & 0 & 2 & 0 & 1 & 0 & 0 & 0 & 0 & 0 & 0 & 0 & 0 & 2 \\ 1 & 0 & 0 & 0 & 0 & 1 & 0 & 0 & 0 & 1 & 1 & 1 & 1 & 1 & 0 & 1 \\ 1 & 0 & 0 & 0 & 1 & 1 & 0 & 0 & 1 & 0 & 0 & 0 & 0 & 1 & 2 & 0 \\ 0 & 1 & 1 & 0 & 0 & 0 & 0 & 0 & 1 & 0 & 2 & 1 & 0 & 0 & 0 & 1 \\ 0 & 1 & 0 & 0 & 1 & 1 & 0 & 0 & 0 & 1 & 0 & 1 & 0 & 1 & 0 & 1 \\ 2 & 1 & 0 & 0 & 0 & 1 & 1 & 1 & 0 & 0 & 0 & 0 & 0 & 0 & 0 & 0 \\ 0 & 0 & 0 & 0 & 0 & 0 & 1 & 0 & 0 & 3 & 1 & 1 & 0 & 1 & 0 & 1 \\ 0 & 2 & 0 & 0 & 1 & 0 & 0 & 1 & 1 & 3 & 2 & 0 & 0 & 0 & 1 & 1 \\ 0 & 0 & 2 & 0 & 0 & 0 & 0 & 0 & 0 & 0 & 0 & 0 & 0 & 0 & 0 & 0 \\ 1 & 0 & 1 & 1 & 0 & 0 & 0 & 0 & 1 & 0 & 0 & 0 & 0 & 0 & 0 & 0 \\ 0 & 0 & 0 & 1 & 0 & 0 & 0 & 2 & 1 & 0 & 1 & 0 & 0 & 0 & 1 & 0 \\ 0 & 2 & 0 & 1 & 0 & 0 & 0 & 0 & 0 & 0 & 1 & 0 & 0 & 1 & 0 & 0 \\ 0 & 0 & 0 & 1 & 0 & 0 & 1 & 0 & 0 & 0 & 0 & 0 & 1 & 0 & 1 & 1 \end{pmatrix}. \quad (3.37)$$

We consider the full strategy space and the following game parameters:

Number of agents	N_{tot}	101
Memory size	m	2
Strategies per agent	s	2
Resource level	ϕ	0.5
Time horizon	H	2
Threshold	τ	0.51

The dimension of the transition matrix is thus $2^{H+m} = 16$.

$$\underline{\underline{T}} = \begin{pmatrix}
0 & 0 & 0 & 0 & 0 & 0 & 0 & 0 & 0 & 0 & 0 & 0 & 0 & 0 & 0 & 0 \\
1 & 0 & 0 & 0 & 0 & 0 & 0 & 0 & 1 & 0 & 0 & 0 & 0 & 0 & 0 & 0 \\
0 & 0.5 & 0 & 0 & 0 & 0 & 0 & 0 & 0 & 0.0625 & 0 & 0 & 0 & 0 & 0 & 0 \\
0 & 0.5 & 0 & 0 & 0 & 0 & 0 & 0 & 0 & 0.9375 & 0 & 0 & 0 & 0 & 0 & 0 \\
0 & 0 & 1 & 0 & 0 & 0 & 0 & 0 & 0 & 1 & 0 & 0 & 0 & 0 & 0 & 0 \\
0 & 0 & 0 & 0 & 0 & 0 & 0 & 0 & 0 & 0 & 0 & 0 & 0 & 0 & 0 & 0 \\
0 & 0 & 0 & 0.1875 & 0 & 0 & 0 & 0 & 0 & 0 & 1 & 0 & 0 & 0 & 0 & 0 \\
0 & 0 & 0 & 0.8125 & 0 & 0 & 0 & 0 & 0 & 0 & 0 & 0 & 0 & 0 & 0 & 0 \\
0 & 0 & 0 & 0 & 0.1875 & 0 & 0 & 0 & 0 & 0 & 0 & 0.1094 & 0 & 0 & 0 & 0 \\
0 & 0 & 0 & 0 & 0.8125 & 0 & 0 & 0 & 0 & 0 & 0 & 0.8906 & 0 & 0 & 0 & 0 \\
0 & 0 & 0 & 0 & 0 & 0 & 0 & 0 & 0 & 0 & 0 & 0 & 0.0312 & 0 & 0 & 0 \\
0 & 0 & 0 & 0 & 0 & 1 & 0 & 0 & 0 & 0 & 0 & 0 & 0.9688 & 0 & 0 & 0 \\
0 & 0 & 0 & 0 & 0 & 0 & 0.75 & 0 & 0 & 0 & 0 & 0 & 0 & 0 & 0.5 & 0 \\
0 & 0 & 0 & 0 & 0 & 0 & 0.25 & 0 & 0 & 0 & 0 & 0 & 0 & 0 & 0.5 & 0 \\
0 & 0 & 0 & 0 & 0 & 0 & 0 & 1 & 0 & 0 & 0 & 0 & 0 & 0 & 0 & 1 \\
0 & 0 & 0 & 0 & 0 & 0 & 0 & 0 & 0 & 0 & 0 & 0 & 0 & 0 & 0 & 0
\end{pmatrix}. \quad (3.38)$$

Each nonzero element in the transition matrix corresponds to a probability function in the output space in the Future–Cast operator matrix. Consider that the initial state is $\Gamma = 10$, i.e., the last 4 bits are $\{1001\}$ (obtained from running the game prior to the Future–Casting process). The initial probability state vector is the point $(0,1)$ in the element of the vector $\underline{\varsigma}$ corresponding to the time-horizon state 10. We can then generate the Future–Cast for given U (shown in Fig. 3.15).

Clearly the probability function in output space becomes smoother as U becomes larger and the number of successive convolutions increases, as highlighted by the probability distribution functions at $U = 15$ and $U = 25$ (Fig. 3.16).

We note the non-Gaussian form of the probability distribution for the Future–Casts, emphasising the fact that such a Future–Cast approach is essential for understanding the system's evolution. An assumption of rapid diffusion toward a Gaussian distribution, and hence the future spread in paths increasing as the square-root of time, would clearly be unreliable.

3.6
Evolution Management via Perturbations to Population's Composition

For less simple parameters, the matrix dimension required for the Future–Cast process become very large very quickly. To generate a Future–Cast appropriate to larger parameters, e.g., $m = 3$, $H = 10$, it is still, however, possible to carry out the calculations numerically quite easily. As an example, we generate a random $\underline{\underline{\Omega}}$ (the form of which is given in Fig. 3.19) and an initial time-horizon appropriate to these parameters. This time-horizon is obtained by allowing the system to run prior to the Future–Cast. For visual representation reasons, the Reduced Strategy Space [20] is employed. The other game

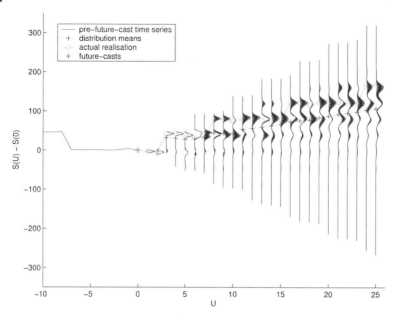

Fig. 3.15 The (non-normalized) evolution of a Future–Cast for the given $\underline{\Omega}$, game parameters and initial state Γ. The figure shows the last 10 time-steps prior to the Future–Cast, the means of the distributions within the Future–Cast itself, and also an actual realization of the game run forward in time. (For a color version please see color plate on page 594.)

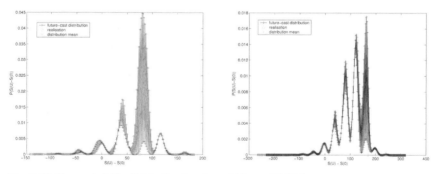

Fig. 3.16 The probability distribution function at $U = 15, 25$ time-steps beyond the present state.

parameters are as previously stated. The game is then instructed to run down every possible winning decision path exhaustively. The spread of output at each step along each path is then convolved with the next spread such that a Future–Cast is built up along each path. Fortunately, not all paths are realized at every time-step since the stochasticity in the winning-decision/state-space results from the condition $N_{ud} \geq D_d$. The Future–Cast as a function of U

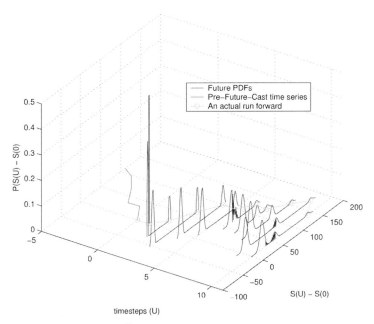

Fig. 3.17 Evolution of $\Pi^U(S)$ for a typical quenched disorder matrix $\underline{\underline{\Omega}}$. (For a color version please see color plate on page 595.)

and S, can thus be built up for a randomly chosen initial quenched disorder matrix (QDM), see Fig. (3.17).

We now wish to consider the situation where it is required that the system should not behave in a certain manner. For example, it may be desirable that it avoids entering a certain regime characterized by a given value of $S(t)$. Specifically, we consider the case where there is a barrier in the output space that the game should avoid, as shown in Fig. 3.18.

The evolution of the spread (i.e., standard deviation) of the distributions in time, confirms the non-Gaussian nature of the system's evolution – we note that this spread can even decrease with time[8]. In the knowledge that this barrier will be breached by this system, we therefore perturb the quenched disorder at $U = 0$. This perturbation corresponds in physical terms to an adjustment of the composition of the agent population. This could be achieved by "rewiring" or "reprogramming" individual agents in a situation in which the agents were accessible objects, or introducing some form of communication

---

8) This feature can be understood by appreciating the multi-peaked nature of the distributions in question. The peaks correspond to differing paths travelled in the Future–Cast, the final distribution being a superposition of these. If these individual path distributions mean-revert, the spread of the actual Future–Cast can decrease over short time scales.

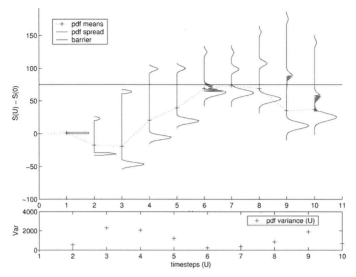

Fig. 3.18 The evolution of the Future–Casts, and the barrier to be avoided. For simplicity the barrier is chosen to correspond to a fixed $S(t)$ value of 110, although there is no reason that it could not be made time-dependent. Superimposed on the (un-normalized) distributions, are the means of the Future–Casts, while their variances are shown below.

channel, or even a more "evolutionary" approach whereby a small subset of species are removed from the population and a new subset added in to replace them. Interestingly we note that this "evolutionary" mechanism need neither be completely deterministic (i.e., knowing exactly how the form of the QDM changes) nor completely random (i.e., a random perturbation to the QDM). In this sense, it seems tantalisingly close to some modern ideas of biological evolution, whereby there is some purpose mixed with some randomness.

Figure 3.20 shows the impact of this relatively minor microscopic perturbation on the Future–Cast and global output of the system. In particular, the system has been steered away from the potentially harmful barrier into "safer" territory.

This set of outputs is specific to the initial state of the system. More typically, we may not know this initial state. Fortunately, we can make use of the characteristic Future–Casts to make some kind of quantitative assessment of the robustness of the quenched disorder perturbation in avoiding the barrier, since this procedure provides a picture of the range of possible future scenarios.

This evolution of the characteristic Future–Casts, for both the initial and perturbed quenched disorder matrices, is shown in Fig. 3.21. A quantitative evaluation of the robustness of this barrier avoidance could then be calculated using traditional techniques of risk analysis, based on knowledge of the distribution functions and/or their low-order moments.

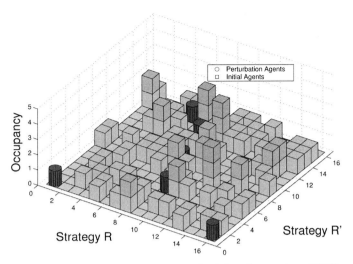

Fig. 3.19 The initial and resulting quenched disorder matrices (QDM), shown in schematic form. The x-y axes are the strategy labels for the two strategies. The absence of a symbol denotes an empty bin (i.e., no agent holding that particular pair of strategies).

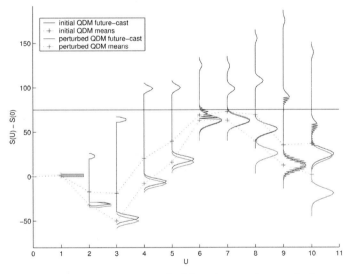

Fig. 3.20 The evolution as a result of the microscopic perturbation to the population's composition (i.e., the QDM). (For a color version please see color plate on page 595.)

3.7
Reducing the Future–Cast Formalism

We introduced the Future–Cast formalism to map from the internal state space of a complex system to the observable output space. Although the formalism

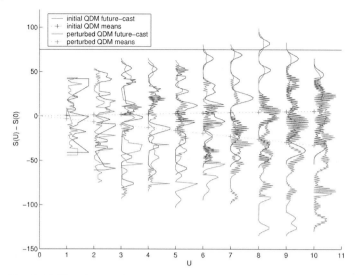

Fig. 3.21 The characteristic evolution of the initial and perturbed QDMs. (For a color version please see color plate on page 596.)

exactly generates the probability distributions of the subsequent output from the system, its implementation is far from trivial. This involves keeping track of numerous distributions and performing appropriate convolutions between them. Often, however it is only the lowest-order moments which are of immediate concern to the system designer. Here, we show how this information can be generated without the computational exhaustion previously required. We demonstrate this procedure for a very simple two-state system, although the formalism is general for a system of any number of states governed by a Markov chain.

Recall the toy model comprizing the two dice of Section 3.2. We previously broke down the possible outputs of each according to the state transition as shown in Fig. 3.22. These distributions were used to construct the matrix $\underline{\underline{Y}}$ to form the Future–Cast process as denoted in Eq. 3.3. This acted on vector $\underline{\varsigma}^U$ to generate $\underline{\varsigma}^{U+1}$. The elements of these vectors contain the partial distribution of outputs which are in the state denoted by the element number at that particular time, so for the two-dice model, $\varsigma_1^U(S)$ contains the distribution of output values at time $t + U$ (or U time-steps beyond the present) which corresponds to the system being in state A and $\varsigma_2^U(S)$ contains those for state B. To reduce the calculation process, we will consider only the moments of each of these individual elements about zero. Therefore we construct a vector, $^n\underline{x}_U$, which takes the form:

$$^n\underline{x}_U = \begin{pmatrix} \sum_{S=-\infty}^{\infty} \varsigma_1^U(S) S^n \\ \sum_{S=-\infty}^{\infty} \varsigma_2^U(S) S^n \end{pmatrix} \qquad (3.39)$$

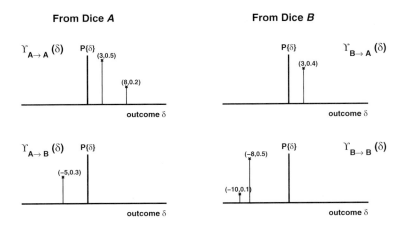

Fig. 3.22 The state transition distributions as prescribed by our dice.

The elements are just the nth moments about zero of the partial distributions within the appropriate state. For $n = 0$ this vector merely represents the probabilities of being in either state at some time $t + U$. We note that for the $n = 0$ case, $^0\underline{x}_{U+1} = \underline{\underline{T}}\,^0\underline{x}_U$ where $\underline{\underline{T}}$ is the Markov Chain transition matrix as in Eq. 3.8. We also note that the transition matrix $\underline{\underline{T}} = ^0\underline{\underline{X}}$ where we define the (static) matrix $^n\underline{\underline{X}}$ in a similar fashion using the partial distributions (described in Fig. 3.22) to be

$$^n\underline{\underline{X}} = \begin{pmatrix} \sum_{\delta=-\infty}^{\infty} \Upsilon_{A \to A}(\delta)\delta^n & \sum_{\delta=-\infty}^{\infty} \Upsilon_{B \to A}(\delta)\delta^n \\ \sum_{\delta=-\infty}^{\infty} \Upsilon_{A \to B}(\delta)\delta^n & \sum_{\delta=-\infty}^{\infty} \Upsilon_{B \to B}(\delta)\delta^n \end{pmatrix} \quad (3.40)$$

Again, this contains the moments (about zero) of the partial distributions corresponding to the transitions between states. The evolution of $^0\underline{x}$, the statewise probabilities with time is trivial as described above. For higher orders, we must consider the effects of superposition and convolution on their values. We know that for the superposition of two partial distributions, the resulting moments (any order) about zero will be just the sum of the moments of the individual distributions, i.e., it is just a summation. The effects of convolution, however, must be considered more carefully. The elements of our vector $^1\underline{x}_U$ are the first-order moments of the values associated with either of the two states at time $t + U$. The first element of which corresponds to those values of output in state A at time $t + U$. Consider that element one step later, $^1x_{1,U+1}$. This can be written as the superposition of the two required convolutions.

$$\sum_S \varsigma_1^{U+1} S = \sum_\delta \sum_S \varsigma_1^U \Upsilon_{A \to A}(S+\delta) + \sum_\delta \sum_S \varsigma_2^U \Upsilon_{B \to A}(S+\delta)$$

$$^1x_{1,U+1} = {^0x_{1,U}}\,^1X_{1,1} + {^0X_{1,1}}\,^1x_{1,U} + {^0x_{2,U}}\,^1X_{1,2} + {^0X_{1,2}}\,^1x_{2,U} \quad (3.41)$$

In simpler terms,

$$^1\underline{x}_{U+1} = {}^0\underline{\underline{X}}\,{}^1\underline{x}_U + {}^1\underline{\underline{X}}\,{}^0\underline{x}_U \tag{3.42}$$

The $(S + \delta)$ term in the expression relates to the nature of series generating algorithm, $S_{t+1} = S_t + \delta_t$. If the series updating algorithm were altered, this would have to be reflected in this convolution.

The overall output of the system is the superposition of the contributions in each state. So, the resulting first moment about zero (the mean) for the overall output at U time-steps into the future is simply $^1\underline{x}_{U+1} \cdot \underline{1}$ where $\underline{1}$ is a vector containing all ones and $\cdot$ is the familiar dot product.

The other moments about zero can be obtained similarly.

$$\begin{aligned}
^0\underline{x}_{U+1} &= {}^0\underline{\underline{X}}\,{}^0\underline{x}_U \\
^1\underline{x}_{U+1} &= {}^0\underline{\underline{X}}\,{}^1\underline{x}_U + {}^1\underline{\underline{X}}\,{}^0\underline{x}_U \\
^2\underline{x}_{U+1} &= {}^0\underline{\underline{X}}\,{}^2\underline{x}_U + 2\,{}^1\underline{\underline{X}}\,{}^1\underline{x}_U + {}^2\underline{\underline{X}}\,{}^0\underline{x}_U \\
^3\underline{x}_{U+1} &= {}^0\underline{\underline{X}}\,{}^3\underline{x}_U + 3\,{}^1\underline{\underline{X}}\,{}^2\underline{x}_U + 3\,{}^2\underline{\underline{X}}\,{}^1\underline{x}_U + {}^3\underline{\underline{X}}\,{}^0\underline{x}_U \\
&\vdots \qquad \vdots
\end{aligned} \tag{3.43}$$

More generally

$$^n\underline{x}_{U+1} = \sum_{\gamma=0}^{n} {}^nC_\gamma\,{}^\gamma\underline{\underline{X}}\,{}^{n-\gamma}\underline{x}_U \tag{3.44}$$

where $^nC_\gamma$ is the conventional *choose* function.

To calculate time-averaged properties of the system, for example, the one-time-step mean or variance, we set the initial vectors such that

$$^0\underline{x}_0 = {}^0\underline{\underline{X}}\,{}^0\underline{x}_0 \tag{3.45}$$

and $^\beta\underline{x}_0 = \underline{0}$ for $\beta > 0$. The moments about zero can then be used to calculate the moments about the mean. The mean of the one time-step increments in output averaged over an infinite run will then be $^1\underline{x}_1 \cdot \underline{1}$ and σ^2 will be

$$\sigma^2 = {}^2\underline{x}_1 \cdot \underline{1} - ({}^1\underline{x}_1 \cdot \underline{1})^2 \tag{3.46}$$

These can be calculated for any size of rolling window. The mean of all U-step increments, $S(t + U) - S(t))$ or conversely the mean of the Future–Cast U steps into the future from unknown current state is simply $^1\underline{x}_U \cdot \underline{1}$ and σ_U^2 will be

$$\sigma_U^2 = {}^2\underline{x}_U \cdot \underline{1} - ({}^1\underline{x}_U \cdot \underline{1})^2 \tag{3.47}$$

again with initial vectors calculated from $^0\underline{x}_0 = {^0\underline{\underline{X}}}\,{^0\underline{x}_0}$ and $^\beta\underline{x}_0 = \underline{0}$ for $\beta > 0$. Examining this explicitly for our two-dice model, the initial vectors are:

$$^0\underline{x}_0 = \begin{pmatrix} \frac{4}{7} \\ \frac{3}{7} \end{pmatrix}$$

$$^1\underline{x}_0 = \begin{pmatrix} 0 \\ 0 \end{pmatrix}$$

$$^2\underline{x}_0 = \begin{pmatrix} 0 \\ 0 \end{pmatrix} \tag{3.48}$$

and the (static) matrices are:

$$^0\underline{\underline{X}} = \begin{pmatrix} 0.7 & 0.4 \\ 0.3 & 0.6 \end{pmatrix}$$

$$^1\underline{\underline{X}} = \begin{pmatrix} 3.1 & 1.2 \\ -1.5 & -5.0 \end{pmatrix}$$

$$^2\underline{\underline{X}} = \begin{pmatrix} 17.3 & 3.6 \\ 7.5 & 42. \end{pmatrix} \tag{3.49}$$

These are all we require to calculate the means and variances for our system's potential output at any time in the future. To check that all is well, we employ the Future–Cast to generate the possible future distributions of output up to 10 time-steps, Π^1_{char} to Π^{10}_{char}. The means and variances of these are compared to the reduced Future–Cast formalism and also a numerical simulation. This is a single run of the game over 100 000 time-steps. The means and variances are then measured over rolling windows of between 1 and 10 time-steps in length. The comparison is shown in Fig. 3.23. Fortunately, they all concur. The Reduced Future–Cast formalism and the moments about either the mean or zero from the distributions generated by the Future–Cast formalism are identical. Clearly numerically simulations require progressively longer run times to investigate the properties of distributions further into the future, where the total number of possible paths becomes large.

3.8
Concluding Remarks and Discussion

We have presented an analytical formalism for the calculation of the probabilities of outputs from the BAR system at a number of time-steps beyond the present state. The construction of the (static) Future–Cast operator matrix allows the evolution of the systems output, and other macroscopic quantities of the system, to be studied without the need to follow the microscopic details of each agent or species. We have demonstrated a technique to investigate the

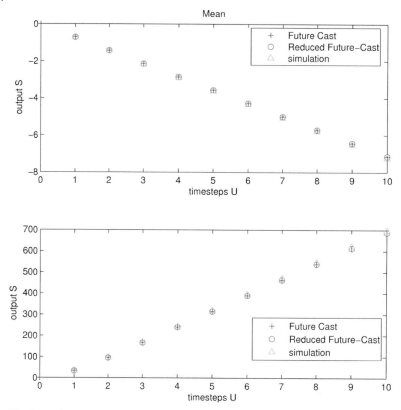

Fig. 3.23 The means and variances of the characteristic distributions Π_{char}^{1} to Π_{char}^{10} as compared to a numerical evaluation and the reduced Future–Cast.

macroscopic effects of population perturbations but it could also be used to explore the effects of exogenous noise or even news in the context of financial markets. We have concentrated on single realizations of the quenched disorder matrix, since this is appropriate to the behavior and design of a particular realization of a system in practice. An example could be a financial market model based on the BAR system whose derivatives could be analyzed quantitatively using expectation values generated with the Future–Casts. We have also shown that through the normalized eigenvector solution of the Markov Chain transition matrix, we can use the Future–Cast operator matrix to generate a characteristic probability function for a given game over a given time period. The formalism is general for any time-horizon game and could, for example, be used to analyze systems (games) where a level of communication between the agents is permitted, or even linked systems (i.e., linked games or "markets"). In the context of linked systems, it will then be interesting to pursue the question as to whether adding one "safe" complex system to another "safe" complex system, results in an "unsafe" complex system. Or thinking more optimistically, when can we put together two or more "unsafe" systems and get a "safe" one?

We have also presented a simplified and altogether more usable interpretation of the Future–Cast formalism for tracking the evolution of the output variable from a complex system whose internal states can be described as a Markov process. We have illustrated the application of the results for an example case both for the evolution from a known state, or when the present state is unknown, to give characteristic information about the output series generated by such a system. The formalism is generalizable to Markov Chains whose state transitions are not limited to just two possibilities and also to systems whose mapping from state transitions to output-space are governed by continuous probability distributions.

Future work will focus on the "reverse problem" of generating a specific system to behave in a desired fashion. As such, the effects of any perturbation to the system's heterogeneity could be pre-engineered. A future application might include the global control problem with discrete actuating controllers [27]. In light of the formalisms presented, we ask the question of whether a complex system could then constructed to replicate real-world systems whose output variables exhibit behavior inconsistent with popular analysis.

References

1 See BOCCARA, N., *Modeling Complex Systems*, Springer, New York, **2004**, and references within, for a thorough discussion

2 WOLPERT, D. H., WHEELER, K., TUMER, K., *Europhys. Lett.* 49 (**2000**), p. 6

3 STROGATZ, S. H., *Nonlinear Dynamics and Chaos*, Addison-Wesley, Reading, **1995**

4 ISRAELI, N., GOLDENFELD, N., *Phys. Rev. Lett.* 92 (**2004**), p. 74105

5 CASTI, J. L., *Would-be Worlds*, Wiley, New York, **1997**

6 BRIAN ARTHUR, W., *Amer. Econ. Rev.*, 84 (**1994**), 406; *Science* 284 (**1999**), p. 107

7 JOHNSON, N. F., JARVIS, S., JONSON, R., CHEUNG, P., KWONG, Y. R., HUI, P. M., *Physica A* 258 (**1998**), p. 230

8 ANGHEL, M., TOROCZKAI, Z., BASSLER, K. E., KRONISS, G., *Phys. Rev. Lett.* 92 (**2004**), p. 58701

9 GOURLEY, S., CHOE, S. C., HUI, P. M., JOHNSON, N. F., *Europhys. Lett.* 67 (**2004**), pp. 867–873

10 CHOE, S. C., JOHNSON, N. F., HUI, P. M., *Phys. Rev. E* 70 (**2004**), p. 55101

11 LO, T. S., CHAN, H. Y., HUI, P. M., JOHNSON, P. M., *Phys. Rev. E* 70 (**2004**) p. 56102

12 CHALLET, D., ZHANG, Y.-C. *Physica A* 246 (**1997**), p. 407

13 CHALLET, D., MARSILI, M., OTTINO, G. cond-mat/0306445

14 JOHNSON, N. F., CHOE, S. C., GOURLEY, S., JARRETT, T., HUI, P. M., in *Advances in Solid State Physics* 44 (**2004**) p. 427, Springer, Heidelberg

15 We originally introduced the finite time-horizon MG, plus its "grand canonical" variable-N and variable-L generalizations, to provide a minimal model for financial markets. See HART, S. C., JEFFERIES, P., JOHNSON, N. F. *Physica A* 311 (**2002**), p. 275; HART, S. C., LAMPER, D., JOHNSON, N. F., *Physica A* 316 (**2002**), p. 649; LAMPER, D., HOWISON, S. D., JOHNSON, N. F., *Phys. Rev. Lett.* 88 (**2002**), p. 017902; JOHNSON, N. F., JEFFERIES, P., HUI, P. M.,

Financial Market Complexity, Oxford University Press, **2003**. See also CHALLET, D., GALLA, T. cond-mat/0404264, which uses this same model

16 CHALLET, D., ZHANG, Y.-C. *Physica A* 256 (**1998**), p. 514

17 CHALLET, D., MARSILI, M., ZECCHINA, R. *Phys. Rev. Lett.* 82 (**1999**), p. 2203

18 CHALLET, D., MARSILI, M., ZECCHINA, R., *Phys. Rev. Lett.* 85 (**2000**), p. 5008

19 See http://www.unifr.ch/econophysicsforMinorityGameliterature.

20 JOHNSON, N. F., HART, M., HUI, P. M., *Physica A* 269 (**1999**), p. 1

21 HART, M., JEFFERIES, P., JOHNSON, N. F., HUI, P. M., *Physica A* 298 (**2001**), p. 537

22 JOHNSON, N. F., HUI, P. M., ZHENG, D., HART, M., *J. Phys. A: Math. Gen.* 32 (**1999**), p.L427

23 HART, M. L., JEFFERIES, P., JOHNSON, N. F., HUI, P. M., *Phys. Rev. E* 63 (**2001**), p. 017102

24 JEFFERIES, P., HART, M., JOHNSON, N. F., HUI, P. M., *J. Phys. A: Math. Gen.* 33 (**2000**), p.L409

25 JEFFERIES, P., HART, M. L., JOHNSON, N. F., *Phys. Rev. E* 65 (**2002**), p. 016105

26 JOHNSON, N. F., HUI, P. M., (**2003**), cond-mat/0306516v1

27 BIENIAWSKI, S., KROO, I. M., *AIAA Paper* (**2003**) pp. 2003–1941

28 SOBOL, I. M., *A Primer for the Monte Carlo Method* Boca Raton, FL: CRC Press, **1994**

4
Growth of Firms and Networks
Yoshi Fujiwara, Hideaki Aoyama, and Wataru Souma

4.1
Introduction

If a firm's size were to be regarded as human height, anyone could easily see that a population of firms would look quite peculiar and striking. From a distance, only a few giants can be identified, while a large number of dwarves are visible only faintly. Removing the giants from the population and looking closely, one would have a similar landscape with a rescaled dimension.

This phenomenon, first observed for personal income by Vifredo Pareto more than a century ago, is now known as Pareto's law. In mathematical terms, high-income distribution follows a power law: the probability $P_>(x)$ that a given individual has an income equal to, or greater than x, obeys

$$P_>(x) \propto x^{-\mu} \tag{4.1}$$

with μ as a constant, which is about 2 for Japanese personal income, for example. See Fig. 4.1(a). There have been many observations in different countries [4,8,11,12,29,34,47] about this fact. On the other hand, low and middle-income distribution has been considered to obey log-normal distribution, the Boltzman distribution or another functional form.

It transpires that (4.1) also describes the distribution of *firm size* with μ close to 1 as we shall see in this chapter. The case $\mu = 1$ is customarily called Zipf's law. Let us call it Pareto–Zipf's law for the distribution with a generic value of μ. Reference [7] gives a striking example with firm size measured by the number of employees. Figure 4.1(b) is Japanese firms' income. These examples show that the Pareto–Zipf's law is valid over several orders of magnitude in a large-firm region.

No less interesting than such a static snapshot of distribution is the way in which each individual changes its income and size with time. Since there are possible scenarios that bring about a power-law distribution [10,24,25,32,33, 44,46], it would be highly desirable to have direct observation of the dynam-

Econophysics and Sociophysics: Trends and Perspectives.
Bikas K. Chakrabarti, Anirban Chakraborti, Arnab Chatterjee (Eds.)
Copyright © 2006 WILEY-VCH Verlag GmbH & Co. KGaA, Weinheim
ISBN: 3-527-40670-0

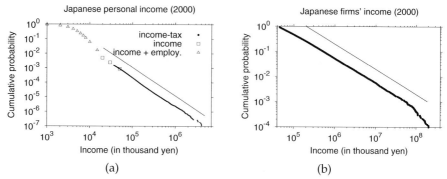

Fig. 4.1 (a) Cumulative probability distribution of Japanese personal income in the year 2000. The line is a guide with $\mu = 2.0$ in (4.1). Note that the dots are income tax data of about 80 000 taxpayers. (See [4] and [34] for the details). (b) Cumulative probability distribution of Japanese firms' income in the year 2000. Data consist of the top 78 314 firms. The line corresponds to $\mu = 1.0$ (See also [5, 6, 27, 28]).

ical process of growth and fluctuation of firm size [1, 2, 41–43] (see [13] and review [17, 40] for personal income). Again, the reader will see, in this chapter, that the dynamics of firms' (and individual) growth has quite interesting statistical properties.

Understanding the dynamics of a firm's growth has importance in economics, because even if the range for which (4.1) is valid is a few percent in the upper tail of the distribution, it is often observed that such a small fraction of firms occupies a large amount of the total sum. Small idiosyncratic shock can bring about a considerable macro-economic impact (see [3, 18]).

In Section 4.2, by employing exhaustive lists of large firms in European countries, we show that the upper tail of the distribution of firm size obeys the Pareto–Zipf law, and that in this region the growth rate of each firm satisfies what is called Gibrat's law. Additionally, we will see that detailed balance holds in the power-law region. In Section 4.3, we prove several relationships including the Pareto–Zipf law, Gibrat's law and the condition of detailed balance, which are in good agreement with the data. We also give a conjecture on the origin of $\mu = 1$. In Section 4.4, we show that small-business firms have qualitatively different characteristics of firm-size growth from those for large firms. By employing Japanese large dataset for small-business firms, we show that Gibrat's law breaks down for the small and mid-sized companies corresponding to the non-power-law region, while the law asymptotically holds in the larger-sized region, for all the variables examined. In Section 4.5, we turn our attention to the aggregate behavior of firms that are linked with each other. Specifically, we examine two kinds of firm networks: ownership and production networks. We find scale-free properties for both of these networks, and attempt to explain how this fact is related to a firm's growth and failure through stochastic multiplicative process with a reset event. In addition, the structure of a production network is crucial in the understanding of bankruptcy, which also obeys a Zipf law, and the chain of firms failure.

4.2 Growth of Firms

4.2.1 Dataset of European Firms

For the study of the growth of firms[1], we use the dataset, Bureau van Dijk's AMADEUS, which contains financial data for about 260 000 firms of 45 European countries for the years 1992–2001. For every firm there are reported descriptive data (such as bankruptcy) and a series of data drawn from its balance and normalized. It reports the current values (in several currencies) of balance sheet (BS), profit and loss account (P/L), financial ratios and stocktaking. To be included in the data set, firms must satisfy at least one of these three criteria as follows. For UK, France, Germany, Italy, Russia and Ukraine, operating revenue equal to at least 15 million Euros; total assets equal to at least 30 million Euros; number of employees equal to at least 150. For the other countries, operating revenue equal to at least 10 million Euros; total assets equal to at least 20 million Euros; number of employees equal to at least 100.

As a proxy for firm size, we utilize one of the financial and fundamental variables; total assets, sales and number of employees. We use the number of employees as a complementary variable so as to check the validity and robustness of our results. Note that the dataset includes firms with smaller total assets, simply because either the number of employees or the operating revenue (or both of them) exceeds the corresponding threshold. We thus focus on complete sets of those firms that have a larger amount of total assets than the threshold, and similarly those for the number of employees. For sales, we assume that our dataset is nearly complete since a firm with a small amount of total assets and a small number of employees is unlikely to make a large amount of sales. For our purposes, therefore, we discard all the data below each corresponding threshold for each measure of firm size. This procedure makes the number of data points much smaller. However, for several developed countries, we have enough data for the study of Gibrat's law. In what follows, our results are shown for the UK and France, although we obtained similar results for other developed countries. The threshold for total assets in these two countries is 30 million Euros, and that for the number of employees

---

1) This section is based on the work of [15]. The authors are indebted to Mauro Gallegati for the use of this dataset.

is 150 persons, as described above. For sales, we used 15 million Euros per year as a threshold. We will also show the results for Italy and Spain in addition to the UK and France only when examining the annual change of Pareto indices. Our study should be understood as an analysis conditional on the survival of firms.

A Caveat: Exhaustive Data vs Biased Sampling

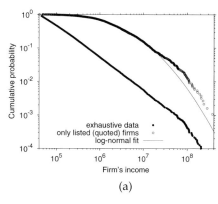

(a)

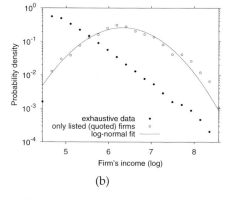

(b)

Fig. 4.2 (a) Cumulative probability distribution $P_>(x)$ for tax income of firms x. (b) Probability density $P(y)$ for $y = \log_{10} x$. In both plots, x is the annual taxable income (Japan in the year 2001), in units of thousand yen (roughly ten Euros), declared to and investigated by the national tax administration. Exhaustive dataset comprizes more than 70 000 top taxable income earners (filled circle), from which only the firms that are listed in stock markets, and that are traded in over-the-counter markets of Japan are selected (open circle). The line is a log-normal fit for the latter dataset. Obviously, the actual distribution does not fit with a log-normal, but rather a power law.

In this chapter, we will show several distributions for firms' stock variables and flow variables. These variables basically represent a firm's size. In the literature, a firm-sized distribution is quite often determined based on a dataset, which is obtained by some biased sampling. The sampling is usually done in a way that is of limited relevance to the population of firm size. For example, a set of firms is selected as a dataset, simply because the firms are listed (quoted) in particular financial markets. Many firms in such a dataset, of course, are outstanding, known to some extent and usually large in size. Nevertheless, they are by no means an unbiased sampling from the population. Usage of a biased dataset for determining a firm size distribution, however the size is measured, can potentially yield a misleading result.

To illustrate this point, which the authors consider sometimes crucial in order to know the form of the firm sized distribution, let us take a digression and compare two distributions, one based on a biased sampling and the other based on an exhaustive dataset.

To be brief, for taxation, the Japanese tax administration investigates firms' income as strictly as possible, every year. Tax income is a flow variable, which is basically revenue *minus* costs (legally admitted) of a firm. This variable can be used as a proxy that represents the firm's size in terms of flow. Since the administration compiles an exhaustive list of firms' income annually, the income of firms exceeding a certain threshold are exhaustively listed. As an alternative dataset, we intentionally select the firms that are listed in stock markets, Tokyo (TSE) and others, and that are traded in over-the-counter markets of Japan. For the year 2001, the exhaustive dataset comprizes the largest income firms, top 70 000 and more, while the latter sampling comprizes approximately 10 000 of them.

For each of the two datasets, we plot the distribution of firms' income in Fig. 4.2, in both a cumulative probability distribution and a probability density distribution. Suppose the biased dataset is the only available one. Then one may attempt, by taking a look at the cumulative or the probability density distribution, to fit the data with a log-normal distribution. As shown in Fig. 4.2, the fit turns out to be quite satisfactory. However good the fitting appears, it is obvious that this result makes little sense, because the population observed from the exhaustive dataset has completely different distribution in the range of firm size examined. Namely, the distribution obeys a power law over four orders of magnitude in exactly the same range.

It is, therefore, very crucial to have an exhaustive dataset in order to talk about the functional form of a firm size distribution and to draw conclusions. Alternatively, one could use a sample survey, or a tabulated dataset. Unfortunately, such tabulated data does not usually cover the upper tail of the distribution (or the lower region), so one cannot often draw a decisive conclusion from it. In this chapter, we will use an exhaustive dataset wherever it is available, and augment it with tabulated data wherever it is necessary.

4.2.2
Pareto–Zipf's Law for Distribution

First we show that the distribution of firm size obeys a power law in the range of our observation whatever we take as a variable for firm size. Figure 4.3 depicts the cumulative distributions for total assets in France (a), sales in France (b), and number of employees in UK (c). The number of data points is respectively (a) 8313, (b) 15776 and (c) 15055.

Pareto–Zipf law states that the cumulative distribution $P_>(x)$ for firm size x follows (4.1). The power-law fit for $x \geq x_0$, where x_0 denotes the threshold mentioned above for each measure of firm size, gives the values of μ; (a) 0.886±0.005, (b) 0.896±0.011, (c) 0.995±0.013 (standard error at 99% significance level). μ is close to unity. Note that the power-law fit is quite well nearly three orders of magnitude in size of firms.

The Pareto index is surprisingly stable in its value. Figure 4.4 is a panel for the annual change of Pareto indices for four countries, Italy, Spain, France and the UK estimated from total assets, number of employees and sales (except UK). Different measures of firm size give approximately the same behavior. It is observed that the value μ is quite stable being close to unity in all the countries.

4.2.3
Gibrat's Law for Growth

Let us denote a firm's size by x and its two values at two successive points in time (i.e., two consecutive years) by x_1 and x_2. The growth rate is given by

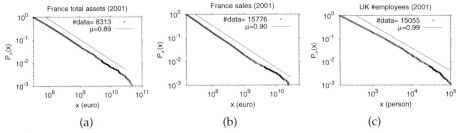

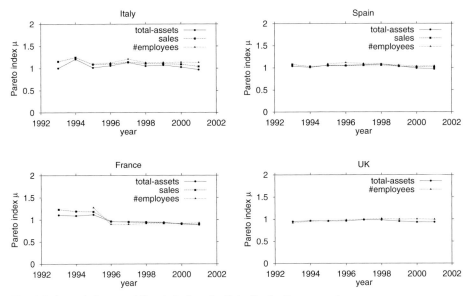

Fig. 4.3 Cumulative probability distribution $P_>(x)$ for firm size x. (a) Total-assets in France (2001) greater than 30 million Euros, (b) sales in France (2001) greater than 15 million Euros. (c) number of employees in UK (2001) greater than 150 persons. Guidelines are the power law with Pareto index corresponding to, 0.89, 0.90 and 0.99, respectively (obtained by least-square-fit in the logarithmic scale).

Fig. 4.4 Annual change of Pareto indices for Italy, Spain, France and the UK from 1993 to 2001 for total assets, number of employees, and sales (except UK). The estimate of Pareto index in each year was done by extracting a range of distribution corresponding to large-sized firms, which is common to different countries but different for a different measure of size, and by least-squares-fit in logarithmic scales of rank and size.

$R \equiv x_2/x_1$. We also express the rate in terms of its logarithm, $r \equiv \log_{10} R$. We examine the probability density for the growth rate $P(r|x_1)$ on the condition that the firm size x_1 in an initial year is fixed.

For the conditioning we divide the range of x_1 into logarithmically equal bins. For the total assets in the dataset (Fig. 4.5(a)), the bins are taken as

$x_1 \in 3 \times [10^{7+0.4(n-1)}, 10^{7+0.4n}]$ (Euros) with $n = 1, \cdots, 5$. For the sales in (b), $x_1 \in 1.5 \times [10^{7+0.4(n-1)}, 10^{7+0.4n}]$ (Euros) with $n = 1, \cdots, 5$. For the number of employees in (c), $x_1 \in 1.5 \times [10^{2+0.4(n-1)}, 10^{2+0.4n}]$ (persons) with $n = 1, \cdots, 5$. In all the cases, the range of conditioning covers two orders of magnitude in each variable. We calculated the probability density function for r for each bin, and checked the statistical dependence on x_1 by a graphical method.

Figure 4.5 is the probability density function $P(r|x_1)$ for each case. It should be noted that, due to the limit $x_1 > x_0$ and $x_2 > x_0$, the data for large negative growth are not available. In all the cases, it is obvious that the function $P(r|x_1)$ has little statistical dependence on x_1, since all the curves for different n collapse on a single curve. This means that the growth rate is *statistically independent* of firm size in the initial year. This is known as the *law of proportionate effect* or Gibrat's law (see [45] for a review).

4.2.4
Detailed Balance

The validity of Gibrat's law in the Pareto–Zipf regime appears to be in disagreement with recent literature on firm growth. In the next section, we will show that this is not actually the case by proving that the two laws, Pareto–Zipf and Gibrat, are kinematically related to each other. In fact, Pareto–Zipf follows from Gibrat under an assumption. The assumption is detailed balance, the validity of which is checked here.

Let us denote the joint probability distribution function for the variable x_1 and x_2 by $P_{12}(x_1, x_2)$. *Detailed balance* is the assumption that $P_{12}(x_1, x_2) = P_{12}(x_2, x_1)$. The joint probabilities for our datasets are depicted in Fig. 4.6 as scatter plots of individual firms.

We used two different methods to check the validity of detailed balance. One is an indirect way to check a nontrivial relationship between the growth-rate on the positive side ($r > 0$) and that on the negative ($r < 0$). That is, as we shall prove in the next section, the probability density distribution in positive and negative growth rates must satisfy what we call the reflection law, (4.7), if the detail balance holds. We fitted the cumulative distribution only for positive growth rate with a nonlinear function, converted to a density function, and predicted the form of distribution for negative growth rate by (4.7) so as to compare with the actual observation (see Appendix in [15] for details). In each plot of Fig. 4.5, a solid line in the $r > 0$ side is such a fit, and a broken line in the $r < 0$ side is our prediction. The agreement with actual observation is quite satisfactory, thereby supporting the validity of time-reversal symmetry.

The other way we can proceed is a direct statistical test for symmetry in the two arguments of $P_{12}(x_1, x_2)$. This can be done by two-dimensional Kolmogorov–Smirnov (K–S) test, which is not widely known but was devel-

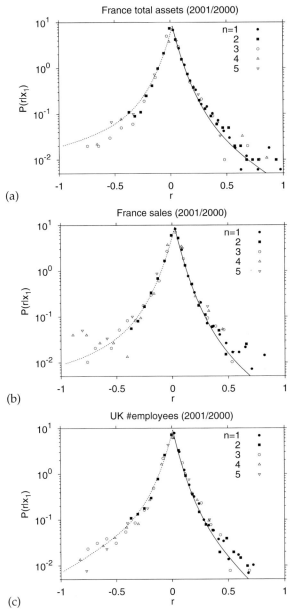

Fig. 4.5 Probability density $P(r|x_1)$ of growth rate $r \equiv \log_{10}(x_2/x_1)$ for the two years, 2000/2001. The data from top to bottom correspond to those in Fig. 4.3. Different bins of initial firm size with equal magnitude in logarithmic scale were taken over two orders of magnitude as described in the main text. The solid line in the portion of positive growth ($r > 0$) is a nonlinear fit. The dashed line ($r < 0$) in the negative side is calculated from the fit by the reflection law given in Eq.(4.7).

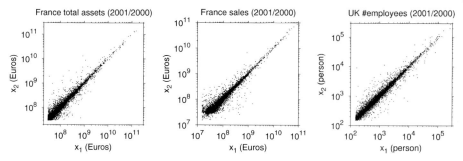

Fig. 4.6 Scatter-plot of all firms whose size exceeds a threshold. The data from left to right correspond to those in Fig. 4.3. Thresholds are 30 million Euros for total assets, 15 million Euros for sales, and 150 persons for number of employees. The number of such large firms is respectively 6969, 13 099 and 12 716.

oped by astrophysicists to test uniform distribution of galaxies appearing in the sky (see references in [15]). This statistical test is not strictly nonparametric (like the well-known one-dimensional K–S test), but has little dependence on parent distribution except through the coefficient of correlation. We compare the scatter-plot sample for $P_{12}(x_1, x_2)$ with another sample for x_1 and x_2 interchanged by making the null hypothesis that these two samples are taken from the same parent distribution. We used the logarithms $\zeta_1 = \log x_1$ and $\zeta_2 = \log x_2$, and added constants to ζ_1 and ζ_2 so that the average growth rate is zero. This addition (or multiplication in x_1 and x_2) is simply subtracting the nominal effects due to inflation, etc. We applied a two-dimensional K–S test to the resulting samples. The null hypothesis is not rejected in a 95% significance level in all the cases we studied.

Summarizing this section, we have observed three phenomenological properties. Namely, (A) detailed balance, (B) the Pareto–Zipf law and (C) Gibrat's law. These three properties are observed for personal income as well as firm size. See the recent review [17] for a study of personal income.

4.3
Pareto–Zipf and Gibrat under Detailed Balance

The probability distribution for the growth rate, such as the one observed in Fig. 4.5, contains information on dynamics. One can see that it has a skewed and heavy-tailed shape with a peak at $R = 1$. How is such a functional form consistent with the detailed balance shown in Fig. 4.6? And how are these phenomenological facts consistent with Pareto's law in Fig. 4.3?

4.3.1
Kinematics

Let x be a personal income or a firm size, and let its values at two successive points in time (i.e., two consecutive years) be denoted by x_1 and x_2. We denote the joint probability distribution for the variables x_1 and x_2 by $P_{12}(x_1, x_2)$. The joint probability distribution of x_1 and $R = x_2/x_1$ is denoted by $P_{1R}(x_1, R)$. We define conditional probability by $P_{1R}(x_1, R) = Q(R|x_1)P_1(x_1)$, where $P_1(x_1)$ is marginal, i.e., $P_1(x_1) = \int_0^\infty P_{1R}(x_1, R)dR = \int_0^\infty P_{12}(x_1, x_2)dx_2$.

The phenomenological properties can be summarized as follows.

(a) *Detailed balance*:
$$P_{12}(x_1, x_2) = P_{12}(x_2, x_1) \tag{4.2}$$

(b) *Pareto–Zipf law*:
$$P_1(x) \propto x^{-\mu-1} \tag{4.3}$$
for $x \to \infty$ with $\mu > 0$.

(c) *Gibrat's law*: The conditional probability $Q(R|x)$ is independent of x:
$$Q(R|x) = Q(R) \tag{4.4}$$

We note here that this holds only for x larger than a certain value. All the arguments below are restricted in this region.

Now we prove that the properties (a) and (c) lead to (b). Under the change of variables from (x_1, x_2) to (x_1, R), since $P_{12}(x_1, x_2) = (1/x_1)P_{1R}(x_1, R)$, one can easily see that $P_{1R}(x_1, R) = (1/R)P_{1R}(R x_1, R^{-1})$. It immediately follows from the definition of $Q(R|x)$ that

$$\frac{Q(R^{-1}|x_2)}{Q(R|x_1)} = R\frac{P_1(x_1)}{P_1(x_2)} \tag{4.5}$$

This equation is thus equivalent to the detailed balance condition.

If Gibrat's law holds, $Q(R|x) = Q(R)$, then

$$\frac{P_1(x_1)}{P_1(x_2)} = \frac{1}{R}\frac{Q(R^{-1})}{Q(R)} \tag{4.6}$$

Note that while the left-hand side of (4.6) is a function of x_1 and $x_2 = Rx_1$, the right-hand side is a function of the ratio R only. It can be easily shown that the equality is satisfied by and only by a power-law function (4.3)[2].

---
[2] Expand (4.6) with respect to R around $R = 1$ and to equate the first-order term to zero, which gives an ordinary differential equation for $P_1(x)$.

As a bonus, by inserting (4.3) into (4.6), we have a nontrivial relation:

$$Q(R) = R^{-\mu-2} Q(R^{-1}) \tag{4.7}$$

which relates the positive and negative growth rates, $R > 1$ and $R < 1$, through the Pareto index μ. We call this relation (4.7) a "reflection law". It should be noted that the above relation is true for the conditional probability for large x. Contrastingly, the marginal probability for the growth-rate, $P_R(R) = \int_0^\infty P_{1R}(x_1, R) dx_1$ obeys a similar but different relation.

$$P_R(R) = R^{-2} P_R(R^{-1}) \tag{4.8}$$

which can be obtained by inserting its definition in (4.5).

One can also show that $Q(R)$ has a cusp at $R = 1$; $Q'(R)$ is discontinuous at $R = 1$. Explicitly, $[Q^{+\prime}(1) + Q^{-\prime}(1)]/Q(1) = -\mu - 2$, where we denote the right and left-derivative of $Q(R)$ at $R = 1$ by the signs $+$ and $-$ in the superscript, respectively. If the probability density $Q(R)$ has a cusp at $R = 1$, as we have shown in our data analysis, then this relation states that the shape of cusp is constrained in the way described by this relation which depends on the value of Pareto index μ.

Summarizing this subsection, we have proved that under the condition of detailed balance (a), Gibrat's law (c) implies Pareto–Zipf law (b). The opposite (b) → (c) is not true. See [15] for several kinematic relations, and also [5,13,14] for the validity of our findings in personal income and firms' data.

4.3.2
Growth of Firms and Universality of Zipf's Law

The relations (4.7) and (4.8) lead to several intriguing relations between moments of x. Let us discuss the latter first. We denote the average of a function $F(R)$ of the growth rate R by $\langle f(R) \rangle$;

$$\langle f(R) \rangle \equiv \int_0^\infty f(R) P_R(R) \, dR \tag{4.9}$$

By using the relation (4.8) and changing the integration variable to R^{-1}, we find that

$$\langle f(R) \rangle = \langle f(R^{-1}) \rangle \tag{4.10}$$

By choosing $F(R) = \ln R$, we find an interesting equality;

$$\langle \ln R \rangle = 0 \tag{4.11}$$

which one might think is intuitively true, or even obvious, since the law of detailed balance is assumed. Such an intuitive argument, however, is mistaken.

In fact, for $F(R) = R$, we find that $\langle R \rangle = \langle R^{-1} \rangle$ from (4.10), which leads to the following inequality;

$$\langle R \rangle = \frac{1}{2} \langle R + R^{-1} \rangle \geq \langle 1 \rangle = 1 \tag{4.12}$$

Since the latter holds if and only if $P_R(R) = \delta(R-1)$, we find that the average growth rate $\langle R \rangle$ is greater than one in a real economic situation. We stress that this is a direct consequence of the the detailed balance (4.2) alone.

Next, let us look at the consequences of the relation (4.7). We define the average of $f(R)$ for large x_1 by the following;

$$\langle\!\langle f(R) \rangle\!\rangle \equiv \int_0^\infty f(R)\, Q(R \mid x_1)\, dR = \int_0^\infty f(R)\, Q(R)\, dR \tag{4.13}$$

where we used the fact that x_1 is large enough for the Pareto–Zipf law (4.3) to holds The relation (4.7) leads to the following;

$$\langle\!\langle f(R) \rangle\!\rangle = \langle\!\langle R^\mu f(R^{-1}) \rangle\!\rangle \tag{4.14}$$

which is analogous to, but different from the equality (4.10). With $F(R) = R^\mu$, we find that

$$\langle\!\langle R^\mu \rangle\!\rangle = 1 \tag{4.15}$$

More useful relations may be obtained by changing the integration variable to R^{-1} for $R \in [1, \infty]$ in (4.13);

$$\langle\!\langle f(R) \rangle\!\rangle = \int_0^1 \left(f(R) + R^\mu f(R^{-1}) \right) Q(R)\, dR \tag{4.16}$$

For $F(R) = \ln R$, we can use the above to find that

$$\langle\!\langle \ln R \rangle\!\rangle = \int_0^1 \left((1 - R^\mu) \ln R \right) Q(R)\, dR \leq 1 \tag{4.17}$$

where we used the fact that $\mu > 0$, as is required for the normalizability of $P_1(x)$ in (4.3). By substituting $F(R) = R - 1$ in (4.16), we find the following;

$$\langle\!\langle R - 1 \rangle\!\rangle = -\int_0^1 (1-R)(1 - R^{\mu-1})\, Q(R)\, dR \begin{cases} > 0 & \text{for } \mu < 1 \\ = 0 & \text{for } \mu = 1 \\ < 0 & \text{for } \mu > 1 \end{cases} \tag{4.18}$$

In the above, we have derived several inequalities and equalities from the phenomenological properties (a), (b) and (c). We stress that this is a very powerful argument, in the sense that no dynamic assumptions are needed. No matter what the fundamental (stochastic or deterministic) laws governing the behavior of large firms, as long as these phenomenological properties are satisfied, these inequalities and equalities hold.

The inequality (4.18) offers an insight into the apparent universality of $\mu = 1$ in the actual data (see Fig. 4.4). It is most natural for the firms to try to grow through their economic activities; they were born to do just that. Such effort by the large (or huge) firms manifests itself as a "force" that pushes the value of μ downward, since the smaller μ, the fewer firms have a larger share of the market. According to (4.18), if μ is larger than one, the average growth rate for large firms, $\langle\!\langle R \rangle\!\rangle$, is less than one when the law of detailed balance holds. Since the efforts of the large firms to grow involves making $\langle\!\langle R \rangle\!\rangle$ larger, which corresponds to smaller μ; then, why does μ stay close to one and not become much smaller? There are a number of causes which attempt to push μ upward: inefficiency that comes simply from the size in the huge firms; higher competition; a social outcry against monopoly and oligopoly and the resulting governmental restrictions and measures such as the antitrust laws. We propose that the last is the major cause. Figure 4.7 shows several plots of

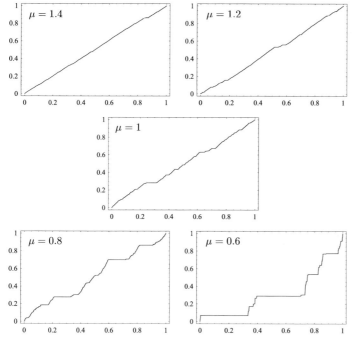

Fig. 4.7 Sample of randomly-generated distribution of 100 000 ($\equiv \mathcal{N}$) firms that follow the Pareto–Zipf law (4.3). Each step on these staircases represents a firm. For the nth firm, the vertical coordinate is $n/\mathcal{N}$ and the horizontal width is the ratio of share, $x_n / \sum_{k=1}^{\mathcal{N}} x_k$. In other words, the horizontal axis represents the cumulative ratio $\sum_{k=1}^{n} x_k / \sum_{k=1}^{\mathcal{N}} x_k$. Note that the firms are not ordered in any particular way.

the simulated distribution of the firm size for 100 000 firms. When $\mu = 1.4$, the distribution is close to the straight diagonal line, although there are some large firms. As μ becomes smaller, the zig-zagging becomes more apparent. When μ becomes lower than one, significant large steps, that is, huge monopolistic firms begin to appear, and the fractal nature of the whole distribution as a Devil's staircase becomes apparent, with much lower fractal dimension than for $\mu \geq 1$. The lower the value of μ, the more significant is this feature. It is natural to expect that the inequality of the shares, the oligopolistic nature, apparent for $\mu < 1$, would drive the social/governmental force to reduce it. Once the value of μ is one or larger, the oligopoly becomes tame; the inequality is there, but is not apparent. So, the driving force could disappear, or at least become significantly reduced.

This argument shows that $\mu = 1$ is much like a critical point, or a balancing point, of the driving forces. The economic efforts by large firms pushes μ downward but is met by the government measures below $\mu = 1$. Since the average growth rate is equal to one at $\mu = 1$, it is the point at which large firms could be content. Although this argument, illustrated in Fig. 4.8 is not supported by a solid, rigorous evidence, we stress that this is a simple but rather strong explanation of the universal nature of $\mu = 1$. Many stochastic models of firms that yield the Pareto–Zipf law have been constructed. Some of them can yield $\mu = 1$, with or without choice of the values of the parameters. We have, then, no reason to believe one over the other. In contrast, the explanation we offered here is a simple one based on a few general properties of the distribution and we believe that it is therefore a convincing one. Models of the firm dynamics with social/governmental background forces may be constructed along this line of argument.

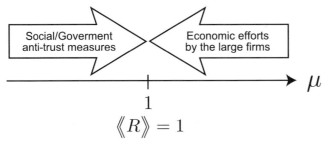

Fig. 4.8 Pictorial representation of our explanation of the universality of $\mu = 1$.

4.4 Small and Mid-sized Firms

While company growth in the power-law regime has been extensively studied, as we have seen, little has been known about the growth and fluctuations of small-business firms. Such a study is important because most firms are present in the non-power-law regime; therefore, financing those small-business firms is quite often one of the biggest financial problems for financial institutions in a country. The problem concerns how a financial institution should finance small and mid-sized companies with an appropriately determined interest rate for a certain period of time. It would be easier to understand firms' growth and failure by using an applicable database for quantitative study.

The Credit Risk Database (CRD) is the largest database of Japanese small and mid-sized companies, which covers nearly 1 million small-business firms, more than 60% of all companies in Japan. By employing this database, in this chapter, we show that small-business firms have qualitatively different characteristics of firm-size growth from those for large firms.

4.4.1 Data for Small and Mid-sized Firms

According to a survey by the statistics bureau of a Japanese ministry, the number of Japanese companies is approximately 1.6 million in the year 2001. Data in the CRD is sampled annually by a credit guarantee association, government-affiliated institutions and private-sector financial institutions all over Japan since the year 1997. It mainly covers small and mid-sized companies, the definition of which can be stated as follows. In accordance with the Japanese Small and Medium Enterprise Basic Law, if either the number of employees or amount of shareholders fund when established is less than threshold values, then the firm is said to be a small or mid-sized firm. The threshold value depends on the business sector which the firm belongs to in its main business. For wholesale's, it is 100 for the number of employees and 0.1 billion yen for the capital, when established. For retail, 50 and 50 million yen; for services, 100 and 50 million yen; for manufacturing and other sectors, 300 and 0.3 billion yen. The data coverage is more than 60 % of such firms in the year 2001. The database includes financial statements, nonfinancial facts (establishment etc.) and default facts.

Firm size distribution with respect to the number of employees is depicted in Fig. 4.9. This is the cumulative probability distribution of all firms according to a sample survey by a domestic census. One can observe that there exists

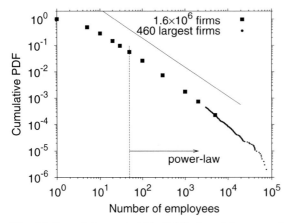

Fig. 4.9 Cumulative probability distribution of Japanese firm size in the year 2001. The squares correspond to tabulated data by a sample survey (Establishment and Enterprise Census 2001 by Ministry of Internal Affairs and Communications). The dots show an exhaustive list of the largest 460 firms (database by Diamond, Inc.). The line is simply a guide with $\mu = 1.34$ in the power-law $P_>(x) \propto x^{-\mu}$.

a transition from a non-power-law regime to a power-law regime[3]. The transition occurs at around a few tens in terms of the number of employees. The CRD which covers a large fraction of small and midsize firms as explained above, therefore, covers the non-power-law regime and the transition region. In what follows, we shall examine stock variables of total assets and total debts, and flow variable of sales. Because our database covers non-power-law regime and transition region, it is expected that the fluctuation of growth has a transition region for all of these variables. This will be shown to be true as we see in the next section.

4.4.2
Growth and Fluctuations

Let x be a variable which measures firm size (such as total assets and sales), and x_1 and x_2 be the variables measured on two successive years. The joint probability $P(x_1, x_2)$ is shown in Fig. 4.10 for the total assets, total debts, and sales on two successive years.

Our concern is the annual change of individual firm size, namely its growth. Exactly in the same way as was done for large firms, growth rate is defined as $R = x_2/x_1$. It is customary to use the logarithm of R, $r \equiv \log_{10} R$. We examine

---

[3] It should be mentioned that our database may have poor sampling for extremely small firms, e. g., those with less than 10 employees. See also other countries data such as [7, 21] for comparison. However, our study does not depend on extremely small firms.

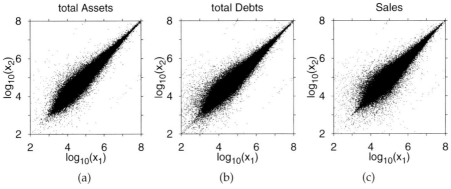

Fig. 4.10 Scatter plots corresponding to joint probabilities $P(x_1, x_2)$ for total-assets (a), total-debts (b) and sales (c). The values of x_1 and x_2 are in units of 10^3 yen.

the probability density for the growth rate $P(r|x_1)$ on the condition that the size x_1 in an initial year is fixed. If $P(r|x_1)$ does not depend on x_1, Gibrat's law holds. For large firms in the power-law regime, we showed in Section 4.3.1 that Gibrat's law holds and that Gibrat's law implies the existence of a power-law in the firm size distribution under detailed balance conditions. However, for small and mid-sized ones, we find that Gibrat's law breaks down as shown below.

Figure 4.11 shows the breakdown of Gibrat's law by depicting the probability density function $P(r|x_1)$ for logarithmic growth rate r. The probability density has an explicit dependence on x_1 showing the breakdown of Gibrat's law. In order to quantify the dependence, we examine how the standard deviation of r in the ensemble defined in each bin of x_1 scales, as x_1 becomes larger. Let the standard deviation of r be denoted by σ. Figure 4.11 (d)–(f) show how σ scales as a function of x_1.

We find that two regimes exist for all the variables we have studied. Namely, there is present a transition point x_1^* such that for $x_1 < x_1^*$,

$$\sigma \propto x_1^{-\beta} \tag{4.19}$$

with a power exponent β. x_1^* for different variables of asset, debt and sales all correspond to the transition of firm-size distribution that was shown in Fig. 4.9. Since $\beta > 0$, in this regime, the growth-rate fluctuation is larger for smaller firms.

For $x_1 > x_1^*$, we see that σ asymptotically approaches a nonscaling regime ($\sigma \sim$ const). This is the region in which Gibrat's law holds. This fact is consistent with what we have shown in Section 4.3.1.

Furthermore, we also examine the growth rate for different time-scales Δt. Let us define $r_{\Delta t} = \log R_{\Delta t} = \log(x_{t+\Delta t}/x_t)$, and examine the dependence

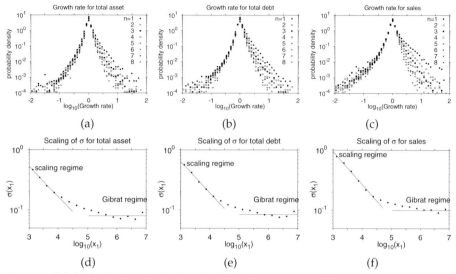

Fig. 4.11 (a)–(c) Probability density function $P(r|x_1)$ for $r = \log_{10}(R)$. For conditioning x_1, we use different bins of initial firm size with an equal interval in the logarithmic scale as $x_1 \in [10^{4+0.25(n-1)}, 10^{4+0.25n}]$ ($n = 1, \cdots, 8$) for total assets, total debts and sales. (d)–(f) Standard deviation σ of r as a function of initial year's firm size x_1 for total assets, total debts and sales.

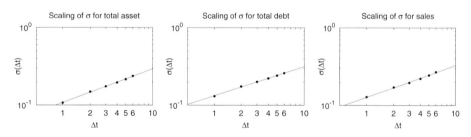

Fig. 4.12 Standard deviation $\sigma_{\Delta t}$ of $r_{\Delta t} = \log_{10} R_{\Delta t}$. The growth rate $R_{\Delta t} = x_{t+\Delta t}/x_t$ is defined for different time-scales Δt which range from 1 year to 6 years. The scaling relation in the form $\sigma_{\Delta t} \propto (\Delta t)^{-\gamma}$ is evident for total assets, total debts and sales, from left to right. The values of γ are respectively 0.44, 0.38 and 0.41.

of the standard deviation $\sigma_{\Delta t}$ of $r_{\Delta t}$ by changing the time-scale Δt from 1 to 6 years.

Figure 4.12 shows $\sigma_{\Delta t}$ as a function of Δt. It is obvious that there is a scaling relation in $\sigma_{\Delta t}$ for each of the variables we studied. Namely, we find that

$$\sigma_{\Delta t} \propto (\Delta t)^{-\gamma} \qquad (4.20)$$

with a power exponent γ.

This finding can be considered to be important in order to understand a temporal change in the financial state of stock (balance-sheet) and flow (profit-and-loss), because it is relevant to the growth and default process of small and mid-sized firms. Moreover, this study of fluctuation in growth rate would be of practical importance to financial institutions. Suppose a financial institution has a "portfolio" of credits (lending money) to a group of small and mid-sized firms. Then the short-term behavior of total assets and debts for several years of time-scale would be quite important to know quantitatively. Application to practical problems in this direction is in progress by collaborating with practitioners in the CRD Association.

4.5 Network of Firms

Recently, many studies have revealed true structures of real-world networks [9, 30, 48]. This development has been done also in the field of econophysics [23, 35–39]. Such studies have investigated business networks, shareholding networks, world trade networks, and corporate board networks. As is common practice, in order to discuss networks, we must define their nodes and edges. Edges represent some relationship between the nodes. In the following, we shall first consider firms as nodes, and the shareholding relationship between them as edges. Then we shall briefly take a look at a business network in relation to a firm's failure; bankruptcy.

4.5.1 Shareholding Networks

4.5.1.1 Degree Distribution

We consider Japanese shareholding networks as they existed in the years 1985, 1990, 1995, 2000, 2002 and 2003 (for other countries, see [23], for example, which studies the shareholding networks of firms quoted in the stock markets of MIB, NYSE and NASDAQ). We use data published by Toyo Keizai, Inc. This data source provides lists of shareholders for firms listed on the stock market or on the over-the-counter market in Japan. The number of shareholders in a firm's list varies from one firm to another. The data before the year 2000 contain information on the top 20 shareholders for each firm. On the other hand, the data for 2002 and 2003 contain information on the top 30 shareholders for each firm. In order to analyze the data uniformly, we consider the top 20 shareholders for each firm.

Types of shareholders include listed firms, nonlisted financial institutions (commercial banks, trust banks and insurance companies), officers and other individuals. For our study we do not consider officers and other individuals,

Tab. 4.1 Change in the size of a shareholding network N, the total number of edges K, and the exponent γ of the outgoing degree distribution $p(k_{\text{out}}) \propto k_{\text{out}}^{-\gamma}$.

Year	1985	1990	1995	2000	2002	2003
N	2 078	2 466	3 006	3 527	3 727	3 770
K	23 916	29 054	33 860	32 586	30 000	26 407
γ	1.68	1.67	1.72	1.77	1.82	1.86

so the shareholding networks are constructed only from firms. The number of nodes, N, and the total number of edges, K, vary with the years, and these are summarized in Table 4.1.

If we draw arrows from shareholders to stock corporations, we can represent a shareholding network as a directed graph. If we count the number of incoming edges and also of outgoing edges for each node, we can obtain the degree distribution for incoming degree, k_{in}, and that for outgoing degree, k_{out}. However, as explained above, we consider only the top 20 shareholders for consistency. Therefore, the incoming degree has an upper bound, $k_{\text{in}} \leq 20$, while the outgoing degree has no bound.

The log-log plot of k_{out} is shown in Fig. 4.13(a). In this figure, the horizontal axis corresponds to k_{out}, and the vertical axis corresponds to the cumulative probability distribution $P(k_{\text{out}} \leq)$, that is defined by the probability distribution function $p(k_{\text{out}})$,

$$P(k_{\text{out}} \leq) = \int_{k_{\text{out}}}^{\infty} dk'_{\text{out}} \, p(k'_{\text{out}})$$

in the continuous case. We can see that the distribution follows the power-law function, $p(k_{\text{out}}) \propto k_{\text{out}}^{-\gamma}$, except for the tail part. The exponent γ depends on the year, as summarized in Table 4.1. It has also been reported that the degree distributions of shareholding networks for firms listed on the Italian stock market (Milano Italia Borsa; MIB), the New York Stock Exchange (NYSE) and the National Association of Security Dealers Automated Quotations (NASDAQ), each follow the power-law distribution [23]. The exponents are $\gamma_{\text{MIB}} = 1.97$ in 2002, $\gamma_{\text{NYSE}} = 1.37$ in 2000, and $\gamma_{\text{NASDAQ}} = 1.22$ in 2000. These values are similar to the Japanese case.

The semi-log plot is shown in Fig. 4.13(b), and the meaning of the axes is the same as in (a). We can see that the tail part of the distribution approximately follows the exponential function. The exponential part of the distribution mainly consists of financial institutions. On the other hand, almost all of the power-law part of the distribution consists of nonfinancial institutions. This result suggests that different mechanisms work in each range of the distribution.

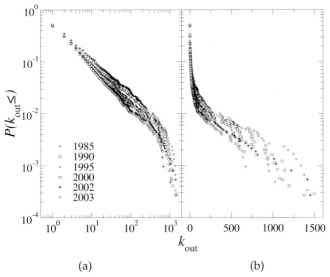

Fig. 4.13 (a) Log-log plot. (b) semi-log plot of the cumulative probability distribution of the outgoing degree.

4.5.1.2 Correlation Between Degree and a Firm's Growth

Knowledge of the characteristics of the node could be potentially useful for constructing a model explaining the dynamical growth of the network. However, for many types of complex network, it is often difficult to measure quantitative characteristics of the node. On the other hand, for networks in economics, especially networks of firms, we can obtain characteristics of the node quantitatively by using a balance sheet and an income statement, for example. We believe that this allows us to understand business networks in terms of firm growth.

We consider the correlation between outgoing degree and a firm's age. This is because ageing of nodes often plays an important role in many phenomena of growing networks and models explaining their generation mechanism. We also consider the correlation between the outgoing degree and a firm's total assets. We consider that creating a new link produces costs, and the costs that can be borne depend on the total assets. Hence it would be natural to assume that a firm with a large amount of total assets has a relatively large number of outgoing edges.

In the following, we measure a firm's age in units of months. Figure 4.14(a) is the semi-log plot of the distribution of a firm's age at the end of March 2002. The horizontal axis is the firm's age, T, and the vertical axis is the rank. In this figure, the dotted line corresponds to fitting by the exponential function, denoting that the distribution of a firm's age approximately follows an exponential distribution. The inner small panel shows a log-log plot of the correlation between the outgoing degree and a firm's age in units of months at the end of March 2002. This figure shows that these quantities have little correlation. Kendall's rank correlation τ can quantitatively clarify this correlation. The result is $\tau = 0.203$, meaning that there is no significant correlation between outgoing degree and a firm's age.

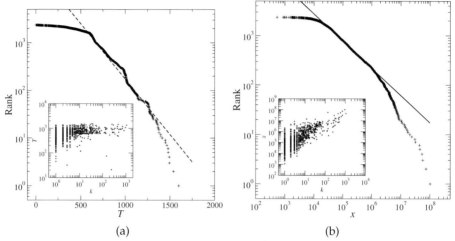

Fig. 4.14 (a) Semi-log plot of rank distribution for the age of a firm T. Dotted line in the left panel is the exponential function $p(T) \propto \exp(-T)$. Inner small figure is a scatter plot in the log-log scale of the k_{out} and T. (b) Log-log plot of rank distribution of total assets A. Solid line is a power low function, $p(x) \propto x^{-b}$ with $b = 1.6$. Inner small figure is the scatter plot in the log-log scale of the k_{out} and x.

Figure 4.14(b) is the log-log plot of the distribution of a firm's total assets at the end of March 2002. In this figure, the horizontal axis is the total assets x in units of million yen, and the vertical axis is the rank. This figure shows that the level of total assets in the middle range follows the power-law distribution. The solid line corresponds to fitting by a power-law function $p(x) \propto x^{-b}$ with exponent $b = 1.6$. In addition, this figure shows that the tail part of the distribution deviates from the power-law distribution. The inner small panel shows the log-log plot for the correlation between the outgoing degree and total assets in units of million yen at the end of March 2002. This figure shows that these two quantities correlate with each other. In this case, Kendall's rank correlation is $\tau = 0.530$. This means that there is significant correlation between outgoing degree and total assets.

4.5.1.3 Simple Model of Shareholding Network

Many models have been proposed to explain the dynamical growth of scale-free networks. However, many of them do not include characteristics of the node under consideration. On the other hand, several models have been proposed to explain the power law distribution of the firm's size. One of these models is based on a stochastic multiplicative process (SMP) with reset event [26]. Let us regard bankruptcy as a reset event.

We denote an initial distribution function of $x_i(0)$ as $p_0(x_i(0))$. Here i distinguishes entity and $i = 1, 2, \cdots, N$, and $x_i(t)$ is the size of firm i. The SMP with reset events is given as follows. At each time step, $x_i(t)$ is reset with a probability q to a new value $x_i(0)$. Here $x_i(0)$ follows a probability density function $p_0(x_i(0))$ that gives the initial distribution of $x_i(0)$. If the reset event does not occur, $x_i(t)$ is multiplied by a random positive factor $a_i(t)$ with a probability density function $p(a_i(t))$. Namely,

$$x_i(t+1) = \begin{cases} x_i(0) & \text{with probability } q \\ a_i(t)x_i(t) & \text{with probability } 1-q \end{cases} \quad (4.21)$$

We regard $x_i(t)$ in (4.21) as firm i's total assets, and assume that it is multiplied by a factor $a_i(t)$ with probability $1 - q$ and that a firm can go bankrupt with probability q. In addition, we fix the total number of companies, N, and replace bankrupted firms with new ones with total assets which are sampled from $p_0(x(0))$.

In the following, we fix $N = 50\,000$, $x_i(0) = 1$, and $q = 0.005$. We assume that $a_i(t)$ is uniformly distributed, i.e., $a_i(t) \in [0.5, 1.5)$. However, even if we use a different distribution function for $p_0(x_i(0))$, the tail part of the stationary distribution does not change. On the other hand, the reset probability q and the distribution function $P(a_i(t))$ change the value of exponents in the stationary distribution of $x_i(t)$ (see [26]).

Figure 4.15(a) represents the convergence of the simulation. In this figure, the horizontal axis is firm's total assets $x_i(t)$ and the vertical axis is the rank. This figure shows that the distribution rapidly converges after $t = 200$, and the converged distribution follows a power-law distribution with exponent $b = 2.0$, which is greater than in the case of Figure 4.14(b). The small figure in Figure 4.15(a) is the semi-log plot of the distribution of the firm's age at $t = 1000$. This figure shows that T_i follows an exponential distribution, a result almost the same as in Figure 4.14(b), except for the magnitude of the age. However, this difference can be absorbed by redefinition of the units with which age is measured.

In the following we simply denote the outgoing degree as k_i. We assume that k_i is defined by $k_i \equiv x_i/r_i$, where r_i is the uniformly distributed random variable, i.e., $r_i \in [1,5)$. Figure 4.15(b) represents the stationary distribution for k_i at $t = 1000$ in different runs of simulation. In this plot, the horizontal axis is k_i and the vertical axis is the rank. This figure shows that the outgoing degree follows a power-law distribution with $\gamma \approx 2.0$. This value is slightly greater than the exponents shown in Table 4.1.

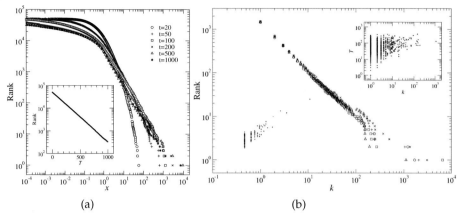

Fig. 4.15 (a) Convergence of simulation for a firm's total assets and a firm's age. Inner small figure is the semi-log rank plot of T_i at $t = 1000$. (b) Simulation results for outgoing degree distribution, correlation between outgoing degree and firm's age, and correlation between outgoing degree and a firm's total assets.

The scatter plot of k_i and T_i in the log-log scale is shown in the inner right small panel of Fig. 4.15(b). In this case, Kendall's rank correlation is $\tau = 0.210$, which is only slightly greater than the actual data shown in Fig. 4.14. The scatter plot of k_i and x_i in the log-log scale is shown in the inner left small panel of Fig. 4.15(b). In this case, $\tau = 0.60$, which is almost the same as the actual data shown in Fig. 4.14. Hence our model can explain correlation between outgoing degree, a firm's age, and a firm's total assets.

4.5.2
Business Networks

Firms do transaction with other firms. A firm buys intermediate goods and services from *upstream* firms, and sells products, goods and services to *downstream* firms. The firm puts on added value during the production process, potentially making profit as a result. This is a chain of firms who are putting on added values. Such a production network or business network is another important relationship between firms. Here we focus our attention on an extreme state of a firm's growth, namely failure or bankruptcy, and we shall see how the structure of a business network is crucial in understanding a chain of a firm's failure. Let us take a look at bankruptcy at an individual level.

4.5.2.1 Bankruptcy of Firms

Our bankruptcy data, provided by a firms databank company (Tokyo Shoko Research, Ltd.), is an exhaustive list of Japanese bankrupted firms in the year 1997. The data is exhaustive in the sense that any bankrupted firm with a total debt exceeding 10 million yen (roughly 80 000 Euros) is listed in it. The number of such bankrupted firms in the year is 16 526. The data contains information about debt, sales, business sector, number of employees, primary bank, when bankrupted, dates of establishment and failure, and amount of shareholders' fund when established. Two remarks immediately follow.

First, *bankruptcy* or business failure is not a legal term but a general term that represents a state of critical financial insolvency of a debtor. Classification of the different types, according to legal and private procedures taken by bankrupted firms, are different from one case to another. What is important for our purpose here, however, is the fact that most of the cases in bankruptcy, whatever the formal types may be, are caused by the financially insolvent state of the firm. Indeed, among all cases (16 526), the most frequent is suspension of bank transactions (13 850). That is, even short-term financing (for payment of labor and short-term borrowing, for example) becomes difficult, which is the situation common to almost all cases of bankruptcy. The causes for such a critical state of financial insolvency will be mentioned later, in relation to the problem of the bankruptcy chain.

Second, *debt* refers to the total sum of liabilities in the credit ("right-hand") side of the balance-sheet. This quantity, when a firm goes bankrupt, is relatively easy to measure in comparison with other balance-sheet quantities such as total assets and equity[4]. Since each item in the debt has a creditor outside of the bankrupted firm, the amount of debt is not hidden inside but is made public. Actual measurement of debt is done either by legal documents or by investigation of current and noncurrent liabilities in the balance sheet *plus* the amount of bills discounted and transfers by endorsement, according to whether the bankrupted firm takes legal procedures or private procedures, respectively.

For our study we used the entire dataset, irrespective of the formal types and procedure taken by bankrupted firms.

4.5.2.2 Distribution of Debt

Left panel of Fig. 4.16 shows the cumulative probability distribution of debt when the firm was bankrupted. Let x be the amount of debt in million yen (roughly ten thousand Euros). For x exceeding 10^2, we can observe a power-law distribution for three orders of magnitude or even more. The probability that a given bankrupted firm has a debt equal to or greater than x, denoted by $P_>(x)$, obeys $P_>(x) \propto x^{-\mu}$ with a constant μ, the Pareto index for bankruptcy. The index can be estimated as $\mu = 0.911 \pm 0.008$ (by least-squares-fit for sam-

---

4) Here debt is not equal to net liabilities, which is defined by the total liabilities *minus* the amount of assets not hypothecated. Such quantity would represent liabilities that the firm has in relation to other firms and banks, but is difficult to know.

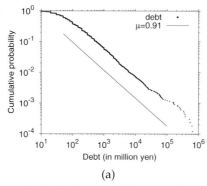 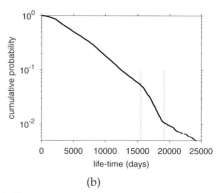

Fig. 4.16 (a) Distribution of debt when bankrupted. The number of data amounts to 16 526 firms which were bankrupted in the year 1997. A Guide is given by the line which corresponds to $\mu = 0.91$. (b) Cumulative probability distribution for lifetime of bankrupted firms. Time is given in units of days. The two dotted lines show 10 years right after the World War II, during which the birth-rate of firms was very much higher than at other epochs.

ples equally spaced in a logarithm of rank in the power-law region; the error is at 90% level). This finding is what we call the Zipf law for the debt of firms when bankrupted. It was previously pointed out in [4] that debt may obey the Zipf law, but this may be inconclusive due to the limited number of data (the data corresponded only to the tail with x larger than 10^5).

Since it has been known that firm size follows the Zipf law [1,6,7,22,31], it would be useful to know the relation between firm size and firm debt right before bankruptcy. As shown in [16], there is a strong correlation between debt and firm size (sales, number of employees and so on). The ratio between debt and size is statistically independent of firm size. This finding provides the picture that the Zipf law for firm size and that for firm debt when bankrupted are two sides of the same coin.

4.5.2.3 Lifetime of Bankrupted Firms

As interesting as how a large firm can fail is the problem of how long a failed firm can live. Figure 4.16(b) gives the cumulative distribution of lifetime τ, $P_>(\tau)$, for the same set of bankrupted firms. With this plot being semi-log, we can observe that the entire distribution follows an exponential distribution.

We also see that the distribution has a sudden drop and kink in a particular epoch. This coincides with about 10 years right after World War II, during which the birth-rate of firms was very much higher than at other epochs. In fact, as shown in [16], one can show that the process of survival is basically homogeneous in epochs from the historical data of *entry of firms* and by observing that a hazard-rate function can be explained with good agreement by entry rate alone. This can be considered to be brought about by the dissolution of the *zaibatsu* (corporate alliances) monopoly under the government by US General Headquarters (from 1945 to 1952).

These stylized facts about a firm's bankruptcy can be explained by a multidimensional stochastic model that includes the balance-sheet dynamics of firms and a financial institute (bank), which have been proposed by economists [19, 20]. See [16] for the simulation and results.

4.5.2.4 Chain of Bankruptcy and Business Network

During the last ten years, the annual number of bankrupted firms is 10 000 or more. (Remember that the total number of Japanese corporations is typically a few millions). For the year 2001, the number is 19 991, and the total amount of debt in bankruptcy is 16.28 trillion yen (roughly 160 billion Euros), which amounts to more than 3% of the nominal GDP in the same year. The different causes of bankruptcy are investigated and classified. The most frequent cause of bankruptcy is poor performance in sales (with the number 11 290 in total). However, the cases due to failure of other business-related firms are not negligible. Specifically, the influence by another firm's bankruptcy (1 731 in total) and the failure of commercial credit contracts in the transaction. Interestingly, as the amount of debt when bankrupted becomes larger, the number of these cases related to a bankruptcy chain and critical influence is observed to be a larger fraction. In fact, for the 208 bankruptcies each with a debt of more than 10 billion yen (roughly 0.1 billion Euros)[5], the cases of influence by another firm's bankruptcy is 62 in number, while the cases of poor performance amount to 48, even less. Other years have a similar qualitative feature. Moreover, it is reported that the total amount of debt due to a bankruptcy chain is more than 20% of the entire debt involved in all cases of bankruptcy. Thus bankruptcy of other firms in a business network has large effect which cannot be underestimated.

By looking at individual examples of bankruptcy, the amount of debt borrowed by a bankrupted firm is largest for banks and financial institutions as lending entities. But the number of related institutions is largest for nonfinancial firms who are lending as a payment of a business transaction. In other words, pay-back scheduled in the next period, that is, payment by firm A to B, was not actually performed in the commercial credit contract after the bankruptcy of firm A. If B is bankrupted at a later time, this process forms a bankruptcy chain described above. Such a process is considered to propagate from downstream to upstream and is mainly caused by failure in commercial

---

5) The debt amount of 10 billion yen is in the power-law region as we saw in Fig. 4.16. The number of such large bankruptcy is small, because it is in the tail of distribution. But the total amount of debt, so its impact to related firms and industries, can be huge.

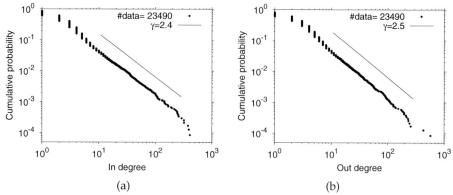

Fig. 4.17 Cumulative degree distributions for in-degree (a) and out-degree (b). Here a direct link from firm A to B is defined to be present if intermediate products flow from A to B. The data is for Japanese firms (total number 23 490) in the year 2003.

credit contracts. Most vulnerable firms are those who have a small number of links for selling products. Because a firm which has a single transaction, for example, would experience a critical insolvency state when the single firm in its downstream goes into bankruptcy.

The actual business network has a heavy-tailed distribution of degree, that is, the numbers of firms in the upstream side and the downstream side. Figure 4.17 shows cumulative distributions of degree (a) in-degree, (b) out-degree) for Japanese firms (total number 23 490) in the year 2003. Here a directed edge or link is defined to be present from firm A to firm B, if A sells its product to B (equivalently, if B buys an intermediate product from A). In each plot, a power-law distribution $P_>(k) \propto k^{-\gamma+1}$, in the same notation γ as we used for shareholding network, is shown where the degree is denoted by k, simply as guide for eyes.

Firms with the largest out-degree are trading and distribution companies which include Mitsubishi Corporation, Itochu and Marubeni, as well as manufactures which include electronics companies Fujitsu and Matsushita Electric (Panasonic). Those with the largest in-degree are the same firms as well as automobile companies such as Toyota and Nissan, and construction companies such as the Taisei Corporation.

However, as described above, the firms that play an influential role in the bankruptcy chain are those which are linked to the firms downstream with relatively low out-degree. That is because those firms depend crucially on a small number of downstream firms to sell their products.

4.6
Conclusion

We have demonstrated the following stylized facts concerning the distribution of firm sizes, their growth and fluctuations, by studying exhaustive lists of firm sizes in Europe and in Japan.

- In the power-law regime, detailed balance and Gibrat's law hold.

- Under the condition of detailed balance, Gibrat's law implies the Pareto–Zipf law (but not vice versa).

- Growth-rate distribution has a nontrivial relation between its positive and negative growth sides through the Pareto index (reflection law).

- For firm size in a non-power-law regime corresponding to small and mid-sized firms, Gibrat's law does not hold. Instead, there is a scaling relation of variance in the growth rates of those firms with respect to firm size, which asymptotically approaches the non-scaling region as the firm size approaches a power-law regime.

We examined two kinds of firm's networks, ownership and production relationships between firms. We found scale-free properties for both of these networks, and have attempted to explain how this fact is related to a firm's growth and failure through a stochastic multiplicative process with reset event. In addition, the structure of a production network is crucial in the understanding of bankruptcy, which also obeys a Zipf law, and the chain of firms' failure.

These stylized facts would serve to test models that explain the dynamics of firm size, which is the origin of the Pareto–Zipf law, the breakdown of these laws, a firm's failure, and the aggregate dynamics of firms in networks.

Acknowledgment

We would like to thank Masano Aoki, Mauro Gallegati, Corrado Di Guilmi, Yuichi Ikeda, Hiroshi Iyetomi and Taisei Kaizoji for collaboration and discussions. We would also like to thank Shigeru Hikuma and the CRD Association (CRD: Credit Risk Database) in Japan for the small-business firms dataset, and Tokyo Shoko Research, Ltd. for firm's bankruptcy data. We are grateful to the editors of this book for the invitation to write this chapter. This research was supported in part by the National Institute of Information and Communications Technology.

References

1 AMARAL, L. A. N., BULDYREV, S.V., HAVLIN, S., LESCHHORN, H., MAASS, P., SALINGER, M. A., STANLEY, H. E., STANLEY, M. H. R., *J Phys (France) I* 7 (**1997**), pp. 621–633

2 AMARAL, L. A. N., BULDYREV, S. V., HAVLIN, S., SALINGER, M. A., *Phys Rev Lett* 80 (**1998**), pp. 1385–1388

3 AOKI, M., *Modeling Aggregate Behaviour and Fluctuations in Economics*, Cambridge University Press, New York, **2002**

4 AOYAMA, H., SOUMA, W., NAGAHARA, Y., OKAZAKI, M. P., TAKAYASU, H., TAKAYASU, M., *Fractals* 8 (**2000**), pp. 293–300

5 AOYAMA, H., SOUMA, W., FUJIWARA, Y., *Physica A* 324 (**2003**), pp. 352–358

6 AOYAMA, H., FUJIWARA, Y., SOUMA, W., in *The Application of Econophysics: Proceedings of the Second Nikkei Econophysics Symposium*, Ed. H. Takayasu, Springer-Verlag, Tokyo, **2003**, pp. 268–273

7 AXTELL, R. L., *Science* 293 (**2001**), pp. 1818–1820

8 BADGER, W. W., in: *Mathematical Models as a Tool for the Social Science*, Ed. B. J. West, Gordon and Breach, New York, **1980**, pp. 87–120

9 BARABÁSI, A.-L., *Linked*, Perseus Publishing, Cambridge, **2003**

10 BOUCHAUD, J.-P., MÉZARD, M., *Physica A* 282 (**2000**), pp. 536–545

11 CHAMPERNOWNE, D. G., *Econometric Journal* 63 (**1953**), pp. 318–351

12 DRĂGULESCU, A. A., YAKOVENKO, V. M., *Physica A* 299 (**2001**), pp. 213–221

13 FUJIWARA, Y., SOUMA, W., AOYAMA, H., KAIZOJI, T., AOKI, M., *Physica A* 321 (**2003**), pp. 598–604

14 FUJIWARA, Y., AOYAMA, H., SOUMA, W., in *The Application of Econophysics: Proceedings of the Second Nikkei Econophysics Symposium*, Ed. H. Takayasu, Springer-Verlag, Tokyo, **2003**, pp. 262–267

15 FUJIWARA, Y., DI GUILMI, C., AOYAMA, H., GALLEGATI, M., SOUMA, W., *Physica A* 335 (**2004**), pp. 197–216

16 FUJIWARA, Y., *Physica A* 337 (**2004**), pp. 219–230

17 FUJIWARA, Y., in *Econophysics of Wealth Distributions*, Eds. A. Chatterjee, S. Yarlagadda, B. K. Chakrabarti, Springer-Verlag, Italia, **2005**, pp. 24–33

18 GABAIX, X., *Power laws and the origins of the business cycle*, MIT preprint, Department of Economics, **2002**

19 DELLI GATTI, D., GALLEGATI, M., PALESTRINI, A., in *Interaction and Market Structure*, Eds. D. Delli Gatti, M. Gallegati, A. Kirman, Springer-Verlag, Berlin, **2000**

20 GALLEGATI, M., GIULIONI, G., KICHIJI, N., in *Proceedings of 2003 International Conference in the Computational Sciences and Its Applications, Lecture Notes in Computer Science 2667*, Eds. V. Kumar, M. L. Gavrilova, C. J. K. Tan, P. L'ecuyer, Springer-Verlag, **2003** pp. 770–779

21 HART, P. E., OULTON, N., *Applied Economics Letters* 4 (**1997**), pp. 205–206

22 IJIRI, Y., SIMON, H. A., (Eds.), Skew Distributions and the Sizes of Business Firms. North-Holland, New York, **1977**

23 GARLASCHELLI, D., ET AL. *Physica A* 350 (**2005**), pp. 491–499

24 LEVY, M., SOLOMON, S., *Int J Mod Phys C* 7 (**1996**), pp. 595–601

25 MANDELBROT, B. B., *Econometrica* 29 (**1961**), pp. 517–543

26 MANRUBIA, S. C., ZANETTE, D. H. *Phys. Rev. E* 59 (**1999**), pp. 4945–4948

27 MIZUNO, T., KATORI, M., TAKAYASU, H., TAKAYASU, M., in *Empirical Science of Financial Fluctuations: The Advent of Econophysics*, Ed. H. Takayasu, Springer-Verlag, Tokyo, **2002**, pp. 321–330

28 MIZUNO, T., KATORI, M., TAKAYASU, H., TAKAYASU, M., in *The Application of Econophysics: Proceedings of the Second Nikkei Econophysics Symposium*, Ed. H. Takayasu, Springer-Verlag, Tokyo, **2004**, pp. 256–261

29 MONTROLL, E. W., SHLESINGER, M. F., *J Stat Phys* 32 (**1983**), pp. 209–230

30 NEWMAN, M. E. J., *SIAM Review* 45 (**2003**), pp. 167–256

31 OKUYAMA, K., TAKAYASU, M., TAKAYASU, H., *Physica A* 269 (**1999**), pp. 125–131

32 SOLOMON, S., RICHMOND, P., in *Economics with Heterogeneous Interacting Agents*, Eds. A. Kirman, J.-B. Zimmermann, Springer-Verlag, Berlin, **2001**

33 SORNETTE, D., *Critical Phenomena in Natural Sciences*, Springer-Verlag, Berlin, **2000**

34 SOUMA, W., *Fractals* 9 (**2001**), pp. 463–470

35 SOUMA, W., FUJIWARA, Y., AOYAMA, H., *Physica A* 324 (**2003**), pp. 396–401

36 SOUMA, W., FUJIWARA, Y., AOYAMA, H., *Physica A* 344 (*2004*), pp. 73–76

37 SOUMA, W., FUJIWARA, Y., AOYAMA, H., in *Science of Complex Networks: From Biology to the Internet and WWW CNET2004 (API Conference Proceedings, Vol. 776)*, Eds. J. F. F. Mendes, et al., Melville, New York, **2005**, pp. 298–307

38 SOUMA, W., FUJIWARA, Y., AOYAMA, H., in *Practical Fruits of Econophysics: Proceedings of The Third Nikkei Econophysics Symposium*, Ed. H. Takayasu, Springer-Verlag, Tokyo, **2005**, pp. 307–311

39 SOUMA, W., FUJIWARA, Y., AOYAMA, H., in *The Complex Networks of Economic Interactions: Essays in Agent-Based Economics and Econophysics (Lecture Notes in Economics and Mathematical Systems, Vol. 567)*, Eds. A. Namatame, et al., Springer-Verlag, Tokyo, **2006**, pp. 79–92

40 SOUMA, W., NIREI, M., in *Econophysics of Wealth Distributions*, Eds. A. Chatterjee, S. Yarlagadda, B. K. Chakrabarti, Springer-Verlag, Italia, **2005**, pp. 34–42

41 STANLEY, H. E., AMARAL, L. A. N., GOPIKRISHNAN, P., PLEROU, V., *Physica A* 283 (**2000**), pp. 31–41

42 STANLEY, M. H. R., BULDYREV, S. V., HAVLIN, S., MANTEGNA, R. N., SALINGER, M. A., STANLEY, H. E., *Economics Letters* 49 (**1995**), pp. 453–457

43 STANLEY, M. H. R., AMARAL, L. A. N., BULDYREV, S. V., HAVLIN, S., LESCHHORN, H., MAASS, P., SALINGER, M. A., STANLEY, H. E., *Nature* 379 (**1996**), pp. 804–806

44 STEINDL, J., *Random Processes and the Growth of Firms: A Study of the Pareto Law*, Griffin, London, **1965**

45 SUTTON, J., *J Economic Literature* 35 (**1997**), pp. 40–59

46 TAKAYASU, H., OKUYAMA, K., *Fractals* 6 (**1998**), pp. 67–79

47 YAKOVENKO, V. M., SILVA, A. C., in *Econophysics of Wealth Distributions*, Eds. A. Chatterjee, S. Yarlagadda, B. K. Chakrabarti, Springer-Verlag, Italia, **2005**, pp. 15–23

48 WATTS, D. J., *Small-worlds*, Princeton University, Princeton, **2003**

5
A Review of Empirical Studies and Models of Income Distributions in Society

Peter Richmond, Stefan Hutzler, Ricardo Coelho, and Przemek Repetowicz

5.1
Introduction

The statement that wealth is not distributed uniformly in society appears obvious. However, this immediately leads to several nontrivial questions. How is wealth distributed? What is the form of the distribution function? Is this distribution universal or does it depend on the individual country? Does it depend on time or history? These questions were first studied by Vilfredo Pareto in 1896/97 who noticed that *the rich end* of the distribution was well described by a *power law*. Ever since these early studies of Pareto, economists, and more recently physicists, have tried to first of all infer the exact shape of the entire distribution from economic data, and secondly, to design theoretical models that can reproduce such distributions.

Progress has been made on both fronts, but questions remain. How can wealth be measured? Gross salary income may be a good indicator for low-to-medium income earners, but how do the *super-rich*, many of whom are not employees, fit into this picture?

In this article we review both historical and current data which support the thesis that certain features of the wealth distribution are indeed universal. In our discussion we shall restrict ourselves to *income*. We shall not concern ourselves with the question of *wealth*. However, we shall see that the observed Pareto power-law tail is a feature even of income data. Whilst it is difficult to be precise, it does appear that the reported values of the exponents are changing with time and we might speculate as to why this should be. Does it relate to economic and taxation policy? These are questions for the economic community to consider.

We begin our review by giving some background on the thinking of Pareto, a researcher who was ahead of his time with regards to scientific research, but who also mirrored the zeitgeist of his period. We then summarize early work by Gibrat, Champernowne and Mandelbrot, before reporting more recent approaches. In particular, we feature a number of different agentbased models: the Family Network model, the Generalized Lotka Volterra model and other models where agents exchange money by pairwise transactions in analogy with the exchange of momentum by colliding molecules in a gas. These *collision models* which are conceptually extremely simple, is also accessible to analytical theory and thus currently *en vogue*. Finally we present a case study using UK data for incomes over the period 1992–2002, highlighting the successes and failures of current research.

5.2
Pareto and Early Models of Wealth Distribution

5.2.1
Pareto's Law

The distribution of wealth or money in society has proved to be of great interest for many years. Based on the numerical analysis of an impressive amount of economic data, Italian economist Vilfredo Pareto [1] was the first to suggest that it followed a *natural law* now often simply termed *Pareto's law*. Sketches of both income distribution and cumulative distribution are shown in Figs 5.1 and 5.2. However, there are a number of different forms of Pareto's law quoted in the literature. Mandelbrot [2] distinguishes between three different versions: if $\Pi(m)$ is the percentage of individuals with an income greater than m, the *strong Pareto's law* states that

$$\Pi(m) = \begin{cases} (m/m_0)^{-\alpha} & \text{for } m > m_0 \\ 1 & \text{for } m < m_0 \end{cases} \quad (5.1)$$

Here m_0 is a scale factor and the value of the exponent α is not determined. In the strongest form of *Pareto's law*, $\alpha = 3/2$, which is the average value of α in Pareto's original data (see Table 5.1 and Fig. 5.3). Recent studies show that this average is now close to 2, with a wider distribution (Table 5.2 and Fig. 5.3). It is thus better to refer to the *weak Pareto law* [2], which states that the power law only holds in the limit $m \to \infty$ or, in other words ([2], p. 81):

$$\Pi(m)/(m/m_0)^{-\alpha} \to 1, \text{ as } m \to \infty \quad (5.2)$$

where the value of α remains unspecified. In the remainder of this article we mean this form of the law when we write *Pareto's law*.

Today, the law of Pareto is usually quoted in terms of the distribution density function, $P(m)$, rather than the cumulative distribution function, $\Pi(m) = \int_m^\infty P(m)dm$, viz:

$$P(m) \sim m^{-(1+\alpha)}, \text{ for large } m. \quad (5.3)$$

5.2 Pareto and Early Models of Wealth Distribution

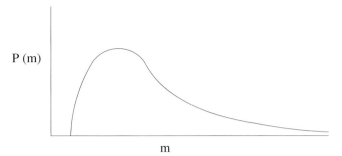

Fig. 5.1 Representation of the distribution of income. For large values of income this follows a power law.

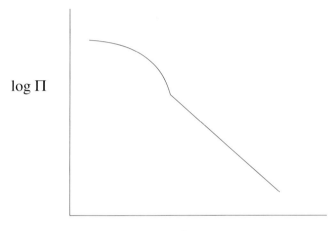

Fig. 5.2 Representation of the cumulative distribution of income. In this log–log plot the power-law regime results in a straight line with slope $-\alpha$, see Eq. (5.1).

Tab. 5.1 Table, taken from Pareto's book [1], showing the exponent α for a number of different data sets. Note that this is only a small extract of all the data that Pareto analyzed.

Country	Year	α	Country	Year	α
England	1843	1.50	Perouse, village		1.69
	1879–80	1.35	Perouse, campagne		1.37
Prussia	1852	1.89	a		1.32
	1876	1.72	Italian villages		1.45
	1881	1.73	Basle	1887	1.24
	1886	1.68	Paris		1.57
	1890	1.60	Augsburg	1471	1.43
	1894	1.60		1498	1.47
Saxony	1880	1.58		1512	1.26
	1886	1.51		1526	1.13
Florence		1.41	Peru	End of 18th century	1.79

a) Some of the Italian main cities: Ancone, Arezzo, Parma and Pisa

Tab. 5.2 Table of empirical data. In column Source: S.H.–Size of Houses; I.–Income; I.T.–Income Tax; Inhe. T.–Inheritance Tax; W.–Wealth. In column Distributions: Par.–Pareto tail; LN–Log-normal; Exp.–Exponential; D. Par.–Double Pareto Log-normal; G.–Gamma.

Country	Source	Distributions	Pareto Exponents	Ref.
Egypt	S.H. (14^{th}B.C.)^a	Par.	$\alpha = 1.59 \pm 0.19$	[34]
Japan	I.T. (1992)	Par.	$\alpha = 2.057 \pm 0.005$	[42–44]
	I. (1998)	Par.	$\alpha = 1.98$	
	I.T. (1998)	Par.	$\alpha = 2.05$	
	I. / I.T. (1998)	Par.	$\alpha = 2.06$	
	I. (1887-2000)	LN / Par.	$\bar{\alpha} \sim 2.0^b$	
USA	I.T. (1997)	Par.	$\alpha = 1.6$	[24]
Japan	I.T. (2000)	Par.	$\alpha = 2.0$	
USA	I. (1998)	Exp. / Par.	$\alpha = 1.7 \pm 0.1$	[10, 14, 45]
UK	Inhe. T. (1996)	Exp. / Par.	$\alpha = 1.9$	
Italy	I. (1977-2002)	LN / Par.	$\alpha \sim 2.09 - 3.45$	[12]
	I. (1987)	LN / Par.	$\alpha = 2.09 \pm 0.002$	
	I. (1993)	LN / Par.	$\alpha = 2.74 \pm 0.002$	
	I. (1998)	LN / Par.	$\alpha = 2.76 \pm 0.002$	
Australia	I. (1993-97)	Par.	$\alpha \sim 2.2 - 2.6$	[20]
USA	I. (1997)	D. Par.^{c}	$\alpha = 22.43 \ / \ \beta = 1.43$	[46]
Canada	I. (1996)	D. Par.	$\alpha = 4.16 \ / \ \beta = 0.79$	
Sri-Lanka	I. (1981)	D. Par.	$\alpha = 2.09 \ / \ \beta = 3.09$	
Bohemia	I. (1933)	D. Par.	$\alpha = 2.15 \ / \ \beta = 8.40$	
USA	1980	G. / Par.	$\alpha = 2.2$	[47]
	1989	G. / Par.	$\alpha = 1.63$	
	2001	G. / Par.		
UK	1996	Par.	$\alpha = 1.85$	
	1998-99	Par.	$\alpha = 1.85$	
USA	I. (1992)	Exp. / LN	^{d}	[31]
UK	I. (1992-2002)	Exp. / LN		
India	W. (2002-2004)^e	Par.	$\alpha \sim 0.81 - 0.92$	[48]
	I. (1997)	Par.	$\alpha = 1.51$	
USA	W. (1996)^f	Par.	$\alpha = 1.36$	[19, 49, 50]
	W. (1997)^g	Par.	$\alpha = 1.35$	
UK	W. (1970)	Par.		
	W. (1997)^h	Par.	$\alpha = 1.06$	
Sweden	W. (1965)	Par.	$\alpha = 1.66$	
France	W. (1994)	Par.	$\alpha = 1.83$	
UK	Inhe. T. (2001)	Par.	$\alpha = 1.78$	[21]
Portugal	I.T. (1998-2000)	Par.	$\alpha \sim 2.30 - 2.46$	[51]

a) Related to the size of houses found in an archaeological study.
b) This value is an average Pareto exponent.
c) α and β are Pareto exponents for the richest and poorest part, respectively.
d) Both distributions are a good fit of the data.
e) 125 wealthiest individuals in India.
f) 400 wealthiest people, by Forbes.
g) Top wealthiest people, by Forbes.
h) Top wealthiest people, by Sunday Times.

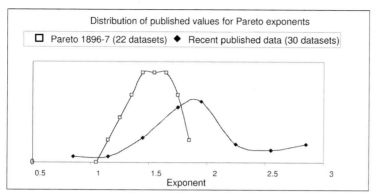

Fig. 5.3 Distribution of values for Pareto exponents α (Eq. (5.2)) taken from Pareto's original data ([1] and Table 5.1) and recently published values (see Table 5.2).

5.2.2
Pareto's View of Society

It is interesting to re-read the writings of Pareto at the time. These reveal more about the process by which he arrived at his conclusions. He opens the chapter on *Population* in his *Manual of Political Economy* [3], which summarizes his findings and thoughts in a mainly nonmathematical fashion, by stating that "society is not homogeneous" (p. 281). The existing "economic and social inequalities" correspond to the "inequalities of human beings *per se*" with respect to "physical, moral, and intellectual viewpoints" (p. 281). Thus an excellent indicator of nonhomogeneity in society is, according to Pareto, the distribution of income in society, as sketched in Fig. 5.1.

Pareto's main achievement, as seen from the perspective of today's econophysicists and economists, is the observation that this distribution is *universal*, i.e., that it "varies very little in space and time; different peoples and different eras yield very similar curves. There is a remarkable stability of form in this curve." (p. 285). Furthermore Pareto discovered that the form of the curve "does not correspond at all to the curve of errors, that is, to the form the curve would have if the acquisition and conservation of wealth depended only on chance".

This *non-Gaussian* character of the curve is obvious from its lack of symmetry about its peak and the pronounced tail at the *rich end* of the distribution, although Pareto does not dwell on the concept of power laws in his 1927 *Manual* [3].

Pareto notes that the poor end of the wealth distribution cannot be fully characterized, due to a lack of data. He stresses, however, the existence of a "minimum income [...] below which men cannot fall without perishing of poverty and hunger" (p. 286).

Finally he notes the *stability* of the distribution: "If, for example, the wealthiest citizens were deprived of all their income [...] sooner or later [the curve] would re-establish itself in a form similar to the initial curve. In the same way, if a famine [...] were to wipe out the lower parts of the population [...] the figure [...] would return to a form resembling the original one." (p. 292).

For Pareto the wealth distribution of Fig. 5.1 "gives a picture of society" (p. 286) and thus forms the basis of his *theory of society*. In using arguments based on Darwin's ideas of social selection, in common with many of his contemporaries, and calling feminism a *malady* and referring to women as "objects of luxury who consume but do not produce" (p. 297), he also paints a picture of society at his time.

In Pareto's view people are, in principle, free to move along the wealth axis in the course of their lifetime, in both directions, but this movement is determined by "whether they are or are not well fitted for the struggle of life" (p. 287). If they drop below the minimum income they "disappear" (p. 286) or are "eliminated" (p. 287). In the region of low incomes "people cannot subsist, whether they are good or bad; in this region selection operates only to a very small extent because extreme poverty debases and destroys the good elements as well as the bad." (p. 287). Pareto views the process of selection to be most important in the area around the peak of the distribution. Here the incomes are "not low enough to dishearten the best elements". He continues with the following statement that reflects views that were probably widely held at that time. "In this region, child mortality is considerable, and this mortality is probably a powerful means of selection" (p. 287–8). In Pareto's ideology this region forms a future aristocracy which will eventually rise to the rich end of the distribution and form the leadership of the country. Since selection does not apply to the rich, this will, however, lead to degeneration in this "social stratum". If this is paired with an "accumulation in the lower strata of superior elements which are prevented from rising" (p. 288), a revolution is unavoidable.

Pareto's ideas for changes in society, based on *social Darwinism*, are today no longer acceptable to the majority of people. They have been completely and utterly discredited by the German NSDAP who used them as a basis for their justification of concentration camps. However, Pareto's idea that a static distribution of wealth does not imply a static society, holds true. People are able to move along the wealth axis in both directions, although in some societies, this movement appears not to be too prevalent. Indeed, it is often found that being born to parents *at the poor end* of the wealth distribution greatly reduces the chances of obtaining a university education, which may form the basis for a high income in later life.

Since the distribution of wealth appears fixed, the main indicator for the degree of development of a society, according to Pareto, is the amount of wealth per person. If this increases, as in Pareto's example of England in the 19th century, it provides "individuals with good opportunities to grow rich and rise to higher levels of society" (p. 296). This is essentially today's neo-liberal argument that a well run and strong economy will *automatically* abolish poverty in society. Others, who are often in close contact with those at the lower reaches of society, such as the homeless, dispute this argument and point to other studies that suggest some forms of intervention are required in order to provide *social justice* across society.

The kind of thinking and explanation based on opinion is not one followed by those physicists who have begun to examine income distribution data. Physicists are basically driven by empiricism, an approach exemplified by Kepler who as a result of rather painstaking observations of the motion of planets, proposed his law of planetary motion. In similar vein, physicists and some economists have begun to construct models based on some underlying mechanisms that allow money to flow throughout a system and, in so doing, link these microscopic mechanisms to the overall distribution of income. Some of the models may be criticized as naïve by the economics community. However, as we shall see, at least the models that do emerge seem capable of predicting distributions of money that are observed to a greater or lesser degree. These advances allow for a rational debate, and through further research, advances to be made in prediction.

5.2.3
Gibrat and Rules of Proportionate Growth

The French economist, Robert Gibrat realized that the power law distribution did not fit all the data and proposed a law of proportionate effect, "la loi de l'effet proportionnel" [4]. This states that a small change in a quantity is independent of the quantity itself. The quantity $dz = dx/x$ should therefore be Gaussian distributed. Hence x should be distributed according to a log normal distribution. As a result of studying the empirical data, he generalizes the statement and concludes that $z = a \ln(x - x_0) + b$. This leads him to the distribution:

$$F(x) \propto \exp\left(-(a\ln(x - x_0) + b)^2\right) \qquad (5.4)$$

Since this was based on the statistics of Gauss, Gibrat felt this was a better approach than that of Pareto. Gibrat defined $100/a$ to be an inequality index. The parameter a is today related to the Gibrat index.

Recently, Fujiwara [5] has shown, using very detailed Japanese data where the variation of individual incomes can be identified over time, how in the power law region, Gibrat's law and the condition of detailed balance, both hold. He then shows how the condition of detailed balance together with Gibrat's law implies Pareto's law (but not vice versa).

5.2.4
The Stochastic Model of Champernowne

An early stochastic model which reproduces Pareto's law is due to Champernowne [6]. The model dates back to his King's College Fellowship thesis of 1937, but was presented in full detail only in 1953 in the Economic Journal. The basic idea is that an individual's income in one year "may depend on what it was in the previous year and on a chance process" (p. 319). Based on the definition of certain ranges of income, Champernowne specifies the probability for the income of an individual to change with time from one income *range* to another. Mathematically such a process may be expressed in terms of a vector $X_r(t)$, specifying the number of income-receivers in the income range r at time t, and a set of stochastic transition matrices $T(r,s|t)$ that represent the proportion of individuals in income range r at time t which move to income range s in the year $t+1$:

$$X_s(t+1) = \sum_r X_r(t) T(r,s|t) \tag{5.5}$$

Champernowne was able to show that provided "the stochastic matrix is assumed to remain constant throughout time [...] the distribution will tend towards a unique equilibrium distribution depending on the stochastic matrix but not on the initial distribution" (p. 318) of income. These equilibrium distributions are described by a Pareto law.

Obviously, the details of the transition matrix $T(r,s|t)$ are crucial. In the simplest case considered by Champernowne, income increases are allowed only by one range each year whereas decreases may occur over many ranges. Furthermore, the transition probabilities are treated as independent of the present income. With these assumptions and empirical data provided by the Institute of Statistics at Oxford, "regarded as unreliable by the authors of the survey" (p. 322), as noted by Champernowne, he determined a transition matrix and deduced that his model exhibited an equilibrium solution. The form was given by the strong Pareto law, and thus was not in agreement with actual data which shows power-law behavior only for high incomes.

However, two generalizations of the model led Champernowne to the weak form of Pareto's law. First he allowed annual income increases over more than one range, and secondly he linked the possible range of jumps to the income before the jump, except for high incomes. Further modifications by Champernowne that took account of the age of an individual, and occupation dependent prospects did not change the result. Champernowne concludes his paper with the observation that his models "do not throw much light on the mechanism that determines the actual observed values for Pareto's alpha" (p. 349). This makes it impossible to draw "any simple conclusions about the effect on Pareto's alpha of various redistribution policies" (p. 351). This remark is interesting in the sense that it is clear that Champernowne was already thinking of how to engineer specific income distributions within society.

5.2.5
Mandelbrot's Weighted Mixtures and Maximum Choice

While Champernowne presents a particular model with some variations that reproduces an income distribution which follows Pareto's law, Benoit Mandelbrot comes to a more general conclusion in his 1960 article [7] "The Pareto–Lévy Law and the Distribution of Income" and his "informal and non mathematical" paper "New Methods in Statistical Economics" [2]. According to Mandelbrot: "random variables with an infinite population variance are indispensable for a workable description of price changes, the distribution of income, and firm sizes, etc."([7], p. 421). This statement is based on Mandelbrot's observation that "essentially the same law continues to be followed by the distribution of income, despite changes in the definition of these terms." ([2], p. 85.)

To understand the relevance of the statement one needs to consider the so-called *stability* of a probability distribution. In the case of a Gaussian distribution it is known that numbers made up from the sum of independent Gaussian variables are again Gaussian distributed. This stability under summation (or invariance under aggregation) is, however, not restricted to Gaussian. It also holds for *Pareto–Lévy* distributions with index α between 1 and 2. (Mandelbrot introduces the term *Pareto–Lévy* in honor of his former supervisor Lévy, who studied their properties.) Such distributions had generally not been considered in the economical literature. The fact that their moments may be infinite led the physics community to ignore them for many years, partly on the grounds that they did not seem to correspond to physical reality. However, Mandelbrot dedicates a long section in his 1963 paper to scale invariance of such distributions. Specifically, he considers the stability of the Pareto–Lévy distribution under three different transformations.

Invariance under aggregation or simple addition may be met by using Gaussian as well as Pareto–Lévy distributions. Mandelbrot notes that the common belief at that time that only the Gaussian is invariant under aggregation is correct only if one excludes random variables with infinite moments.

The idea behind his concept of so-called *"weighted mixtures"* covers the case where the origin of an income data set is not known. If this is the case, one may consider that it was picked at random from a number of possible basic distributions. The distribution of observed incomes would then be a mixture of the basic distributions. With price data, for example, they often refer to various grades of commodity that may not be known; with companies or firms "the very notion is to some extent indeterminate, as can be seen in the case of almost wholly owned, but legally distinct subsidiaries". The situation is analogous to measuring the distribution of molecules in a box over a period of time where each day, for example, and unknown to the person making the measurements, a demon shifts the temperature and allows a new thermal equilibrium to come about prior to new measurements being made. Thus the data obtained for the distribution may be considered to be from a set of different Boltzmann distributions characterized by different temperatures. Such an approach leads to the so-called *super-statistics* discussed recently by, for example, Bek [8]. Gaussian distributions are not invariant under this type of transformation whereas Levy distributions are.

The third property Mandelbrot termed *"maximizing choice, the selection of the largest or smallest quantity in the set"*. "It may be that all we know about a set of quantities is the size of the one chosen by a profit maximizer. Similarly, using historical data, one may expect to find only fully reported events such as droughts, floods or famines. At best, data might be a mixture of complete reporting and reporting limited to extreme cases." ([7], p. 424.)

Such transformations need not be the only ones of interest. However, they are so important that they should characterize the laws they leave invariant. In this sense, the observation that income distributions are the same whatever the definition of income, is used by Mandelbrot to support the claim that they are Paretian. Mandelbrot summarizes ([7], p. 425): "It is true that incomes (or firm sizes) follow the law of Pareto; it is not true that the distributions of income are very sensitive to the methods of reporting and of observation.".

5.3
Current Studies

Since the contributions of Mandelbrot, numerous more recent empirical studies, for example Levy and Solomon [9], Dragulescu [10] and Souma [11] have all shown that the power-law tail is an ubiquitous feature of income distributions. The value of the exponent may vary with time and depends on the source of the data. However, over 100 years after Pareto's observation, a complete understanding of the shape and dynamics of wealth distribution is still evasive. This is partly due to incomplete data, but may also reflect the fact that there might indeed be two distributions, one for the *super rich*, one for the *low to medium rich*, with some intermediate region in-between. For the US, for example, only the top three percent of the population follow Pareto's weak law, the vast majority of people appear to be governed by a completely different law. The distribution function for the majority of the population seems to fit a different curve.

Clementi and Gallegati in a recent publication [12] have examined data from the US, UK and Germany and reaffirmed support for a two-parameter log-normal function to fit the data outside the power law regime and suggest that different mechanisms are working in the low and high-income regimes.

Other fitting functions have also recently been proposed. Mimkes and Willis [13] assert that for people in low to middle wage classes in US manufacturing, a Boltzmann–Gibbs distribution fits the data just as well as the log-normal distribution. Such a function was also found by Willis to fit data from the UK, in the low to middle income regime. This form of fitting function has been confirmed by Yakovenko [14]. More recently, Ferrero [15] has, using data for Argentina, shown how an entire income distribution including even the power-law tail of income may be fitted using a Tsallis distribution function. The Tsallis distribution reduces to the BG distribution for low income values yet yields a power law for high values of income. Anteneodo and Tsallis [16] have shown that the Tsallis distribution can be obtained from a simple type of multiplicative stochastic process.

Table 5.2 summarizes recent empirical studies according to the type of data, the models used for their analysis, and the value of the power-law exponent. While in the empirical studies made by Pareto α is around 1.5 (Table 5.1), in the later studies, exponents less than 1.5 only feature for the top wealthiest in the society and are generally obtained from a list of *super rich* people, published in magazines (as can be seen in Fig. 5.4). For all the other studies, which are mainly obtained from income distributions, the exponent is bigger than this (see also Fig. 5.3). From this one might conclude that income does not show the real exponent for the richest part.

In the next section we will discuss three particular models for wealth distributions in greater detail. This will give a good indication, on the one hand, that already very basic models yield key features of wealth distributions. On the other hand, it also highlights how hard it is to link the model parameters to economic parameters.

5.3.1
Generalized Lotka–Volterra Model

The Generalized Lotka–Voterra model of Solomon, Richmond and colleagues [17–19] is based on the re-distribution of wealth of a total number N of agents in a society. The governing equation is given by:

$$m_{i,t+1} = (1+\xi_t)m_{i,t} + \frac{a}{N}\sum_j m_{j,t} - c\sum_j m_{i,t}m_{j,t} \qquad (5.6)$$

and combines a multiplicative random process with an auto-catalytic process.

In the above equation, $m_{i,t}$ represents the amount of money assigned to agent i at time t. The random numbers ξ_t are chosen from a probability distribution with variance D. The second term on the RHS of Eq. (5.6) redistributes at each time-step a fraction of the total money, to ensure the money possessed

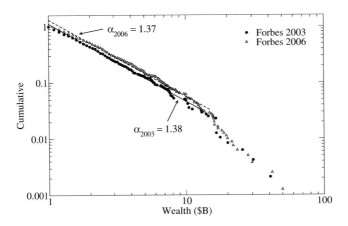

Fig. 5.4 Cumulative distribution of wealth for the top richest people in the World. The data was taken from Forbes Magazine for the years 2003 (476 individuals) and 2006 (793 individuals). The wealth is in billions of dollars.

by any agent is never zero as a result of the random process. One interpretation of this is that it simulates in a simplistic way the effect of a tax or social security policy. The parameter c in the final term on the RHS controls the overall growth of the system. It represents external limiting factors: finite amount of resources and money in the economy, technological inventions, wars, disasters, etc. It also includes internal market effects, competition between investors, adverse influence of bids on prices (such as when large investors sell assets to realize their profits and thereby cause prices/ profits to fall). This term has the effect of limiting the growth of the total amount of money, m_t, in the system, to values sustainable for the current conditions and resources. Clearly the equation has no stationary solution and the total money within this system can change with time.

However, it turns out that the relative value of money possessed by an agent, $x_{i,t} = m_{i,t}/\langle m_t \rangle$, is independent of the function c. The distribution function for the relative value of money, $x_{i,t}$, depends only on a and the variance D of the random variables and if the ratio a/D is constant, the dynamics of the relative money are independent of time. Consequently, even in a nonstationary system, after some time, the relative wealth distribution will eventually converge to a time-invariant form.

A mean-field approximation leads to a stationary distribution function $P(x)$ of the form:

$$P(x) = \exp\left(-(\alpha - 1)/x\right)/x^{1+\alpha} \tag{5.7}$$

The positive exponent, α, is a ratio of parameters of the model that are related to the redistribution process and volatility of the random process. For large values of income, x, this expression indeed exhibits a Pareto-like behavior. Numerical simulations by Malcai et al. [9] of the complete dynamics of this model, show this solution is essentially exact.

A particular form of an agent exchange model was proposed by Di Matteo et al. [20]. In this model, agents exchange money according to the following rule:

$$m_{l,t+1} = m_{l,t} + \sum_{j \neq l} m_{j,t} T(j,l|t) - m_{l,t} \sum_{j \neq l} T(l,j|t) + \xi_t \qquad (5.8)$$

Numerical calculations show that the solution of this equation yields a one-agent distribution function that agrees qualitatively with empirical income data for Australia in both the low and high-income regions.

5.3.2
Family Network Model

Coelho and colleagues [21] have recently published the so-called *family network model*, which not only yields realistic wealth distributions, but also a topology of wealth. This is a model of asset exchange where the main mechanisms of wealth transfer are *inheritance* and *social costs* associated with raising a new family. The structure of the network of social (economic) interactions is not predefined but emerges from the asset dynamics. These evolve in discrete time-steps in the following manner. For each time-step:

- From the initial configuration (step I in Fig. 5.5), the oldest family (node) is taken away, and its assets are uniformly distributed between the families linked to it (neighbors) (step II in Fig. 5.5).

- A new family (node) is added to the system and linked to two existing families (nodes), that have wealth greater than a minimum value q (step III in Fig. 5.5).

- The small amount q, is subtracted from the wealth of the selected families (nodes) and redistributed in a preferential manner in the society. (This process aims to model the wealth needed to raise a child. The preferential redistribution is justified by the fact that wealthier families control more business and benefit more from the living costs of a child.).

- A portion p of the remaining wealth of each of the two families is donated to the new family as *start-up money* (step IV in Fig. 5.5).

The total wealth and the number of families are conserved after each time-step. Numerical calculations yield the cumulative wealth distribution in line

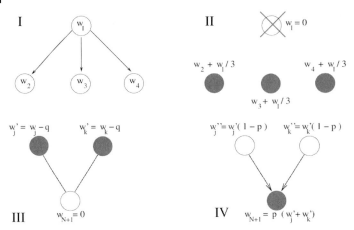

Fig. 5.5 Schematic representation of a time-step in the Family Network Model. Circles represent families and arrows the links between them.

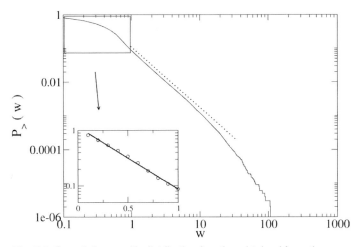

Fig. 5.6 Cumulative wealth distribution function obtained from the Family Network Model ($p = 0.3$, $q = 0.7$, $N = 10\,000$ and results after 10 MCS). The tail is approximated by a power-law with exponent $\alpha = 1.80$, and the initial part of the curve follows an exponential. The inset shows this initial trend on a log-linear scale.

with empirical data. For reasonable values of p and q we observe, for the upper 10% of the society, the scale-free Pareto distribution with Pareto exponents that lie between 1.8 and 2.7. The Pareto tail forms relatively quickly, usually after less than two generations. The degree-distribution of the family-network converges also rapidly to an exponential form. Interesting correlations between wealth, connectivity, and wealth of the first neighbors are revealed. Wealthiest families are linked together and have a higher number of links compared with the poorest families. These correlations yield new insights into the way the Pareto distribution arises in society.

5.3.3
Collision Models

In 1960, Mandelbrot wrote "There is a great temptation to consider the exchanges of money which occur in economic interaction as analogous to the exchanges of energy which occur in physical shocks between molecules. In the loosest possible terms, both kinds of interactions *should* lead to *similar* states of equilibrium. That is, one *should* be able to explain the law of income distribution by a model similar to that used in statistical thermodynamics: many authors have done so explicitly, and all the others of whom we know have done so implicitly." [2]. Unfortunately Mandelbrot does not provide any references to this bulk of material!

In analogy to two-particle collisions with a resulting change in their individual momenta, income exchange models may be based on two-agent interactions. Here two randomly picked agents exchange money by some predefined mechanism. Assuming the exchange process does not depend on previous exchanges, the dynamics follows a Markovian process as follows:

$$\begin{pmatrix} m_{i,t+1} \\ m_{j,t+1} \end{pmatrix} = \mathbf{M} \begin{pmatrix} m_{i,t} \\ m_{j,t} \end{pmatrix} \tag{5.9}$$

where $m_{i,t}$ is the income of agent i at time t and the collision matrix $\mathbf{M}$ defines the exchange mechanism.

The simplest case was considered by the economist Angle who proposed that exchange occurs by money from both agents being aggregated and then split randomly between them [22, 23]. Chatterjee, Chakrabarti, Chakraborti, Kaski, Patriarka and co-workers [24–27] and Kar Gupta [28] have studied this model using computational methods. Moreover, they have extended the model beyond that proposed by Angle and introduced the concept of *saving*. Thus, agents i and j continue to exchange money by analogy with an ensemble of gas molecules that exchange momentum. However, the agents are now also allowed to save a fraction λ_i of their money prior to an interaction.

The governing equations for the evolution of wealth m_i and m_j of agents i and j respectively are:

$$\begin{pmatrix} m_{i,t+1} \\ m_{j,t+1} \end{pmatrix} = \begin{pmatrix} \lambda_i + \epsilon(1-\lambda_i) & \epsilon(1-\lambda_j) \\ (1-\epsilon)(1-\lambda_i) & \lambda_j + (1-\epsilon)(1-\lambda_j) \end{pmatrix} \begin{pmatrix} m_{i,t} \\ m_{j,t} \end{pmatrix} \tag{5.10}$$

ϵ is a random number between zero and 1, and the set of numbers λ_i are savings parameters associated with the set of agents i. An important feature of this model is that the total money held between two agents remains conserved during the interaction process.

In the absence of savings ($\lambda_i = 0$ for all agents) the solution was shown to follow the Gibbs rule, $P(m) \sim \exp(-m/\langle m \rangle)$ where $\langle m \rangle$ is the average value of m. In fact this solution can be readily obtained from the maximum entropy approach familiar to physicists. One only has to propose the existence of the Boltzmann Gibbs entropy $S = \int dm P(m) \ln P(m)$ and maximize this function subject to the constraint that money is conserved, i.e, $M = \int dm P(m) m$.

The problem for the situation where all agents save the same fixed percentage of their money at time t was solved analytically by the present authors, Repetowicz, Hutzler and Richmond [29, 30] who using the Boltzmann equation, obtained a relation between the one and two agent distribution functions. Invoking a mean-field approximation this approach allowed the solution to be solved via a moment expansion of the one agent distribution function, $P(m)$. It was demonstrated that to third order, the moments agreed with a heuristic solution proposed in reference [26] as a result of numerical calculations:

$$P(m) = \frac{n^n}{\Gamma(n)} m^{n-1} \exp(-nm) \tag{5.11}$$

The Gamma function, $\Gamma(n)$, and parameter, n, are related to the savings propensity, λ, as follows:

$$n(\lambda) = 1 + \frac{3\lambda}{1-\lambda} \tag{5.12}$$

An extensive assessment of how this function fits UK empirical data and a comparison with the log-normal has been made by Willis and Mimkes [31]. In order to obtain a good fit, they found it necessary to introduce an arbitrary factor that diverts both these functions in the direction of positive values of income. They argued this reflects a minimum income or wage level below which "life is not possible". The need to assume a minimum value for m is reminiscent of the redistribution process encapsulated naturally in the approach using the Lotka–Volterra model discussed earlier.

If the savings of agents i and j are not equal but determined by some distribution, $\rho(\lambda)$, a moment expansion of the distribution function is not possible since the moments may not exist. Numerical solutions suggested that the Pareto exponent for this model was close to one. Analytic calculations performed by Repetowicz, Hutzler and Richmond [29, 30] demonstrated clearly that, assuming such a solution existed, the Pareto exponent for this model is exactly one! This result appeared to hold regardless of the form of the savings distribution function, $\rho(\lambda)$. Allowing simultaneous exchange between three or more agents seemed to make no difference to this result; nor did allowing random times between collisions. The Pareto index remained unity.

Repetowicz, Hutzler and Richmond [32] have shown how the model can be generalized and a Pareto exponent greater than unity obtained. The dynamics is assumed to be governed by the following process:

$$\begin{pmatrix} m_{i,t+1} \\ m_{j,t+1} \end{pmatrix} = \begin{pmatrix} \lambda_i + \epsilon(1-\lambda_i) & \epsilon(1-\lambda_j) \\ (1-\epsilon)(1-\lambda_i) & \lambda_j + (1-\epsilon)(1-\lambda_j) \end{pmatrix} \begin{pmatrix} \hat{m}_{i,t} \\ \hat{m}_{j,t} \end{pmatrix} \quad (5.13)$$

On the face of things, this looks to be as before. However, notice that the starting position is for agent i is $\hat{m}_{i,t}$. This is assumed to be the money possessed by agent i at a previous time in the past:

$$\hat{m}_{i,t} = m_{i,t} + \gamma m_{i,q} \quad (5.14)$$

The time, q, is a random time in the past that precedes t and determines a kind of memory horizon for the agents. γ is a "money accumulation" parameter. The approach is, as before to formulate the Boltzmann equation and invoke the mean-field approximation. However, in this case one has to deal with two time correlation functions. The outcome admits Pareto exponents greater than unity. The penalty, however, is that money is no longer conserved in the model.

Angle developed the idea for this type of model within a philosophical framework where agents were thought to exchange money via a hierarchy of exchange processes ranging from theft and burglary through to philanthropy. Not all economists, however, seem comfortable with this idea. For example, macro-economist, Makoto Nirei, has said [33]: "The model seems to me not like an economic exchange process, but more like a burglar process. People randomly meet and one just beats up the other and takes their money.".

A different viewpoint is to assume that all agents are providing some kind of service; a haircut, for example. It seems a simple matter then to imagine one agent interacting in some way with another to receive the service, in this case a *haircut*. The criticism by some economists that the model is not valid because money is conserved and other levels of money, such as credit, are not accounted for, seems to these authors to be ill-founded. It is certainly possible to develop any model to include, for example, debt. This could be simply a matter of allowing the money held by an individual to take negative values. Furthermore, the distribution of wealth away from the Pareto regime is now known to have followed the form currently observed even in ancient times [34] and would seem to predate considerably the development of modern concepts of different kinds of money. The key issue from our perspective is that this model leaves open the not-unimportant issue of how the price of the service is decided.

Lux [35] has also criticized the model for these reasons. Lux also noted that economists prefer to search for models based on the use of *utility* or *preference* functions and pointed to a recent paper by Silver, Slud and Takamoto [36] that describes an exchange market consisting of many agents with stochastic preferences for two goods. If the price is fixed this model can be shown to map onto the simple Angle exchange model and the wealth follows a simple exponential distribution. $P(m) \sim \exp(-am)$ which does not fit the empirical data at the low end where there is clearly a maximum away from $m = 0$. Of course in reality, the price is not constant. Lux suggests a self consistent approach but full details of these calculations remain to be evaluated.

A further variant of the collision model has been proposed by Slanina [37]. In Slanina's model, agents i and j exchange money according to the following rule:

$$\begin{pmatrix} m_{i,t+1} \\ m_{j,t+1} \end{pmatrix} = \begin{pmatrix} 1-\beta+\epsilon & \beta \\ \beta & 1-\beta+\epsilon \end{pmatrix} \begin{pmatrix} m_{i,t} \\ m_{j,t} \end{pmatrix} \qquad (5.15)$$

Thus agent i gives a fraction, β, of its money to agent j and vice versa. In addition it is assumed that additional money, $\epsilon(m_i + m_j)$, is created in the exchange via some sort of wealth creating process. Technically this additional term is required to obtain a solution to the mathematical model thus posed which turns out to be singular at $\epsilon = 0$. Since money is not conserved in this model, there is no stationary solution for the distribution of money, m. However, as with the GLV model, there is a stationary distribution function for the relative value: $x(t) = m(t)/\langle m(t) \rangle$. The solution is then obtained by solving the associated Boltzmann equation within a mean-field approximation. Taking the limits $\beta \to 0$ and $\epsilon \to 0^+$ whilst keeping the assumed power law, α, constant, yields:

$$P(x) = \frac{\exp\left(-(\alpha-1)/x\right)}{x^{1+\alpha}} \qquad (5.16)$$

where $\alpha - 1 = 2\beta/\epsilon^2$. It is interesting to see that this result is identical in form to that given by the GLV Eq. (5.7) [5, 12]. However, the origin of the Pareto exponent arises now from *a pair exchange process* controlled by the exchange parameter, β, together with the parameter, ϵ, that determines the wealth creation process.

5.4
A Case Study of UK Income Data

There is a widely held view in the economics community that the log-normal distribution, as originally proposed by Gibrat [4], is the appropriate function to use when fitting data and it is clearly of interest to see how well the models that have been developed and discussed above fit the empirical data which is now available.

The UK National Statistics Office (*http://www.statistics.gov.uk/*) runs an annual survey of weekly income of employees in the UK – the so-called "New Earnings Survey" or NES. The NES takes a 1% sample of all employees in

Great Britain (i.e., the United Kingdom excluding Northern Ireland). This includes full and part-time employees. Individuals are identified through national insurance numbers together with surveys of major employers. The resulting survey sample is large and of high quality and provenance. However, it should be noted that the survey only includes wage earners; it excludes the unemployed, the self-employed, those earning less than the lowest tax threshold, and those living on private income. (These are collectively referred to as "Unwaged income" for the purposes of this paper). The data sets analyzed here span the years 1992–2002 inclusive and show weekly incomes before tax.

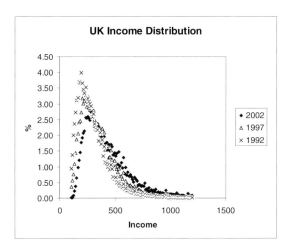

Fig. 5.7 Distribution functions for UK income for 2002, 1997 and 1992.

Figure 5.7 shows the probability distributions constructed using the data for 1992, 1997 and 2002. Points from the data set running from the £101–£110 band to the £1191–£1200 band have been shown. Data above and below these levels have been lumped together in the data sets, and have not been included in the figure. The peak of the graph moves to the right, reflecting increasing overall income, as time progresses. The height of the peak drops to compensate for the increasing width of the distribution. The data is shown on a log-linear plot in Fig. 5.8.

Much more insight into the nature of the data can, however, be seen when we construct the cumulative distribution functions. These are much smoother curves that can be examined in much more detail. Figure 5.9 shows the cumulative distribution functions for 1992, 1997, 2000 and 2002. A simple step is to adjust the incomes by an annual inflation factor. By assuming an inflation

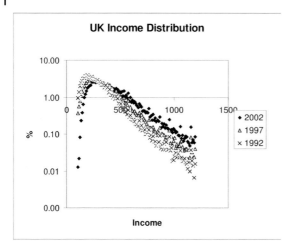

Fig. 5.8 Distribution functions for UK income for 2002, 1997 and 1992 (Log-linear plot).

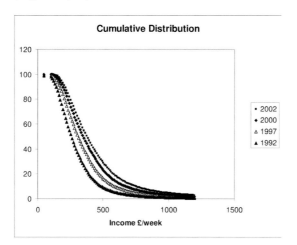

Fig. 5.9 Cumulative distribution functions for UK income over 1992, 1997, 2000 and 2002.

of 4% per annum over the decade 1992–2002 we obtain the results shown in Fig. 5.10.

The superposition looks on the face of things to be surprisingly good suggesting that there is some universal function that should describe the data.

One point that can be made without any further ado is that the simple exponential function proffered by Yakovenko [38] and recently discussed in the context of Australian income data does not fit this data. The slope of the cumulative distribution appears to be essentially zero for some distance away from the origin (zero income). This suggests some mechanism, either savings

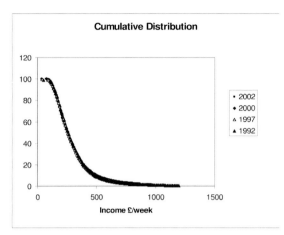

Fig. 5.10 UK income distributions for 1992 to 2002 scaled using a scaling factor for incomes of 4% per annum across the entire decade.

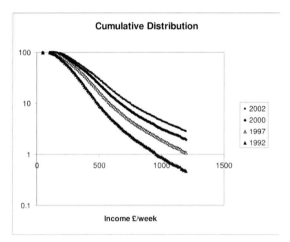

Fig. 5.11 Cumulative distribution functions for UK income over 1992, 1997, 2000 and 2002 plotted on a log linear scale.

as proposed in reference [26] or some social dynamics implemented by a taxation mechanism.

To proceed further we can examine the raw data in the form of a log-linear plot. This is shown in Fig. 5.11. It is immediately evident that for large income value the curve does not follow a simple exponential decay. The data deviates from a straight line. Furthermore, it can be seen from Fig. 5.12 that the curves actually do not quite overlap at the higher income end, even when the incomes are scaled by the 4% inflation factor.

We have examined the nature of this high end tail for each of the data sets over the 11 year period. For more than half of the range (72 out of the 112 data

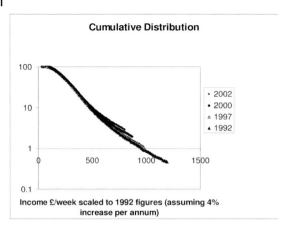

Fig. 5.12 *Scaled* income distributions for the years 2002, 2000, 1997 and 1992 plotted on a log linear scale.

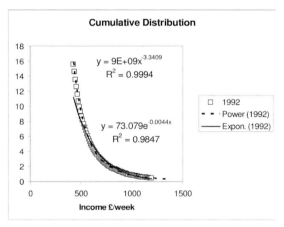

Fig. 5.13 The higher income part of the cumulative distribution for data for 1992 illustrating the comparison of power-law and exponential fits.

points) whilst an exponential function gives a reasonable fit with a correlation coefficient of 0.985, a power law gives a better fit with a correlation coefficient of 0.999. This is true of each of the 11 data sets. Figure 5.13 illustrates the point for the year 1992.

This data is re-plotted on a log-log scale in Fig. 5.14 where it is abundantly clear that the power law fits better than the exponential function.

One might still be sceptical of this interpretation since power laws are notoriously difficult to infer from data such as this. However, there is one other set of UK data that was obtained from the UK Revenue commissioners by Cranshaw who showed it in a poster at APFA3 in London in 2000 [39]. There is an amusing story surrounding the data. Cranshaw was given the low-end data

5.4 A Case Study of UK Income Data

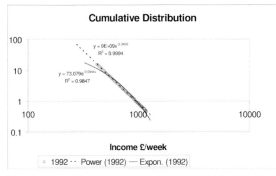

Fig. 5.14 Log-log plot of income data for 1992 showing trend lines for exponential and power laws.

set after an initial inquiry to the Revenue. However, on noticing that there did not seem to be good evidence of the power law at the high end, Cranshaw returned to the revenue and asked if there were more data points. These data were finally produced but only after a fee was paid! Figure 5.15 reproduces the NES data set for 1995 together with the data set taken off the graph published by Cranshaw.

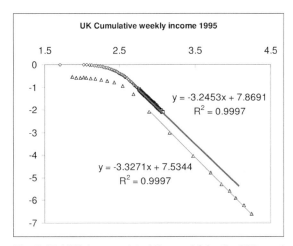

Fig. 5.15 NES income data (diamonds) for the UK together with points taken from the publication of Cranshaw (triangles) shown as a log10-log10 plot.

There is a small shift between the two curves in the ordinate which we are unable to explain. It is probably associated with the normalization of the distribution function. Certainly shifting the ordinate in the Cranshaw data by ~ 0.33 which is the difference between the two curves will bring the two curves into coincidence. The interesting point is that the power laws for the data sets which are over quite different ranges of income essentially coincide.

Tab. 5.3 Power laws for UK income data over the period 1992–2002.

Year	Power law Index	Year	Power law Index
1992	3.34	1998	3.15
1993	3.25	1999	3.15
1994	3.21	2000	3.15
1995	3.21	2001	2.68
1996	3.15	2002	2.70
1997	3.15		

In the Table 5.3 we summarize the values now obtained from all the NES data over the whole decade from 1992 to 2002.

The steady decrease of the power-law index across the decade might be interpreted from an economic perspective as *the rich getting richer*. It is interesting to note that during the first five years, the UK was ruled by a Conservative administration, whereas during the latter five years a Labour government was in power. Clearly it would be interesting to see what has happened in the more recent period 2002–2006 as the Labour government consolidated its grip on UK economic policy. Have the rich continued to get richer or has this phenomenon been reversed?

From the perspective of a physicist an interesting feature is not only the presence of power laws greater than two but also the fact that the actual number, ~ 3, almost coincides with the value obtained by Gopikrishnan, Stanley and collaborators for stock returns [40].

It is possible that there are a few people with private earnings or income earners in an even higher range that are not recorded in these data sets and who will lead to a smaller cut-off value. Queen Elizabeth II did not pay income tax until around this time and may not be in the data set. Perhaps other notable people such as pop stars and a few soccer players are also not included. These results should prompt further studies to see if an even higher different power-law regime exists.

Let us finally see whether we are able to fit the cumulative distribution over the *whole* range of data to one particular function. We have done this for the UK-NES data set of 1995 using the following normalized distributions:

- The log-normal function:

$$P_{LN}(x) = \frac{1}{x\sigma\sqrt{2\pi}} \exp\left(-\frac{(\ln x - \mu)^2}{2\sigma^2}\right) \tag{5.17}$$

$$\Pi_{LN}(x) = \frac{1}{2}\left(1 - \mathrm{erf}\left(\frac{1}{\sqrt{2}}\frac{(\ln x - \mu)}{\sigma}\right)\right) \tag{5.18}$$

- The Boltzmann function:

$$P_{MB}(x) = \frac{\beta^\beta}{\Gamma(\beta)} x^{\beta-1} \exp(-\beta x) \tag{5.19}$$

$$\Pi_{MB}(x) = \frac{\Gamma(\beta, \beta x)}{\Gamma(\beta)} \tag{5.20}$$

- The Generalized Lotka–Volterra function (GLV):

$$P_{GLV}(x) = \frac{(\alpha-1)^\alpha}{\Gamma(\alpha)} \frac{\exp\left(-\frac{(\alpha-1)}{x}\right)}{x^{1+\alpha}} \tag{5.21}$$

$$\Pi_{GLV}(x) = 1 - \frac{\Gamma\left(\alpha, \frac{\alpha-1}{x}\right)}{\Gamma(\alpha)} \tag{5.22}$$

Each of these contain one free parameter (we have set $\mu = 0$ in the log-normal function) and in order to compare with our data we need a scaling parameter for the abscissa. Figures 5.16, 5.17 and 5.18 show the results of our two-parameter least-square fits. It is clear that the function that offers the most potential for describing the data is the GLV function. This exhibits a power law for high values of income. It also exhibits the characteristics required for low values of the income. However, there is a clear deviation in the intermediate regime (£500–900), just prior to the power-law end.

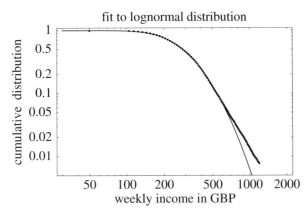

Fig. 5.16 Least-squares fit of the log-normal cumulative distribution function Π_{LN} (Eq. (5.18)) to the UK NES data set for 1995.

From analysis of US data Yakovenko and Silva [41] assert that there are two quite different regimes. A Boltzmann regime where the majority of the population is in statistical equilibrium, and a power-law regime, corresponding to a minority population out of equilibrium. However, the Lotka–Volterra model provides a single mechanism that allows one to see the entire population as one *complex* system. The analogy is with Tsallis statistics rather than Boltzmann statistics.

Nevertheless we do recognize that the simple GLV model requires some refinement to account properly for the intermediate region in the same way as a Lennard–Jones potential requires some refinement to deal with the interaction potential between atoms.

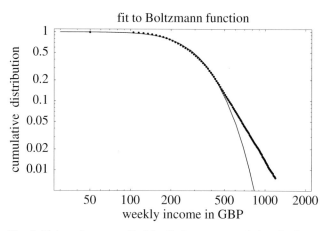

Fig. 5.17 Least-squares fit of the Boltzmann cumulative distribution function Π_{MB} (Eq. (5.20)) to the UK NES data set for 1995.

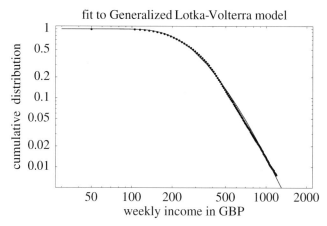

Fig. 5.18 Least-squares fit of the Generalized Lotka–Volterra cumulative distribution function Π_{GLV} (Eq. (5.22)) to the UK NES data set for 1995. Although this clearly offers the best description of the data, there is a noticeable deviation for the income between 500 and 900 GBP.

5.5 Conclusions

A renewed interest in studying the distribution of income in society has emerged over the last 10 years, driven principally by the new interest of physicists in the areas of economy and sociology. This has resulted in the development of a number of theoretical models, based on concepts of statistical physics. It now seems clear that some of these models, despite their simplicity, can reproduce key features of income distribution data. In particular, the Generalized Lotka Volterra model with its simple dynamics essentially describes the full range of data from the low to medium income region through to the high income Pareto power-law regime. There are some problems with the intermediate regime; however, we argue that these may be resolved by extensions to the model and do not reflect a breakdown of the basic model.

In the course of our review of published work we have noticed that the quality of available data is still not of the standard that would normally be available to a physicist. We would like to emphasize that this is not due to data not being monitored; simply that data is not published in a detailed form. In particular it would be of great benefit if standard datasets with good resolution over both low and high income regions were more widely available to allow models to be tested.

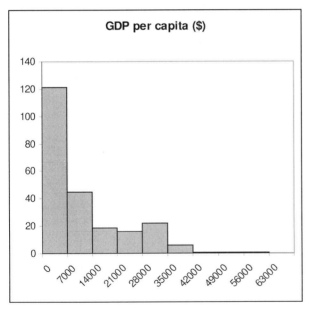

Fig. 5.19 Histogram of GDP per capita for 233 countries across the world showing a second maximum between 28 000 and 35 000 USD. The EU average is $28 000 and the US $41 800. (Data is taken from *www.cia.gov*).

Finally, we note that deviations from the kind of distributions discussed in this review do exist. Figure 5.19 shows the distribution of GDP per capita for 233 countries. The existence of two maxima, one at the very poor end and the other encompassing mainly Western countries is very interesting. To a physicist it would suggest two different population groups in different equilibrium states. This suggests that coupling between the two populations is weaker than the intra-population coupling. In economic terms this implies that the money flows between the two populations is weak. Perhaps this will be obvious to an economist; however, by examining data in this way one can monitor changes over time and begin to uncouple in more detail the impact of external activity such as aid programmes and other forms of economic intervention. The approach may therefore be useful for policy development.

Acknowledgment

This publication has emanated from research conducted with the financial support of Science Foundation Ireland (04/BR/P0251).

References

1 PARETO, V., *Cours d'Économie Politique*, Libraire Droz (Genève), **1964**, new edition of the original from **1897**

2 MANDELBROT, B. B., *International Economic Review* 1 (**1960**), p. 79–106

3 PARETO, V., *Manual of Political Economy*, Macmillan (London), **1971**, translated from the original edition of **1906**

4 GIBRAT, R., *Les Inégalites Économiques*, Libraire du Recueil Sirey (Paris), **1931**

5 FUJIWARA, Y., in *Econophysics of Wealth Distributions*, Eds. A. Chatterjee, S. Yarlagadda, B. K. Chakrabarti, Springer-Verlag Italia, Milan, **2005**, p. 24–33

6 CHAMPERNOWNE, D. G., *The Economic Journal* 63 (**1953**), p. 318–351

7 MANDELBROT, B. B., *J. Political Economy* 71 (**1963**), p. 421–440

8 BEK, C., COHEN, E. G. D., *Physica A* 322 (**2003**), p. 267–275

9 MALCAI, O., BIHAM, O., RICHMOND, P., SOLOMON, S., *Phys. Rev. E* 66 (**2002**), p. 031102

10 DRĂGULESCU, A. A., YAKOVENKO, V. M., *Physica A* 299 (**2001**), p. 213–221

11 AOYAMA, H., SOUMA, W., FUJIWARA, Y., *Physica A* 324 (**2003**), p. 352–358

12 CLEMENTI, F., GALLEGATI, M., *Physica A* 350 (**2005**), p. 427–438

13 MIMKES, J., WILLIS, G., in *Econophysics of Wealth Distributions*, Eds. A. Chatterjee, S. Yarlagadda, B. K. Chakrabarti, Springer-Verlag Italia, Milan, **2005**, p. 61–69

14 DRĂGULESCU, A. A., YAKOVENKO, V. M., *Eur. Phys. J. B* 20 (**2001**), p. 585–589

15 FERRERO, J. C., in *Econophysics of Wealth Distributions*, Eds. A. Chatterjee, S. Yarlagadda, B. K. Chakrabarti, Springer-Verlag Italia, Milan, **2005**, p. 159–167

16 ANTENEODO, C., TSALLIS, C., *J. Math. Phys.* 44 (**2003**), p. 5194–5203

17 SOLOMON, S., RICHMOND, P., *Eur. Phys. J. B* 27 (**2002**), p. 257–261

18 RICHMOND, P., SOLOMON, S., *Int. J. Mod. Phys. C* 12 (**2001**), p. 333–343

19 LEVY, M., SOLOMON, S., *Physica A* 242 (**1997**), p. 90–94

20 DI MATTEO, T., ASTE, T., HYDE, T., in *The Physics of Complex Systems (New Advances and Perspectives)*, Eds. F. Mallamace and H. E. Stanley, Amsterdam **2004**, p. 435

21 COELHO, R., NÉDA, Z., RAMASCO, J. J., SANTOS, M. A., *Physica A* 353 (**2005**), p. 515–528

22 ANGLE, J., *Social Forces* 65 (**1986**), p. 293–326

23 ANGLE, J., *Journal of Mathematical Sociology* 18 (**1993**), p. 27–46

24 CHATTERJEE, A., CHAKRABARTI, B. K., MANNA, S. S., *Physica Scripta T* 106 (**2003**), p. 36

25 CHATTERJEE, A., CHAKRABARTI, B. K., MANNA, S. S., *Physica A* 335 (**2004**), p. 155–163

26 PATRIARCA, M., CHAKRABORTI, A., KASKI, K., *Phys. Rev. E* 70 (**2004**), p. 016104

27 CHATTERJEE, A., CHAKRABARTI, B. K., STINCHCOMBE, R. B., *Phys. Rev. E* 72 (**2005**), p. 026126

28 KAR GUPTA, A., *Physica A* 359 (**2006**), p. 634–640

29 REPETOWICZ, P., HUTZLER, S., RICHMOND, P., *Physica A* 356 (**2005**), p. 641–654

30 RICHMOND, P., REPETOWICZ, P., HUTZLER, S., in *Econophysics of Wealth Distributions*, Eds. A. Chatterjee, S. Yarlagadda, B. K. Chakrabarti, Springer-Verlag Italia, Milan, **2005**, p. 120–125

31 WILLIS, G., MIMKES, J., *Microeconomics 0408001, Economics Working Paper Archive EconWPA* **2004**

32 REPETOWICZ, P., RICHMOND, P., HUTZLER, S., NI DHUINN, E., in *The Logistic Map and the Route to Chaos*, M. Ausloos and M. Dirickx Editors, Springer-Verlag (Berlin Heidelberg), **2006**, p. 259–272

33 NIREI, M., *New Scientist* 2490 (**2005**), p. 6

34 ABUL-MAGD, A. Y., *Phys. Rev. E* 66 (**2002**), p. 057104

35 LUX, T., in *Econophysics of Wealth Distributions*, Eds. A. Chatterjee, S. Yarlagadda, B. K. Chakrabarti, Springer-Verlag Italia, Milan, **2005**, p. 51–60

36 SILVER, J., SLUD, E., TAKAMOTO, K., *J. Economic Theory* 106 (**2002**), p. 417–435

37 SLANINA, F., *Phys. Rev. E* 69 (**2004**), p. 046102

38 BANERJEE, A., YAKOVENKO, V. M., DI MATTEO, T., *preprint physics/0601176*

39 CRANSHAW, T., Presented as a poster within *Applications of Physics to Financial Analysis*, London, **2000**

40 GOPIKRISHNAN, P., PLEROU, V., AMARAL, L. A. N., MEYER, M., STANLEY, H. E., *Phys. Rev. E* 60 (**1999**), p. 5305–5316

41 YAKOVENKO, V. M., SILVA, A. C., in *Econophysics of Wealth Distributions*, Eds. A. Chatterjee, S. Yarlagadda, B. K. Chakrabarti, Springer-Verlag Italia, Milan, **2005**, p. 15–23

42 AOYAMA, H., SOUMA, W., NAGAHARA, Y., OKAZAKI, M. P., TAKAYASU, H., TAKAYASU, M., *Fractals* 8 (**2000**), p. 293–300

43 FUJIWARA, Y., SOUMA, W., AOYAMA, H., KAIZOJI, T., AOKI, M., *Physica A* 321, (**2003**), p. 598–604

44 SOUMA, W., *Fractals* 9 (**2001**), p. 463–470

45 DRĂGULESCU, A. A., YAKOVENKO, V. M., in *AIP Conference Proceedings*, 661 (**2003**), p. 180–183

46 REED, W. J., *Physica A* 319 (**2003**), p. 469–486

47 SCAFETTA, N., PICOZZI, S., WEST, B. J., *Quantitative Finance* 4 (**2004**), p. 353–364

48 SINHA, S., *Physica A* 359 (**2006**), p. 555–562

49 LEVY, S., "Wealthy People and Fat Tails: An Explanation for the Lévy Distribution of Stock Returns", **1998**, Finance p. 30–98

50 LEVY, M., *J. Econ. Theory* 110 (**2003**), p. 42–64

51 COELHO, R., *MSc. Thesis "Modelos de Distribuição de Riqueza"*, **2004**

6
Models of Wealth Distributions – A Perspective
Abhijit Kar Gupta

6.1
Introduction

Wealth is accumulated in various forms and factors. The continual exchange of wealth (a value assigned) among the agents in an economy gives rise to interesting and often universal statistical distributions of individual wealth. Here the word "wealth" is used in a general sense for the purpose and the spirit of the review (in spite of separate meanings attached to the terms "money", "wealth" and "income"). Econophysics of wealth distributions [1] is an emerging area where mainly the ideas and techniques of statistical physics are used in interpreting real economic data of wealth (available mostly in terms of income) of all kinds of people or other entities (e.g., companies) for that matter, pertaining to different societies and nations. Literature in this area is growing steadily (see an excellent website [2]). The prevalence of income data and apparent interconnections of many socio-economic problems with physics have inspired a wide variety of statistical models, data analysis and other works in econophysics [3], sociophysics and other emerging areas [4] in the last decade or more (see an enjoyable article by Stauffer [5]).

The simple approach of *agent-based models* has been able to bring out all kinds of wealth distributions that open up a whole new way of understanding and interpreting empirical data. One of the most important and controversial issues has been to understand the emergence of *Pareto's law*:

$$P(w) \propto w^{-\alpha} \tag{6.1}$$

where $w \geq w_0$, w_0 being some value of wealth beyond which the power law is observed (usually towards the tail of the distributions). Pareto's law has been observed in income distributions among the people of almost all kinds of social systems across the world in a robust way. This phenomenon has now been known for more than a century and has been discussed at great length in innumerable works in economics, econophysics, sociophysics and physics dealing with power-law distributions. In many places, while mentioning *Pareto's law*, the power-law is often written in the form: $P(w) \propto w^{-(1+\nu)}$, where ν is referred to as "Pareto index". This index is usually found between 1 and 2 from empirical data fitting. Power laws in distributions appear in many other cases [6–8] like that of computer file sizes, the growth of sizes of business firms and cities, etc. Distributions are often referred to as "heavy tailed" or "fat tailed" distributions [9]. The smaller the value of α, the fatter the tail of the distribution as it may easily be understood (the distribution is more spread out).

Some early attempts [10] have been made to understand the income distributions which follow Pareto's law at the tail of the distributions. Some of them are stochastic logistic equations or some related generalized versions of that which have been able to generate power laws. However, the absence of direct interactions of one agent with any other often carries less significance in terms of interpreting real data.

Some part of this review is centered around the concept of emergence of Pareto's law in the wealth distributions, especially in the context of the models that are discussed here. However, a word of caution is in order. In the current literature as well as in the historical occurrences, the power-law distribution has often been disputed with a closely related log-normal distribution [6]. It is often not easy to distinguish between the two. Thus a brief discussion is given here on this issue. Let us consider the probability density function of a log-normal distribution:

$$p(w) = \frac{1}{\sqrt{2\pi}\sigma w} \exp[-(\ln w - \overline{w})^2/2\sigma^2] \tag{6.2}$$

The logarithm of the above can be written as:

$$\ln p(w) = -\ln w - \ln \sqrt{2\pi}\sigma - \frac{(\ln w - \overline{w})^2}{2\sigma^2} \tag{6.3}$$

If now the variance σ^2 in the log-normal distribution is large enough, the last term on the right-hand side can be very small so that the distribution may appear linear on a log-log plot. Thus the cause of concern remains, particularly when one deals with real data.

In the literature, sometimes one calculates a *cumulative distribution* function (to show the power law in a more convincing way) instead of plotting an ordinary distribution from a simple histogram (probability density function). The cumulative probability distribution function $P(\geq w)$ is such that the argument has a value greater than or equal to w:

$$P(\geq w) = \int_w^\infty P(w')dw' \tag{6.4}$$

If the distribution of data follows a power law $P(w) = Cw^{-\alpha}$, then

$$P(\geq w) = C \int_w^\infty w'^{-\alpha} dw' = \frac{C}{\alpha - 1} w^{-(\alpha-1)} \tag{6.5}$$

When the ordinary distribution (found from just histogram and binning) is a power law, the cumulative distribution thus also follows a power-law with the exponent 1 less: $\alpha - 1$, which can be seen from a log-log plot of data. An extensive discussion on power laws and related matters can be found in [7].

Besides power laws, a wide variety of wealth distributions from exponential to something like Gamma distributions are all reported in recent literature in econophysics. Exchange of wealth is considered to be a primary mechanism behind all such distributions. In a class of wealth exchange models [11–14] that follow, the economic activities among agents have been assumed to be analogous to random elastic collisions among molecules as considered in kinetic gas theory in statistical physics. An analogy is drawn between wealth (w) and Energy (E), where the average individual wealth ($\overline{w}$) at equilibrium is equivalent to temperature (T). Wealth (w) is assumed to be exchanged between two randomly selected economic agents like the exchange of energy between a pair of molecules in kinetic gas theory. The interaction is such that one agent wins and the other loses the same amount so that the sum of their wealth remains constant before and after an interaction (trading): $w_i(t+1) + w_j(t+1) = w_i(t) + w_j(t)$; each trading increases the time t by one unit. Therefore, it is basically a process of zero sum exchange between a pair of agents; the amount won by one agent is equal to the amount lost by another. This way wealth is assumed to be redistributed among a fixed number of agents (N) and the local conservation ensures that the total wealth ($W = \sum w_i$) of all the agents will remain conserved.

Random exchange of wealth between a randomly selected pair of agents may be viewed as a *gambling process* (with zero sum exchange) which leads to Boltzmann–Gibbs type exponential distribution in individual wealth ($P(w) \propto \exp(-w/\overline{w})$). However, a little variation in the mode of wealth exchange can lead to a distribution distinctly different from exponential. A number of agent-based conserved models [12, 15–21], invoked in recent times, are essentially variants of a gambling process. A wide variety of distributions evolve out of these models. There has been a renewed interest in such two-agent exchange models in the present scenario while dealing with various problems in social systems involving complex interactions. A good insight can be drawn by looking at the 2×2 transition matrices associated with the process of wealth exchange [22].

In this review, the aim would be to arrive at some understanding of how wealth exchange processes in a simple working way may lead to a variety of distributions within the framework of the conserved models. A fixed number of N agents in a system are allowed to interact (trade) stochastically and thus wealth is exchanged between them. The basic steps of such a wealth exchange model are as follows:

$$w_i(t+1) = w_i(t) + \Delta w, \quad (6.6)$$

$$w_j(t+1) = w_j(t) - \Delta w,$$

where $w_i(t)$ and $w_j(t)$ are wealths of ith and jth agents at time t and $w_i(t+1)$ and $w_j(t+1)$ are that at the next time step $(t+1)$. The amount Δw (to be won or to be lost by an agent) is determined by the nature of the interaction. If the agents are allowed to interact for a long enough time, a steady state equilibrium distribution for individual wealth is achieved. The equilibrium distribution does not depend on the initial configuration (initial distribution of wealth among the agents). A single interaction between a randomly chosen pair of agents is referred to here as one "time step". In some simulations, N, such interactions are considered as one time step. This, however, does not matter as long as the system is evolved through enough time steps to come to a steady state and then data is collected to make equilibrium probability distributions. For all the numerical results presented here, data have been produced following the available models, conjectures and conclusions. Systems of $N = 1000$ agents have been considered in each case. In each numerical investigation, the system is allowed to equilibrate for a sufficient time that ranges between 10^5 and 10^8 time steps. Configuration averaging has been done over 10^3–10^5 initial configurations in most cases. The average wealth (averaged over the agents) is kept fixed at $\overline{w} = 1$ (by taking the total wealth, $W = N$) for all the cases. The wealth distributions that are dealt with in this review are ordinary distributions (probability density function) and not the cumulative ones.

6.2
Pure Gambling

In a pure gambling process (usual kinetic gas theory), the entire sum of the wealths of two agents is available for gambling. Some random fraction of this sum is shared by one agent and the rest goes to the other. The randomness or stochasticity is introduced into the model through a parameter ϵ which is a random number drawn from a uniform distribution in [0, 1]. (Note that ϵ is independent of a pair of agents i.e., a pair of agents is not likely to share the same fraction of aggregate wealth in the same way when they interact repeatedly). The interaction can be seen through:

$$w_i(t+1) = \epsilon[w_i(t) + w_j(t)] \quad (6.7)$$

$$w_j(t+1) = (1-\epsilon)[w_i(t) + w_j(t)]$$

where the pair of agents (indicated by i and j) are chosen randomly. The amount of wealth that is exchanged is $\Delta w = \epsilon[w_i(t) + w_j(t)] - w_i(t)$. The individual wealth distribution ($P(w)$ vs. w) at equilibrium turns out to be a Boltzmann–Gibbs distribution, like an exponential. Exponential distribution of personal income has in fact been shown to appear in real data [13, 14]. In the kinetic theory model, the exponential distribution is found by standard formulation of the master equation or by the entropy maximization method; the latter has been discussed later in brief in Section 6.7. A normalized exponential distribution obtained numerically out of this pure gambling process is shown in Fig. 6.1 in a semi-logarithmic plot. The high end of the distribution appears noisy due to sampling of data. The successive bins on the right-hand side of the graph contain a smaller and smaller number of samples in them so the fractional counts in them are seen to fluctuate more (finite-size effect). One way to get rid of this sampling error to a great extent is by using logarithmic binning [7]. Here it is not important to do so as the idea is to show the nature of the curve only. (In the case of a power-law distribution, an even better way to demonstrate and extract the power-law exponent is to plot the cumulative distribution as already discussed.)

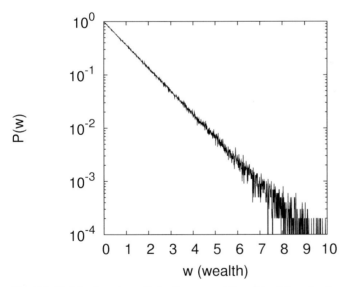

Fig. 6.1 Distribution of wealth for the case of Pure Gambling: the linearity in the semi-log plot indicates an exponential distribution.

If one intends to take the time average of wealth of a single agent over a sufficient time, it comes out to be equal for all the agents. Therefore, the distribution of individual *time averaged wealth turns out to be a delta function* which can be checked from numerical data. This is because the fluctuation of the wealth of any agent over time is statistically no different from that of any other. The same is true in case of the distribution of wealth of a single agent over time. However, when the average of the wealth of any agent is calculated over a short time period, the delta function broadens, the right end part of which decays exponentially. The distribution of individual wealth at a certain time turns out to be purely exponential as mentioned earlier. This may be thought of as a 'snap shot' distribution.

6.3
Uniform Saving Propensity

Instead of random sharing of their aggregate wealth during each interaction, if the agents decide to save (keep aside) a uniform (and fixed) fraction (λ) of their current individual wealth, then the wealth exchange equations appear as the following:

$$w_i(t+1) = \lambda w_i(t) + \epsilon(1-\lambda)[w_i(t) + w_j(t)] \tag{6.8}$$

$$w_j(t+1) = \lambda w_j(t) + (1-\epsilon)(1-\lambda)[w_i(t) + w_j(t)]$$

where the amount of wealth that is exchanged is $\Delta w = (\epsilon - 1)(1 - \lambda)[w_i(t) + w_j(t)]$. The concept of saving as introduced by Chakrabarti and group [12] in an otherwise gambling type of interaction results in distinctly different distributions. A number of numerical works followed [23–25] in order to understand the emerging distributions to some extent. Saving induces the accumulation of wealth. Therefore, it is expected that the probability of finding agents with zero wealth may be zero, unlike in the previous case of pure gambling where, due to the unguarded nature of exchange, many agents are likely to go nearly bankrupt! (It is to be noted that for an exponential distribution, the peak is at zero.) In this case the most probable value of the distribution (peak) is somewhere else than at zero (the distribution is right-skewed). The right end, however, decays exponentially for large values of w. It has been claimed through heuristic arguments (based on numerical results) that the distribution is a close approximate form of the Gamma distribution [23]:

$$P(w) = \frac{n^n}{\Gamma(n)} w^{n-1} e^{-nw} \tag{6.9}$$

where the Gamma function $\Gamma(n)$ and the index n are understood to be related to the saving propensity parameter λ by the following relation:

$$n = 1 + \frac{3\lambda}{1-\lambda} \tag{6.10}$$

The emergence of a probable Gamma distribution is also subsequently supported by numerical results in [24]. However, it has later been shown in [26] by considering the moments equation, that moments up to third order agree with that obtained from the above form of distribution subject to the condition stated in Eq. (6.10). Discrepancies start to appear only from fourth order onwards. Therefore, the actual form of distribution still remains an open question.

In Fig. 6.2, two distributions are shown for two different values of saving propensity factor: $\lambda = 0.4$ and $\lambda = 0.8$. The smaller the value of λ, the lesser the amount one is able to save. This in turn means that more wealth is available in the market for gambling. In the limit of zero saving ($\lambda = 0$) the model reduces to that of pure gambling. In the opposite extent of large saving, only a small amount of wealth is available for gambling. Then the exchange of wealth will not be able to drastically change the amount of individual wealth. This means the width of distribution of individual wealth will be narrow. In the limit of $\lambda = 1$, all the agents save all of their wealth and thus the distribution never evolves. The concept of "saving" here is of course a little different from that in real life where people do save some amount to be kept in a bank and the investment (or business or gambling) is done generally not with the entire amount (or a fraction) of the wealth that one holds at one time.

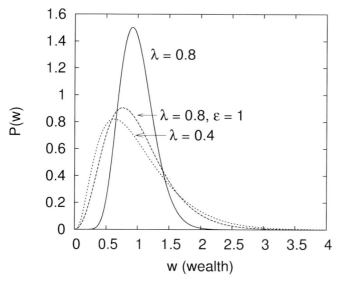

Fig. 6.2 Wealth distribution for the model of uniform and fixed saving propensity. Two distributions are shown with $\lambda = 0.4$ and $\lambda = 0.8$ where the stochasticity parameter ϵ is drawn randomly and uniformly in [0, 1]. Another distribution is plotted with $\lambda = 0.8$ but with a fixed value of the stochasticity parameter, $\epsilon = 1$.

Stochastic evolution of individual wealth is also examined without the inclusion of the stochastic parameter ϵ. The stochasticity seems to be automatically introduced anyway through the random selection of a pair of agents (and the random choice of the winner or loser as well) each time. Therefore, it is interesting to see how the distributions evolve with a fixed value of ϵ. As an example, the equations in (6.8) reduce to the following with $\epsilon = 1$:

$$w_i(t+1) = w_i(t) + (1-\lambda)w_j(t) \tag{6.11}$$

$$w_j(t+1) = \lambda w_j(t)$$

The above equations indicate that the randomly selected agent j keeps (saves) an amount $\lambda w_j(t)$ which is proportional to the wealth he currently has, and transfers the rest to the other agent i. This is indeed a stochastic process and is able to produce Gamma-type distributions in wealth as observed. However, a distribution with random ϵ and that with a fixed ϵ are different. Numerically, it has been observed, that the distribution with $\lambda = 0.8$ and with $\epsilon = 1$ is very close to that with $\lambda = 0.5$ and with random ϵ. In Fig. 6.2 the distribution with fixed $\lambda = 0.8$ and fixed $\epsilon = 1$ is plotted, along with two other distributions with random ϵ. It should also be noted that, while with fixed ϵ one does not get Gamma-type distributions for all values of λ, particularly for low values of λ the distributions come close to exponential, as observed. However, this is not clearly understood.

It has recently been pointed out in [27] that a very similar kind of agent-based model was proposed by Angle [28] (see other references cited in [27]) in sociological journals several years ago. The pair of equations in Angle's model are as follows:

$$w_i(t+1) = w_i(t) + D_t \omega w_j(t) - (1-D_t)\omega w_i(t) \tag{6.12}$$

$$w_j(t+1) = w_j(t) + (1-D_t)\omega w_i(t) - D_t \omega w_j(t)$$

where ω is a fixed fraction and the winner is decided through a random toss D_t which takes a value of either 0 or 1. Now, the above can be seen as the more formal way of writing the pair of equations (6.11) which can be arrived at by choosing $D_t = 1$ and identifying $\omega = (1-\lambda)$.

It can, in general, be said that within the framework of this kind of (conserved) model, different ways of incorporating wealth-exchange processes may lead to drastically different distributions. If the gamble is played in a *biased way*, then this may lead to a distinctly different situation from the case when it is played in a normal unbiased manner. Since in this class of models negative wealth or debt is not allowed, it is desirable that in each wealth exchange, the maximum that any agent may invest is the amount that he has at that time. Suppose, the norm is set for an "equal amount investment" where the amount to be deposited by an agent for gambling is decided by the amount the poorer agent can afford and, consequently, the same amount is agreed to be deposited by the richer agent. Let us suppose $w_i > w_j$. Now, the poorer agent (j) may invest a certain fraction of his wealth, an amount λw_j and the rest $(1-\lambda)w_j$ is saved by him. Then the total amount $2\lambda w_j$ is up for gambling and as usual a fraction of this, $2\epsilon\lambda w_j$ may be shared by the richer agent i where the rest $(1-\epsilon)\lambda w_j$ goes to the poorer agent j. This may appear fine, however, it leads to the "rich gets richer and the poor get poorer". The richer agent draws more and more wealth in his favour in the successive encounters and the poorer agents are only able to save less and less and finally there is a condensation of wealth in the hands of the richest person. This is more apparent when one considers an agent with $\lambda = 1$ where it can easily be checked that the richer agent automatically saves an amount equal to the difference of their wealth $(w_i - w_j)$ and the poorer agent ends up saving zero amount. Eventually, poorer agents get extinct. This is the "minimum exchange model" [21].

6.4 Distributed Saving Propensity

The distributions emerge out to be dramatically different when the saving propensity factor (λ) is drawn from a uniform and random distribution in [0,1] as introduced in a model proposed by Chatterjee, Chakrabarti and Manna [15]. Randomness in λ is assumed to be quenched (i.e., remains unaltered in time). Agents are indeed heterogeneous. They are likely to have different (characteristic) saving propensities. The two wealth exchange equations are now written as:

$$w_i(t+1) = \lambda_i w_i(t) + \epsilon[(1-\lambda_i)w_i(t) + (1-\lambda_j)w_j(t)] \qquad (6.13)$$

$$w_j(t+1) = \lambda_j w_j(t) + (1-\epsilon)[(1-\lambda_i)w_i(t) + (1-\lambda_j)w_j(t)]$$

A power law with exponent $\alpha = 2$ (Pareto index $\nu = 1$) is observed at the right end of the wealth distribution for several decades. Such a distribution is plotted in Fig. 6.3 where a straight line is drawn in the log-log plot with slope $= -2$ to illustrate the power law and the exponent. Extensive numerical results with different distributions in the saving propensity parameter λ are reported in [18]. The power law (with exponent $\alpha = 2$) is found to be robust. The value of the Pareto index obtained here ($\nu = 1$), however, differs from what is generally extracted (1.5 or above) from most of the empirical data of income distributions (see discussions and analysis on real data by various authors in [1]). The present model is not able to resolve this discrepancy and it is not expected at the outset either. The simple approach of incorporating the concept of distributed and quenched saving propensities within a population, that brings out a power law in wealth distribution, is itself an interesting point. There have been attempts to modify the model by introducing random waiting time in the interactions of agents in order to have a justification for a larger value of the exponent ν [29].

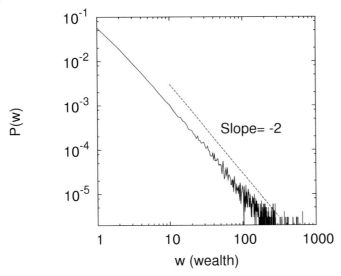

Fig. 6.3 Wealth distribution for the model of random saving propensity plotted in a log-log scale. A straight line with slope $= -2$ is drawn to demonstrate that the power law exponent is $\alpha = 2$.

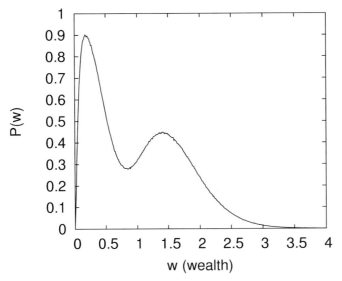

Fig. 6.4 Bimodal distribution of wealth (w) with fixed values of saving propensities, $\lambda_1 = 0.2$ and $\lambda_2 = 0.8$. Emergence of two economic classes are apparent.

The distributed saving gives rise to an additional interesting feature when a special case is considered where the saving parameter λ is assumed to have only two fixed values, λ_1 and λ_2 (preferably widely separated). A bimodal distribution in individual wealth results in [22]. This can be seen from Fig. 6.4. The system evolves towards a robust and distinct two-peak distribution as the difference in λ_1 and λ_2 is increased systematically. Later it is seen that one still gets a two-peak distribution even when λ_1 and λ_2 are drawn from narrow distributions centered around two widely separated values (one large and one small). Two economic classes seem to persist until the distributions in λ_1 and λ_2 have got sufficient widths. A population can be imagined to have two distinctly different kinds of people: some of them tend to save a very large fraction (fixed) of their wealth and the others tend to save a relatively small fraction (fixed) of their wealth. Bimodal distributions (and a polymodal distribution, in general) are, in fact, reported with real data for the income distributions in Argentina [30]. The distributions were derived at a time of political crisis and thus they may not be regarded as truly equilibrium distributions. However, it remains an interesting possibility from a simple model of wealth exchange.

6.4.1
Power Law from Mean-field Analysis

One can have an estimate of an ensemble averaged value of wealth [31] using one of the above equations (6.13) in Section 6.4. Emergence of a power law in the wealth distribution can be established through a simple consideration as follows. Taking the ensemble average of all the terms on both sides of the first equation (6.13), one may write:

$$\langle w_i \rangle = \lambda_i \langle w_i \rangle + \langle \epsilon \rangle [(1 - \lambda_i) \langle w_i \rangle + \langle \frac{1}{N} \sum_{j=1}^{N} (1 - \lambda_j) w_j \rangle] \tag{6.14}$$

The last term on the right-hand side is replaced by the average over agents where it is assumed that any agent (here the ith agent), on average, interacts with all other agents of the system, allowing sufficient time to interact. This is basically a *mean-field approach*. If ϵ is assumed to be distributed randomly and uniformly between 0 and 1 then $\langle \epsilon \rangle = \frac{1}{2}$. The wealth of each individual keeps on changing due to interactions (or wealth exchange processes that take place in a society). No matter what personal wealth one begins with, the time evolution of wealth of an individual agent at the steady state is independent of that initial value. This means that the distribution of wealth of a single agent over time is stationary. Therefore, the time averaged value of wealth of any agent remains unchanged whatever the amount of wealth one starts with. In the course of time, an agent interacts with all other agents (presumably repeatedly) given sufficient time. One can thus think of a number of ensembles (configurations) and focus attention on a particular tagged agent who eventually tends to interact with all other agents in different ensembles. Thus the time averaged value of wealth is equal to the ensemble averaged value in the steady state.

Now if one writes

$$\overline{\langle (1-\lambda)w \rangle} \equiv \langle \frac{1}{N}\sum_{j=1}^{N}(1-\lambda_j)w_j \rangle \qquad (6.15)$$

Equation (6.14) reduces to:

$$(1-\lambda_i)\langle w_i \rangle = \overline{\langle (1-\lambda)w \rangle)} \qquad (6.16)$$

The right-hand side of the above equation is independent of any agent-index and the left-hand side is referred to any arbitrarily chosen agent i. Thus, it can be argued that the above relation can be true for any agent (for any value of the index i) and so it can be equated to a constant. We now consider $C = \overline{\langle (1-\lambda)w \rangle}$, a constant which is found by averaging over all the agents in the system and which is further averaged over ensembles. Therefore, one arrives at a unique relation for this model:

$$w = \frac{C}{(1-\lambda)} \qquad (6.17)$$

where one can get rid of the index i and may write $\langle w_i \rangle = w$ for brevity. The above relation is also verified numerically which is obtained by many authors in their numerical simulations and scaling of data [18,24]. One can now derive $dw = \frac{w^2}{C}d\lambda$ from the above relation (6.17). An agent with a (characteristic) saving propensity factor (λ) ends up with wealth (w) such that one can in general relate the distributions of the two:

$$P(w)dw = g(\lambda)d\lambda \qquad (6.18)$$

If now the distribution in λ is considered to be uniform then $g(\lambda)$ = constant. Therefore, the distribution in w is bound to be of the form:

$$p(w) \propto \frac{1}{w^2} \qquad (6.19)$$

This may be regarded as Pareto's law with index $\alpha = 2$ which is already numerically verified for this present model. The same result has also been obtained recently in [32] where the treatment is argued to be exact.

6.4.2
Power Law from Reduced Situation

From numerical investigations, it seems that the stochasticity parameter ϵ is irrelevant as long as the saving propensity parameter λ is made random. It has been tested that the model is still able to produce a power law (with the same exponent, $\alpha = 2$) for any fixed value of ϵ. As an example, the case for $\epsilon = 1$ is considered. The pair of wealth exchange equations (refakg:eqn:ransave) now reduce to the following:

$$w_i(t+1) = w_i(t) + (1-\lambda_j)w_j(t) = w_i(t) + \eta_j w_j(t) \tag{6.20}$$
$$w_j(t+1) = w_j(t) - (1-\lambda_j)w_j(t) = (1-\eta_j)w_j(t)$$

The exchange amount, $\Delta w = (1-\lambda_j)w_j(t) = \eta_j w_j(t)$ is now regulated by the parameter $\eta = (1-\lambda)$ only. If λ is drawn from a uniform and random distribution in [0, 1], then η is also uniformly and randomly distributed in [0, 1]. *To achieve a power law in the wealth distribution it seems essential that randomness in η has to be quenched.* For an "annealed" type of disorder (i.e., when the distribution in η varies with time) the power law gets washed away (which is observed through numerical simulations). It has also been observed that a power law can be obtained when η is uniformly distributed between 0 and some value less than or equal to 1. As an example, if η is taken in the range between 0 and 0.5, a power law is obtained with the exponent around $\alpha = 2$. However, when η is taken in the range $0.5 < \eta < 1$, the distribution clearly deviates from a power law which is evident from the log-log plot in Fig. 6.5. *Thus there seems to be a crossover from the power law to some distribution with an exponentially decaying tail as one tunes the range in the quenched parameter η.*

At this point, *two important criteria may be identified to obtain a power law* within this reduced situation:

- The disorder in the controlling parameter η has to be quenched (fixed set of η's for a configuration of agents).

- It is required that η, when drawn from a uniform distribution, should have a lower bound of 0.

The above criteria may appear ad hoc, but have been checked by extensive numerical investigations. It was further checked that the power law exponent does not depend on the width of the distribution in η (as long as it is between 0 and something less than 1). This claim is substantiated by taking various ranges of η in which it is uniformly distributed. Systematic investigations are made for the cases where η is drawn in [0, 0.2], [0, 0.4], ... ,[0, 1]. A power laws result in in all the cases with the exponent around $\alpha = 2$.

6.5
Understanding by Means of the Transition Matrix

The evolution of wealth in the kind of two-agent wealth-exchange process can be described by the following 2×2 transition matrix (T) [22]:

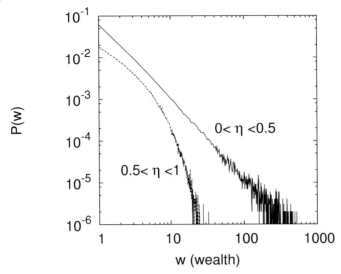

Fig. 6.5 Wealth distributions (plotted in the log-log scale) for two cases of the "reduced situation": (i) $0 < \eta < 0.5$ and (ii) $0.5 < \eta < 1$. In one case, the distribution follows a power law (with exponent around $\alpha = 2$) and in the other case, it is seen to be clearly deviating from a power law.

$$\begin{pmatrix} w'_i \\ w'_j \end{pmatrix} = T \begin{pmatrix} w_i \\ w_j \end{pmatrix}$$

where it is written, $w'_i \equiv w_i(t+1)$ and $w_i \equiv w_i(t)$ and so on. The transition matrix (T) corresponding to a *pure gambling* process (in Section 6.2) can be written as:

$$T = \begin{pmatrix} \epsilon & \epsilon \\ 1-\epsilon & 1-\epsilon \end{pmatrix}$$

In this case the above matrix is *singular* (determinant, $|T| = 0$) which means the inverse of this matrix does not exist. This in turn indicates that an evolution through such transition matrices is bound to be *irreversible*. This property is connected to the emergence of the exponential (Boltzmann–Gibbs) wealth distribution. The same may also be perceived in a different way. When a product of such matrices (for successive interactions) is taken, the leftmost matrix (of the product) itself returns:

$$\begin{pmatrix} \epsilon & \epsilon \\ 1-\epsilon & 1-\epsilon \end{pmatrix} \begin{pmatrix} \epsilon_1 & \epsilon_1 \\ 1-\epsilon_1 & 1-\epsilon_1 \end{pmatrix} = \begin{pmatrix} \epsilon & \epsilon \\ 1-\epsilon & 1-\epsilon \end{pmatrix}$$

The above signifies the fact that during the repeated interactions of the same two agents (via this kind of transition matrix), the last of the interactions is what matters (the last matrix of the product survives) $[T^{(n)}.T^{(n-1)} \ldots T^{(2)}.T^{(1)} = T^{(n)}]$. This "loss of memory" (random history of collisions in the case of molecules) may here be attributed to the path towards irreversibility in time.

The singularity can be avoided if one considers the following general form:

$$T_1 = \begin{pmatrix} \epsilon_1 & \epsilon_2 \\ 1 - \epsilon_1 & 1 - \epsilon_2 \end{pmatrix}$$

where ϵ_1 and ϵ_2 are two different random numbers drawn uniformly from [0, 1]. (This ensures the transition matrix to be nonsingular.) The significance of this general form can be seen through the wealth exchange equations in the following way. The ϵ_1 fraction of wealth of the first agent (i) added to the ϵ_2 fraction of wealth of the second agent (j) is retained by the first agent after the trade. The rest of their total wealth is shared by the 2nd agent. This may happen in a number of ways which can be related to the detailed considerations of a model. The general matrix T_1 is nonsingular as long as $\epsilon_1 \neq \epsilon_2$ and then the two-agent interaction process remains reversible in time. Therefore, it is expected to have a steady state equilibrium distribution of wealth which may deviate from an exponential distribution (as in the case with a pure gambling model). When one considers $\epsilon_1 = \epsilon_2$, one again gets back to the pure exponential distribution. A trivial case is obtained for $\epsilon_1 = 1$ and $\epsilon_2 = 0$. The transition matrix then reduces to the identity matrix $I = \begin{pmatrix} 1 & 0 \\ 0 & 1 \end{pmatrix}$ which trivially corresponds to no interaction and no evolution.

It may be emphasized that any transition matrix $\begin{pmatrix} t_{11} & t_{12} \\ t_{21} & t_{22} \end{pmatrix}$, for such conserved models is bound to be of the form such that the sum of two elements of either of the columns has to be *unity by design*: $t_{11} + t_{21} = 1$, $t_{12} + t_{22} = 1$. It is important to note that no matter what extra parameter one incorporates within the framework of the conserved model, the transition matrix has to retain this property.

In Fig. 6.6 three distributions (with $\epsilon_1 \neq \epsilon_2$) are plotted where ϵ_1 and ϵ_2 are drawn randomly from uniform distributions with different ranges. It is demonstrated that qualitatively different distributions are possible as the parameter ranges are tuned appropriately.

Now let us compare the above situation with the model of equal saving propensity as discussed in Section 6.3. With the incorporation of a saving propensity factor λ, the transition matrix now appears as:

$$\begin{pmatrix} \lambda + \epsilon(1 - \lambda) & \epsilon(1 - \lambda) \\ (1 - \epsilon)(1 - \lambda) & \lambda + (1 - \epsilon)(1 - \lambda) \end{pmatrix}$$

The matrix elements can now be rescaled by assuming $\tilde{\epsilon}_1 = \lambda + \epsilon(1 - \lambda)$ and $\tilde{\epsilon}_2 = \epsilon(1 - \lambda)$ in the above matrix. Therefore, the above transition matrix

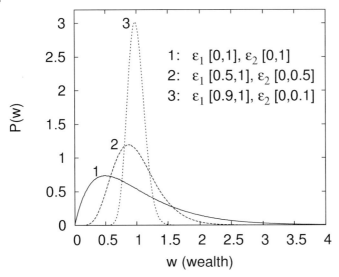

Fig. 6.6 Three normalized wealth distributions are shown corresponding to the general matrix T_2 (in general with $\epsilon_1 \neq \epsilon_2$) as discussed in the text. Curves are marked by numbers (1, 2 and 3) and the ranges of ϵ_1 and ϵ_2 are indicated within which they are drawn uniformly and randomly.

reduces to

$$T_2 = \begin{pmatrix} \tilde{\epsilon}_1 & \tilde{\epsilon}_2 \\ 1 - \tilde{\epsilon}_1 & 1 - \tilde{\epsilon}_2 \end{pmatrix}$$

Thus the matrix T_2 is of the same form as T_1. The distributions due to the above two matrices of the same general form can now be compared if one can correctly identify the ranges of the rescaled elements. In the model of uniform saving: $\lambda < \tilde{\epsilon}_1 < 1$ and $0 < \tilde{\epsilon}_2 < (1-\lambda)$ as the stochasticity parameter ϵ is drawn from a uniform and random distribution in $[0, 1]$. As long as $\tilde{\epsilon}_1$ and $\tilde{\epsilon}_2$ are different, the determinant of the matrix is nonzero ($|T_2| = \tilde{\epsilon}_1 - \tilde{\epsilon}_2 = \lambda$). Therefore, the incorporation of the saving propensity factor λ results in *two effects*:

- The transition matrix becomes nonsingular.
- The matrix elements t_{11} (= $\tilde{\epsilon}_1$) and t_{12} (= $\tilde{\epsilon}_2$) are now drawn from truncated domains (somewhere in $[0, 1]$).

Hence it is clear from the above discussion that the wealth distribution with uniform saving is likely to be qualitatively no different from what can be achieved with general transition matrices having different elements, $\epsilon_1 \neq \epsilon_2$. The distributions obtained with a different λ may correspond to that with appropriately chosen ϵ_1 and ϵ_2 in T_1.

In the next stage, when the saving propensity factor λ is distributed as in Section 6.4, the transition matrix between any two agents having different λ's (say, λ_1 and λ_2) now appears as:

$$\begin{pmatrix} \lambda_1 + \epsilon(1-\lambda_1) & \epsilon(1-\lambda_2) \\ (1-\epsilon)(1-\lambda_1) & \lambda_2 + (1-\epsilon)(1-\lambda_2) \end{pmatrix}$$

Again the elements of the above matrix can be rescaled by putting $\tilde{\epsilon}'_1 = \lambda_1 + \epsilon(1-\lambda_1)$ and $\tilde{\epsilon}'_2 = \epsilon(1-\lambda_2)$. Hence the transition matrix can again be reduced to the same form as that of T_1 or T_2:

$$T_3 = \begin{pmatrix} \tilde{\epsilon}'_1 & \tilde{\epsilon}'_2 \\ 1-\tilde{\epsilon}'_1 & 1-\tilde{\epsilon}'_2 \end{pmatrix}$$

The determinant here is $|T_3| = \tilde{\epsilon}'_1 - \tilde{\epsilon}'_2 = \lambda_1(1-\epsilon) + \epsilon\lambda_2$. Here also the determinant is ensured to be nonzero as all the parameters ϵ, λ_1 and λ_2 are drawn from the same positive domain: [0, 1]. This means that each transition matrix for two-agent wealth exchange remains nonsingular, which ensures that the interaction process is reversible in time. Therefore, it is expected that *qualitatively different distributions are possible when one appropriately tunes the two independent elements in the general form of the transition matrix (T_1 or T_2 or T_3)*. However, the emergence of a power-law tail (*Pareto's law*) in the distribution cannot be explained by this. Later, it is found that, to obtain a power law in the framework of present models, it is essential that the distribution in λ should be quenched (frozen in time) which means that the matrix elements in the general form of any transition matrix have to be quenched. In Section 6.4.2, it has been shown that the model of distributed saving (Section 6.4) is equivalent to a reduced situation where one needs only one variable η. In this case the corresponding transition matrix looks even simpler:

$$T_4 = \begin{pmatrix} 1 & \eta \\ 0 & 1-\eta \end{pmatrix}$$

where a nonzero determinant ($|T_4| = 1-\eta \neq 0$) is ensured, among other things.

6.5.1
Distributions from the Generic Situation

From all the previous discussions, it is clear that the the transition matrix (for zero sum wealth exchange) is bound to be of the following general form:

$$\begin{pmatrix} \epsilon_1 & \epsilon_2 \\ 1-\epsilon_1 & 1-\epsilon_2 \end{pmatrix}$$

The matrix elements, ϵ_1 and ϵ_2 can be appropriately associated with the relevant parameters in a model. A generic situation occurs where one can generate all sorts of distributions by controlling ϵ_1 and ϵ_2.

As long as $\epsilon_1 \neq \epsilon_2$, the matrix remains nonsingular and one achieves Gamma-type distributions. In a special case, when $\epsilon_1 = \epsilon_2$, the transition matrix becomes singular and a Boltzmann–Gibbs type of exponential distribution results. It has been numerically checked that a power law with exponent $\alpha = 2$ is obtained with the general matrix when the elements ϵ_1 and ϵ_2 are of the same set of quenched random numbers drawn uniformly in $[0, 1]$. The matrix corresponding to the reduced situation in the Section 6.4.2, as discussed, is just a special case with $\epsilon_1 = 1$ and $\epsilon_2 = \eta$, drawn from a uniform and (quenched) random distribution. Incorporation of any parameter in an actual model (saving propensity, for example) results in the adjustment or truncation of the full domain $[0, 1]$ from which the element ϵ_1 or ϵ_2 is drawn. Incorporating distributed λs in Section 6.4 is equivalent to considering the following domains: $\lambda_1 < \epsilon_1 < 1$ and $0 < \epsilon_2 < (1 - \lambda_2)$.

A more general situation occurs when the matrix elements ϵ_1 and ϵ_2 are of two sets of random numbers drawn separately (one may identify them as $\epsilon_1^{(1)}$ and $\epsilon_2^{(2)}$ to distinguish them) from two uniform and random distributions in the domain: $[0, 1]$. In this case a power law is obtained with the exponent $\alpha = 3$ which is, however, distinctly different from that obtained in the "distributed saving model" in Section 6.4. To test the robustness of the power law, the distributions in the matrix elements are taken in the following truncated ranges: $0.5 < \epsilon_1 < 1$ and $0 < \epsilon_2 < 0.5$ (widths are narrowed down). A power law is still obtained with the same exponent (α close to 3). These results are plotted in Fig. 6.7.

It is possible to achieve distributions other than power laws as one draws the matrix elements, ϵ_1 and ϵ_2 from different domains within the range between 0 and 1. There is indeed a *crossover from power-law to Gamma-like distributions* as one tunes the elements. It appears from extensive numerical simulations that the power law disappears when both the parameters are drawn from some ranges that do not include the lower limit 0. For example, when $0.8 < \epsilon_1 < 1.0$ and $0.2 < \epsilon_2 < 0.4$, the wealth distribution does not follow a power law. In contrast, when ϵ_1 and ϵ_2 are drawn from the ranges, $0.8 < \epsilon_1 < 1.0$ and $0 < \epsilon_2 < 0.1$, the power law distribution reappears.

It now appears that *to achieve a power law in such a generic situation, the following criteria must be fulfilled*:

- It is essential to have the randomness or disorder in the elements ϵ_1 and ϵ_2 quenched,

- In the most general case, ϵ_1 should be drawn from a uniform distribution whose upper bound has to be 1 and for ϵ_2 the lower bound has to

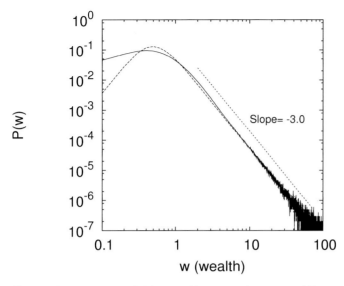

Fig. 6.7 Distribution of individual wealth (w) for the most general case with random and quenched ϵ_1 and ϵ_2. The elements are drawn from two separate distributions where $0 < \epsilon_1 < 1$ and $0 < \epsilon_2 < 1$ in one case and in the other case, they are chosen from the ranges, $0.5 < \epsilon_1 < 1$ and $0 < \epsilon_2 < 0.5$. Both show power laws with the same exponent around 3.0 (the two distributions almost superpose). A straight line (with slope -3.0) is drawn to demonstrate the power law in the log-log scale.

be 0. Then a power law with higher exponent $\alpha = 3$ is achieved. To have a power law with exponent $\alpha = 2$, the matrix elements are to be drawn from the same distribution. (These choices automatically make the transition matrices nonsingular.)

The above points are not supported analytically at this stage. However, the observation seems to bear important implications in terms of generation of power-law distributions.

When the disorder or randomness in the elements ϵ_1 and ϵ_2 changes with time (i.e., is not quenched) unlike the situation just discussed above, the problem is perhaps similar to the mass diffusion and aggregation model by Majumdar, Krishnamurthy and Barma [33]. The mass model is defined on a one dimensional lattice with periodic boundary condition. A fraction of mass from a randomly chosen site is assumed to be continually transported to any of its neighboring sites at random. The total mass between the two sites is then unchanged (one site gains mass and the other loses the same amount) and thus the total mass of the system remains conserved. The mass of each site evolves as

$$m_i(t+1) = (1 - \eta_i)m_i(t) + \eta_j m_j(t) \tag{6.21}$$

Here it is assumed that η_i fraction of mass m_i is dissociated from the site i and joins either of its neighboring sites $j = i \pm 1$. Thus $(1 - \eta_i)$ fraction of mass m_i remains at that site whereas a fraction η_j of mass m_j from the neighboring site joins the mass at site i. Now, if we identify $\epsilon_1 = (1 - \eta_i)$ and $\epsilon_2 = \eta_j$ then this model is just the same as described by the general transition matrix as discussed so far. If the η_is are drawn from a random and uniform distribution in $[0, 1]$ then a mean-field calculation (which turns out to be exact in the thermodynamic limit), as shown in [33], brings out the stationary mass distribution $P(m)$ as a Gamma distribution:

$$P(m) = \frac{4m}{\overline{m}^2} e^{-2m/\overline{m}} \qquad (6.22)$$

where $\overline{m}$ is the average mass of the system. It has been numerically checked that there seems to be no appreciable change in the distribution even when the lattice is not considered. Lattice seems to play no significant role in the case of kinetic theory, the same as wealth distribution models. Incidentally, this distribution with $\overline{m} = 1$ is exactly the same as the Gamma distribution [Eq. (6.9)], mentioned in Section 6.3 when one considers $n = 2$. The index n equals 2 when one puts $\lambda = \frac{1}{4}$ in the relation (6.10).

In the general situation ($\epsilon_1 \neq \epsilon_2$), when both the parameters are drawn from a random and uniform distribution in $[0, 1]$, the emerging distribution very nearly follows the above expression (6.22). Only when the randomness in them is quenched (fixed in time), is there a possibility of getting a power law as already mentioned. The Gamma distribution (Eq. (6.22)) and the numerically obtained distributions for different cases (as discussed in the text) are plotted in Fig. 6.8 in support of the above discussions.

6.6
Role of Selective Interaction

So far only those models of wealth exchange processes have been considered where a pair of agents is selected randomly. However, interactions or trade among agents in a society are often guided by personal choice or some social norms or perhaps other reasons. Agents may like to interact selectively and it would be interesting to see how the Individual wealth distribution is influenced by selection [19]. The concept of selective interaction is already there when one considers the formation of a family. The members of a same family are unlikely to trade (or interact) between each other. It may be worth examining the role played by the concept of "family" in wealth distributions of families: "family wealth distribution" for brevity. A family in a society usually consists of more than one agent. In a computer simulation, the agents belonging to the same family are colored to keep track of them. To find wealth

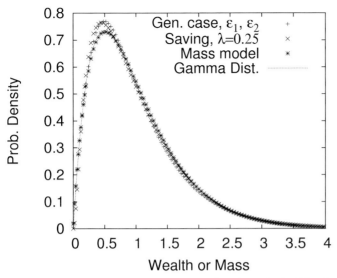

Fig. 6.8 Normalized probability distribution functions obtained for three different cases: (i) Wealth distribution with random and uniform ϵ_1 and ϵ_2 in [0, 1], (ii) Wealth distribution with uniform and fixed saving propensity, $\lambda = \frac{1}{4}$, (iii) Mass distribution for the model [33] in one dimensional lattice (as discussed in text). The theoretical Gamma distribution [the eqn. (6.22)] is also plotted (line draw) to have a comparison.

distributions of families, the contributions of the same family members are added up. In Fig. 6.9 family wealth distributions are plotted for three cases: (i) families consisting of two members each, (ii) families consisting of four members each and (iii) families of mixed sizes between one and four. The distributions are clearly not purely exponential, but modified exponential distributions (Gamma-type distributions) with different peaks and different widths. This is quite expected as the probability of zero income for a family is zero. The modified exponential distribution for family wealth is also supported by fitting real data [14].

Some special way of incorporating selective interaction is seen to have a drastic effect in individual wealth distribution. To implement the idea of "selection", a "class" of an agent is defined through an index ϵ. The class may be understood in terms of some sort of efficiency in accumulating wealth or some other closely related property. Therefore, ϵs are assumed to be quenched. It is assumed that, during the interactions, the agents may convert an appropriate amount of wealth proportional to their efficiency factor in their favor or against. Now, the model can be understood in terms of the general form of

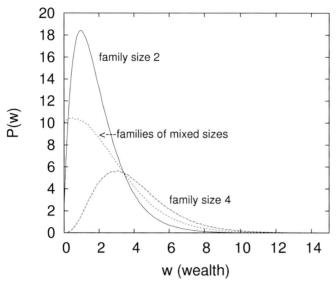

Fig. 6.9 Family wealth distributions: two curves are for families consisting of all equal sizes of 2 and 4. One curve is for a system of families consisting of various sizes between 1 and 4. The distributions are not normalized.

equations:

$$w_i(t+1) = \epsilon_i w_i(t) + \epsilon_j w_j(t)$$
$$w_j(t+1) = (1-\epsilon_i) w_i(t) + (1-\epsilon_j) w_j(t)$$
(6.23)

where the ϵ_is are quenched random numbers between 0 and 1 (randomly assigned to the agents at the beginning). Now the agents are supposed to make a choice of whom not to trade with. This option, in fact, is not unnatural in the context of a real society where individual or group opinions are important. There has been a lot of work on the process and dynamics of opinion formations [4, 34] in model social systems. In the present model it may be imagined that the "choice" is simply guided by the relative class index of the two agents. It is assumed that an interaction takes place when the ratio of two class factors remain within a certain upper limit. The requirement for interaction (trade) to happen is then $1 < \epsilon_i/\epsilon_j < \tau$, where $\epsilon_i > \epsilon_j$. Wealth distributions for various values of τ are numerically investigated. Power laws in the tails of the distributions are obtained in all cases. In Fig. 6.10 the distributions for $\tau = 2$ and $\tau = 4$ are shown. Power laws are clearly seen with an exponent, $\alpha = 3.5$ (a straight line with slope around -3.5 is drawn) which means the Pareto index ν is close to 2.5. It is not investigated further whether the exponent (α) actually differs in a significant way for different choices of τ. It has been shown that preferential behavior [20] generates a power law in money distribution with some imposed conditions which allows the rich to have a higher probability of getting richer. The rich are also favored in a model with some kind of asymmetric exchange rules as proposed in [21] where a power law results. The dice seem to be loaded in favor of the rich otherwise the rich cannot be the rich!

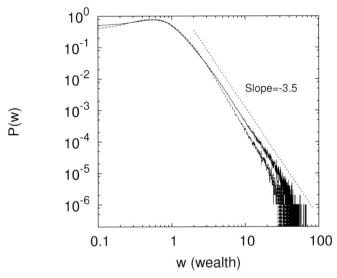

Fig. 6.10 Distribution of individual wealth with selective interaction. A power law is evident in the log-log plot where a straight line is drawn with slope $= -3.5$ for comparison.

6.7
Measure of Inequality

The emergence of Pareto's law signifies the existence of inequality in wealth in a population. Inequality or disparity in wealth or that of income is known to exist in almost all societies. To have a quantitative idea of inequality one generally plots a Lorenz curve and then calculates the Gini coefficient. Here the entropy approach [35] is considered. The time evolution of an appropriate quantity is examined which may be regarded as a measure of wealth inequality.

Let us consider $w_1, w_2, \ldots, w_N$ to be the wealths of N agents in a system. Let $W = \sum_{i=1}^{N} w_i$ be the total wealth of all the agents. Now $p_i = w_i/W$ can be considered as the fraction of wealth the ith agent shares. Thus each of $p_i > 0$ and $\sum_{i=1}^{N} p_i = 1$. Thus the set of $p_1, p_2, \ldots, p_N$ may be regarded as a probability distribution. The well-known Shannon entropy is defined as the

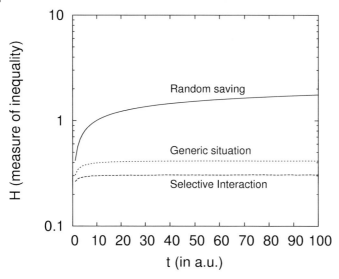

Fig. 6.11 Comparison of time evolution of the measure of inequality (H) in wealth for different models. Each "time step" (t) is equal to a single interaction between a pair of agents. Data is taken after every 10^4 time steps to avoid clumsiness and each data point is obtained by averaging over 10^3 configurations. The Y-axis is shown in log scale for a fair comparison.

following:

$$S = -\sum_{i=1}^{N} p_i \ln p_i \quad (6.24)$$

From the maximization principle of entropy it can be easily shown that the entropy (S) is a maximum when

$$p_1 = p_2 = \cdots = p_N = \frac{1}{N} \quad (6.25)$$

giving the maximum value of S to be $\ln N$ where it is a limit of equality (everyone possesses the same wealth). A measure of inequality should be something which measures a deviation from the above ideal situation. Thus one can have a measure of wealth inequality as

$$H = \ln N - S = \ln N + \sum_{i=1}^{N} p_i \ln p_i = \sum_{i=1}^{N} p_i \ln(N p_i). \quad (6.26)$$

The greater the value of H, the greater the inequality.

It can be seen that the wealth exchange algorithms are so designed that the resulting disparity or variance (or measure of inequality), in effect, increases

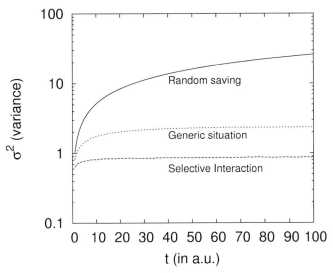

Fig. 6.12 Evolution of variance (σ^2) with "time" (t) for different models. The Y-axis is shown in a log scale to accommodate three sets of data in a same graph. Data is taken after every 10^4 time steps to avoid clumsiness and each data point is obtained by averaging over 10^3 configurations.

with time. Wherever a power law in distribution results, the distribution naturally broadens which indicates that the variance (σ^2) or the inequality measure [H in Eq. (6.26)] should increase. In Figs 6.11 and 6.12 the time evolution of inequality measure H and variance σ^2 respectively are plotted with time for three models to have a comparison. It is apparent that the measure of inequality in the steady state attains different levels due to different mechanisms of wealth-exchange processes, giving rise to different power law exponents. The growth of variance is seen to be different for different models considered, which is responsible for power laws with different exponents, as discussed in the text. The power-law exponents (α) appear to be related to the magnitudes of variance that are attained in equilibrium in the finite systems.

6.8
Distribution by Maximizing Inequality

It is known that the probability distribution of the wealth of the majority is different from that of a small minority (rich people). Disparity is more or less a reality in all economies. A wealth-exchange process can be thought of within the present framework where the interaction among agents eventually leads to increasing variance. It is numerically examined [19] whether the process of

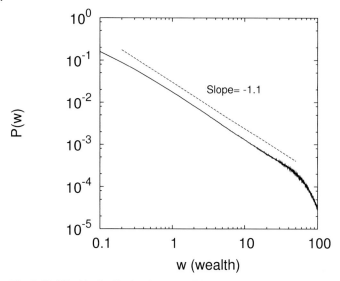

Fig. 6.13 Wealth distribution by maximizing the variance in the pure gambling model. A power law is clearly seen (in the log-log plot) and a straight line is drawn with slope $= -1.1$ for comparision.

forcing the system to have an ever-increasing variance (measure of disparity) leads to a power law, as it is known that the power law is usually associated with infinite variance. Evolution of variance, $\sigma^2 = \langle w^2 \rangle - \langle w \rangle^2$ is calculated after each interaction in the framework of a pure gambling model (the pair of equations (6.7)) and it is then forced to increase monotonically by comparing this to the previously calculated value (the average value $\overline{w}$ is fixed by virtue of the model). This results in a very large variance under this imposed condition. The inequality factor H likewise increases monotonically and attains a high value. A power-law distribution is obtained with the exponent α close to 1. None of the available models brings out such a low value of the exponent. The variance in any of the usual models generally settles at a level much lower than that obtained in this way. The resulting distribution of wealth is plotted as a log-log scale in Fig. 6.13.

A power law, however, could not be obtained in the same way in the case of a nonconserved model like the following: $w_i(t+1) = w_i(t) \pm \delta$, where the increase or decrease (δ) in wealth (w) of any agent is independent of any other.

It has also been noted, considering some of the available models, that the larger the variance, smaller the exponent one gets. For example, the variance is seen to attain higher values (with time) in the model of distributed (random) saving propensities [18] compared to the model of selective interaction [19] and the resulting power law exponent α is close to 2.0 in the former case whereas it is close to 3.5 in the later. In the present situation the variance attains even higher values and the exponent α seems to be close to 1, the lowest among all.

6.9 Confusions and Conclusions

As it is seen, the exchange of wealth in various modes generates a wide variety of distributions within the framework of simple wealth-exchange models as discussed. In this review, some basic structures and ideas of interaction are looked at which seem in fundamental bringing out the desired distributions. In this kind of agent-based model (for some general discussions, see [36]) the division of labour, demand and supply and human qualities (selfish acts or altruism) and efforts (investments, business) which are essential ingredients in classical economics, are not considered explicitly. What causes the exchange of wealth of a specific kind amongst agents is not important in this discussion. Models are considered to be conserved (no inflow or outflow of money/wealth into or out from the system). It is not essential to look for inflation, taxation, debt, investment returns, etc., in an economic system at the outset for the kind of questions that are addressed here. The essence of the complexity of interaction leading to distributions can be understood in terms of the simple (microscopic) exchange rules much in the same way as the simple logistic equations that went on to construct "roads to chaos" and opened up a new horizon of thinking about a complex phenomenon like turbulence [37].

Some models of zero-sum wealth exchange are examined here in this review. One may start thinking in a new way of how the distributions emerge out of the kind of algorithmic exchange processes that are involved. The exchange processes can be understood in a general way by looking at the structure of associated 2×2 transition matrices. Wealths of individuals evolve to have a specific distribution in a steady state through the kind of interactions which are basically stochastic in nature. The distributions shift away from Boltzmann–Gibbs like exponential to Gamma-type distributions and in some cases distributions emerge with power-law tails known as Pareto's law ($P(w) \propto w^{-\alpha}$). It is also seen that the wealth distributions appear to be influenced by personal choice. In a real society, people usually do not interact arbitrarily but rather do so with purpose and thought. Some kind of personal preference is always there which may be incorporated in some way or other. A power law with a distinctly different exponent ($\alpha = 3.5$, Pareto exponent $\nu = 2.5$) is achieved by a certain method of selective interaction. The value of the Pareto index ν does not correspond to what is generally obtained empirically. However, the motivation is not to attach much importance to the numerical value at the outset rather than to focus on the fact of how power laws emerge with distinctly different exponents governed by the simple rules of wealth exchange.

The fat-tailed distributions (power laws) are usually associated with large variance, which can be a measure of disparity. Economic disparity usually exists within a population. The detailed mechanism leading to disparity is not always clear but it can be said to be associated with the emergence of power law tails in wealth distributions. Monotonically increasing variance (with time) can be associated with the emergence of a power-law in individual wealth distributions. The mean and variance of a power-law distribution can be analytically derived [7] to see that they are finite when the power law exponent α is greater than 3. For $\alpha \leq 3$, the variance diverges but then the mean is finite. In the case of the models discussed here, the mean is kept fixed but large or enhanced variance is observed in different models whenever there results in a power law. It remains a question of what the mechanisms can be (in the discrete and conserved models) that generate a large variance and power-law tails. Large and increasing variance is also associated with log-normal distributions. A simple multiplicative stochastic process like $w(t+1) = \epsilon(t)w(t)$ can be used to explain the emergence of a log-normal distribution and indefinite increase in variance. However, empirical evidence shows that the Pareto index and some other appropriate indices (the Gibrat index, for example), generally dwindle within some range [38] indicating that the variance (or any other equivalent measure of inequality) does not increase forever. It seems to attain a saturation, given sufficient time. This is indeed the case which the numerical results suggest. Normally there occurs a simultaneous increase in the variance and mean in statistical systems (in fact, the relationship between mean and variance goes by a power law as $\sigma^2 \propto \overline{w}^b$ know as Taylor's power law [39] as curiously observed in many natural systems). In this conserved model the mean is not allowed to vary as it is fixed by virtue of the model. It may be the case that σ^2 then has to have a saturation. The limit of σ^2 is found by an artificial situation in which the association of the power law with large variance is tested in a reverse way.

Understanding the emergence of the power law [7, 8] itself has been of great interest for decades. There is usually no accepted framework which may explain the origin and wealth of varieties of its appearance. It is often argued that the dynamics which generate power laws are dominated by multiplicative processes. It is true that, in an economy, wealth (or money) of an agent multiplies and that is coupled to the large number of interacting agents. The generic stochastic Lotka–Volterra systems like $w_i(t+1) = \epsilon w_i(t) + a\overline{w}(t) - bw_i(t)\overline{w}(t)$ have been studied [34, 40] to achieve power-law distributions in wealth. However, this kind of models is not discussed in this review as the basic intention is to understand the ideas behind models of conserved wealth of which the above is not an example.

In a change of thinking, let us imagine a distribution curve which can be stretched in any direction which one wishes to have, keeping the area under it invariant. If now the curve is pulled too high around the left region then the right-hand side will fall off too quickly, and an exponential decay is then a possible option. On the other hand, if the width is stretched too far (the distribution becomes fat) on the right-hand side, it should then decay fairly slowly giving rise to a possible power-law fall at the right end, while preserving the area under the curve. What makes such a stretching possible? This review has been an attempt to integrate some ideas regarding models of wealth distribution and also to reinvent things in a new way. During the process some confusion, conjectures and conclusions have emerged and many questions have possibly been answered with further questions and doubts. At the end of the day, the usefulness of this review may be measured by the further interest and extra attention which it may generate on the subject.

Acknowledgment

The author is grateful to Dietrich Stauffer for many important comments and criticisms, at different stages of publication on the results that are incorporated in this review.

References

1 *Econophysics of Wealth Distributions*, (Ed.: A. Chatterjee, S.Yarlagadda, B. K. Chakrabarti), Springer-Verlag Italia, Milan, **2005**

2 Econophysics Forum: *www.unifr.ch/econophysics*

3 MANTEGNA, R. N., STANLEY, H. E., *An Introduction to Econophysics*, Cambridge Univ. Press, Cambridge, UK, **2000**

4 STAUFFER, D., MOSS DE OLIVEIRA, S., DE OLIVEIRA, P. M. C., SA MARTINS, J. S., *Biology, Sociology, Geology by Computational Physicists*, Elsevier, Amsterdam, **2006**; MOSS DE OLIVEIRA, S., DE OLIVEIRA, P. M. C., STAUFFER, D., *Evolution, Money, War and Computers*, B.G. Tuebner, Stuttgart, Leipzig, **1999**

5 STAUFFER, D., *Physica A* 336 (**2004**), p. 1

6 MITZENMACHER, M., *Internet Mathematics* 1 (**2004**), p. 226 www.internetmathematics.org

7 NEWMAN, M. E. J., *Contemporary Physics* 46 (**2005**), p. 323

8 REED, W. J., HUGHES, B. D., *Phys. Rev. E* 66 (**2002**), p. 067103

9 MANDELBROT, B. B., *Scaling and Fractals in Finance*, Springer, Berlin, **1997**

10 SOLOMON, S., LEVY, M., *Int. J. Mod. Phys. C* 7 **1996**; LEVY, M. AND SOLOMON, S., *Physica A* 242 (**1997**), p. 90; ISPOLATOV, S., KRAPIVSKY, P. L., REDNER, S., *Eur. Phys. J. B* 2 (**1998**), p. 267; PODOBNIK, B., IVANOV, P. CH., LEE, Y., CHESSA, A., STANLEY, H. E., *Europhys. Lett.* 50 (**2000**), p. 711

11 CHAKRABARTI, B. K., MARJIT, S., *Ind. J. Phys. B* 69 (**1995**), p. 681

12 CHAKRABORTI, A., CHAKRABARTI, B. K., *Eur. Phys. J. B* 17 (**2000**), p. 167

13 DRĂGULESCU, A. A., YAKOVENKO, V. M., *Eur. Phys. J. B* 17 (**2000**), p. 723

14 Drăgulescu, A. A., Yakovenko, V. M., *Physica A* 299 (**2001**), p. 213

15 Chakrabarti, B. K., Chatterjee, A., in *Application of Econophysics, Proc. 2nd Nikkei Econophys. Symp. (Tokyo, 2002), ed. H. Takayasu*, Springer, Tokyo, (**2004**), p. 280

16 Chatterjee, A., Chakrabarti, B. K., Manna, S. S., *Physica Scripta T* 106 (**2003**), p. 36; Chatterjee, A., Chakrabarti, B. K. and Manna, S. S., *Physica A* 335 (**2004**), p. 155

17 Silver, J., Slud, E., Takamoto, K., *J. Eco. Th.* 106 (**2002**), p. 417; González, M. C., Lind, P. G. and Herrmann, H. J., *Phys. Rev. Lett.* 96 (**2005**), p. 088702; Xie, Y.-B., Wang, B.-H., Hu, S., Zhou, T., *Phys. Rev. E* 71 (**2005**), p. 046135

18 Chatterjee, A., Chakrabarti, B. K., *in [1]*, p. 79; Chatterjee, A., Chakrabarti, B. K., Manna, S. S., *Physica A* 335 (**2004**), p. 155; Chatterjee, A., Chakrabarti, B. K., Stinchcombe, R. B., *Phys. Rev. E* 72 (**2005**), p. 026126

19 Kar Gupta, A., *Commun. Comp. Phys.* 1 (**2006**), p. 505; also in *arXiv: physics/0509172*

20 Ding, N., Wang, Y., Xu, J., Xi, N., *Int. J. Mod. Phys. B* 18 (**2004**), p. 2725; Wang, Y. and Ding, N., [1], p. 126

21 Sinha, S., in [1], p. 177

22 Kar Gupta, A., *Physica A* 359 (**2006**), p. 634; also in *arXiv: physics/0505115*

23 Patriarca, M., Chakraborti, A., Kaski, K., *Physica A* 340 (**2004**), p. 120; Patriarca, M., Chakraborti, A., Kaski, K., *Phys. Rev. E* 70 (**2004**), p. 016104

24 Bhattacharya, K., Mukherjee, G., Manna, S. S., in [1], p. 111

25 Das, A., Yarlagadda, S., *Phys. Scr. T* 106 (**2003**), p. 39

26 Repetowicz, P., Hutzler, S., Richmond, P., *Physica A* 356 (**2005**), p. 641

27 Lux, T., in [1], p. 51

28 Angle, J., *Social Forces* 65 (**1986**), p. 294; Angle, J., *Physica A* (**2006**), in press

29 Richmond, P., Repetowicz, P. and Hutzler, S., in [1], p. 120

30 Ferrero, J. C., in [1], p. 159

31 Kar Gupta, A. (**2006**), unpublished

32 Mohanty, P. K., *arXiv:physics/0603141*

33 Majumdar, S. N., Krishnamurthy, S., Barma, M. *J. Stat. Phys.* 99 (**2000**), p. 1

34 Biham, O., Malcai, O., Levy, M., Solomon, S., *Phys. Rev. E* 58 (**1998**), p. 1352

35 Kapur, J. N., *Maximum-Entropy Models in Science and Engineering*, Wiley, New York, **1989**

36 Leombruni, R., Richiardi, M., *Physica A* 355 (**2005**), p. 103

37 Kadanoff, L. P., *Physics Today* Dec. (**1983**), p. 46

38 Souma, W., Nirei, M., *Empirical Study and Model of Personal Income* in [1], p. 34

39 Taylor, L. R., *Nature* 189 (**1961**), p. 732; also see www.zoo.ufl.edu/bolker/emd/notes/taylor-pow.html

40 Solomon, S., Richmond, P., *Eur. Phys. J. B* 27 (**2002**), p. 257

7
The Contribution of Money-transfer Models to Economics
Yougui Wang, Ning Xi, and Ning Ding

7.1
Introduction

Recently some physicists have shifted their research interests to economics, regarding an economy as a complex system, and have formed a new branch which is named econophysics [1]. In this interdisciplinary subject, several topics including financial markets, wealth distribution and so on, have attracted much attention. Since econophysicists view the economy as a collection of particles that interact with each other, they have applied methods and techniques of statistical physics to economic and financial problems. With these tools of statistical analysis and multi-agent modeling, they have made some meaningful contributions to economics.

To explore the mechanism behind Pareto-law income and wealth distributions [2], several models have been developed [3]. S. Ispolatov et al. proposed a model where assets are exchanged between two traders randomly chosen according to the trading rule, and total wealth is conserved [4]. Their work was followed by some research [5,6]. In reality, wealth does not follow the conservation law and can be produced and destroyed. Thus, using the analogy between asset exchange and inelastic scattering in a granular gas, F. Slanina put forward a new approach in order to understand these cases [7]. Different from the nonconservation of wealth, the amount of money in an economy always remains constant, which is similar to the energy in an ideal gas. Focusing on the statistical mechanics of money, V. M. Yakovenko's group investigated the case of the random transfer of money [8]. Further, B. K. Chakrabarti and some other researchers considered trading rules with a saving factor and put much effort into mathematical analysis [9–14]. Taking the role of preference in a real economy into account, we introduced preferential dispensing behavior into the trading process and exposed its effect [15]. In a modern economy, money creation plays an important role. Recognizing its significance, R. Fischer and D. Braun analyzed the process of creation and annihilation of money from a

Econophysics and Sociophysics: Trends and Perspectives.
Bikas K. Chakrabarti, Anirban Chakraborti, Arnab Chatterjee (Eds.)
Copyright © 2006 WILEY-VCH Verlag GmbH & Co. KGaA, Weinheim
ISBN: 3-527-40670-0

mechanical perspective by proposing analogies between assets and the positive momentum of particles and between liabilities and their negative momentum [16–18].

These models mentioned above consist of some common factors, such as a group of traders, a pile of money and a trading rule. In the process of evolution, money is always held by traders and transferred from one trader to another continually; and the transferred amount of money is determined by a specific trading rule. Thus the trading rule is an essential determinant to the statistical properties of money in an economy. Since this kind of model has the property of continual money transfers between traders, it could be referred to as a money transfer model. As mentioned above, the prime theme of constructing such models is to explore the mechanism of statistical distribution. In fact, in the evolution process of these simple models, one can conveniently record the amount of money held by any given trader at any given moment and easily trace the motion of any given unit of money. Thus these models can be applied to some other economic issues, such as monetary circulation, money creation, and economic mobility. In this chapter, we will show how to modify the money transfer models to fulfill corresponding goals and what discover the results which we can obtain from these applications.

We first apply the money transfer models to the understanding of monetary circulation. In reality, money is transferred from hand to hand consecutively. This phenomenon is called circulation of money in economics. The scientific term usually used to describe the circulation is the velocity of money, which can be computed by the ratio of total transaction volume to the money stock. In traditional economics, the velocity has been investigated based on exchange equation and money demand theory. In fact, it measures how fast money moves between agents. This rate can be observed by recording the time intervals of each unit of money that the agents hold. As virtual economies, money transfer models can be used to carry out such an observation.

The time interval mentioned above is defined as the holding time of money. With the help of money transfer models, a steady distribution of holding time can be observed as the virtual economy reaches its equilibrium state [19, 20]. The velocity of money is also found to have an inverse relationship with the average holding time of money. Since the average holding time is governed by the agents in the models, this relation suggests that the velocity is determined by the behavior patterns of economic agents. Employing a simple version of the life-cycle model in economics, we demonstrate that the average holding time of money can be obtained from the individual's optimal choice. When the circulation process is in the form of a Poissonian, the probability distribution of the holding time of money takes an exponential form [20]. Under this assumption, the relation between the Keynsian Multiplier and the velocity of money is deduced, and it is found that the Multiplier is just a tortuous representation of the velocity of money. These investigations provide a new insight into the velocity of money circulation and the relevant concepts.

In order to discuss the impact of money creation on the monetary aggregate, we extend money transfer models by introducing a banking system, where money creation is achieved by bank loans and the monetary aggregate is determined by the monetary base and the required reserve ratio. The dependence of the monetary aggregate on the required reserve ratio has been illustrated by a simplified multiplier model in traditional economics. However, this model only presents the final result without demonstrating the process. Instead, we examine the evolutionary process of a money transfer model to see how money is created by classifying traders according to the amount of money they hold. By distinguishing different roles they play in the money-transferring process, we find that the monetary aggregate increases proportionally with the number of the traders with no monetary wealth.

In a similar framework, we also study how money creation affects the velocity of money. We know different definitions of money lead to different velocities of money. Each kind of velocity can be computed by dividing the total volume of transactions by the corresponding money stock. Based on modified transfer models, we examine the impacts of money creation on two types of velocity for narrow money and broad money using the methods of statistical physics and circuit dynamics.

Besides making contributions to research on money circulation and creation, money transfer models are also helpful in the study of economic mobility. During the simulations of money transfer models, the amount of money held by agents varies over time, meanwhile the rank of each agent shifts from one position to another. This phenomenon is called mobility in economics. Just as money circulation and creation is the dynamics of money, mobility is the dynamics of income or wealth. Thus the analysis of mobility is very helpful in comprehending the dynamic mechanism behind the distribution.

To show the mobility phenomenon with clarity, we perform the simulations on the transfer model with uniform saving rate and record the time series of the agent's rank. With the displacement index proposed by economists, we measure the mobility for different saving rates. It is found that the measurement results of mobility in the transfer model depend on the sampling interval. This defect stems from the framework of mobility indices where the contributions of the process to the mobility and the time length of the process are neglected. Starting from this idea, we modify the index by taking account of process and time, and remeasure the mobility in the transfer model. The results show that the measurement is dependent on the sampling interval no longer.

7.2
Understanding Monetary Circulation

Money does matter in the performance of an economy. As a medium of exchange, the money moves as it transfers between agents. This kind of movement from hand to hand in sequence is usually figured out as monetary circulation. The circulation process is depicted by two variables: money stock and velocity. In comparison with money stock, the velocity is more significant and complicated in the economy. As indicated by our previous work [19–21], with the help of money transfer models, the velocity of monetary circulation can be understood deeply. Starting from those primary results, we go further in the theoretical study on this topic.

7.2.1
Velocity of Money Circulation and its Determinants

The velocity of money circulation is a central matter in monetary theory which has attracted much attention for hundreds of years [22]. Although exploration of the velocity can be traced backwards to earlier works in the 1660s, most of the current investigation of velocity is commonly attributed to Irving Fisher who presented an influential exchange equation [23]:

$$MV = PY \qquad (7.1)$$

where M is the amount of money in circulation, V is the velocity of circulation of money, P is the average price level and Y is the level of real income. From this equation, the velocity of money can be expressed as follows

$$V = \frac{PY}{M} \qquad (7.2)$$

This result indicates the velocity can be computed as the ratio of transaction volume or aggregate income to money stock.

Based on this equation, many theoretical and empirical research work on the velocity have been carried out which gives the equation a good reputation [24–27]. Nevertheless, it also has had many critics. First, the exchange equation is only an arithmetic description of the relationship between money flow and product flow. This equation is an identity that arises from the value-equality of any trade. The amount of money that the buyer pays out in each transaction is always equal to the value of the goods that the seller has sold to the buyer. All individual exchanges that have been performed during one unit of time can be aggregated to an exchange equation for the whole economy. The equality still holds after aggregation, where the sum of money paid out corresponds to the total money flow (left side of Eq. (7.1)), and the sum of the value of all sales is nominal GDP (right side of Eq. (7.1)). Any one variable of the identity is governed by some quantities which may not be variables in the equation. Just as Rothbard [28] argues, "it is absurd to dignify any quantity with a place in an equation unless it can be defined independently of the other terms in the equation." There is no exception for the velocity of money. It must have a "life of its own".

Like all other identities, it says nothing about the causality between the left side and the right side of the equation unless we add more assumptions. In some cases, production flow drives money to circulate which means the direction of causation runs from right to left. In other cases, the change in money flow may be the cause of production which corresponds to the reverse of causation. In traditional monetary theories, the quantity theory of money is a representative of left-to-right causation, while the Cambridge Equation makes the opposite proposal.

With respect to which factors govern the velocity of money, there have been considerable debates over hundreds of years. To explore the determinants of the velocity, much effort has been put into research, theoretically and empirically. An overall review has been presented by T. M. Humphrey [29]. A prominent explorer is M. Friedman, who is the leader of the monetarism school. His theoretical analysis on the velocity of money is a combination of money demand theory and portfolio theory. Money is regarded as one sort of asset. The representative agent decides to hold the amounts of money and other assets, such as bonds and stocks, based on an equilibrium benefit–cost analysis [26]. Dividing the amount of money which results from the balance by the given income yields the reciprocal of the velocity. The final conclusion shows that the determinants of the velocity include mainly income, interest rates of bonds and stocks, inflation rate, and so on. However, as indicated by an empirical work of R. T. Selden, which is contained in the same book, the theoretical prediction of the correlation between the velocity and the interest rate could not be verified [26].

The fatal defect of this kind of analysis is that it derives a macroeconomic conclusion from an individual optimal choice, leading to a synthetic fallacy. According to the money demand theory, as long as all individuals intend to hold more money, the aggregate amount of money will increase automatically. Actually, in order to increase the quantity of holding money, an individual has either to reduce his expense as income is given, or to increase his income as expense is given. The most common case is the former, but not the latter, since expense is more easy to control than income. According to the identity of income and expense, when all individuals would like to reduce some of their expenses monetary, the gross income will drop equivalently. The intention of holding more money leads to less money flow in the system instead of a larger aggregate of money. In other words, when people face a change in money demand, the aggregate outcome of individual choices is characterized by a variation in money flow. If the total amount of money is fixed, this can finally be marked by a change in the velocity of money.

In brief, since the exchange equation at aggregate level fails to specify what factors govern the velocity of money, a micro approach is needed. However, the money demand theory which has long been used to solve this problem at micro level is found to have a weakness beyond retrieval. This caused us to seek an alternative approach which will be presented in the following subsections.

7.2.2
Holding Time versus Velocity of Money

7.2.2.1 From Holding Time to Velocity of Money

The way in which money moves in an economy is different from that in which commodities move. Any commodity will undergo a process of being produced, transmitted, consumed or used. Sooner or later, they will disappear out of our world. This is true in particular for perishable consumption goods. Money is not like them. The purpose of people in holding and using money is to exchange it for what they need, instead of consuming it. Although money might be destroyed due to abrasion, it cannot be intentionally used up in an economic sense. In this way, money is always transferred from one hand to another. The money one receives today must be spent on something sooner or later. Nevertheless, after one gets an amount of money, it will not be paid out until the proper time. In other words, once a unit of money mediates an exchange, it will be held by its owner for a certain period, instead of taking part in another one immediately. During this period, money may exist in various forms, such as cash or a deposit in bank. In any event, the money idles as long as its holder does not take part in trade, rendering no service to the transaction. The time interval between two consecutive trades involved with one unit of money represents the activity of this unit of money in participating in the trade. The longer the time interval, the less active will be the unit of money, and vice versa. It follows that the time interval is an important index which measures how frequently money participates in trade and how deeply it makes its effect on an economy.

To our knowledge, Wicksell was the first to propose this concept and called it the "average period of idleness" or the "interval of rest" of money [30]. When we initially presented this concept we were not aware of Wicksell's work [19]. We named it "holding time" in order to place emphasis on its passive position. Obviously, this kind of time interval is not determined by money itself, but its holder's behavior. To make it clear, it is worth noting Mises' statement about this character [31]: "money can be in the process of transportation, it can travel in trains, ships, or planes from one place to another. But it is in this case, too, always subject to somebody's control, is somebody's property." Later we recognized Wicksell's denomination. But we argue that holding time is well-founded and should take over.

The length of holding time is determined by the holder's motives to use or store money. According to the money demand theory in economics, the motives can be sorted into three types: the transactions-motive; the precautionary-motive; and the speculative-motive. We can further divide the motives into more categories as Keynes did [32]. However, this is impracticable and there is no need to know of the true purpose of the holders individually. In fact, what is clear is that the holding times of different units of money are different at the same moment, and the holding time of any unit of money always varies over time. At micro level, any one unit of money can be viewed to be transferred at random. Nevertheless, the holding time of money might have a stationary probabilistic nature as a whole, which is what we should be concerned about. We can imagine that as an economy reaches its equilibrium state, any single amount of money's holding time continues to vary over time, but the overall probability character does not change any further. The term we use to depict this character is the probability distribution of money over the holding time.

In a closed economy, the amount of money M is fixed. The portion of money whose holding time lies between τ and $\tau + d\tau$ is $MP(\tau)\,d\tau$, where $P(\tau)$ is defined as the probability distribution function of money over the holding time. If $P(\tau)$ is given, the money flow generated by the fraction of money $MP(\tau)\,d\tau$ can be expressed by the following

$$F(\tau) = MP(\tau)\frac{1}{\tau}d\tau \tag{7.3}$$

Adding up the contributions of all parts leads to the total money flow in the economy,

$$F = M\int_0^\infty P(\tau)\frac{1}{\tau}d\tau \tag{7.4}$$

From Eq. (7.2), we have

$$V = \int_0^\infty P(\tau)\frac{1}{\tau}d\tau \tag{7.5}$$

This is the statistical expression of the velocity of money in terms of holding time. The result shows the velocity V is the mathematical expectation of $1/\tau$.

Given the probability function $P(\tau)$, we can also express the average holding time of money $\bar{\tau}$ as the following

$$\bar{\tau} = \int_0^\infty P(\tau)\tau\,d\tau \tag{7.6}$$

From Eq. (7.5) and (7.6), it is obvious that the velocity and the average holding time are inversely related. On average, the longer the holding time of money, the slower the money circulates in the economy, and vice versa. Particularly when the probability function $P(\tau)$ is of the Gamma form, the velocity of money and the average holding time have the following relation

$$V = \frac{2}{\bar{\tau}} \tag{7.7}$$

This simple relation indicates explicitly that the velocity of money is determined by its average holding time.

7.2.2.2 Calculation of Average Holding Time

Obviously, to explore the determinants of velocity, we need to make clear what decides the length of the average holding time. As mentioned above, the holding time is not governed by money itself, but by the money holder's behavior. Its length is determined by various economic choices of the money holders. There is a great variety of economic choices which are involved with money, for money permeates the whole economy. Economic agents may use money to invest, or to consume, or to save for the future, or to speculate in financial markets. Although these choices treat money in diverse ways, all these choices have one thing in common, that is, money is received and then paid out. Whatever an economic choice may be, the length of holding time of money in this choice is eventually characterized by the receipt-payment pattern.

Consider a complete receipt-payment process of money during a given period T, where at any time t, the inflow of money is $Y(t)$, the outflow is $C(t)$ and the accumulative money balance of an agent is $W(t)$. Assume both the initial and final balances are zero, that is, $W(0) = W(T) = 0$. Thus the average holding time of money in this process can be computed by the following form [21]

$$\bar{\tau} = \frac{\int_0^T [C(t) - Y(t)] t\, dt}{\int_0^T Y(t)\, dt} \tag{7.8}$$

To illustrate this calculation, we employed a very simple version of a life-cycle model which was developed by Franco Modigliani, Richard Brumberg and Albert Ando and has been adopted in many macroeconomics textbooks. The simple version of the model considers a representative individual who will live T years more, and work for only T_0 years. During his life, he earns income $Y(t) = Y$ within the working period, but $Y(t) = 0$ in the remaining period. Thus, he has a balance of $W = YT_0$ that can be spent in his entire life. The optimal choice for this agent is to spend $C = YT_0/T$ each year. From

Eq. (7.8), the average holding time of money in this model can thus be obtained as follows

$$\bar{\tau} = \frac{1}{2}\left(1 - \frac{T_0}{T}\right)T \tag{7.9}$$

This result is straightforward. The more years one lives, the longer one tends to hold one's money. And the longer the working period, the shorter the holding time.

The life-cycle model conveys to us that as one's income pattern is given, the time pattern of consumption expenditure can be obtained by maximizing the representative individual's goal. Thus we have the average holding time of money for a representative consumer. Similarly, with respect to other kinds of economic choices, as long as the time patterns of receipt and expenditure are specified, the corresponding average holding time can be calculated from Eq. (7.8).

7.2.3
Keynesian Multiplier versus Velocity of Money

We have progressed simulations of holding time distribution based on a random-transfer model and an expense-preferential model [20]. We found that when the money is randomly received and paid out, the process of money flow exhibits a stationary Poisson nature. As the system reaches its equilibrium state, if we set a certain moment as a starting point in time, then for any one unit of money at time t the probability of participating in trade for the first time can be given by

$$P(t) = Ve^{-Vt} \tag{7.10}$$

where V denotes the velocity of money in the system.

In this process, we take a period of the continuous transaction T as a sampling time interval and examine how money transfers between agents during this period. We define the amount of money received by agents as total income, and that paid out as total expenditure. Due to the conservation of money in trade, the total income is always equal to the total expenditure. It is worth noting that, within this period, some expenditure comes from current income, while other comes from past incomes. Along the line of income-expenditure theory which was proposed by Keynes and has become a staple in macroeconomic education at the introductory and higher levels, we call the expenditure that comes from current income "deduced expenditure", and the others "autonomous expenditure".

Given the amount of money M circulating in the economy, the total expenditure generated within this period can be computed as the following

$$E_t = MVT \tag{7.11}$$

The deduced expenditure is that part of the transaction volume in which the money involved has taken part in trade at least twice during the reference period. On the contrary, the autonomous expenditure is that part of the transaction volume in which the money involved has taken part in trade only once during the period. We can express the autonomous expenditure in terms of the probability distribution function as the following

$$E_a = \int_0^T \int_{T-t_1}^\infty MP(t_1)P(t_2)dt_1 dt_2 \tag{7.12}$$

Substituting (7.10) into (7.12), we obtain

$$E_a = MVTe^{-VT} \tag{7.13}$$

It is obvious that the deduced expenditure plus the autonomous expenditure equals the total expenditure, that is,

$$E_t = E_a + E_d \tag{7.14}$$

Combining Eq. (7.11), (7.13) and (7.14), we get the deduced expenditure

$$E_d = MVT(1 - e^{-VT}) \tag{7.15}$$

Following the definition of the marginal propensity to expend in economics, we write it in the following form

$$e_{mp} = \frac{E_d}{Y} \tag{7.16}$$

where Y corresponds to the total income, which is equal to the total expenditure. From Eq. (7.11) and (7.15), we immediately obtain

$$e_{mp} = 1 - e^{-VT}. \tag{7.17}$$

Similarly, the marginal propensity to save is related to the marginal propensity to expend in the following form

$$s_{mp} = 1 - e_{mp}. \tag{7.18}$$

Thus, we have

$$s_{mp} = e^{-VT}. \tag{7.19}$$

This result indicates that the marginal propensity to save s_{mp} depends on the length of the time interval T. For the two extreme cases $T \to 0$ and $T \to \infty$, we get

$$\lim_{T \to 0} s_{mp} = 1 \tag{7.20}$$

$$\lim_{T \to \infty} s_{mp} = 0 \tag{7.21}$$

The former equation tells us that, when the time interval is short enough, all current income is saved. In the opposite case, when the time interval is long enough, all expenditure comes from income in this period. There is no saving at all.

The Keynesian Multiplier is the most basic concept in macroeconomics, but it has long given rise to much controversy. In a simplified model it states that, as the autonomous expenditure increases, one additional unit, the equilibrium output, will increase more than one unit. This multiplier effect can also be seen in money transfer models. Under the assumptions given above we can see that an amount of money is an exogenous force which can lead to a change in autonomous expenditure and equilibrium output. Suppose there is a small change in the amount of money δM. So the corresponding variations in the autonomous expenditure and the equilibrium output are respectively

$$\delta E_a = VTe^{-VT}\delta M \tag{7.22}$$

and

$$\delta Y = VT\delta M \tag{7.23}$$

Let k denote the multiplier, and we then get

$$k = \frac{\delta Y}{\delta E_a} = e^{VT} = \frac{1}{s_{mp}}. \tag{7.24}$$

This corresponds to the inference of the income-expenditure theory. From Eq. (7.24) we immediately have

$$\ln k = VT \tag{7.25}$$

This equation shows us that the multiplier is also dependent on the length of the sampling time interval. If we define the time interval as one unit of time, then the multiplier is obviously equivalent to the velocity of money. In related economics literature, the Keynesian Multiplier is usually deduced in a logical time framework. Due to the neglect of real time, it reflects the velocity of money in a tortuous way. This being the case, the multiplier should be discarded in economics and totally replaced by the velocity of money.

7.3
Inspecting Money Creation and its Impacts

Since the largest part of the monetary aggregate that circulates in modern economies is created by debts through the banking system, money creation has an important influence on the monetary economic system. Recently, some investigations on this aspect have been carried out, mainly focusing on the impact of money creation on monetary distribution [8, 16]. In this section, we will illuminate how money creation affects money circulation from two aspects, which are the monetary aggregate and the velocity of money, by employing money transfer models.

7.3.1
Money Creation and Monetary Aggregate

The modern banking system is a fractional reserve banking system, which absorbs savers' deposits and loans to borrowers. Since they are purchasing, the public can pay in currency or in deposits. In this sense, currency held by the public and deposits in a bank can both play the role of exchange media. Thus, in economics, the monetary aggregate is measured by the sum of the currency held by the public and the deposits in the bank. When the public places a part of their currency into commercial banks, this part of the currency turns into deposits. Such a transformation does not change the monetary aggregate. Once commercial banks loan to borrowers, usually in deposit form, deposits in the bank increase while currency held by the public remains constant. So the loaning behavior of commercial banks increases the monetary aggregate and achieves money creation.

7.3.1.1 A Simplified Multiplier Model

Economists have developed a model to figure out this mechanism of money creation. It is called the multiplier model of money [33–35]. Here we introduce its simplified version. In the economy, commercial banks are required to keep a percentage of their deposits in currency form as required reserves, which is determined by the central bank and named as the required reserve ratio. The simplified multiplier model assumes that all the currency issued by the central bank is held by commercial banks as required reserves. In order to fulfill the requirement of reserves, the commercial banks must grant enough loans such that the monetary aggregate satisfies the following relation

$$M = \frac{M_0}{r} \qquad (7.26)$$

where M denotes the monetary aggregate, M_0 the monetary base and r the required reserve ratio. The volume of outstanding debts is the difference between the monetary aggregate and the monetary base, which can be written as

$$L = \frac{M_0}{r} - M_0 \qquad (7.27)$$

From these two equations we can see that this model provides only the final outcomes of money creation, but fails to demonstrate any detail of the whole process.

7.3.1.2 A Modified Money-transfer Model

To represent the process of money creation, econophysicists have also proposed an approach (called double entry book-keeping) of money transfer, from a physical perspective [16–18]. In this approach, a money transfer from trader P to trader R triggers one of the four different cases below:

1. Money creation if P has liabilities and R has assets. P will increase its liabilities and R will increase its assets.

2. Money transfer if both P and R have assets. P will decrease its assets and R will increase its assets.

3. Liability transfer if both P and R have liabilities. P will increase its liabilities and R will decrease its liabilities.

4. Money annihilation if P has assets and R has liabilities. P will decrease its assets and R will decrease its liabilities.

Let F denote the transaction volume in one round, and let F_{la}, F_{aa}, F_{ll} and F_{al} in turn denote transaction volumes which are contributed by the four cases above, respectively. Thus a quantitative equation is obtained,

$$F = F_{la} + F_{aa} + F_{ll} + F_{al} \tag{7.28}$$

Combining the approach and money-transfer model, we can better understand what is the source of money creation. The model we propose here is an extension of the model in [16]. The economy consists of N traders and a bank. At the beginning, a constant monetary base M_0 is equally allocated to these traders and all the monetary base is saved in the bank. In the transferring process of money, each of the traders chooses his partner randomly in each round, and yields N trade pairs. Then one is chosen as "payer" randomly and the other as "receiver" in each trade pair. If the payer has deposits in the bank, he pays one unit of money to the receiver in deposit form. If the payer has no deposit and the bank has excess reserves, the payer borrows one unit of money from the bank and pays it to the receiver. But if the bank has no excess reserve, the trade is cancelled. After receiving one unit of money, if the receiver has loans, he repays his loans. Otherwise the receiver holds this unit of money in deposit form.

For convenience in manipulation, we classify traders into three groups: the traders with positive monetary wealth, the ones with no monetary wealth and the ones with negative monetary wealth. From the trading mode of our

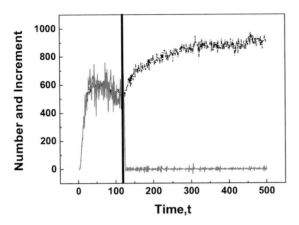

Fig. 7.1 The evolutions of the increment of the monetary aggregate (solid) and the number of traders with no monetary wealth (dash dot) over time for required reserve ratio $r = 0.8$. The vertical line denotes the moment at which the monetary aggregate reaches the steady value.

model, it can be reckoned out that each trader participates in trade on average twice in one round. When a trader with positive monetary wealth makes an exchange, he neither borrows nor repays. When a trader with negative monetary wealth does so, he either borrows or repays. When a trader has no monetary wealth, he has only the possibility of borrowing but never repays. Thus, money creation originates from the borrowing behavior of both the traders with negative monetary wealth and those with no monetary wealth. On the other hand, money annihilation is caused only by the traders with negative monetary wealth. When the bank is not constrained by the requirement of reserves, it can grant loans freely. In this case each trader with negative monetary wealth has the same probability of 0.5 to get loans and to repay the debts, leading to no contribution to a variation in the monetary aggregate. Therefore, only the traders with no monetary wealth contribute to the money creation in the whole economy. According to the setting of our model, the increment of monetary aggregate in one round is computed to be on average equal to the number of traders with no monetary wealth.

This prediction can be verified by computer simulations of this money-transfer model. Setting various required reserve ratios, we performed the simulations and recorded the data of monetary aggregate and the number of traders with no monetary wealth. Figure 7.1 shows the evolutions of the number of traders with no monetary wealth and the increment of the monetary aggregate over time when the required reserve ratio is 0.8. The curve of the increment fluctuates about that of the number of the traders with no monetary wealth before the monetary aggregate reaches the steady value. From this figure we can see that the theoretical deduction is in good agreement with the simulation results.

7.3.2
Money Creation and Velocity of Money

Two kinds of definition for velocity of money are usually used in economics and these are the velocity of narrow money and that of broad money. Both velocities are computed by dividing the total transaction volume by the corresponding money stock, where the stock of narrow money is the monetary base and that of broad money corresponds to the monetary aggregate. When we employ the method of statistical physics to study how money creation affects the velocity of narrow money, we tend to highlight the influence of loan size. On the other hand, when we use the circuit approach to see how it affects the velocity of broad money, we place more emphasis on the influence of the circular rate of loan.

7.3.2.1 Velocity of Narrow Money

As mentioned above, the velocity of narrow money is equal to the value of the transaction volume in one round divided by the monetary base, that is,

$$V_0 = \frac{F}{M_0} \tag{7.29}$$

To expose the impact of money creation on this velocity, we continue to use the model proposed in the last subsection. Since the monetary base is given, what we need to do is to analyze the volume F and express it in terms of the required reserve ratio.

In contrast to the classification of the traders in the previous subsection; traders are now sorted into two groups: the traders with positive monetary wealth and the ones with nonpositive monetary wealth, whose numbers are denoted by N_+ and N_-, respectively. It is well known that each trader participates in trade on average twice in one round. In each transfer of money, the probability of paying one unit of money is $1/2$ for the traders with positive monetary wealth, and it must be less than $1/2$ for the traders with nonpositive monetary wealth, as borrowing may fail due to the limitation of required reserves. Let ω denote this unknown probability, and from the detailed balance condition which holds in the steady state, we have

$$\omega p_-(m_1)p_+(m_2) = \tfrac{1}{2}p_-(m_1 - 1)p_+(m_2 + 1) \tag{7.30}$$

where $p_-(m)$ and $p_+(m)$ denote the monetary wealth distribution, whose expression can be obtained by the method of the most probable distribution [36,37]. The formula for the stationary distribution of monetary wealth in terms of the required reserve ratio is as follows

$$p_+(m) = \frac{N_0}{N} e^{-\frac{m}{\overline{m}_+}} \quad \text{for } m \geq 0$$
$$p_-(m) = \frac{N_0}{N} e^{\frac{m}{\overline{m}_-}} \quad \text{for } m < 0 \tag{7.31}$$

where N_0 denotes the number of the traders with no monetary wealth, $\overline{m}_+$ is equal to the average amount of positive monetary wealth and $\overline{m}_-$ is equal to the average amount of negative monetary wealth. The expressions of $\overline{m}_+$ and $\overline{m}_-$ can be written as

$$\overline{m}_+ = \frac{1 + \sqrt{1-r}}{r} \frac{M_0}{N} \tag{7.32}$$

and

$$\overline{m}_- = \frac{1 - r + \sqrt{1-r}}{r} \frac{M_0}{N} \tag{7.33}$$

Substituting the expression for monetary wealth distribution (7.31) into Eq. (7.30), we obtain

$$\omega = \frac{1}{2} e^{-\frac{1}{\overline{m}_+} - \frac{1}{\overline{m}_-}} \tag{7.34}$$

Thus the total average transaction volume in each round can be expressed as

$$F = N_+ + N_- e^{-\frac{1}{\overline{m}_+} - \frac{1}{\overline{m}_-}} \tag{7.35}$$

Substituting Eq. (7.35) into (7.29), the velocity of narrow money can be given by

$$V_0 = \frac{N_+}{M_0} + \frac{N_-}{M_0} e^{-\frac{1}{\overline{m}_+} - \frac{1}{\overline{m}_-}} \tag{7.36}$$

Since in the steady state the number of traders whose monetary wealth changes from 0 to 1 is equal to that of the traders whose monetary wealth changes from 1 to 0, we have the following approximate relation

$$\omega N_0 \frac{N_-}{N} + \frac{1}{2} N_0 \frac{N_+}{N} = \frac{1}{2} N_1 \frac{N_+}{N} + \frac{1}{2} N_1 \frac{N_-}{N} \tag{7.37}$$

where $N_1 = N_0 e^{-\frac{1}{\overline{m}_+}}$ is the number of traders with a monetary wealth of 1. The left-hand side of Eq. (7.37) represents the number of traders whose monetary wealth changes from 0 to 1 and the right-hand side denotes the number of traders whose monetary wealth changes from 1 to 0. Substituting Eq. (7.34) into (7.37) and taking $N = N_+ + N_-$ into account, yields

$$N_+ = \frac{e^{\frac{1}{m_-}} - 1}{e^{\frac{1}{m_+} + \frac{1}{m_-}} - 1} N \qquad (7.38)$$

and

$$N_- = \frac{e^{\frac{1}{m_+} + \frac{1}{m_-}} - e^{\frac{1}{m_-}}}{e^{\frac{1}{m_+} + \frac{1}{m_-}} - 1} N \qquad (7.39)$$

Combining Eqs (7.36), (7.38) and (7.39), we can obtain

$$V_0 = \frac{N}{M_0} e^{-\frac{1}{m_+}} \qquad (7.40)$$

From Eq. (7.40), it is clear that the velocity of narrow money has an inverse relation with the required reserve ratio. This can be interpreted in the following way. In each round, if every pair of traders could fulfill their transfer of money, the transaction volume would be N in our model. However, in each round some transfers are cancelled because the payers with nonpositive monetary wealth may not get loans from the bank. As indicated by Eq. (7.40), the average realized transfer ratio can be expressed in the form of $e^{-\frac{1}{m_+}}$, which decreases as the required reserve ratio increases. Thus the transaction volume in each round decreases, and as a result the velocity of narrow money decreases.

To verify the theoretical relationship, we performed several simulations for different required reserve ratios. Since the initial settings of the amount of money and the number of traders have no impact on the final results, the parameters M_0 and N are set to 2.5×10^5 and 2.5×10^4, respectively. We are fully convinced that the whole system has reached a stationary state by 8×10^5 rounds. Thus, after that moment we collected latency time which is defined as the time interval between the sampling moment and the moment when money takes part in trade for the first time after the sampling moment. Please note that each transfer of the deposits can be regarded as that of currency chosen randomly from reserves in the bank, equivalent to collecting the data of latency time. It is seen that latency time follows an exponential law, which indicates that the transferral process of currency is of Poisson type. In this case, the velocity of narrow money for different required reserve ratios can be calculated by the data of latency time [19]. Figure 7.2 shows the relationship between the velocity of narrow money and the required reserve ratio, from simulation results and also from Eq. (7.40), respectively. It can be seen that the theoretical results are in good agreement with the simulation ones.

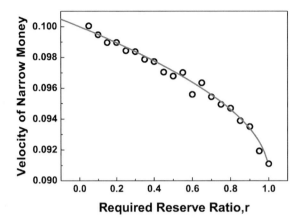

Fig. 7.2 The velocity of narrow money versus the required reserve ratio obtained from simulation results (dots) and from the corresponding analytical formula (continuous curve) given by Eq. (7.40), respectively.

7.3.2.2 Velocity of Broad Money

In Section 7.2.2, we have calculated the velocity based on the holding time distribution, where money creation is not taken into account. In this subsection, we intend to express the velocity of broad money in the same way. However, when money creation is involved, what is circulating in the economy includes not only money, but also liability. To describe these two kinds of circuit, we should specify the holding time of money in this case, and introduce another time interval related to liability, which is called the loaning time of liability. Under the assumption that all of the monetary base is saved in the bank, the holding time τ_m is defined as the time interval between the appearance and disappearance of a unit of deposit. Similarly, the loaning time τ_l is defined as the time interval between the appearance and disappearance of a unit of liability.

By analogy with Eq. (7.4), the transaction volume contributed by broad money circulation can be expressed by

$$F_m = \int_0^\infty M P_m(\tau_m) \frac{1}{\tau_m} d\tau_m \qquad (7.41)$$

where $P_m(\tau_m)$ denotes the steady distribution of holding time τ_m. Similarly, the transaction volume contributed by liability circulation is given by

$$F_l = \int_0^\infty L P_l(\tau_l) \frac{1}{\tau_l} d\tau_l \qquad (7.42)$$

where $P_l(\tau_l)$ denotes the steady distribution of loaning time τ_l. For simplicity, letting

$$V_m = \int_0^\infty P_m(\tau_m) \frac{1}{\tau_m} d\tau_m \qquad (7.43)$$

and

$$V_l = \int_0^\infty P_l(\tau_l) \frac{1}{\tau_l} d\tau_l \qquad (7.44)$$

we can rewrite Eqs (7.41) and (7.42) as the following, respectively

$$F_m = MV_m \qquad (7.45)$$

and

$$F_l = LV_l \qquad (7.46)$$

Relating to the four components in Eq. (7.28), we find that F_m is generated by money transfer and money annihilation and F_l is generated by liability transfer and liability annihilation. Since the liability annihilation is the same process as money annihilation, we then have

$$F_m = F_{aa} + F_{al} \qquad (7.47)$$

and

$$F_l = F_{ll} + F_{al} \qquad (7.48)$$

When the system reaches a steady state, the volume of money creation is equal to that of money annihilation, that is, $F_{al} = F_{la}$. Thus, combining Eqs (7.28), (7.47) and (7.48), we have

$$F = F_m + F_l \qquad (7.49)$$

We know, that the velocity of broad money is equal to the value of the transaction volume in one round divided by the monetary aggregate. From Eqs (7.45), (7.46), (7.49), (7.26) and (7.27), we have the expression for the velocity of broad money as

$$V = V_m + (1-r)V_l \qquad (7.50)$$

To verify Eq. (7.50), we further modify the model so that it is one with some variables being easily controlled. Four parameters p_{la}, p_{aa}, p_{ll} and p_{al} are added to represent the probabilities at which feasible transfers that correspond to the four cases mentioned above, are achieved. Simulations were performed

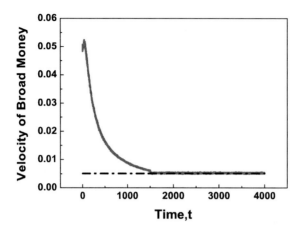

Fig. 7.3 The evolution of velocity over time (solid curve) and its estimated value (dash dot line). The parameters $p_{la} = 0.9$, $p_{aa} = 0.5$, $p_{ll} = 0.5$, $p_{al} = 0.6$ and $r = 0.1$.

for different settings of parameters with $M_0 = 5 \times 10^4$ and $N = 5 \times 10^3$. We collected the total transaction volume and the monetary aggregate in each round, and gave the time series of the velocity according to its definition. The evolution of velocity over time in a specific setting is shown in Fig. 7.3 (solid curve). After 3.5×10^5 rounds, by which the whole system has reached a steady state, we collected the holding time of money and the loaning time of liability, and calculated V_m and V_l respectively according to Eqs (7.43) and (7.44). Substituting these values and the given required reserve ratio into Eq. (7.50), we get the estimated value of the velocity of broad money, which is also shown in Fig. 7.3 (dash dot line). It can be seen that, after the system gets to its steady state, the velocity remains at its estimated value all the time.

7.4 Refining Economic Mobility

The study on economic mobility started in the 1960s [38]. Gradually this issue has become a branch of welfare economics. Economists augured that analysis on mobility is a supplement to distribution when exploring the generating mechanism of income and measuring the degree of inequality [39, 40]. A series of empirical analyses on mobility show that income mobility is essentially due to "exchange" [41–43]. "Exchange" mobility means that persons just shift their ranks while the distribution remains unchanged. This phenomenon can be observed easily from a money-transfer model. When the virtual economy reaches its steady state, the distribution is unchanged, while the rank of any individual varies over time. So we use this kind of model to examine economic mobility. This investigation reveals that there are some defects in the indices proposed by economists. We further modify the mobility indices and verify the new index, based on the money-transfer model.

7.4.1
Concept and Index of Measurement

In this burgeoning field there is still little consensus on the concept and measurements of mobility. Some researchers deem mobility as a re-ranking phenomenon which is a purely relative concept [38, 44, 45]. While some others assert that mobility is an absolute concept, and any variation of anyone's income or wealth should be taken into account [46, 47]. Correspondingly, the researchers construct distinct indices to measure mobility according to different concepts which is well reviewed in [48].

In fact, all of these indices are constructed in the same framework, which is called the two time periods framework, according to economics practice. Please note that data analyzed and compared are collected at the beginning period and ending period but not within the two periods. In this framework, the base and final periods are labelled 0 and 1 respectively. Assume there are N persons in the economy who are labelled by an index set $\{1, 2, ..., N\}$. At time k ($k \in \{0, 1\}$), the quantity measured of person i is denoted by x_i^k and the distribution is a vector $x^k = (x_1^k, ..., x_n^k)'$. Thus, all of the data at the heart of analysis form a matrix $x = (x^0, x^1)$. Let A be the domain of x^0 and x^1 and A^2 be the domain of x, therefore a particular mobility index becomes a map $M : A^2 \to R$. x_i denotes the rank of person i when measuring the relative mobility; the income or wealth when measuring the absolute mobility.

As an illustrative example, let us see the index proposed by G. S. Fields and E. A. Ok [47] which has recently been widely accepted and applied [41]. The index is constructed as follows

$$M_b = \frac{1}{N} \sum_{i=1}^{N} |\log x_i^0 - \log x_i^1| \tag{7.51}$$

It can easily be seen that the index is just a particular map $M_b : R_+^{2n} \to R_+$. Since there are N persons in the economy, any given state of this economy can be regarded as a point in an N-dimensional space. In this view, any index is actually one kind of displacement between the two points (or economy states). So we call them displacement-like indices.

7.4.2
Mobility in a Money-transfer Model

To show the "exchange" mobility, we employ the transfer model proposed by Chakraborti and Chakrabarti [9]. The virtual economy is composed of N agents and M units of money, where the money is held by these agents individually. All the agents keep part of their money as savings when participating in trade. The ratio of the savings to the money held is called the saving rate and denoted by s. Every agent has the same saving rate which is set at the beginning of the simulations and remains unchanged. Time is discrete. In each round, a pair of agents are chosen randomly to trade with each other. Suppose that at the tth round, agent i and j take part in trade, so at the $t+1$th round their money $m_i(t)$ and $m_j(t)$ changes to

$$m_i(t+1) = m_i(t) + \Delta m; m_j(t+1) = m_j(t) - \Delta m \tag{7.52}$$

where

$$\Delta m = (1-s)[(\varepsilon - 1)m_i(t) + \varepsilon m_j(t)]$$

and ε is a random fraction. It can be seen that if Δm is positive, agent i is the receiver of the trade, otherwise it is the payer.

The simulations continued to proceed so that a steady Gamma distribution appeared, which indicates the system is approaching a steady state. After this, we carried out the measurement of mobility. The sampling interval was set to be 1000 rounds in our simulations. The amount of money which each agent heldsat the end of each sampling period was recorded. Correspondingly, the ranks of the agents were obtained by sorting all of the agents according to their money. In this way, we get the time series of rank for all agents. Some typical rank time series for $s = 0$ and $s = 0.5$ are shown in Fig. 7.4. The value of rank is inversely related to the amount of money held, so the more money an agent holds, the lower is his position in the panels.

From the figure it can be seen that all of the agents can be either rich or poor. The rich have the probability of becoming poor and the poor also may be lucky enough to become rich. The economy in this model is quite fair where every agent has equal opportunity. This is easy to understand because agents are homogenous. There is no reason why any agent should always excel. From the figure, it can also be seen that the lower the saving rate, the higher the fluctuation frequency. In other words, the lower saving rate leads to a greater mobility in the transfer model. The intuition for this result is straightforward. The more money that agents keep when trading, the less money they take out to participate in trade, and the less probability of a change in rank. Therefore, the higher the saving rate, the lower the mobility.

After qualitative analysis, we turned to the quantitative measurement of mobility. We adopt the index proposed by Fields and Ok (see Eq. (7.51)) be-

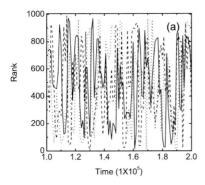 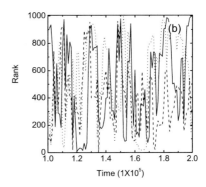

Fig. 7.4 The typical time series of rank from the Uniform Saving Rate Model with saving rates $s = 0$ (a) and $s = 0.5$ (b).

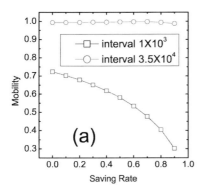 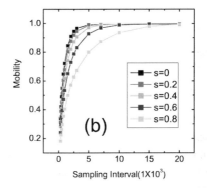

Fig. 7.5 (a) The relation between saving rate and mobility measured by employing displacement mobility index: (□) the sampling interval 1000, (○) the sampling interval 3.5×10^4; (b) The relation between sampling interval and mobility measured by employing displacement-like mobility index. (For a color version please see color plate on page 596.)

cause it embodies all variations in income at microlevel. The measurement results are obtained in the following way. First, more than 9000 samples are recorded continuously after the distribution becomes steady. Secondly, the "displacement" between any two consecutive samples is calculated according to Eq. (7.51). Finally, these "displacements" are averaged over 9000 samples as a final value of M_b. The relation between mobility and saving rate is plotted at the bottom of Fig. 7.5(a). This measurement result verifies our qualitative analysis above to some degree.

However, as the sampling interval is lengthened, the inverse relationship between mobility and saving rate becomes indistinct. When the sampling interval is 3.5×10^4 rounds, the measurement result becomes almost independent of the saving rate (see the upper curve in Fig. 7.5(a)). To reveal this phenomenon more clearly, we present the relationship between the mobility index and sampling interval in Fig. 7.5(b). It can be seen that as the sampling interval becomes long enough, the measurements reach the same maximum (about 1) for the different saving rates. The value of the saving rate only has an effect on the convergence speed. This phenomenon indicates that there is something wrong with the index which we applied and causes us to rethink the measurement framework of mobility proposed by economists. In the next section, we will propose a new approach to measuring the mobility.

7.4.3
Modification in the Measurement Index

Since the measurement results depend on the sampling interval, this displacement-like index, we think, is just a distorted reflection of the mobility which cannot accurately depict the picture of mobility. The culprit, we argue, is the absence of the contribution of the process in the two-periods framework of mobility. Suppose an agent's income is $1 in the first year, $10 in the second year and $1 in the third year. No matter which index we use to measure the mobility during these three years, the result is 0. From this example we can see that the variations in income or wealth during the sampling interval may differ a lot, but are neglected by the measurement index. The longer the sampling interval, the more the information on the mobility is neglected. So for this reason, the measurements highly depend on the sampling interval. Due to this defect, we cannot affirm that the degree of the mobilities in two countries are still the same, even if the measurements in practice, yield the same value. To make the measurement more useful, the contribution of the process should be introduced into the measurement.

So in order to modify the framework of mobility, we are assisted by a simple analogical example. Assume that two drunken guys, who cannot run straight but only in a zigzag, have a race. We would not be surprised at the result that the tortoise is faster than Achilles, if displacement is employed as the measurement index. Apparently, to solve this problem, the contribution of the process needs to be taken into account. However, this is not enough. The time spent on the whole process should also be considered. In other words, the "speed" should be measured in terms of accumulated "distance" and length of time. In this way, the modifications do not just turn the displacement-like indices into distance-like ones, they turn them into speed-like ones. Taking the index proposed by Fields and Ok (see Eq. (7.51)) as an example, we modify

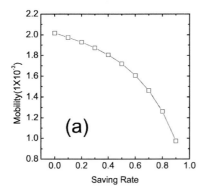
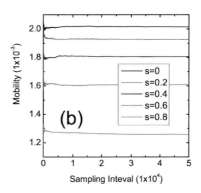

Fig. 7.6 (a) The relation between saving rate and mobility measured by employing a speed mobility index with the sampling interval 3.5×10^4; (b) The relation between sampling interval and mobility measured by employing the speed mobility index. (For a color version please see color plate on page 597.)

this displacement-like index in the following way

$$l(T, T') = \frac{1}{NK} \sum_{i=1}^{N} \sum_{t=1}^{K} |\log x_i(T_0 + t \cdot \Delta t) - \log x_i(T_0 + (t-1)\Delta t)| \quad (7.53)$$

where

$$K = \frac{T - T_0}{\Delta t}$$

and $x_i(t)$ is the quantity of agent i at t, such as rank, income or wealth; T_0 and T are the starting period and ending period of the sampling interval, respectively; Δt is the time of one period, which is one round in the model. This equation shows that the displacement-like index is turned into a speed-like index by dividing the aggregate of all the variations by the time. In fact, the new index is the average contribution of the process to the mobility in one period.

With this new index, we remeasure the mobility in the transfer model. To draw a comparison with Fig. 7.5, we show the relationship between the mobility and saving rate at the sampling interval of 3.5×10^4 rounds and the dependence of the measurement on the sampling interval in Fig. 7.6(a,b) respectively. These results indicate that the new index can reflect the mobility properly no matter how long the sampling interval may be. It is obvious that the sampling interval no longer has an effect, and the measurement results for different sampling intervals can be compared with each other.

Although the modification is only verified in a virtual economy, it has changed the traditional framework of mobility measurement and can be applied in practice. To make the measurement valid, the relevant data in all periods should be collected and the time of one period needs to be calibrated so that the loss of information is as small as possible. This calls for the reform in relevant data collection in the future.

7.5 Summary

Money-transfer models have been developed to explore the mechanism behind income and wealth distribution. Due to their simplicity and elegance, this kind of model can be applied to a wide range of economic issues. In this chapter, we present the applications of the money-transfer model on monetary circulation, money creation, and economic mobility. This study brings some new insights to some essential economic issues, such as the determinants of the velocity of money and the measurement indices of economic mobility. We believe that they still have the potential to contribute more to economics.

Acknowledgments

This research work is supported by the National Nature Science Foundation of China under Grant Nos. 73071072 and 73071073.

References

1 MANTEGNA, R. N., STANLEY, H. E., *An Introduction to Econophysics*, Cambridge University Press, Cambridge, **2000**

2 PARETO, V., *Cours d'Économie Politique*, Macmillan, London, **1897**

3 HAYES, B., *American Scientist* 90 (**2002**), p. 400

4 ISPOLATOV, S., KRAPIVSKY, P. L., REDNER, S., *The European Physical Journal B* 2 (**1998**), p. 267

5 SINHA, S., *Physica Scripta T* 106 (**2003**), p. 59

6 SINHA, S., in: *Econophysics of Wealth Distributions* (Ed.: A. Chatterjee, S. Yarlagadda, and B. K. Chakrabarti), Springer-Verlag, Italy, **2005**, p. 177

7 SLANINA, F., *Physical Review E* 69 (**2004**), 046102

8 DRĂGULESCU, A. A., YAKOVENKO, V. M., *The European Physical Journal B* 17 (**2000**), p. 723

9 CHAKRABORTI, A., CHAKRABARTI, B. K., *The European Physical Journal B* 17 (**2000**), p. 167

10 CHATTERJEE, A., CHAKRABARTI, B. K., MANNA, S. S., *Physica A* 335 (**2004**), p. 155

11 KAR GUPTA, A., *Physica A* 359 (**2006**), p. 634

12 PATRIARCA, M., CHAKRABORTI, A., KASKI, K., *Physical Review E* 70 (**2004**), 016104

13 PATRIARCA, M., CHAKRABORTI, A., KASKI, K., *Physica A* 340 (**2004**), p. 334

14 CHATTERJEE, A., CHAKRABARTI, B. K., STINCHCOMBE, R. B., *Physical Review E* 72 (**2005**), 026126

15 DING, N., WANG, Y., XU, J., XI, N., *International Journal of Modern Physics B* 18 (**2004**), p. 2725

16 FISCHER, R., BRAUN, D., *Physica A* 321 (**2003**), p. 605

17 BRAUN, D., *Physica A* 290 (**2001**), p. 491

18 FISCHER, R., BRAUN, D., *Physica A* 324 (**2003**), p. 266

19 WANG, Y., DING, N., ZHANG, L., *Physica A* 324 (**2003**), p. 665

20 DING, N., XI, N., WANG, Y., *The European Physical Journal B* 36 (**2003**), p. 149

21 WANG, Y., QIU, H., *Physica A* 353 (**2005**), p. 493

22 BARNETT, W. A., FISHER, D., SERLETIS, A., *Journal of Economic Literature* 30 (**1992**), p. 2086

23 FISHER, I., *The Purchasing Power of Money*, Macmillan, New York, **1911**

24 LAIDLER, D., *The Golden Age of the Quantity Theory*, Princeton University Press, Princeton, N.J., **1991**

25 BRIDEL, P., *Cambridge Monetary Thought*, St. Martin's Press, NewYork, **1987**

26 FRIEDMAN, M., *Studies in the Quantity Theory of Money*, University of Chicago Press, Chicago, **1956**

27 FRIEDMAN, M., SCHWARTZ, A. J., *A Monetary History of the United States, 1867-1960*, Princeton University Press for NBER, Princeton, **1963**

28 ROTHBARD, M. N., *Man, Economy, and State*, Ludwig Von Mises Inst., Auburn, Alabama, **2004**, p. 841

29 HUMPHREY, T. M., *Federal Reserve Bank of Richmond Economic Quarterly* 79(4) (**1993**), p. 1

30 WICKSELL, K., *Interest and Prices*, **1898**, Translated by R. F. Kahn, with an introduction by Bertil Ohlin, Macmillan, London, **1936**

31 VON MISES, L., *Human Action: A Treatise on Economics*, 4th Revised Edition, Fox & Wilkes, San Francisco, **1996**, p. 402

32 KEYNES, J. M., *The General Theory of Employment, Interest and Money*, the Royal Economic society, Macmillan Press, London **1973**

33 BRUNNER, K., *International Economic Review* January (**1961**), p. 79

34 BRUNNER, K., MELTZER, A. H., *Journal of Finance* May (**1964**), p. 240

35 GARFINKEL, M. R., THORNTON, D. L., *Federal Reserve Bank of St. Louis Review* 73 (**1991**), p. 47

36 GASKELL, D. R., *Introduction to the Thermodynamics of Materials*, 4th edn., Taylor & Francis, New York, **2003**

37 XI, N., DING, N., WANG, Y., *Physica A* 357 (**2005**), p. 543

38 PRAIS, S. J., *Journal of the Royal Statistical Society A, Part I* 118 (**1955**), p. 56

39 KUZNETS, S. S., *Modern Economic Growth: Rate, Structure and Spread*, Yale University, New Haven, **1966**, p. 203

40 JARVIS, S., JENKINS, S. P., *Economic Journal* 108 (**1998**), p. 1

41 KERM, P. V., *Economica* 71 (**2004**), p. 223

42 MARKANDYA, A., *Scottish Journal of Political Economy* 29 (**1982**), p. 75

43 MARKANDYA, A., *Economica* 51 (**1984**), p. 457

44 SHORROCKS, A. F., *Econometrica* 46 (**1978**), p. 1013

45 SOMMERS, P. S., CONLISK, J., *Journal of Mathematical Sociology* 6 (**1979**), p. 253

46 FIELDS, G. S., OK, E. A., *Journal of Economic Theory* 71 (**1996**), p. 349

47 FIELDS, G. S., OK, E. A., *Economica* 66 (**1999**), p. 455

48 FIELDS, G. S., OK, E. A., in: *Handbook of Inequality Measurement* (Ed.: J. Silber), Kluwer Academic Publishers, Dordrecht, **1999**, p. 557

8
Fluctuations in Foreign Exchange markets
Yukihiro Aiba and Naomichi Hatano

8.1
Introduction

In this chapter, we discuss fluctuations in foreign exchange markets. We all know that the price in a market with many traders fluctuates strongly. It is, however, quite difficult to explain the fluctuation. The purpose of this chapter is to give an explanation from the viewpoint of statistical mechanics. We here focus on the fluctuation in foreign exchange markets partly because the data are abundant and partly because there are so many traders involved that we expect it to be typical of all markets.

Classical economics argues that a foreign exchange rate should be determined by fundamental elements such as interest rates and GDPs. One might also think that foreign exchange rates are affected by news. Foreign exchange rates, however, strongly fluctuate every minute although the fundamental elements do not change from day to day and no "big news" is coming in every hour.

Financial engineering basically claims that the fluctuation is simply due to the fact that many traders are involved in the market. Each trader, although fed the same information on interest rates and GDPs, may judge differently; the different response of each trader can yield a strong collective fluctuation. It may hence be quite natural to assume the fluctuation to be a Gaussian. All important conclusions in financial engineering are based on this assumption. Note that this assumption is valid when the response of the traders is independent from each other.

It is, however, quite easy to show from the actual data that the fluctuation of foreign exchange markets is not a Gaussian; it typically exhibits a sharper central peak and fatter tails than does a Gaussian distribution. Detailed analyses of high-frequency data have revealed that the fluctuation of foreign exchange rates is quite well approximated by a power-law distribution. This observation tells us that the foreign-exchange market is a strongly correlated many-body system.

Econophysics and Sociophysics: Trends and Perspectives.
Bikas K. Chakrabarti, Anirban Chakraborti, Arnab Chatterjee (Eds.)
Copyright © 2006 WILEY-VCH Verlag GmbH & Co. KGaA, Weinheim
ISBN: 3-527-40670-0

It is interesting to liken the above development to the development of statistical physics. Classical economics takes the standpoint of Newtonian physics, which claimed that the dynamics of any system was deterministic. Then there is statistical physics, which treats a many-body system statistically. We, however, had to wait until the 1960s, when theories on the critical phenomena of strongly correlated systems started to blossom.

The analogue encourages us to apply methods of statistical physics to analyses of fluctuations in foreign exchange markets. Economic systems obviously consist of a large number of interacting units. Systems consisting of many interacting units such as strongly correlated many-body systems are of great interest in statistical physics. In such systems, exotic phenomena like phase transitions occur, but we cannot see them emerging if we look at each unit separately. Statistical physics treats the interacting units as a whole and thereby has successfully elucidated the mechanism of the phenomena. We can hence expect it to be possible that methods and concepts developed in the study of strongly correlated systems may yield new results in the analysis of financial data.

We first show in Section 8.2 that statistical physical modeling in fact reproduces empirical laws found in the data of foreign exchange rates. We further discuss correlations inherent in the data. Nowadays, correlations in financial fluctuations are becoming a considerable interest. For example, we can find that the fluctuations of foreign exchange rates have negative autocorrelation over a short time scale [1,2]. This fact means that foreign exchange rates tend to move in the opposite direction, against that of the previous movement. We here focus our attention on correlation among multiple markets. The correlation is analyzed and reproduced first macroscopically and then microscopically in Sections 8.3–8.5.

8.2
Modeling Financial Fluctuations with Concepts of Statistical Physics

There are many models aiming to reproduce price fluctuations in the financial market (for example, Refs [3–8]). Here we first review the Sznajd model of price formation proposed by Sznajd-Weron and Weron [8]. Next, we review Sato and Takayasu's dealer model [5]. These models reproduce the power-law behavior of the price fluctuation very well using quite different approaches.

8.2.1
Sznajd Model

The time evolution of the Sznajd model is as follows. Prepare an Ising chain consisting of N spins S_i with a periodic boundary condition. Regard the spins

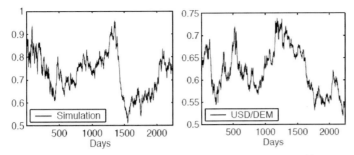

Fig. 8.1 A typical fluctuation of the simulated price process $m(t)$ on the left with the USD/DEM exchange rate on the right. In the simulation, eight simulation steps are regarded as one day. This graph is adapted from [8].

as traders in a financial market. The directions of the spins represent the traders' actions: if the ith spin is up, the ith trader wants to buy; if the ith spin is down, the ith trader wants to sell.

We now define the rule of opinion formation. Select a pair of consecutive traders S_i and S_{i+1} at random. If $S_i S_{i+1} = 1$, then make the directions of S_{i-1} and S_{i+2} the direction of $S_i (= S_{i+1})$. If $S_i S_{i+1} = -1$, then change the directions of S_{i-1} and S_{i+2} to ± 1 at random. Let the magnetization

$$m(t) = \frac{1}{N} \sum_{i=1}^{N} S_i(t) \tag{8.1}$$

be the price of the market, which is the normalized difference between demand and supply.

Obviously, the above model has two stable states, all spins up and all spins down. They are, however, not the states that we want to reproduce. In order to avoid this problem, let one of the N traders be a fundamentalist. The fundamentalist changes his/her direction depending on the price m. The fundamentalist buys, or takes the value 1 at time t with probability $|m(t)|$ if $m(t) < 0$ and sells, or takes the value -1 with probability $m(t)$ if $m(t) > 0$. This rule means that if the system comes close to the stable state "all up," the fundamentalist will place a sell order, take the value -1 almost certainly and hence the system will start to reverse. When the price $m(t)$ is close to the other stable state "all down," on the other hand, the fundamentalist will place a buy order, take the value 1, and the price will start to grow. Thus the ferromagnetic states are made unstable states.

The dynamics of the price $m(t)$ simulated by the Sznajd model is shown in Fig. 8.1 together with the USD/DEM exchange rate. The returns $r(t) \equiv m(t) - m(t-1)$ are compared to the USD/DEM exchange rate in the top panels of Fig. 8.2(a,b) and the normal probability plot of $r(t)$ are compared to the USD/DEM returns in Fig. 8.2(c,d). Sznajd-Weron et al. concluded that this

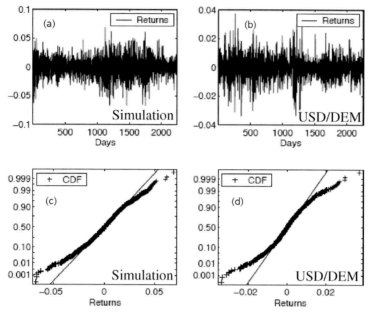

Fig. 8.2 The returns $r(t)$ of the simulated price process $m(t)$ and daily returns of the USD/DEM exchange rate during the last decade, respectively (a),(b). The normal probability plots of $r(t)$ and USD/DEM returns, respectively, clearly show fat tails of the price-return distributions (c),(d). This graph is adapted from [8].

simple model is a good first approximation of a number of real financial markets, because the results show good agreement with the actual market data.

This model is very simple at first sight; there is no connection between an Ising-like spin system and a financial market. Nonetheless, the model very well reproduces the statistics of the price change in foreign exchange markets including the fat-tail behavior of the fluctuations. It is interesting that two systems having no connection at first sight behave in a similar way.

8.2.2
Sato and Takayasu's Dealer Model

We next review Sato and Takayasu's dealer model (the ST model) briefly [5] (Fig. 8.3). The ST model also reproduces the power-law behavior of the price fluctuations. The basic assumption of the ST model is that dealers want to buy stocks or currencies at a lower price and sell them at a higher price. There are N dealers; the ith dealer has bidding prices to buy, $B_i(t)$, and to sell, $S_i(t)$, at time t. We assume that the difference between the buying price and the selling price is a constant $\Lambda \equiv S_i(t) - B_i(t) > 0$ for all i, in order to simplify the model.

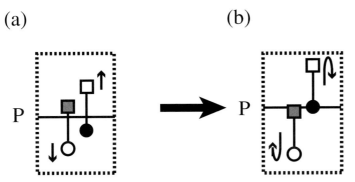

Fig. 8.3 A schematic image of a transaction of the ST model. Only the best bidders are illustrated, in order to simplify the image. The circles denote the dealers' bidding price to buy and the squares denote the dealers' bidding price to sell. The filled circles denote the best bidding price to buy, $\max\{B_i\}$, and the grey circles denote the best bidding price to sell, $\min\{B_i\} + \Lambda$. In (a), the condition (8.3) is not satisfied, and the dealers, following Eq. (8.5), change their relative positions by a_i. Note that the term $c\Delta P$ does not depend on i; hence it does not change the relative positions of dealers but changes the whole of the dealers' positions. In (b), the best bidders satisfy the condition (8.3). The price P is renewed according to Eq. (8.4), and the buyer and the seller, respectively, become a seller and a buyer according to Eq. (8.6).

The model assumes that a trade takes place between the dealer who proposes the maximum buying price and the one who proposes the minimum selling price. A transaction thus takes place when the condition

$$\max\{B_i(t)\} \geq \min\{S_i(t)\} \tag{8.2}$$

or

$$\max\{B_i(t)\} - \min\{B_i(t)\} \geq \Lambda \tag{8.3}$$

is satisfied, where $\max\{\cdot\}$ and $\min\{\cdot\}$, respectively, denote the maximum and the minimum values in the set of the dealers' buying thresholds $\{B_i(t)\}$. The market price $P(t)$ is defined by the mean value of $\max\{B_i\}$ and $\min\{S_i\}$ when the trade takes place. The price $P(t)$ maintains its previous value when the condition (8.3) is not satisfied:

$$P(t) = \begin{cases} (\max\{B_i(t)\} + \min\{S_i(t)\})/2 & \text{if the condition (8.3) is satisfied} \\ P(t-1) & \text{otherwise} \end{cases} \tag{8.4}$$

The dealers change their prices in unit time by the following deterministic rule:

$$B_i(t+1) = B_i(t) + a_i(t) + c\Delta P(t) \tag{8.5}$$

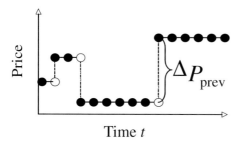

Fig. 8.4 A schematic image of the price difference ΔP. The price difference ΔP is defined as the difference between the present price and the price at the time when the previous trade was done, and it maintains its value until the next trade occurs.

where $a_i(t)$ denotes the ith dealer's characteristic movement in the price at time t, $\Delta P(t)$ is the difference between the price at time t and the price at the time when the previous trade was done (see Fig. 8.4), and $c(>0)$ is a constant which specifies the dealers' response to the market price change, and is common to all of the dealers in the market. The absolute value of a dealer's characteristic movement $a_i(t)$ is given by a uniform random number in the range $[0, \alpha)$ and is fixed throughout the time. The sign of a_i is positive when the ith dealer is a buyer and is negative when the dealer is a seller. The buyer (seller) dealers move their prices up (down) until the condition (8.3) is satisfied. Once the transaction takes place, the buyer of the transaction becomes a seller and the seller of the transaction becomes a buyer; in other words, the buyer dealer changes the sign of a_i from positive to negative and the seller dealer changes it from negative to positive:

$$a_i(t+1) = \begin{cases} -a_i(t) & \text{for the buyer and the seller} \\ a_i(t) & \text{for other dealers} \end{cases} \quad (8.6)$$

The initial values of $\{B_i\}$ are given by uniform random numbers in the range $(-\Lambda, \Lambda)$. We thus simulate this model specifying the following four parameters: the number of dealers, N; the spread between the buying price and the selling price, Λ; the dealers' response to the market price change, c; and the average of the dealers' characteristic movements in a unit time, α.

The ST model reproduces the power-law behavior of the price change well, when the dealers' response to the market change $c > c^*$, where c^* is a critical value of the power-law behavior (Figs 8.5 and 8.6). The critical point depends on the other parameters; e.g. $c^* \simeq 0.25$ for $N = 100$, $\Lambda = 1.0$ and $\alpha = 0.01$ [5]. For $c < c^*$, the probability distribution of the price change ΔP can be approximated by a hybrid distribution in the tails of $|\Delta P|$. For $c > c^*$ the probability distribution is approximated by a power law. As c increases, the distribution has longer tails and the exponent of the power-law distribution

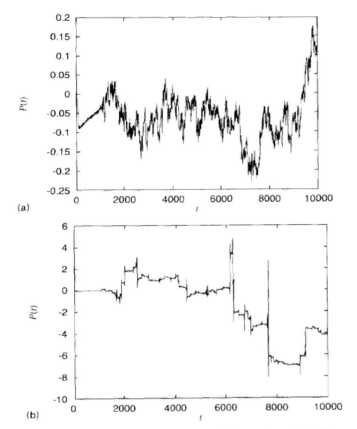

Fig. 8.5 Examples of temporal fluctuations of the market price $P(t)$, simulated by the ST model: (a) $c = 0.0$; (b) $c = 0.3$. This graph is adapted from [5].

is estimated to be smaller. For c greater than 0.45, the price fluctuation is very unstable and diverges quickly; that is, one cannot observe any steady distributions. The probability distribution looks similar to the distribution of price changes for real foreign exchange markets in the case $c \simeq 0.3$ except the tail parts for very large $|\Delta P|$.

8.3
Triangular Arbitrage as an Interaction among Foreign Exchange Rates

Analyzing the correlation in financial time series is a topic of considerable interest [9–25]. In the foreign exchange market, a correlation among the exchange rates can be generated by a triangular arbitrage transaction. The triangular arbitrage is a financial activity that takes advantage of the three ex-

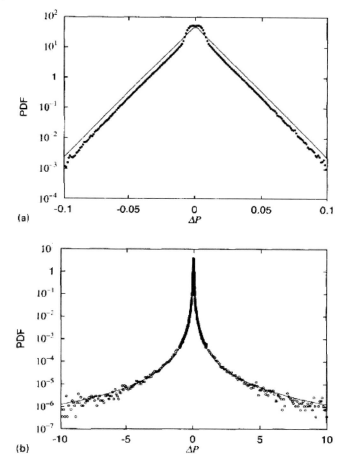

Fig. 8.6 Semi-logarithmic plots of the probability density functions of ΔP, simulated by the ST model: (a) $c = 0.0$; (b) $c = 0.3$. The dots represent results of the numerical simulation and the lines represent theoretical curves: (a) a hybrid of the Gaussian–Laplacian distribution, whose variance is 0.001; (b) a power-law distribution, $f(\Delta P) \propto (\Delta P)^{-2.5}$. This graph is adapted from [5].

change rates between three currencies [2, 26–32]. Suppose that we exchange one US dollar to some number of Japanese yen, exchange the amount of Japanese yen to some number of Euros, and finally exchange the Euros back to US dollars; then how many US dollars do we have? There is a possibility that we will have more than one US dollar. The triangular arbitrage transaction is the trade that uses this type of opportunity.

For those who might doubt the existence of triangular arbitrage transactions in actual markets, we can think of the following transaction. Suppose that a Japanese company earned some number of US dollars and wants to exchange

8.3 Triangular Arbitrage as an Interaction among Foreign Exchange Rates

it to Japanese yen. There can be instances when it is more advantageous to exchange the US dollar to Japanese yen *via Euros* than to exchange it directly to Japanese yen. The condition for such instances to happen is almost the same as the condition for the triangular arbitrage opportunities to appear.

Once there is a triangular arbitrage opportunity, many traders will make the transaction. This makes the product of the three exchange rates converge to a certain value [27], thereby eliminating the opportunity; the triangular arbitrage is thus a form of interaction among currencies. Triangular arbitrage opportunities nevertheless appear, because each rate r_x fluctuates strongly.

The purpose of this section is to show that there are, in fact, triangular arbitrage opportunities in foreign exchange markets and they generate an interaction among foreign exchange rates; the product of three foreign exchange rates has a narrow distribution with fat tails. We use real data of the yen-dollar rate, the yen-euro rate and the dollar-euro rate, taken from January 25 1999 to March 12 1999, except for weekends.

In order to quantify the triangular arbitrage opportunities, we define the quantity

$$\mu(t) = \prod_{x=1}^{3} r_x(t) \tag{8.7}$$

where $r_x(t)$ denotes each exchange rate at time t. We refer to this quantity as the rate product. There is a triangular arbitrage opportunity whenever the rate product is greater than unity.

To be more precise, there are two types of rate product. One is based on the arbitrage transaction in the direction of dollar to yen to Euros to dollar. The other is based on the transaction in the opposite direction of dollar to Euros to yen to dollar. Since these two values show similar behavior, we focus on the first type of $\mu(t)$ in the present and the next sections. Thus, we specifically define each exchange rate as

$$r_1(t) \equiv \frac{1}{\text{yen-dollar ask}(t)} \tag{8.8}$$

$$r_2(t) \equiv \frac{1}{\text{dollar-Euro ask}(t)} \tag{8.9}$$

$$r_3(t) \equiv \text{yen-Euro bid}(t) \tag{8.10}$$

Here, "bid" and "ask," respectively, represent the best bidding prices to buy and to sell in each market. We assume here that an arbitrager can transact instantly at the bid and the ask prices provided by information companies and hence we use the prices at the same time to calculate the rate product.

For later convenience, we also define the logarithm rate product ν as the logarithm of the product of the three rates:

$$\nu(t) = \ln \prod_{x=1}^{3} r_x(t) = \sum_{x=1}^{3} \ln r_x(t) \qquad (8.11)$$

There is a triangular arbitrage opportunity whenever this value is positive.

We can define another logarithm rate product ν', which has the opposite direction of the arbitrage transaction to ν, that is, from Japanese yen to Euros to US dollars back to Japanese yen:

$$\nu'(t) = \sum_{x=1}^{3} \ln r'_x(t) \qquad (8.12)$$

where

$$r'_1(t) \equiv \text{yen-dollar bid } (t) \qquad (8.13)$$
$$r'_2(t) \equiv \text{dollar-Euro bid } (t) \qquad (8.14)$$
$$r'_3(t) \equiv \frac{1}{\text{yen-Euro ask } (t)} \qquad (8.15)$$

This logarithm rate product ν' will appear in Section 8.5.

Figure 8.7(a)–(c) shows the actual changes in the three rates: the yen-Euro ask, the dollar-Euro ask and the yen-Euro bid. Figure 8.7(d) shows the behavior of the rate product $\mu(t)$. We can see that the rate product μ fluctuates around the average

$$m \equiv \langle \mu(t) \rangle \simeq 0.99998. \qquad (8.16)$$

(The average is less than unity because of the spread; the spread is the difference between the ask and the bid prices and is usually of the order of 0.05% of the prices.) The probability density function of the rate product μ (Fig. 8.8) has a sharp peak and fat tails while those of the three rates do not. It means that the fluctuations of the exchange rates have correlation that makes the rate product converge to the average m.

8.4
A Macroscopic Model of a Triangular Arbitrage Transaction

We here introduce a new model that takes account of the effect of the triangular arbitrage transaction in the form of an interaction among the three rates. Many models of price change have been introduced so far: for example, the Lévy-stable non-Gaussian model [9]; the truncated Lévy flight [33]; the

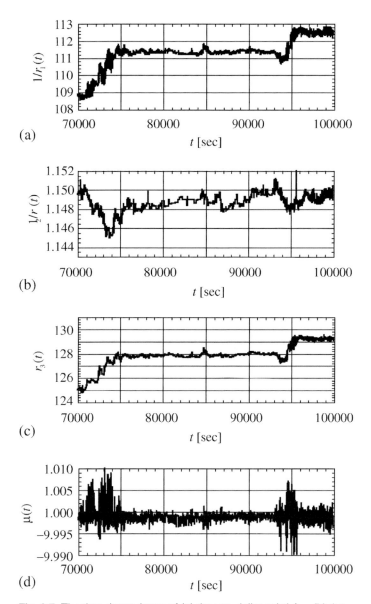

Fig. 8.7 The time dependence of (a) the yen-dollar ask $1/r_1$, (b) the dollar-Euro ask $1/r_2$, (c) the yen-Euro bid r_3 and (d) the rate product μ. The horizontal axis denotes the seconds from 00:00:00, January 12 1999.

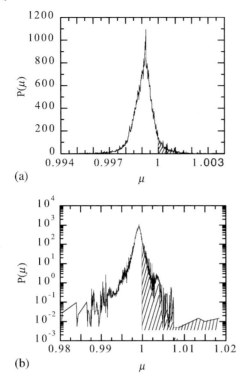

Fig. 8.8 The probability density function of the rate product μ. (b) is a semi-logarithmic plot of (a). The shaded area represents triangular arbitrage opportunities. The data were taken from January 25 1999 to March 12 1999.

ARCH/GARCH processes [34, 35]. However, they discuss only the change in one price. They do not consider an interaction among multiple prices. As we discussed in Section 8.3, however, the triangular arbitrage opportunity exists in the market and is presumed to affect price fluctuations in the way that the rate product tends to converge to a certain value. The purpose of this section is to introduce a macroscopic model that describes the interaction among multiple markets through traiangular arbitrage transactions [2, 28–31].

8.4.1
Basic Time Evolution

The basic equation of our model is a time-evolution equation of the logarithm of each rate [28]:

$$\ln r_x(t+T) = \ln r_x(t) + \eta_x(t) + g(\nu(t)) \qquad (x = 1, 2, 3) \qquad (8.17)$$

where ν is the logarithm rate product (8.11), and T is a time step which controls the time scale of the model; we later use the actual financial data every T[sec].

Just as μ fluctuates around $m = \langle \mu \rangle \simeq 0.99998$, the logarithm rate product ν fluctuates around

$$\epsilon \equiv \langle \ln \mu \rangle \simeq -0.00091 \tag{8.18}$$

(Fig. 8.9(a)). In this model, we focus on the logarithm of the rate-change ratio $\ln(r_x(t+T)/r_x(t))$, because the relative change is presumably more essential than the absolute change. We assumed in Eq. (8.17) that the change of the logarithm of each rate is given by an independent fluctuation $\eta_x(t)$ and an attractive interaction $g(\nu)$. The triangular arbitrage is presumed to make the logarithm rate product ν converge to the average $\epsilon = \langle \nu \rangle$; thus, the interaction function $g(\nu)$ should be negative for ν greater than ϵ and positive for ν less than ϵ:

$$g(\nu) \begin{cases} < 0 & \text{for } \nu > \epsilon \\ > 0 & \text{for } \nu < \epsilon \end{cases} \tag{8.19}$$

As a linear approximation, we define $g(\nu)$ as

$$g(\nu) \equiv -k(\nu - \epsilon) \tag{8.20}$$

where k is a positive constant which specifies the interaction strength.

The time-evolution equation of ν is given by summing Eq. (8.17) over all x:

$$\nu(t+T) - \epsilon = (1 - 3k)(\nu(t) - \epsilon) + F(t) \tag{8.21}$$

where

$$F(t) \equiv \sum_{x=1}^{3} \eta_x(t) \tag{8.22}$$

This is our basic time-evolution equation of the logarithm rate product.

From a physical viewpoint, we can regard the model equation (8.17) as a one-dimensional random walk of three particles with a restoring force, by interpreting $\ln r_x$ as the position of each particle (Fig. 8.10). The logarithm rate product ν is the summation of $\ln r_x$, hence it is proportional to the center of gravity of the three particles. The restoring force $g(\nu)$ makes the center of gravity converge to a certain point $\epsilon = \langle \nu \rangle$. The form of the restoring force (8.20) is the same as that of the harmonic oscillator. Hence we can regard the coefficient k as a spring constant.

8.4.2
Estimation of Parameters

The spring constant k is related to the auto-correlation function of ν as follows:

$$1 - 3k = c(T) \equiv \frac{\langle \nu(t+T)\nu(t) \rangle - \langle \nu(t) \rangle^2}{\langle \nu^2(t) \rangle - \langle \nu(t) \rangle^2} \tag{8.23}$$

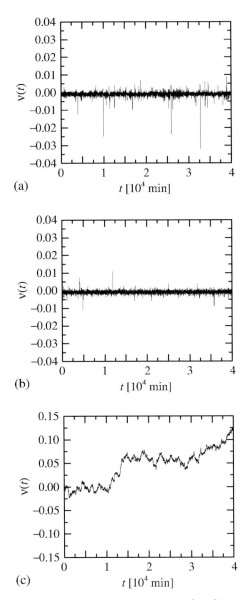

Fig. 8.9 The time dependence of $\nu(t[\min])$ of (a) the real data, (b) the simulation data with the interaction and (c) without the interaction. In (b), ν fluctuates around ϵ like the real data.

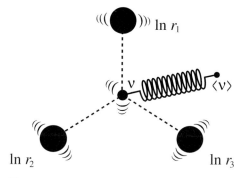

Fig. 8.10 A schematic image of the model. The three random walkers with the restoring force working on the center of gravity.

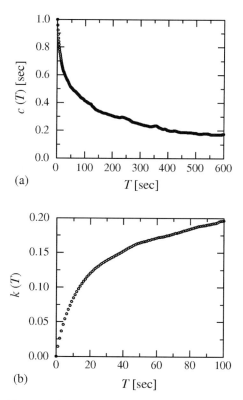

Fig. 8.11 (a) The auto-correlation function of v, $c(T)$. (b) The spring constant k as a function of the time step T. The spring constant k increases with the time step T.

Using Eq. (8.23), we can estimate k from the real data series as a function of the time step T. The auto-correlation function $c(T)$ is shown in Fig. 8.11(a). The estimate of $k(T)$ is shown in Fig. 8.11(b). The spring constant k increases with the time step T. This may be regarded as "renormalization" of a coupling constant. In the present section, we fix the time step at $T = 60[\text{sec}]$ and hence use

$$k(1[\text{min}]) = 0.17 \pm 0.02 \tag{8.24}$$

for our simulation. We will come back to this point in Section 8.5.

On the other hand, the fluctuation of a foreign exchange rate is known to be a fat-tail noise [1,36]. Here we take $\eta_x(t)$ as the truncated Lévy process [33,37]

$$P_{\text{TLF}}(\eta; \alpha, \gamma, l) q P_{\text{L}}(\eta; \alpha, \gamma) \Theta(l - |\eta|) \tag{8.25}$$

where q is the normalization constant, $\Theta(x)$ represents the step function and $P_{\text{L}}(x; \alpha, \gamma)$ is the symmetric Lévy distribution of index α and scale factor γ:

$$P_{\text{L}}(x; \alpha, \gamma) = \frac{1}{\pi} \int_0^\infty e^{-\gamma |k|^\alpha} \cos(kx) dk \quad 0 < \alpha < 2 \tag{8.26}$$

We determine the parameters α, γ and l by using the following relations for $1 < \alpha < 2$ [36,38]:

$$c_2 = \frac{\alpha(\alpha-1)\gamma}{|\cos(\pi\alpha/2)|} l^{2-\alpha} \tag{8.27}$$

$$\kappa = \frac{(3-\alpha)(2-\alpha)|\cos(\pi\alpha/2)|}{\alpha(\alpha-1)\gamma} l^\alpha \tag{8.28}$$

where c_n denotes the nth cumulant and κ is the kurtosis $\kappa = c_4/c_2^2$. The estimates are shown in Table 8.1. The generated noises with the estimated

Tab. 8.1 The estimates of the parameters.

Rate	α	γ	l
r_1 (1/yen-dollar ask)	1.8	7.61×10^{-7}	1.38×10^{-2}
r_2 (1/dollar-Euro ask)	1.7	4.06×10^{-7}	3.81×10^{-2}
r_3 (yen-Euro bid)	1.8	6.97×10^{-7}	7.58×10^{-2}

parameters are compared to the actual data in Fig. 8.12.

We simulated the time evolution equation (8.21) with the parameters given in Eqs (8.18), (8.24) and Table 8.1. The probability density function of the results (Fig. 8.9(b)) is compared to that of the real data (Fig. 8.9(a)) with $T = 1[\text{min}]$ in Fig. 8.13. The fluctuation of the simulation data is consistent with that of the real data. In particular, we see good agreement around $\nu \simeq \epsilon$ as a result of the linear approximation of the interaction function. Figure 8.9(c) shows $\nu(t)$ of the simulation without the interaction, i.e. $k = 0$. The quantity ν fluctuates freely, which is inconsistent with the real data.

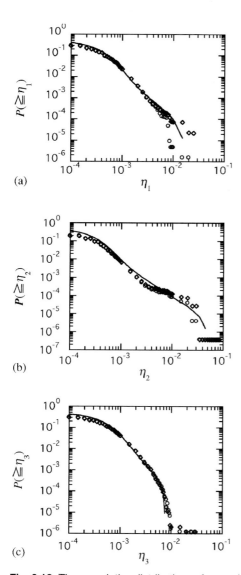

Fig. 8.12 The cumulative distributions of one-minute changes $|\ln r_x(t+1[\min]) - \ln r_x(t)|$ (○ represents upward movements and ◇ represents downward movements) and the generated noise η_x (—): (a) the yen-dollar ask and η_1, (b) the dollar-Euro ask and η_2, and (c) yen-Euro bid and η_3. The real data were taken from January 25 1999 to March 12 1999.

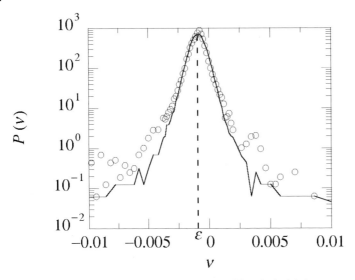

Fig. 8.13 The probability density function of ν. The circle (o) denotes the real data and the solid line denotes our simulation data with the interaction. The simulation data fit the real data well.

8.5
A Microscopic Model of Triangular Arbitrage Transaction

In Section 8.3, we pointed out the existence of the triangular arbitrage opportunity in the foreign exchange market and showed that the triangular arbitrage transaction makes the product of the three foreign exchange rates converge to its average, thereby generating an interaction among the rates. In order to study the effects of the triangular arbitrage on the fluctuations of the exchange rates, in the previous section, we introduced the stochastic model (8.17) describing the time evolution of the exchange rates with an interaction. The model successfully describes the fluctuation of the data of the real market. The model is phenomenological; i.e. it treats the fluctuations of the rates as fluctuating particles and the interaction among the rates as a spring. We refer to this model as the "macroscopic model" hereafter.

The purpose of this section is to understand *microscopically* the effects of the triangular arbitrage on the foreign exchange market. For this purpose, we introduce a new model which focuses on the dynamics of each dealer in the markets; we refer to the new model as the "microscopic model" hereafter. We then show the relation between the macroscopic model and the microscopic model through an interaction strength which is regarded as a spring constant.

In order to describe each foreign exchange market microscopically, we use Sato and Takayasu's dealer model [5] (the ST model; see Section 8.2.2), which reproduces the power-law behavior of price changes in a single market well.

Although we focus on the interactions among three currencies, two of the three markets can be regarded as one effective market [30]; i.e. the yen-Euro rate and the Euro-dollar rate are combined to an effective yen-dollar rate. (This means that each dealer is put in the situation of the Japanese company that we mentioned in Section 8.3.) In terms of the macroscopic model, we can redefine a variable r_2 as the product of r_2 and r_3. Then the renormalized variable r_2 follows a similar time-evolution equation. We therefore describe triangular arbitrage opportunities with only two interacting ST models, in order to simplify the situation.

8.5.1
Microscopic Model of Triangular Arbitrage: Two Interacting ST Models

We describe our microscopic model as a set of the ST models [32]. As is noted above, we prepare two systems of the ST model, the market X and the market Y. Note that we can reproduce the markets, interaction by preparing all of the three markets; see [32].

The dealers in the markets X and Y change their bidding prices according to the ST model as follows:

$$B_{i,X}(t+1) = B_{i,X}(t) + a_{i,X}(t) + c\Delta P_X(t) \quad \text{and} \tag{8.29}$$
$$B_{i,Y}(t+1) = B_{i,Y}(t) + a_{i,Y}(t) + c\Delta P_Y(t) \tag{8.30}$$

where X and Y denote the markets X and Y, respectively. An intra-market transaction takes place when the condition

$$\max\{B_{i,x}(t)\} \geq \min\{S_{i,x}\}, \quad x = X \text{ or } Y \tag{8.31}$$

is satisfied. We assume that Λ is common to the two markets. The price $P_x(t)$ is renewed analogously to the ST model:

$$P_x(t) = \begin{cases} (\max\{B_{i,x}(t)\} + \min\{S_{i,x}(t)\})/2 & \text{if the condition (8.31) is satisfied} \\ P_x(t-1) & \text{otherwise} \end{cases}$$
$$\tag{8.32}$$

where $x = X$ and Y.

We here add a new *inter*-market transaction rule which makes the systems interact. When a dealer cannot find another dealer to transact within his/her market, he/she tries to find one in the other market. An arbitrage transaction can take place when one of the conditions

$$v_X \equiv \max\{B_{i,X}(t)\} - (\min\{B_{i,Y}(t)\} + \Lambda) \geq 0 \tag{8.33}$$
$$v_Y \equiv \max\{B_{i,Y}(t)\} - (\min\{B_{i,X}(t)\} + \Lambda) \geq 0 \tag{8.34}$$

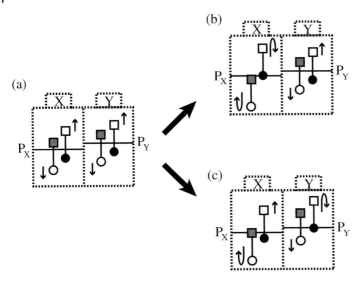

Fig. 8.14 A schematic image of the transactions. Only the best bidders in the markets are illustrated, in order to simplify the image. The circles and the squares denote the dealers' bidding price to buy and to sell. The filled circles denote the best bidding prices to buy in the markets, $\max\{B_{i,X}\}$ and $\max\{B_{i,Y}\}$, and the gray squares denote the best bidding prices to sell in the markets, $\min\{B_{i,X}\} + \Lambda$ and $\min\{B_{i,Y}\} + \Lambda$. In case (a), none of the conditions (8.31), (8.33) and (8.34) are satisfied. The buyers move their prices up, and the sellers move their prices down. In case (b), the dealers in the market X satisfy the condition (8.31); hence the intra-market transaction takes place. The price in the market X, P_X, is renewed, and the buyer and the seller of the transaction become a seller and a buyer, respectively. In case (c), the seller in the market X and the buyer in the market Y satisfy the condition (8.34); hence the arbitrage transaction takes place. The price P_X in the market X becomes $\min\{B_{i,X}\} + \Lambda$, and the price P_Y in the market Y becomes $\max\{B_{i,Y}\}$. The buyer and the seller of the transaction become a seller and a buyer, respectively. The arbitrage transaction thus makes the interaction between the markets X and Y.

is satisfied (see Fig. 8.14). When the conditions (8.31) and (8.33) or (8.34) are both satisfied simultaneously, the condition (8.31) precedes.

Note that the arbitrage conditions $\nu_X \geq 0$ and $\nu_Y \geq 0$ in the microscopic model correspond to the arbitrage condition $\nu \geq 0$ in the actual market, where ν is defined by Eq. (8.11). We assume that the dealers' bidding prices $\{B_i\}$ and $\{S_i\}$ correspond to the logarithm of the exchange rate, $\ln r_i$. Therefore, $\max\{B_{i,X}\}$ may be equivalent to $-\ln(\text{yen-dollar ask})$ while $\min\{S_{i,Y}\}$ may be equivalent to $\ln(\text{dollar-Euro ask}) - \ln(\text{yen-Euro bid})$, and hence ν_X may be equivalent to ν. More precisely, the direction of the arbitrage transaction determines which of the quantities, ν_X or ν_Y, corresponds to the logarithm rate product ν. There are two directions of the triangular arbitrage transaction. The definition (8.11) specifically has the direction of Japanese yen to US dollar to Euro to Japanese yen. As is mentioned in Section 8.3, we can define another logarithm rate product v' in the actual market which has the opposite direction to v, Japanese yen to Euro to US dollar to Japanese yen. Hence, if the logarithm rate product v in the actual market corresponds to v_X in Eq. (8.33), v' corresponds to $-v_Y$ in Eq. (8.34).

The procedures for the simulation of the microscopic model are as follows (Fig. 8.14):

1. Prepare two systems of the ST model, the market X and the market Y, as described in Section 8.2.2. The parameters are common to the two systems.

2. Check the condition (8.31) for each market. If the condition (8.31) is satisfied, renew the price of the market by Eq. (8.32), skip step 3 and proceed to step 4. Otherwise, proceed to step 3.

3. Check the arbitrage conditions (8.33) and (8.34). If the condition (8.33) is satisfied, renew the prices $P_X(t)$ and $P_Y(t)$ to $\max\{B_{i,X}(t)\}$ and $\min\{B_{i,Y}(t)\} + \Lambda$, respectively. If the condition (8.34) is satisfied, renew the prices $P_X(t)$ and $P_Y(t)$ to $\min\{B_{i,X}(t)\} + \Lambda$ and $\max\{B_{i,Y}(t)\}$, respectively. If both of the conditions in Eqs (8.33) and (8.34) are satisfied, choose one of them with the probability of 50% and carry out the arbitrage transaction as described just above. If the arbitrage transaction takes place, proceed to step 4; otherwise skip step 4 and proceed to step 5.

4. Calculate the difference between the new prices and the previous prices, $\Delta P_X(t) = P_X(t) - P_X(t-1)$ and $\Delta P_Y(t) = P_Y(t) - P_Y(t-1)$.

5. If any of the conditions (8.31), (8.33) and (8.34) are not satisfied, maintain the previous prices, $P_X(t) = P_X(t-1)$ and $P_Y(t) = P_Y(t-1)$, as well as the previous price differences, $\Delta P_X(t) = \Delta P_X(t-1)$ and $\Delta P_Y(t) = \Delta P_Y(t-1)$.

6. Change the dealers' bidding prices following Eqs (8.29) and (8.30).

7. Change the buyer and the seller of the transaction to a seller and a buyer, respectively. In other words, change the signs of $a_{i,X}$ and $a_{i,Y}$ of the dealers who transacted.

8. Repeat the steps from 2 to 7.

The quantities v_X and v_Y are shown in Fig. 8.15. (The parameters are common to the two markets X and Y: $N = 100$, $\alpha = 0.01$ and $\Lambda = 1.0$, which follows Ref. [5].) In Fig. 8.15(b) for $c = 0.3$, the fat-tail behavior of the price difference v_X is consistent with the actual data as well as with the macroscopic model. Furthermore, v_X reproduces the skewness of the actual data, which

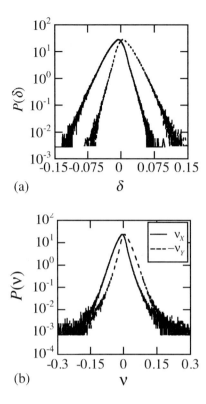

Fig. 8.15 The distributions of v_X and $-v_Y$. The parameters are fixed to $N = 100$, $\alpha = 0.01$, $\Lambda = 1.0$ and (a) $c = 0.0$, (b) $c = 0.3$, and are common to the market X and the market Y. The solid line denotes v_X and the dashed line denotes v_Y in each graph.

cannot be reproduced by the macroscopic model (Fig. 8.16). Note that the skewness of v_Y is consistent with the behavior of $-v'$.

8.5.2
The Microscopic Parameters and the Macroscopic Spring Constant

In this section, we discuss the relation between the macroscopic model and the microscopic model through the interaction strength, or the spring constant k.

In the microscopic model, we define the spring constant k_{micro}, which corresponds to the spring constant k of the macroscopic model, as follows:

$$k_{\text{micro}} \equiv \frac{1}{2}\left(1 - \frac{\langle v_X(t+1)v_X(t)\rangle - \langle v_X(t)\rangle^2}{\langle v_X(t)^2\rangle - \langle v_X(t)\rangle^2}\right) \tag{8.35}$$

Figure 8.17 shows the estimate (8.35) as a function of several parameters.

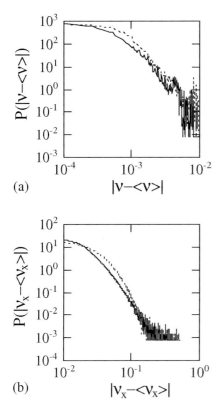

Fig. 8.16 The probability distribution of the difference between the logarithm rate product and its average for (a) the actual data, $|v - \langle v \rangle|$ and for (b) the data of the microscopic model, $|v_x - \langle v_x \rangle|$ for $c = 0.3$. The solid lines represent the part in which the difference is positive and the dotted lines represent the part in which the difference is negative, in both graphs. We can see that the probability distribution of the logarithm rate product v has a skewness around its average, and the microscopic model qualitatively reproduces it well.

Remember that, in the macroscopic model, the spring constant k depends on the time step T (see Fig. 8.11(b)). The spring constant of the microscopic model k_{micro} also depends on a time scale as follows. The time scale of the ST model may be given by the following combination of parameters [5]:

$$\langle n \rangle \simeq \frac{3\Lambda}{N\alpha} \tag{8.36}$$

where n denotes the interval between two consecutive trades. Hence, the inverse of Eq. (8.36),

$$f \equiv \frac{1}{\langle n \rangle} \simeq \frac{N\alpha}{3\Lambda} \tag{8.37}$$

is the frequency of the trades.

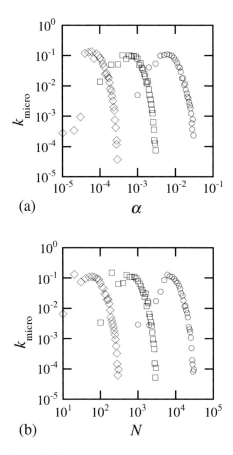

Fig. 8.17 The spring constant k_{micro} as a function of parameters. The panel (a) shows the dependence on N. The other parameters are fixed to $\Lambda = 1.0$ and $\alpha = 0.0001, 0.001$ and 0.01 for the circles, the squares and the diamonds, respectively, and $c = 0.3$ for the all plots. The panel (b) shows the dependence on α. The other parameters are fixed to $\Lambda = 1.0$ and $N = 100, 1000$ and $10\,000$ for the circles, the squares and the diamonds, respectively, and $c = 0.3$ for the all plots.

Although there are four parameters N, α, Λ and c, we change only three parameters N, α, and c and set $\Lambda = 1$, because only the ratios N/Λ and α/Λ are relevant in this system. The ratio N/Λ controls the density of the dealers and α/Λ controls the speed of the dealers' motion on average.

We plot the spring constant k_{micro} as a function of the trade frequency $f \equiv N\alpha/3\Lambda$ in Fig. 8.18. The plots show that the spring constant $k_{\text{micro}}(N, \alpha, \Lambda)$ can be scaled well by the trade frequency f.

In order to determine a reasonable range of the parameters, let us consider the situation in Fig. 8.19, where the arbitrage transaction is about to take place.

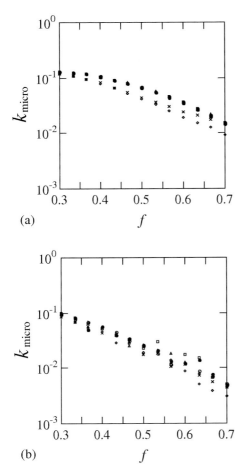

Fig. 8.18 The scaling plot of the spring constant k_{micro} as a function of the trade frequency $f = N\alpha/3\Lambda$. The vertical axes are displayed in the logarithmic scale. The dealers' response to the price change, c, is fixed to 0.0 in (a) and 0.3 in (b). We fix $\alpha = 0.0001, 0.001$ and 0.01 and change N (open circles, squares, and diamonds, respectively) and $N = 100, 1000$ and $10\,000$ and change α (crosses, filled circles and triangles, respectively), while Λ is fixed to 1. Note that all points collapse onto a single curve. The spring constant k_{micro} is scaled by f, and decays exponentially in both of the plots (a) and (b).

At the moment, the positions of the second best bidders (hexagons) in the markets X and Y are, on average, Λ/N away from the prices transacted, P_X and P_Y. In the next step, the second best bidders in the markets X and Y will move by $\alpha/2$ on average toward to the prices P_X and P_Y, respectively. The next transaction will be carried out probably by the second best bidders. For $\alpha/2 > \Lambda/N$, the prices of the transactions may move away from each other. The arbitrage transaction cannot bind the two prices of the markets X and Y

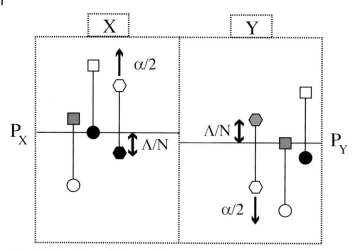

Fig. 8.19 A schematic image of the second best bidders' motion. The circles and the squares denote the dealers' bidding price to buy and to sell. The filled circles and the grey squares represent the best bidding prices to buy and sell, respectively. The hexagons denote the second best bidding prices to buy (the filled one) and to sell (the grey one).

enough and the two prices P_X and P_Y fluctuate rather freely. It is not a realistic situation. Therefore, the condition

$$f = \frac{N\alpha}{3\Lambda} \leq \frac{2}{3} \tag{8.38}$$

should be satisfied for the real market to be reproduced. On the other side, the simulation data have too large errors in the region $f < 1/3$ because the transaction rarely occurs. We hence use the data in the region $1/3 \leq f \leq 2/3$ hereafter.

The spring constant k_{micro} decays exponentially in the range $1/3 \leq f \leq 2/3$ in both of the plots (a) and (b) of Fig. 8.18, having different slopes. Hence we assume that the spring constant decays as

$$k_{\text{micro}} \propto e^{-f/f_0(c)} \tag{8.39}$$

where $f_0(c)$ denotes the characteristic frequency dependent on c. The estimates of the characteristic frequency $f_0(c)$ are shown in Fig. 8.20 as a function of c. The characteristic frequency $f_0(c)$, thus estimated, decays linearly with c. The reason why $f_0(c)$ behaves in this way is an open question.

In Fig. 8.21, we plot the same data as in Fig. 8.11(b), but by making the horizontal axis the trade frequency f_{real}. In order to compare it with Fig. 8.18 quantitatively, we used the time scale $T_{\text{real}} = 7[\text{sec}]$; the interval between two consecutive trades in the actual foreign exchange market is, on average, known [39] to be about 7[sec]. The spring constant in the actual market k

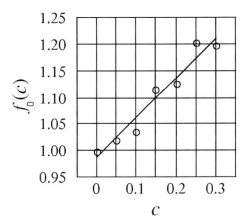

Fig. 8.20 The dependence of $f_0(c)$ on c, estimated by fitting the data in Fig. 8.18 as well as the same plots for different values of c.

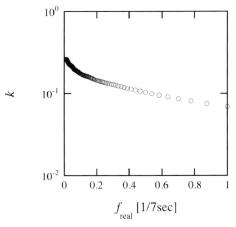

Fig. 8.21 We plotted the same data as in Fig. 8.11(b), but the horizontal axis here is the trade frequency scaled by the realistic time scale of the trades, $T_{\text{real}} = 7[\text{sec}]$.

is of the same magnitude as k_{micro}. It decays exponentially with the trade frequency f_{real}, which is also consistent with that of the microscopic model shown in Fig. 8.18. The real characteristic frequency in Fig. 8.21, however, is quite different from that of the microscopic model plotted in Fig. 8.18. This is also an open question.

8.6
Summary

In Section 8.3 we showed that triangular arbitrage opportunities exist in the foreign exchange market. The probability density function of the rate product μ has a sharp peak and fat tails. This means that the fluctuations of the exchange rates have a correlation that makes the rate product converge to its average $\langle \mu \rangle \simeq 0.99998$. If the rate product μ is greater than unity, the trader can make a profit through the above transaction. This is the triangular arbitrage transaction. Once there is a triangular arbitrage opportunity, many traders will make the transaction. This makes μ converge to a value less than unity, thereby eliminating the opportunity. Triangular arbitrage opportunities nevertheless appear, because each rate r_x fluctuates strongly.

In Section 8.4, we first introduced a model including the interaction caused by the triangular arbitrage transaction. We showed that the interaction is the reason for the sharp peak and the fat tails of the distribution of the logarithm rate product ν.

In Section 8.5, we introduced the microscopic model, which consists of two systems of the ST model. The microscopic model reproduced the actual behavior of the logarithm rate product ν well. The microscopic model can describe more detail than the macroscopic model, in particular, the skewness of the distribution of the logarithm rate product ν. We finally explored the relation between the spring constant of the macroscopic model and the parameters in the microscopic model. The spring constant of the microscopic model k_{micro} can be scaled by the trade frequency f, and it decays exponentially with f, which is consistent with the spring constant k of the actual market.

References

1. TAKAYASU, H., TAKAYASU, M., OKAZAKI, M. P., MARUMO, K., SHIMIZU, T., in *Paradigms of Complexity: Fractals and Structures in the Sciences*, (Ed. M. M. Novak) World Scientific, Singapore, **2000**, pp. 243–258
2. AIBA, Y., HATANO, N., TAKAYASU, H., MARUMO, K., SHIMIZU, T., *Physica A* 324 (**2003**), pp. 253–257
3. TAKAYASU, H., MIURA, H., HIRABAYASHI, T., HAMADA, K., *Physica A* 184 (**1992**), pp. 127–134
4. HIRABAYASHI, T., TAKAYASU, H., MIURA, H., HAMADA, K., *Fractals* 1 (**1993**) pp. 29–40
5. SATO, A.-H., TAKAYASU, H., *Physica A* 250 (**1998**), pp. 231–252
6. CHOWDHURY, D., STAUFFER, D., *Eur. Phys. J. B* 8 (**1999**), pp. 477–482
7. CONT, R., BOUCHAUD, J.-P., *Macroeconomic Dynamics* 4 (**2000**), pp. 170–196
8. SZNAJD-WERON, K., WERON, R., *Int. J. Mod. Phys. C* 13 (**2002**), pp. 115–123
9. MANDELBROT, B. B., *J. Business* 36 (**1963**) pp. 394–419
10. FAMA, E. F. *J. Finance* 25 (**1970**), pp. 383–423
11. DING, Z., GRANGER, C. W. J., ENGLE, R. F., *J. Empirical Finance* 1 (**1993**), pp. 83–106

12 DACOROGNA, M. M., MULLER, U. A., NAGLER, R. J., OLSEN, R. B., PICTET, O. V., *J. Int. Money Finance* 12 (**1993**), pp. 413–438

13 LIU, Y.-H., CIZEAU, P., MEYER, M., PENG, C.-K., STANLEY, H. E., *Physica A* 245 (**1997**), pp. 437–440

14 MANTEGNA, R. N., *Eur. Phy. J. B.* 11 (**1999**), pp. 193–197

15 PLEROU, V., GOPIKRISHNAN, P., ROSENOW, B., AMARAL, L. A. N., STANLEY, H. E., *Phys. Rev. Lett.* 83 (**1999**), pp. 1471–1474

16 PLEROU, V., GOPIKRISHNAN, P., ROSENOW, B., AMARAL, L. A. N., STANLEY, H. E., *Physica A* 287 (**2000**), pp. 374–382

17 KULLMANN, L., KERTESZ, J., MANTEGNA, R. N., *Physica A* 287 (**2000**), pp. 412–419

18 MANDELBROT, B. B., *Quant. Fin.* 1 (**2001**), pp. 113–123

19 MANDELBROT, B. B., *Quant. Fin.* 1 (**2001**), pp. 124–130

20 KULLMANN, L., KERTESZ, J., KASKI, K., *Phys. Rev. E* 66 (**2002**), p. 026125

21 ONNELA, J.-P., CHAKRABORTI, A., KASKI, K., *Phys. Rev. E* 68 (**2003**), 056110

22 MIZUNO, T., KURIHARA, S., TAKAYASU, M., TAKAYASU, H., in H. Takayasu (Ed.) *The Application of Econophysics, Proceedings of the Second Nikkei Symposium*, Springer-Verlag Tokyo, **2004**, pp. 24–29

23 TASTAN, H., *Physica A* 360 (**2006**), pp. 445–458

24 TOTH, B., KERTESZ, J., *Physica A* 360 (**2006**), pp. 505–515

25 JUNG, W.-S., CHAE, S., YANG, J.-S., MOON, H.-T., *Physica A* 361 (**2006**), pp. 263–271

26 MAVIRIDES, M., *Triangular Arbitrage in the Foreign Exchange Market: Inefficiencies, Technology, and Investment Opportunities*, Quorum Books, New York, **1992**

27 MOOSA, I., *Quantitative Finance* 1 (**2001**), pp. 387–390

28 AIBA, Y., HATANO, N., TAKAYASU, H., MARUMO, K., SHIMIZU, T., *Physica A* 310 (**2002**), pp. 467–479

29 AIBA, Y., HATANO, N., TAKAYASU, H., MARUMO, K., SHIMIZU, T., in H. Takayasu (Ed.) *The Application of Econophysics, Proceedings of the Second Nikkei Symposium*, Springer-Verlag Tokyo, **2004**, pp. 18–23

30 AIBA, Y., HATANO, N., *Physica A* 344 (**2004**), pp. 174–177

31 AIBA, Y., HATANO, N. in, A. Chatterjee, B. K. Chakrabarti (Eds.) *Econophysics of Stock Markets and Minority Games*, Proc. Econophys-Kolkata II, Springer-Verlag, Italia, **2006**, to appear

32 AIBA, Y., HATANO, N., *Physica A*, submitted

33 MANTEGNA, R. N., STANLEY, H. E., *An Introduction to Econophysics: Correlations and Complexity in Finance*, Cambridge University Press, Cambridge, **1999**, pp. 64–67.

34 ENGLE, R. F., *Econometrica* 50 (**1982**), pp. 987–1002

35 BOLLERSLEV, T., *J. Econometrics* 31 (**1986**), pp. 307–327

36 BOUCHAUD, J.-P., POTTERS, M., *Theory of Financial Risks: From Statistical Physics to Risk Management*, Cambridge University Press, Cambridge, **2000**, pp. 34, 56-63

37 MANTEGNA, R. N., *Phys. Rev. E* 49 (**1994**), pp. 4677–4683

38 VOIT, J., *The Statistical Mechanics of Financial Markets*, Springer-Verlag, Berlin, **2001**, pp. 101–103

39 OHIRA, T., SAZUKA, N., MARUMO, K., SHIMIZU, T., TAKAYASU, M., TAKAYASU, H., *Physica A* 308 (**2002**), pp. 368–374

9
Econophysics of Stock and Foreign Currency Exchange Markets
Marcel Ausloos

Econophysics is a science in its infancy, born about ten years ago at this time of writing, at the cross-roads of physics, mathematics, computing and of course economics and finance. It also covers human sciences, because all economics is ultimately driven by human decision. Because of this human factor, econophysics has no hope of achieving the status of an exact science, but it is interesting to discover what can be achieved, to find the potential limits and then to try to extend these limits.

Numerous works have inspired physicists and directed them towards the study of financial markets. We could start with Bachelier [1] who introduced what will perhaps remain the simplest and most successful model for price variations but, for the moment, it suffices to mention that it relies on Gaussian statistics. However, Bachelier could have learned from Pareto [2] that power-laws are ubiquitous in nature. In fact, large price variations at high frequency seems to be described by a crossover between two power laws; time correlations in the variance also decay as a power law and so do numerous other quantities related to finance. Mandelbrot [3] and Fama [4] helped move forward from this empirical evidence by proposing to describe price variations using a class of distributions studied by Lévy [5]. Thereafter, these distributions were abruptly truncated by Mantegna and Stanley [6] and exponentially truncated by Koponen [7] to recover a slow convergence towards a Gaussian distribution for low-frequency data, as observed empirically.

From the previous collection of dates, going as far back as 1897 for Pareto, one concludes that the field of research is not new, but interest has been renewed recently as witnessed by the recent excellent works on econophysics, starting with a conference book published by Anderson, Arrow and Pines [8] followed by the books of Bouchaud and Potters [9] and Mantegna and Stanley [10] Paul and Baschnagel [11] and Voit [12] These works show the variety of interest and successes of the approach to finance by physicists. From these works and the many research papers, we gather that there exists a temporary consensus on empirical data, although there is no such unity in modeling. There is indeed a profusion of models.

Econophysics and Sociophysics: Trends and Perspectives.
Bikas K. Chakrabarti, Anirban Chakraborti, Arnab Chatterjee (Eds.)
Copyright © 2006 WILEY-VCH Verlag GmbH & Co. KGaA, Weinheim
ISBN: 3-527-40670-0

First recall that a "price" can only go up or down, that is, is a one-dimensional quantity. Therefore, it can be seen as a point diffusing on a line. The factors driving this diffusive process are numerous, and their nature is still not fully elucidated. Hence, one can only make plausible assumptions and then test their value by comparing the distribution of changes generated by the diffusion process with actual price evolution changes, price being also understood as the value of some financial index or some exchange rate between currencies.

A fundamental problem is the existence or not of long-range power-law correlations in economic systems as well as the presence of economic cycles. Indeed, traditional methods (like spectral methods) have corroborated the fact that there is evidence that the Brownian motion idea is only approximately correct [3,4]. Long-range power-law correlations discovered in economic systems, particularily in financial fluctuations [6] have been examined through the so-called Lévy statistics already mentioned by Stanley et al. [6] who have shown the existence of long-range power-law correlations in the Standard and Poor (S&P500) index. A method based on wavelet analysis has also shown the emergence of hidden structures in the S&P500 index [13]. We have first performed a Detrended Fluctuation Analysis (DFA) of the USD/DEM ratio [14] and of other foreign currency exchanges and have demonstrated the existence of successive sequences of economic activity having different statistical behaviors. They will be mentioned in the relevant section.

A word of caution is at once necessary: this review is by no means objective. It is strongly biased towards studies undergone by a few physicists in the last few years, in order to give an overview of a few results in the physics literature on only two topics : foreign currency exchanges and stock market indices. Even so, it is not possible to cover all papers published or put on arXives. I do apologize for having missed many papers and ideas. Of course, since no full review of the different developments belonging to the physics literature is possible here, this applies even more to the purely economics literature. However, on one hand, each of the research papers presented here has individually justified its sideline propositions which would be tiring to reproduce. On the other hand, there is a huge amount of overlap between the different works summarized here below and the models developed by many authors. Trying to distinguish the main contributions separately is a titanic work. Attempting to compare and discuss details is quasi-impossible. Yet this review of econophysics and data analysis techniques , restricted to stock markets and foreign exchange currency markets should only be seen as opening a gate into huge undiscovered fields.

In the following section a few data-analysis techniques will be described with emphasis on the Detrended Fluctuation Analysis (DFA) and the Zipf Analysis Technique (ZAT). Information about the original data will be sketchy, but the data mainly concerns the foreign currency exchange market. The robustness of the DFA technique will be underlined and additional remarks will be given suggesting further work. In the next section, models about financial index value evolutions will be recalled, again without going into elaborate work discussing typical agent behavior, but rather with hopefully sufficient information such that the basic ingredients can be memorized before reading some of the vast literature on price formation. Crashes being spectacular phenomena, must retain our attention and do so through data analysis and basic intuitive models. A few "more general" microscopic models will also be outlined.

9.1
A Few Robust Techniques

9.1.1
Detrended Fluctuation Analysis Technique

In the basic Detrended Fluctuation Analysis (DFA) technique one divides a time series $y(t)$ of length N into N/τ equal size nonoverlapping boxes [15]. For smoothing out the data, the integration of the raw time series can often be performed first; in so doing one has to remember that such an integration shifts the exponent by one unit. The variable t is discrete and evolves by a single unit at each time step between $t = 1$ and $t = N$. No data point is supposed to be missing, i.e., breaks due to holidays and week-ends are disregarded. Nevertheless, the τ units are said to be *days*: often a week has only 5 days, and a year about 200 days. Let each box contain τ points and N/τ be an integer. The local trend in each τ-size box is assumed to be linear, i.e., it is taken as $z(t) = a\,t + b$. In each τ-size box one next calculates the root-mean-square deviation between $y(t)$ and $z(t)$. The detrended fluctuation function $F(\tau)$ is then calculated as

$$F^2(\tau) = \frac{1}{\tau}\sum_{t=k\tau+1}^{(k+1)\tau}|y(t) - z(t)|^2 \qquad k = 0, 1, 2, \cdots, \left(\frac{N}{\tau} - 1\right) \tag{9.1}$$

Averaging $F^2(\tau)$ over all N/τ box sizes centered on time τ gives the fluctuations $\langle F^2(\tau)\rangle$ as a function of τ. The calculation is repeated for all possible different values of τ. A power-law behavior is expected as

$$\langle F^2(\tau)\rangle^{1/2} \sim \tau^\alpha \tag{9.2}$$

An exponent $\alpha \neq 1/2$ in a certain range of τ values implies the existence of long-range correlations in that time interval as in the fractional Brownian motion [16]. Such correlations are said to be "persistent" or "antipersistent" when they correspond to $\alpha > 1/2$ and $\alpha < 1/2$ respectively, in practice $Hu = \alpha$. A straight line can fit the data between $\log \tau = 1$ and $\log \tau = 2.6$ well. This interval is called the *scaling range*. Outside the scaling range the error bars are larger, often because of so called finite size effects, and/or the lack of numerous data points. The α exponent is directly related to the Hurst exponent [17] and the signal fractal dimension [3].

Most of the time, for the real or virtual foreign exchange currency (FEXC) rates that we have examined [14, 18–27], the scaling range is well defined. Sometimes it readily appears that the data contains two sets of points which can be fitted by straight lines. Usually that describing the "large τ" data has a 0.50 slope, Indicating correlationless fluctuations. Crossovers from fractional to ordinary Brownian motion can be clearly observed. These crossovers suggest that correlated sequences have characteristic durations with well-defined lower and upper scales [14]. The persistence is usually related to free market (and "runaway") conditions while the antipersistence develops due to strict political control allowing for a finite size bracket in which the *FEXC* rates can fluctuate. The case $\alpha = 0.50$ is surely avoided by speculators.

Notice that to consider overlapping boxes might be useful when not many data points are available. However, it was feared that extra insidious correlations would thereby be inserted. Nevertheless the analysis has shown that the value of α is rather insensitive to the way boxes are used. A cubic trend, like $z(t) = ct^3 + dt^2 + et + f$, can be also considered [28]. The parameters a to f are similarly estimated through a best least-square fit of the data points in each box. Following the procedure described above, the value of the exponent α can be obtained. Again the difference is found to be very small. Extensions of DFA to higher moment components have also been investigated, tending toward some sort of multifractal analysis. Moreover, the linear or cubic or other polynomial law fitting the trend can be usefully replaced by some moving average, leading to some further insight into the Hurst exponent value [29].

An interesting observation consists of looking for *local* (more exactly *temporal*) α values [14, 19, 28].

This allows one to probe the existence or not of *locally correlated or decorrelated sequences*. In order to observe the local correlations, a local observation box of a given finite size is constructed. Its size depends on the upper value of τ for which a reasonable power-law exponent is found. It is chosen to be large enough in order to obtain a sufficiently large number of data points. The *local* exponent α is then calculated for the data contained in that finite-sized box, as above. Thereafter the box is moved along the time axis by an arbitrary finite number of points, often depending on the intended strategy.

The *local α exponent* seems rougher, and varies with time around the overall (mean) α value. The variation depends on the box size. Three typical FEXC rate time dependences, i.e. DEM/JPY, DEM/CHF, and DEM/DKK have been shown for various time intervals in [20] and the α value indicated for the scaling range. For example, the DEM/JPY *local* α is consistently above 0.50 indicating a persistent evolution. A positive fluctuation is likely to be followed by another positive one. The case of DEM/CHF is typically Brownian with fluctuations around 0.50. However, in 1994, some drift is observed toward a value ca 0.55 while in 1997–1998 some drift is observed toward a value ca 0.45. It is clear that some economic policy change occurred in 1994, and a drastic one at the end of 1998 in order to render the system more Brownian-like. The same is true for the DEM/DKK where, due to European economic policy, the spread in this exchange rate was changed several times, leading from a Brownian-like situation to a, nowadays antipersistent, behavior, i.e., a positive fluctuation is followed by a negative one, and conversely, such that *local* α becomes quite low.

Therefore, this procedure interestingly leads to a *local* measurement of the degree of long-range correlations. In other investigations, [14] it has been found on the GBL/DEM exchange rate data that the change in slope of *local* α vs. time corresponds to changes in the Bundesbank interest rate increase or decrease. The 1985 Plaza agreement had some influence in order to curb the runaway *local* α value from a persistent 0.60 back to a more 0.50 Brownian-like value. Thus such FEXC rate behaviors indicate a measure of information, an entropy variation indicating how (whatever) "information is managed by the system. This seems to be an "information" to be taken into account when developing Hamiltonian or thermodynamic-like models.

At some time, [21, 22] a search was made for correlations and anticorrelations in exchange rates of pre-Euro moneys with respect to currencies as CHF, DKK, JPY and USD in order to understand the EUR behavior. The power-law behavior describing the rms deviation of the fluctuations as a function of time, whence the Hurst (or α) time-dependent exponent was obtained. We compared the time dependent α exponent of the DFA as in a correlation matrix for estimating respective influences. In doing a similar analysis, it was shown that the fluctuations of the GBP and EUR with respect to the major currencies were very similar, indirectly indicating that the GBP was effectively tied to EUR. The same applies to studies pertaining to Latin American currencies [19].

Since such temporal correlations can be sorted out, a strategy for profit making can be developed. It is easily observed whether there is persistence or anti-persistence in some exchange rate – according to the temporal value of α. Thus some probability of the next fluctuation, being positive or negative can be calculated. Therefore, a buy or sell decision can be taken. In so doing and taking the example of (DEM/BEF), we performed some virtual game and implemented the most simple strategy [30].

Investment strategies should look for Hu values over different time-interval windows which are continuously shifted. In this respect there is a connection to multifractal analysis. The technique consists in calculating the so-called "qth order height-height correlation function"

$$c_q(\tau) = \langle |y(t) - y(t')|^q \rangle_\tau \qquad (9.3)$$

where only nonzero terms are considered in the average $\langle ... \rangle_\tau$ taken over all couples (t, t') such that $\tau = |t - t'|$. The correlation function $c(\tau)$ is supposed to behave as

$$c_1(\tau) = \langle |y(t) - y(t')| \rangle_\tau \sim \tau^{H_1} \qquad (9.4)$$

where H_1 is the *Hu* exponent. [31, 32] This corresponds to obtaining a spectrum of moving averages [33]. Notice that a DFA and a multifractal analysis cost much more CPU time than a moving average method due to multiple loops which are present in the DFA and c_q algorithms [14, 31, 33, 34].

Of course, subtracting some moving average background, instead of the usual linear trend, is of interest in order to implement a strategy based on different horizons.

9.1.2
Zipf Analysis Technique

The same type of consideration for a strategy can be developed from the Zipf analysis technique (ZAT) performed on stock market or FEXC data. In the (DFA) technique the sign of the fluctuations and their persistence, are taken into account, but it falls short of implementing some strategy from the *amplitude* of the fluctuations. The so-called Zipf analysis [35] originally introduced in the context of natural languages can be performed by calculating the frequency of occurence f of each word in a given text. According to their frequency, a rank R can be assigned to each word, with $R = 1$ for the most frequent one. A power law

$$f \sim R^{-\zeta} \qquad (9.5)$$

is expected [35]. A Zipf plot is simply a transformation of the cumulative distribution. However, it accentuates the upper tail of the distribution, which makes it a useful to analyze this part of the distribution.

A simple extension of the Zipf analysis is to consider m-letter words, i.e., the words strictly made of m characters without considering the white spaces. The available number of different letters or characters k in the alphabet should also be specified. A power law in $f(R)$ is also expected for correlated sequences [36, 37]. There is no theory at this time predicting the exponent as a function of (m, k). The technique has a rather weak value when only two characters and short words are considered [36]. An increase in the number of allowed characters for the alphabet allows one to consider different sizes and signs of the fluctuations, i.e., huge, marginal and small (positive or negative) fluctuations can be considered. After having decided on the number (k) of characters of the alphabet, the signal can be transformed into a text and thereafter analyzed with respect to the frequency of words of a given size (m) to be ranked accordingly. In our work, words of equal lengths were always considered.

The above procedure does not take into account the trend. Some work has been done on the matter, but much is still to be performed since the trend definition can be quite different for later strategy considerations. The relevance of this remark should be emphasized: indeed for a positive (or negative) trend over the time box which is investigated, a bias can occur between words. For a two-character alphabet, e.g., u and d, the frequency f of us, i.e., p_u can be larger (smaller) than the frequency of d's, i.e., p_d. Such a *bias* can be taken into account with respect to the equal probability occurrence, e.g. $\epsilon = p_u - 0.5 = p_d + 0.5$. A new ranking procedure can be performed by defining the ratio of the observed frequency of a word divided by the theoretical frequency of occurrence of a word, assuming independence of characters: e.g., if the word uud occurs say p_{uud} times, since the independence of characters would imply that the word would occur $p_u\ p_u\ p_d$ times, a relative frequency f/f_0 can be defined as $p_{uud}/(p_u\ p_u\ p_d)$. A new ranking can be made with a quite different appearance.

A ZAT variant consists in ranking the words according to their relative frequency and relative ("normalized") rank taking into account for the normalization the probability f_M of the word occuring most often [37]. Indeed for m and k large not all words do occur. Even though, e.g., for the ($m = 6, k = 2$) case there are 64 possible words, and the maximum rank is $R_M = 64$, the frequency of the most often observed word is unknown. Another extension has been recently proposed in which time windows are used in order to improve the sorting of the relevant exponents and probability of occurrence [38,39].

In conclusion of this subsection, it is emphasized that different strategies following the Zipf-analysis technique can be implemented, according to (m, k) values, how the trend is eliminated (or not) and how the ranks and frequencies of occurrence are defined.

9.1.3
Other Techniques for Searching for Correlations in Financial Indices

Simultaneously, Laloux et al. [40,41] and Plerou et al. [42,43] analyzed the correlations between stocks traded on financial markets using the theory of random matrices. Laloux et al. considered daily price changes for the 1991–1996 period (1309 days) of 406 of the companies forming the S&P500 index while Plerou et al. analyzed price returns over a 30 minute period of the 1 000 largest US companies for the 1994–1995 period (6448 points for each company). The correlation coefficient between two stocks i and j is defined by

$$\rho_{ij} \equiv \frac{<R_i R_j> - <R_i><R_j>}{\sqrt{(<R_i^2> - <R_i>^2)(<R_j^2> - <R_j>^2)}} \tag{9.6}$$

where R_i is the price of company i for Laloux et al. and the return of the price for Plerou et al. The statistical average is a temporal average performed on all trading days of the investigated periods. The cross-correlation matrix C of elements ρ_{ij} is measured at equal times.

Laloux et al. [40, 41] focussed on the density $\rho_c(\lambda)$ of eigenvalues of C, defined by

$$\rho_c(\lambda) \equiv \frac{1}{N} \frac{dn(\lambda)}{d\lambda} \tag{9.7}$$

where $n(\lambda)$ is the number of eigenvalues of C less than λ and N the number of companies (or equivalently, the dimension of C). The empirically determined $\rho_c(\lambda)$ is compared with the eigenvalue density of a random matrix, that is, a matrix whose components are random variables. Random matrix theory (RMT) isolates universal features, which means that deviations from these features identifies system-specific properties. Laloux et al. found that 94% of the total number of eigenvalues fall inside the prediction of RMT, which means that it is very difficult to distinguish correlations from random changes in financial markets.

The most striking difference is that RMT predicts a maximum value for the largest eigenvalue which is much smaller than that is observed empirically. In particular, a very large eigenvalue is measured, representing the market itself. Plerou et al. [42, 43] observed similar properties in their empirical study. They measured the number of companies contributing to each eigenvector and found that a small number of companies (compared to 1 000) contribute to the eigenvectors associated with the largest and the smallest eigenvalues. For the largest eigenvalues these companies are correlated, while for the smallest eigenvalues, they are uncorrelated. The observation for the largest eigenvalues does not concern the largest one, which has an associated eigenvector representing the market and has an approximately equal contribution from each stock.

Mantegna [44] analyzed the correlations between the 30 stocks used to calculate the Dow Jones industrial average and the correlations between the companies used to calculate the S&P500 index, for the July 1989 to October 1995 time period in both cases. Only the companies which were present in the S&P500 index during the whole period of time were considered, which leaves 443 stocks. The correlation coefficient of the returns for all possible pairs of stocks was computed. A metric distance between two stocks is defined by $d_{ij} = 1 - \rho_{ij}^2$. These distances are the elements of the distance matrix D.

Using the distance matrix D, Mantegna determined the topological arrangement which is present between the different stocks. His study could also give empirical evidence about the existence and nature of common economic factors which drive the time evolution of stock prices, a problem of major interest. Mantegna determined the minimum spanning tree (MST) connecting the different indices, and thereafter, the subdominant ultrametric structure and heriarchical organization of the indices. In fact, the elements $d_{ij}^<$ of the ultrametric distance matrix $D^<$ are determined from the MST. $d_{ij}^<$ is the maximum Euclidian distance d_{lk} detected by moving in single steps from i to j through the path connecting i to j in the MST. Studying the MST and hierarchical organization of the stocks defining the Dow Jones industrial average, Mantegna showed that the stocks can be divided into three groups. Carrying the same analysis for the stocks belonging to the S&P500, Mantegna obtained a grouping of the stocks according to the industry they belong to. This suggests that stocks belonging to the same industry respond in a statistical way to the same economic factors.

Later, Bonanno et al. [45] extended the previous analysis to consider correlations between 29 different stock market indices, one from Africa, eight from America, nine from Asia, one from Oceania and ten from Europe. Their correlation coefficients ρ_{ij} were calculated using the return of the indices instead of the individual stocks. They also slightly modified the definition of the distance, using $d_{ij} = \sqrt{2(1-\rho_{ij})}$. A hierarchical structure of indices was obtained, showing a distinct regional clustering. Clusters for North America, South America and Europe emerge, while Asian indices are more diversified. Japanese and Indian stock markets are pretty distant from the others. When the same analysis is performed over different periods of time, the clustering is still present, with a slowly evolving stucture of the ultrametric distance with time. The taxinomy is stable over a long period of time.

Two key parameters are the length $L^<$ and the proximity P. The length is $L^< \equiv \sum d_{i,j}^<$, where the sum runs over nearest neighboring sites on the MST. It is a kind of average of first-neighbor distances over the whole MST. The proximity is

$$P \equiv \frac{\sum_{i,j} |d_{i,j} - d_{i,j}^<|}{\sum_{i,j} d_{i,j}} \qquad (9.8)$$

where the sums run over all distinct i,j pairs. The proximity characterizes the degree of resemblance of the subdominant ultrametric space to the Euclidian space. For a long time period, $L^< \to 26.9$, while $L^< = 28\sqrt{2} \approx 39.6$ for sequences without correlations. Similarly, for a long time period, $P \to 0.091$, to compare with $P \to 0.025$ when the same data are shuffled to destroy any correlations. The effects of spurious cross-correlations make the previous time analysis relevant for time periods of the order of three months or longer for daily data.

Others [46] analyzed the sign of daily changes of the NIKKEI, the DAX and the Dow Jones industrial average for the Dow Jones only and for the three time series together, reproducing the evolution of the market through the sign of the fluctuation, as if the latter was represented by an Ising spin. The authors studied the frequency of occurrence of triplets of spins for the January 1980 to December 1990 period, after having removed the bias due to a different probability of having a move upward or downward. They showed that the spin series generated by each index separately is comparable to a randomly generated series. However, they emphasized correlations in the spin series generated by the three indices showing that market places fluctuate in a cooperative fashion. Three upward moves in a row are more likely than expected from a series without correlations between successive changes, and the same applies for three downward moves. This behavior seems to be symmetrical with respect to ups and downs in the time period investigated. The strength of the correlations varies with time, with the difference in frequency of different patterns being neatly marked in the two-year period preceding the 1987 crash. Also, during this period, the up-down symmetry is broken.

Finally let us mention the Recurrence Plot (RP) and Recurrence Quantification Analysis (RQA) techniques for detecting, e.g., critical regimes appearing in financial market indices [47]. These are graphical tools elaborated by Eckmann, Kamphorst and Ruelle in 1987, based on Phase Space Reconstruction [48] and extended by Zbilut and Webber [49] in 1992. RP and RQA techniques are usually intended to provide evidence for chaos, but can also be used for their appropriateness in working with nonstationarity and noisy data [50] and in detecting changes in data behavior, in particular in detecting breaks, like a phase transition [51] and other dynamic properties of a time series [48]. It was indicated that they can be used for detecting bubble nucleation and growth and can even indicate bursting. An analysis was made on two time series, NASDAQ and DAX, taken over a time span of six years including the known (NASDAQ) crash of April 2000. It has been shown that a bubble-burst warning could be given, with some delay ($(m-1)d$ days) with respect to the beginning of the bubble, but with enough time before the crash (three months in the mentioned cases) [47].

9.2
Statistical, Phenomenological and "Microscopic" Models

A few functions used to fit empirical data should be mentioned. The point is to try to determine which laws are the best approximation for which data. Some of these laws rely on plausible and fairly convincing arguments. A major step forward to quantify the discrepancies between real time series and Gaussian statistics was made by Mandelbrot [3] who studied cotton prices. In addition to being non Gaussian, Mandelbrot noted that returns respect time scaling, that is, the distribution of returns for various time scales Δt from 1 day up to one month have a similar functional form. Motivated by this scaling and the fact that large events are far more probable than expected, he proposed that the statistics of returns could be described by symmetrical Lévy distributions; other laws can appear to fit the data better, and also rely on practical arguments, but they appear as *ad hoc* modifications, like the truncated Lévy flight.

9.2.1
ARCH, GARCH, EGARCH, IGARCH, FIGARCH Models

The acronym "ARCH" stands for autoregressive conditional heteroscedasticity, a process introduced by Engle [52]. In short, Engle assumed that the price at a given time is drawn from a probability distribution which depends on information about previous prices, what he referred to as a conditional probability distribution (cpd). Typically, in an ARCH(p) process, the variance of the cpd $P(x_t|x_{t-1}, ..., x_{t-p})$ at time t is given by

$$\sigma_t^2 = \alpha_0 + \alpha_1 x_{t-1}^2 + \alpha_2 x_{t-2}^2 + ... \alpha_p x_{t-p}^2 \tag{9.9}$$

where $x_{t-1}, x_{t_2}, ...$ are random variables chosen from sets of random variables with Gaussian distributions of zero mean and standard deviations $\sigma_{t-1}, \sigma_{t-2}, ...$, respectively. $\alpha_0, \alpha_1, ..., \alpha_p$ are control parameters. Locally (in time), σ_t varies but on a long time-scale, the overal process is stationary for a wide range of values of the control parameters.

Bollerslev [53] generalized the previous process by introducing Generalized ARCH or GARCH(p,q) processes. He suggested modeling the time evolution of the variance of the cpd $P(x_t|x_{t-1}, ..., x_{t-p}, \sigma_{t-1}, ...\sigma_{t-q})$ at time t with

$$\sigma_t^2 = \alpha_0 + \alpha_1 x_{t-1}^2 + ...\alpha_n x_{t-n}^2 + \beta_1 \sigma_{t-1}^2 + ... + \beta_q \sigma_{t-q}^2, \tag{9.10}$$

with an added set $\{\beta_1, ..., \beta_q\}$ of q control parameters in addition to the current p control parameters. Using this process, it is now possible to generate time correlations in the variance. The previous processes have been extended, as in EGARCH processes [55], IGARCH processes [54], FIGARCH processes [56], among others.

9.2.2
Distribution of Returns

Mantegna and Stanley [57] analyzed the returns of the S&P500 index from Jan. 1984 to Dec. 1989. They found that it is well described by a Lévy stable symmetrical process of index $\alpha = 1.4$ for time intervals spanning from 1 to 1 000 minutes, except for the most rare events. Typically, a good agreement with the Lévy distribution is observed when $m/\sigma \leq 6$ and an approximately exponential fall-off from the stable distribution is observed when $m/\sigma \geq 6$, with the variance $\sigma^2 = 2.57.10^{-3}$. For $\Delta t = 1$ minute, the kurtosis $\kappa = 43$. Their empirical study relies mainly on the scaling of the "probability of return to the origin" $P(R = 0)$ (not to be confused with the distribution of returns) as a function of Δt, which is equal to

$$P(0) \equiv \mathcal{L}_\alpha(R_{\Delta t} = 0) = \frac{\Gamma(1/\alpha)}{\pi \alpha (\gamma \Delta t)^{1/\alpha}} \quad (9.11)$$

if the process is Lévy, where Γ is the Gamma function. They obtain $\alpha = 1.40 \pm 0.05$. This value is roughly constant over the years, while the scale factor γ (related to the vertical position of the distribution) fluctuates with a burst of activity localized in specific months. Using an ARCH(1) model with σ^2 and κ constrained to their empirical values, Mantegna and Stanley [58] obtained a scaling value close to 2 (~ 1.93), in disagreement with $\alpha = 1.4$. For a GARCH (1,1) process with similar constraint and the choice $\beta_1 = 0.9$, In [57,58] they obtained a very good agreement for the distribution of returns when $\Delta t = 1$ minute, but they estimated that the scaling index α should be equal to 1.88, in disagreement with the observed value.

Gopikrishnan et al. [59] extended the previous study of the S&P 500 to the 1984–1996 period for one-minute frequency records, to the 1962–1996 period for daily records and the 1926–1996 period for monthly records. In parallel, they also analyzed daily records of the NIKKEI index of the Tokyo stock exchange for the 1984–1997 period and daily records of the Hang-Seng index of the Hong Kong stock exchange for the 1980–1997 period. In another study, Gopikrishnan et al. [60] analyzed the returns of individual stocks in the three major US stock markets, the NYSE, the AMEX and the NASDAQ for the January 1994 to December 1995 period. Earlier works by Lux [61] focussed on daily returns of the German stocks making the DAX share index during the period 1988–1994.

To compare the behavior of the different assets on different time scales, a normalized return $g \equiv g_{\Delta t}(t)$ is defined, with

$$g \equiv \frac{R - <R>_T}{\sqrt{<R_T^2> - <R>_T^2}} \quad (9.12)$$

where the average $<\cdots>_T$ is over the entire length T of the time series considered. The denominator of the previous equation corresponds to the time averaged volatility $\sigma(\Delta t)$. For all data and for Δt from 1 minute to 1 day, the distribution of returns is in agreement with a Lévy distribution of index varying from $\alpha \approx 1.35$ to $\alpha \approx 1.8$ in the central part of the distribution, depending on the size of the region of empirical data used for the fit. In contrast to the previously suggested exponential truncation for the largest returns, the distribution is found to scale like a power law with an exponent $\mu \equiv 1 + \alpha \approx 4$. This value is well outside the Lévy stable regime, which requires $0 < \alpha \leq 2$. Hence, these studies point towards a truncated Lévy distribution, with a power-law truncation.

The previous scaling in time is observed for time-scales Δt of up to 4 days. For a larger time-scale, the data are consistent with a slow convergence towards a Gaussian distribution. This convergence is expected as the presence of a power-law cut-off implies that the distribution of returns is in the Gaussian stable regime, where the CLT applies. What is surprising is the existence of a "metastable" scaling regime observed in timescales as long as four days. Gopikrishnan et al. [59] identified time dependencies as one of the potential sources of this scaling regime, by comparing the actual time series first to a randomly generated time series with the same distribution and second to the shuffled original time series. The two latter display a much faster convergence towards the Gaussian statistics, confirming their hypothesis.

One important conclusion of the previous empirical data is that the theory of truncated Lévy distributions cannot reproduce empirical data as it stands, because in the current framework of truncated Lévy distributions, the random variables are still independent, while it has been shown that time dependencies are a crucial ingredient of the scaling. Also, it does not explain the fluctuations of γ.

Plerou et al. [62] considered the variance of individual price changes, $\omega_{\Delta t}^2 \equiv <(\delta p_i)^2>$ for all transactions of the 1 000 largest US stocks for the 1994–1995 two-year period. The cumulative distribution of this variance displays a power-law scaling $P(\omega_{\Delta t} > x) \sim x^{-\gamma}$, with $\gamma = 2.9 \pm 0.1$. Using detrended fluctuation analysis, they obtain a correlation function characterized by an exponent $\mu = 0.60 \pm 0.01$, that is, weak correlations only, independent variables being characterized by $\mu = 1/2$.

Gopikrishnan et al. [59] found that $\sigma(\Delta t) \sim (\Delta t)^\delta$. The exponent δ experiences a crossover from $\delta = 0.67 \pm 0.03$ when $\Delta t < 20$ minutes to $\delta = 0.51 \pm 0.06$ when $\Delta t > 20$ minutes. This is in agreement with the fact that the autocorrelation function of the returns is exponentially decreasing in a characteristic time τ_{ch} of 4 minutes. These results predict that for $\Delta t > 20$ minutes, the returns are essentially uncorrelated. Hence, important scaling properties of the time series originate from time dependencies, but the autocorrelation function of the returns or the time averaged variance, do not deliver the relevant information required to study these dependencies. Higher order correlations need to be analyzed.

Since the time averaged volatility is a power-law, this invalidates the standard ARCH or GARCH processes, because they all predict an exponentially decreasing volatility. In order to explain the long-range persistence of the correlations, one needs to include the entire history of the returns.

Scalas [63] analyzed the statistics of the price difference between the closing price and the opening price of the next day, taking week-ends and holidays as overnight variations. He concentrated his analysis on Italian government bonds (BTP) futures for the 18 Sept 1991–20 Feb 1997 period. As for other time-scale variations, he did not observe short-range nor long-range correlations in the price changes. In fact he was able to reproduce similar results with a numerical simulation of a truncated trinomial random walk.

A general framework for treating superdiffusive systems is provided by the nonlinear Fokker–Planck equation, which is associated with an underlying Ito–Langevin process. This in turn has a very interesting connection to the nonextensive entropy proposed by Tsallis: the nonlinear Fokker–Planck equation is solved by time-dependent distributions which maximize the Tsallis entropy. This unexpected connection between thermostatistics and anomalous diffusion gives an entirely new way, much beyond the Bachelier-like [1] approach, to study the dynamics of a financial market as if there are anomalously diffusing systems.

Whence the intra-day price changes in the S&P500 stock index have been studied within this framework by direct analysis and by simulation in Refs [64–66]. The power-law tails of the distributions, and the index's anomalously diffusing dynamics, are very accurately described. Results show good agreement between market data, Fokker–Planck dynamics, and a simulation. Thus the combination of the Tsallis non-extensive entropy and the nonlinear Fokker–Planck equation unites in a very natural way the power-law tails of the distributions and their superdiffusive dynamics. In our case the shape and tails of partial distribution functions (PDF) for a financial signal, i.e., the S&P500 and the turbulent nature of the markets were linked through the Beck model, originally proposed to describe the intermittent behavior of turbulent flows. Looking at small and large time windows, both for small and large log-returns, the market volatility (of normalized log-returns) distributions fitted well with a χ^2-distribution. The transition between the small-timescale model of nonextensive, intermittent processes and the large-scale Gaussian extensive homogeneous fluctuation picture, was found to be at ca a 200 day time lag. The intermittency exponent (κ) in the framework of the Kolmogorov log-normal model was found to be related to the scaling exponent of the PDF moments, – thereby giving weight to the model. The large value of κ points to a large number of cascades in the turbulent process. The first Kramers–Moyal coefficient in the Fokker–Planck equation is almost equal to zero, indicating "no restoring force". A comparison is made between normalized log-returns and mere price increments.

The case of foreign exchange markets might have not been studied to my knowledge, and might be reported here for the first time (Fig. 9.1). Consider the time series of the normalized log returns $Z(t, \Delta t) = (\tilde{y}(t) - <\tilde{y}>_{\Delta t})/\sigma_{\Delta t}$ for different (selected) values of the time lag $\Delta t = 1, 2, 4, 8, 10, 32$ days. Let $\tau = log_2(32/\Delta t)$,

$$\frac{d}{d\tau} p(Z,\tau) = \left[-\frac{\partial}{\partial Z} D^{(1)}(Z,\tau) + \frac{\partial}{\partial Z^2} D^{(2)}(Z,\tau) \right] p(Z,\tau) \tag{9.13}$$

in terms of a drift $D^{(1)}(Z,\tau)$ and a diffusion coefficient $D^{(2)}(Z,\tau)$ (thus values of τ represent Δt_i, $i = 1, ...$).

The coefficient functional dependence can be estimated directly from the moments $M^{(k)}$ (known as Kramers–Moyal coefficients) of the conditional

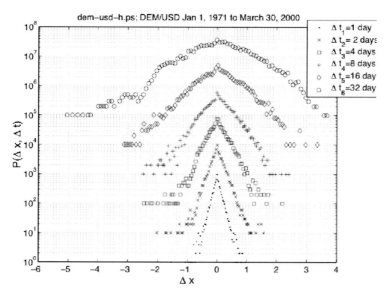

Fig. 9.1 Probability distribution function $P(\Delta x, \Delta t)$ of normalized increments Δx of daily closing price value signal of DEM–USD exchange rate between Jan. 01, 1971 and March 31, 2000, for different time lags: $\Delta t = 1, 2, 4, 8, 16, 32$ days. The normalization is with respect to the width of the PDF for $\Delta t = 32$ days. The PDF symbols for a given Δt are displaced by a factor 10 with respect to the previous Δt; the curve for $\Delta t = 1$ day is unmoved. See the tendency toward a Gaussian for increasing Δt.

probability distributions:

$$M^{(k)} = \frac{1}{\Delta\tau} \int dZ'(Z'-Z)^k p(Z', \tau + \Delta\tau | Z, \tau) \qquad (9.14)$$

$$D^{(k)}(Z,\tau) = \frac{1}{k!} \lim M^{(k)} \qquad (9.15)$$

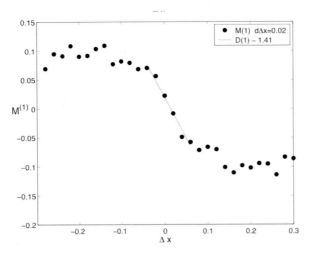

Fig. 9.2 Kramers–Moyal drift coefficient $M^{(1)}$ as a function of normalized increments Δx of daily closing price value of DEM-USD exchange rate between Jan. 01, 1971 and March 31, 2000, with a best linear fit for the central part of the data corresponding to a drift coefficient $D^{(1)} = -1.41$.

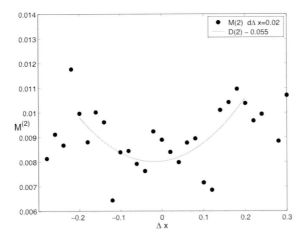

Fig. 9.3 Kramers–Moyal diffusion coefficient $M^{(2)}$ as a function of normalized increments Δx of daily closing price value of DEM–USD exchange rate between Jan. 01, 1971 and March 31, 2000, with a best parabolic fit for the central part of the data corresponding to a diffusion coefficient $D^{(2)} = 0.055$

for $\Delta\tau \to 0$. According to Fig. 9.2 the drift coefficient $D^{(1)} \approx 0$ and the diffusion coefficient $D^{(2)}$ (Fig. 9.3) is not so well represented by a parabola as for the S&P. A greater deviation from regular curves is also seen if one examines specific stock evolutions.

It may be recalled that the observed quadratic dependence of the diffusion term $D^{(2)}$ is essential for the logarithmic scaling of the intermittency parameter in studies on turbulence.

9.2.3
Crashes

Johansen and Sornette [67] stressed that the largest crashes of the twentieth century appear as outliers, not belonging to the distribution of *normal* returns. The claim is easily observable for the three major crashes (downturns of more than 22%), October 1987, World War I and Wall Street 1929, in order of magnitude. It is also supported by the fact that in a simulation of a million-year trading using a GARCH(1,1) model, with a reset every century, three crashes never occurred within the same century. Another important property that points towards a different dynamical process for crashes is that, if anomalously large returns are, for instance, defined as a change of 10% or more of an index over a short time interval (a couple of days at most), one observes only crashes (downturns) and no upsurge. This contrasts with usual returns which have been observed to be symmetric for up and down moves.

In agreement with crashes not being part of the usual distribution of returns, Feigenbaum and Freund [68] and Sornette, Johansen and Bouchaud [69] have independently proposed a picture of crashes as critical points in a hierarchical system with discrete scaling. When a system is at a critical point, it becomes "scale invariant", meaning that it becomes statistically similar to itself when considered at different magnifications. The signature of self-similarity is power laws, which are scale-free laws. For systems with a hierarchical structure, the system is self-similar only under specific scale changes. The signature of this property is that the exponent of the previous power laws has a nonzero imaginary part. Specifically, if $y(t)$ is the price at time t, it is supposed that

$$y(t) = A + B(t_c - t)^\mu [1 + C \cos[\omega \log(t_c - t) + \phi]] \qquad (9.16)$$

in the regime preceeding a crash. The previous equation stipulates that t_c is a critical time where the prices will experience a phase transition. The signature of this transition is a variation of the price as a power law of exponent μ. But because of the hierarchical structure in the system, the divergence is "decorated" by log-periodic oscillations. The phase ϕ defines the chosen unit of timescale. It has appeared that the value of μ is not robust in the nonlinear fits.

In a latter study, the S&P500 and the Dow Jones 1987 crash were examined [70] after subtracting an exponential background due to the time evolution of these indices before a fitting with Eq. (9.16). Next it has been proposed to consider a logarithmic divergence, corresponding to the $\mu = 0$ limit, [71] rather than a power law, i.e.,

$$y(t) = A + B \ln\left(\frac{t_c - t}{t_c}\right)\left[1 + C \sin\left(\omega \ln\left(\frac{t_c - t}{t_c}\right) + \phi\right)\right] \text{ for } t < t_c \quad (9.17)$$

In so doing the analysis of a (closing value) stock market index like the Dow Jones Industrial average (DJIA), the S&P500 [71,72] and DAX [73] leads us to observe the precursor of so-called crashes. This was shown on the Oct. 1987 and Oct. 1997 cases, as was reported in the financial press at the time [74,75]. The prediction of the crash date was made as early as July, in the 1997 case. It should be pointed out that we do not expect any real divergence in the stock market indices, but to give upper bounds are surely of interest. However, the reliability of the method has been questioned by Laloux et al. [41] on the grounds that all empirical fits require a large number of parameters, that the statistical data for crashes is obviously very restricted and that the method predicts crashes that never happened. This can be further debated elsewhere.

Nevertheless, the existence of log-periodic corrections to scaling implies the existence of a hierarchy of time scales $\delta t_n \equiv t_c - t_n$ determined by half-periods of the cosine in Eq. (9.16). These time scales are not expected to be universal, varying from one market to another, but the ratio $\lambda = \delta t_{n+1}/\delta t_n$ could be robust with respect to changes in what is traded and when it is traded, with $\lambda \approx 2.2 - 2.7$, see Sornette [76].

Sornette, Johansen and Bouchaud [69] identify the underlying hierarchical structure as having the currency and trading blocks at its highest level (Euro, Dollar, Yen,...); countries at the level below; major banks and institutions within a country at the level below; the various departments of these institutions at the level below; and so on. Several potential sources can be listed [70] for discrete scaling: hierarchical structure of investors (see Table 9.1), investors with a specific time horizon or periodic events like market closure and opening, quaterly publication of reports, etc.

Feigenbaum and Freund [68] had reported log-periodic oscillations for 1986–1987 weekly data for the S&P500 (October 19, 1987 crash), 1962 monthly data for the S&P500 (NY crash of 1962), 1929 monthly data for the Dow-Jones (NY 1929 crash) and 1990 scanned data for the NIKKEI (Tokyo 1990 crash). Sornette, Johansen and Bouchaud [69] had also reported log-periodic oscillations for July 1985 to the end of 1987 daily data for the S&P500. Since then, this type of oscillation has been detected before many financial index downturns, see [71,72] and [76] for a review.

Tab. 9.1 Different categories of investors and their relative weight on the London Stock Market in 1993 after David [77].

Rank	Investor type	Weight
1	Pension funds	34.2
2	Individuals	17.7
3	Insurance companies	17.3
4	Foreign	16.3
5	Unit trusts	6.6
6	Others	2.3
7	Other personal sector	1.6
8	Industrial and commercial companies	1.5
9	Public sector	1.3
10	Other financial institutions	0.6
11	Banks	0.6

Johansen and Sornette [67] have also analyzed the symmetric phenomenon with respect to time, that is, a power-law decrease of the price after culmination at a maximum value, the power-law being decorated by log-periodic oscillations. Interestingly, they do not obtain empirical evidence of a maximum with log-periodic oscillations before *and* after the maximum. The authors advocate as the main reason the fact that, when the market accelerates to reach a log-periodic regime, it very often ends in a crash, with an associated symmetry breaking. According to this argument, symmetrical phenomena can only be detected whenever a crash had not happened before. They found empirical evidence of "antibubbles" for the NIKKEI, for the 31 Dec. 1989 till 31 Dec. 1998 period and for gold futures after 1980. It seems that λ for decreasing markets could be higher (around 3.4–3.6) than for increasing markets.

9.2.4
Crash Models

Eliezer and Kogan [78] have considered the possibility of the existence of a crashing phase, with different dynamics from the normal or quiescent phase of the market. Their extension is based on a model with two types of trader, but different from the previous noise and rational traders. The distinction comes from the way in which the agents place their orders. Agents that place a limit order are visible to every one on the screen, while agents placing market orders are not. Hence, it is impossible to have an annihilation event between agents placing market orders, as they are "invisible" to each other. This distinction between market-order agents and limit-order agents was emphasized by Cohen et al. [79] as a possible source for widening the spread between ask and bid prices.

9.2.5
The Maslov Model

Maslov [80] has proposed a model similar to the crash model of [78]. He assumes that each agent can either trade a stock at the market price or place a limit order to sell or buy. At each time step, a new trader appears and tries to make a transaction, buying or selling being equally likely to occur. With probability q_{lo}, he places a limit order, otherwise, he trades at the best limit order available, the lowest ask for a buyer or the highest bid for a seller. The price of the last deal is taken as the market price. The price of a new limit order is equal to the market price offset by a random number Δ chosen from a probability distribution, to avoid overlapping buy and sell orders. Agents making a deal are removed from the system. In contrast to the previous models, the number of agents is not fixed and the agents are not able to diffuse. Once they have chosen a price, they can only wait until someone matches their order.

Numerically, it is found that the model displays volatility clustering, a power-law decrease of the autocorrelation function of absolute price change with an exponent $\gamma = 1/2$ (0.3 on real markets), a nontrivial Hurst exponent $H = 1/4$ (0.6–0.7 on real markets) and no correlation on the signs of the price changes. Most importantly, the price increment distribution is non-Gaussian, with a crossover between two power-laws in the wings, from an exponent $1 + \alpha_1$ to $1 + \alpha_2$ for larger changes, with $\alpha_1 = 0.6 \pm 0.10$ and $\alpha_2 = 3.0 \pm 0.2$. In summary, the model is very promising because it has many qualitatively interesting features, but unfortunately a chronic lack of quantitative agreement with real data.

Within a mean-field approximation of the model, Slanina [81] has shown that the stationary distribution for the price changes has power-law tails with an exponent $1 + \alpha = 2$. This result disagrees with numerical simulations of the model, both quantitatively and qualitatively, for reasons that remain unknown.

9.2.6
The Sandpile Model

It can be conjectured that stock markets are hierarchical objects where each level has a different weight and a different characteristic time scale (the horizons of the investors). The hierarchical lattice might be a fractal tree [72] with loops. The geometry might control the type of criticality. This led us to consider the type of avalanches which occur in a tumbling sandpile as an analogy with the spikes in the index variation within which a financial bubble grows.

The Bak, Tang and Wiesenfeld (BTW) [82] in its sandpile version [83] was studied on a Sierpinski gasket [84]. It has been shown that the avalanche dynamics is characterized by a power law with a complex scaling exponent

$\tau + i\omega$ with $\omega = \frac{2\pi}{\ln 2}$. This was understood as the result of the underlying Discrete Scale Invariance (DSI) of the gasket, i.e., the lattice is translation invariant on a log-scale [76]. Such a study of the BTW model was extended to studies in which both the fractal dimension D_f and the connectivity of the lattice were varying. In so doing, connectivity-based corrections to power-law scaling appeared. For most avalanche distributions $P(s) \sim s^{-\tau}$, expressing the scale invariance of the dynamics, we have checked the power-law exponent (τ) as a function of the fractal dimension of RSC lattices and have found that τ seems to be dependent on the real part of the lattice (carpet) fractal dimension $\Re\{D_f\}$. Notice that for $\Re\{D_f\} \to 2$, $\tau = 1.25 \pm 0.03$.

We have observed significant deviations of τ from 1.25, i.e., the $d = 2$ value. It seems that the real part of the fractal dimension of the dissipative system is not the single parameter controlling the value of τ. These oscillations (peaks) can be thought of as originating from the DSI of the RSC lattice as in [84], and mimicking those discussed in the main text for financial indices. We emphasize that the connectivity of the lattice is one of the most relevant parameters. Notice that such log-periodic oscillations and linearly substructured peaks are also observed in the time distribution of avalanche durations $P(t)$.

9.2.7
Percolation Models

Percolation theory was pioneered in the context of gelation, and later introduced in the mathematical literature by Broadbent and Hammersley [85, 86] with the problem of fluid propagation in porous media. Since then, it has been successfully applied in many different area of research : e.g., earthquakes, electrical properties and immunology. Its application to financial market leads to one major conclusion: herding is a likely explanation for the fat tails [87]. Percolation models assume *a contrario* to usual financial theories that the outcomes of decisions of individual agents may not be represented as independent random variables. Such an assumption would ignore an essential ingredient of market organization, namely the *interaction* and *communication* among agents. In real markets, agents may form groups of various sizes which then may share information and act in coordination. In the context of a financial market, groups of traders may align their decisions and act in unison to buy or sell; a different interpretation of a "group" may be an investment fund corresponding to the wealth of several investors but managed by a single fund manager.

To capture the effect of the interactions between agents, it is assumed that market participants form groups or "clusters" through a random matching process but that no trading takes places inside a given group: instead, members of a given group adopt a common market strategy (for example, they decide to buy or sell or not to trade) and different groups may trade with each other through a centralized market process. In the context of a financial market, clusters may represent, for example, a group of investors participating in a mutual fund.

9.2.8
The Cont–Bouchaud model

The Cont–Bouchaud model is an application of bond percolation on a random graph, an approach first suggested by Kirman [88] in the economics literature. Each site of the graph is supposed to be an agent, which is able to make a transaction of one unit on a market. Two agents are connected with each other with probability $p = c/N$. At any time, an agent is either buying with probability a, selling with probability a or doing nothing with probability $1 - 2a$, with $a \in (0, 1/2)$.

The exogenous parameter a controls the transaction rate, or the time scale considered; a close to 0 means a short time horizon of the order of a minute on financial markets, with only a few transactions per time step. The number of traders which are active during one time interval increases with the length of this time interval. All agents belonging to the same cluster are making the same decision at the same time. The cluster distribution represents the network of social interactions, which induces agents to act cooperatively. This network is supposed to model the phenomenon of herding. The aggregate excess demand for an asset at time t is the sum of the decision of all agents,

$$D(t) = \sum_{i=1}^{N} \phi_i(t) \tag{9.18}$$

if the demand of agent i is $\phi_i \in \{-1, 0, +1\}$, $\phi_i = -1$, representing a sell order. $D(t)$ is not directly accessible, so that it has to be related to the returns through some $R = F(D)$.

There is unfortunately no definite consensus about the analytic form of F. This will prevent a convincing quantitative comparison between the model and empirical data. Nevertheless, for the purpose of illustration, we will assume that the price change during a time interval is proportional to the sum of the demand and sell orders from all the clusters which are active during this time interval. That is, we consider $F(D) \sim D$, unless specified otherwise. For $c = 1$ the probability density for the cluster size distribution decreases asymptotically as a power law

$$n_s \sim_{s \to \infty} \frac{A}{s^{5/2}} \tag{9.19}$$

while for $0 < 1 - c \ll 1$, the cluster size distribution is cut off by an exponential tail,

$$n_s \underset{s\to\infty}{\sim} \frac{A}{s^{5/2}} \exp\left(-\frac{(c-1)s}{s_0}\right) \qquad (9.20)$$

For $c=1$, the distribution has an infinite variance while for $c < 1$ the variance becomes finite because of the exponential tail. In this case the average size of a coalition is of order $1/(1-c)$ and the average number of clusters is of order $N(1-c/2)$. Setting the coordination parameter c close to 1 means that each agent tends to establish a link with one other agent. In the limit $N \to \infty$, the number ν_i of neighbors of a given agent i is a Poisson random variable with parameter c,

$$P(\nu_i = \nu) = e^{-c}\frac{c^\nu}{\nu!} \qquad (9.21)$$

The previous results are based in the assumption that either exactly one cluster of traders is making a transaction at each time step, or none. If the time interval of observation is increased enough to allow each cluster to be considered once during a time step, numerous clusters could decide to buy or sell, depending on the value of a. In this case, the probability distribution of the returns will be different from the cluster size distribution. By increasing a from close to 0 to a close to $1/2$, Stauffer and Penna (1998) have shown that the probability distribution of the returns changes from an exponentially truncated power-law distribution to a Gaussian distribution. At intermediate values, a continuous distribution with a smooth peak and a power law in the tails only is obtained, in agreement with a Levy distribution. This situation is similar to a change in time scale. At short time scales, such as every minute, there is either an order that is placed, or none, while for longer time scales, the observed variations on financial markets are the average result of thousands of orders. Increasing a from 0 to $1/2$ is similar to changing the time scale from a short-time scale to a large-time scale. Gopikrishnan et al. [59] have observed empirically this convergence towards a Gaussian distribution by changing the time scale for financial market indices. Recall that Ausloos and Ivanova [65] discussed the fat tails in another way, arguing that they are caused by some "dynamical process" through a hierarchical cascade of short and long-range volatility correlations. Unfortunately, there are no correlations in the Cont–Bouchaud model to compare with this result.

The Cont–Bouchaud model predicts a power-law distribution for the size of associations, with an exponential cut-off. When c reaches one, a finite fraction of the market simultaneously shares the same opinion and this leads to a crash. Unfortunately, this crash lacks any precursor pattern because of the lack of time correlations, amongst other possible shortcomings. From the expected relation $R = F(D)$, it leads, similarly, to power-law tails for the distribution of returns. However, the exact value of the exponent of this power law is still a matter of debate because of the lack of consensus upon $F(D)$. We report in Table 9.2 a summary of the agreement of the model and its limitations.

Tab. 9.2 Summary of the type of agreement between empirical data and the Cont–Bouchaud model.

Property	Agreement
Fat tails	Qualitative
Crossover	No
Symmetric	Yes
Clustering	No
Crashes	Yes
Precursors	No

Eguilúz and Zimmermann [89] introduced a dynamical version of the Cont–Bouchaud model. At each time step, an agent is selected at random. With probability $1 - 2a$, she selects another agent and a link between those two agents is created. Otherwise, this agent triggers her cluster to make a transaction, buying with probability a, selling with probability a. Eguilúz and Zimmermann associated the creation of a link with the exchange of information between agents, but when the agents make a transaction, they make their information public, that is, this information is lost. D'Hulst and Rodgers [90] have shown that this model is equivalent to the Cont–Bouchaud model, except that with the new interpretation, the probability that a cluster of size s makes a transaction is sn_s rather than n_s as in the Cont–Bouchaud model. An extension of the model allowing for the cluster distribution not to change, can be envisaged.

Finally, we should mention that numerical simulations show that it is possible to observe power laws with higher than expected effective exponent. In other words, size effects can modify the exponent. As all empirical data are strongly affected by size effects, looking after a model that reproduces the exact value of the exponent seems less important than trying to justify why the exponent should have a given value.

9.2.9
Crash Precursor Patterns

In the Cont–Bouchaud model for $p \geq p_c$, there is a nonzero probability that a finite fraction of agents decide to buy or sell. These large cooperative events generate anomalous wings in the tails of the distribution, that is, they appear as outliers of the distribution. The same pattern has been observed by Johansen and Sornette for crashes on financial markets, which means that the decision from buying or selling of the percolating cluster can be compared to a crash. These crashes, however, lack any precursor patterns, which have been empirically observed for real crashes, as originally proposed by Sornette, Johansen and Bouchaud.

As explained by Stauffer and Sornette [91] log-periodic oscillations, a crash precursor signature, can be produced with biased diffusion. The sites of a large lattice are randomly initialized as being accessible with probability p, or forbidden, with probability $1 - p$. A random-walker diffuses on the accessible sites only. With probability B, the random-walk moves in one fixed direction, while with probability $1 - B$, it moves to one randomly selected nearest neighbor. In both cases, the move is allowed only if the neighbor is accessible. Writing the time variation of the mean-square displacement from a starting point as $<r^2> \sim t^k$, k changes smoothly as a function of the logarithm of the time. That is, $k(t)$ approaches unity while oscillating according to $\sin(\lambda \ln t)$. Behind these log-periodic oscillations is the fact that the random-walker gets trapped in clusters of forbidden sites. The trapping time depends exponentially on the length of the traps, that are multiples of the lattice mesh size. The resulting intermittent random-walk is thus punctuated by the successive encounters of larger and larger clusters of trapping sites. This result is much more effective in reduced dimensionality, or equivalently, when $B \to 1$.

We have to stress that, even if this model is very closely related to percolation, there exists no explanation of how to relate the distance traveled by a diffusing particle, to transactions on financial markets. Connection to the sandpile model is still to be worked out. Hence, this model is not an extension of the Cont–Bouchaud model of financial markets, but rather a hint of a possible direction for how to implement log-periodic oscillations.

Notice that within the framework of the Langevin equation proposed by Bouchaud and Cont [92], a crash occurs after an improbable succession of unfavorable events, because they are initiated by the noise term. Hence, no precursor pattern can be identified within the original equation. Bouchaud and Cont extended their model and proposed the following mechanism to explain log-periodic oscillations. Every time an anomalously negative value u close to u^* is reached, the market becomes more "nervous". This is similar to saying that the width W of the distribution describing the noise term η increases to $W + \delta W$. Therefore, the time Δt between the next anomalously large negative value will shorten as large fluctuations become more likely. The model predicts

$$\Delta t_{n+1} = \Delta t_n S^{-\delta W/W} \tag{9.22}$$

where, to linear order in δW, S is some constant. This leads to a roughly log-periodic behavior, with the time difference between two events being a geometric series. The previous scenario predicts that a crash is not related to a critical point. That is, there is a crash because u reaches u^*, not because $\Delta t \to 0$.

9.3
The Lux–Marchesi Model

Lux and Marchesi [93] have introduced the model of a financial market where the agents are divided into two groups: fundamentalists and noise traders. Fundamentalists expect the price to follow the so-called fundamental value of the asset (p_f), which is the discounted sum of expected future earnings (for example, dividend payments). A fundamentalist strategy is to buy (sell) when the actual price is believed to be below (above) p_f. The noise traders attempt to identify price trends and patterns and also consider the behavior of other traders as a source of information. The noise traders are also considered as optimistic or pessimistic. The former tend to buy because they think the market will be rising, while the latter bet on a decreasing market and rush out of the market to avoid losses.

The dynamics of the model are based on the movements of agents from one group to another. A switch from one group to the other happens with a certain exponential probability $\nu e^{U(t)} \Delta t$, varying with time. ν is a parameter for the frequency of the revaluation of opinions or strategies by the agents. $U(t)$ is a forcing term covering the factors that are relevant for a change of behavior, and it depends on the type of agent considered. Noise traders use the price trend in the last time steps and the difference between the number of optimistic and pessimistic traders to calculate $U(t)$ corresponding to transitions between the group of optimists and pessimists. For example, observation of a positive trend and more optimists than pessimists is an indication of a continuation of the rising market. This would induce some pessimistic agents to become optimistic. The other type of transition is between fundamentalists and noise traders. The $U(t)$ function corresponding to such transitions is a function of the difference in profits made by agents in each group. An optimistic trader profit consists in short-term capital gain (loss) due to price increase (decrease), while a pessimistic trader gain is given by the difference between the average profit rate of alternative investments (assumed to be constant) minus the price change of the asset they sell. A fundamentalist profit is made from the difference between the fundamental price and the actual price, that is, $|p - p_f|$ is associated to an arbitrage opportunity. In practice, Lux and Marchesi (1999) multiply this arbitrage profit by a factor < 1 to take into account the time it takes for the actual price to reverse to its fundamental value.

To complete the model, it remains to specify how the price and its fundamental value are updated. Price changes are assumed to be proportional to the aggregate excess demand. Optimistic noise traders are buying, pessimistic noise traders are selling and fundamentalists are all buying or all selling depending whether $p_f - p$ is positive or negative, respectively. Finally, relative changes in the fundamental price are Gaussian random variable, $\ln p_{f,t} - \ln p_{f,t-1} = \epsilon_t$, where ϵ_t is a normally distributed random variable, with mean zero and time-invariant variance σ_ϵ^2.

Lux and Marchesi claimed that a theoretical analysis of their model shows that the stationary states of the dynamics are characterized by a price which is on average equal to its fundamental value. This is supported by numerical simulations of the model, where it can be seen that the price tracks the variation of its fundamental value. But comparing price returns, by construction, the fundamental returns are Gaussian, while price returns are not. The distribution of price returns displays fat tails that can be characterized by a power law of exponent $\tau = 1 + \alpha$. Numerically, Lux and Marchesi obtained $\alpha = 2.64$ when sampling at every time step. Increasing the time lag, a convergence towards a Gaussian is observed. They measured a Hurst exponent $H = 0.85$ for the price returns, showing strong persistence in the volatility. The exact value of the exponent is close to empirical data.

The behavior of the model comes from an alternation between quiet and turbulent periods. The switch between the different types of periods are generated by the movements of the agents. Turbulent periods are characterized by a large number of noise traders. There exists a critical value N_c for the number of traders, such as when there are more than N_c traders, the system looses stability. However, the ensuing destabilization is only temporary as the presence of fundamentalists and the possibility to become a fundamentalist ensure that the market always stabilizes again. These temporary destabilizations are known as on-off intermittency in physics.

9.3.1
The Spin Models

There are necessarily crash and financial market aspects which resemble phase transitions; this has led into producing a realm of spin models. The superspin model proposed by Chowdhury and Stauffer [94] is related to the Cont–Bouchaud model presented here above. The cluster of connected agents are replaced by a superspin of size S. $|S_i|$ is the number of agents associated with the superspin i. The value of $|S_i|$ is chosen initially from a probability distribution $P(|S_i|) \sim |S_i|^{-(1+\alpha)}$. A superspin can be in three different states, $+|S_i|$, 0 or $-|S_i|$. Associated to each state is a "disagreement function" $E_i = -S_i(H_i + h_i)$, where $H_i = J\sum_{j \neq i} S_j$ is a local field and h_i an individual bias. E_i represents the disagreement between the actual value of a superspin and two other factors. One of these factors, H_i, is an average over the decision of the other groups of agents. The other factor, h_i, is a personal bias in the character of a group of agents, with groups of optimists for $h_i > 0$, and pessimists for $h_i < 0$. Groups with $h_i = 0$ are pure noise traders. At each time step, every superspin i chooses one of its three possible states, $+|S_i|$ with probability a (buying), $-|S_i|$ with probability a or 0 with probability $1 - 2a$ (doing nothing). A superspin is allowed to change from its present state to the new chosen state with probability $e^{-\Delta E_i/k_b T}$, where ΔE_i is the expected change in its disagreement function. The magnetization M corresponds to the aggregate excess demand, D.

If a linear relation $F(D)$ is assumed between the returns and D, the previous model is characterized by a probability distribution for the prices with power-law tails with the same exponent $1 + \alpha$, as the distribution of spins. In contrast to the Cont–Bouchaud model, this result stays true for all values of a, that is, there is no convergence towards a Gaussian for any value of a, contrary to what is observed on financial markets. Using h_i, it is possible to divide the population in noise traders for $h_i \ll H_i$ and fundamentalists for $h_i \gg H_i$. It is also possible to introduce a dynamics in h_i to reflect the different opinions in different states of the market. No complete study of these properties has been done to date.

Acknowledgments

The author thanks René D'Hulst for an important contribution to this work. MA would also like to thank the many coworkers who helped to make this review possible, i.e., during the last few years, Ph. Bronlet and K. Ivanova. Part of this work has been supported through the Ministry of Education under contract ARC (94-99/174) and (02/07-293) of ULg.

References

1 BACHELIER, L., *Ann. Sci. Ecole Norm. Sup.* 3 (**1900**), p. 21

2 PARETO, V., *Cours d'Economie Politique*, Lausanne and Paris, **1897**

3 MANDELBROT, B. B., *J. Business* 36 (**1963**), p. 294–298

4 FAMA, E. F., *J. Business* 35 (**1963**), pp. 420–429

5 LÉVY, P., *Théorie de l'Addition des Variables Aléatoires*, Gauthier-Villars, Paris, **1937**

6 MANTEGNA, R. N., STANLEY, H. E., *Phys. Rev. Lett.* 73 (**1994**), pp. 2946–2949

7 KOPONEN, I., *Phys. Rev. E* 52 (**1995**), pp. 1197–1199

8 ANDERSON, P. W., ARROW, K., PINES, D., *The Economy as an Evolving Complex System*, Addison-Wesley, Reading, MA, **1988**

9 BOUCHAUD, J.-P., POTTERS, M., *Théorie des Risques Financiers*, Aléa-Saclay, Eyrolles, **1997**; *Theory of Financial Risks*, Cambridge University Press, Cambridge, **2000**

10 MANTEGNA, R. N., STANLEY, H. E., *An Introduction to Econophysics: Correlations and Complexity in Finance* Cambridge University Press, Cambridge, **1999**

11 PAUL, W., BASCHNAGEL, J., *Stochastic Processes from Physics to Finance*, Springer, **1999**

12 VOIT, J., *The Statistical Mechanics of Financial Markets*, Springer Verlag, **2001**

13 RAMSDEN, J. J., KISS-HAYPÁL, G., *Physica A* 277 (**2000**), pp. 220–227

14 VANDEWALLE, N., AUSLOOS, M., *Physica A* 246 (**1997**), pp. 454–459

15 BULDYREV, S., DOKHOLYAN, N. V., GOLDBERGER, A. L., HAVLIN, S., PENG, C. K., STANLEY, H. E., VISWANATHAN, G. M., *Physica A* 249 (**1998**), pp. 430–438

16 WEST, B. J., DEERING, B., *The Lure of Modern Science: Fractal Thinking*, World Scientific, Singapore, **1995**

17 MALAMUD, B. D., TURCOTTE, D. L., *J. Stat. Plan. Infer.* 80 (**1999**), pp. 173–196

18 IVANOVA, K., AUSLOOS, M., *Physica A* 265 (**1999**), pp. 279–286

19 AUSLOOS, M., IVANOVA, K., *Braz. J. Phys.* 34 (**2004**), pp. 504–511

20 AUSLOOS, M. *Physica A* 285 (**2000**), pp. 48–65

21 VANDEWALLE, N., AUSLOOS, M., *Int. J. Phys. C* 9 (**1998**), pp. 711–720

22 AUSLOOS, M., VANDEWALLE, N., BOVEROUX, PH., MINGUET, A., IVANOVA, K., *Physica A* 274 (**1999**), pp. 229–240

23 AUSLOOS, M., IVANOVA, K., *Physica A* 286 (**2000**), pp. 353–366

24 AUSLOOS, M., IVANOVA, K., *Int. J. Mod. Phys. C* 12 (**2001**), pp. 169–196

25 AUSLOOS, M., IVANOVA, K., *Eur. Phys. J. B* 27 (**2002**), pp. 239–247

26 IVANOVA, K., AUSLOOS, M., *Eur. Phys. J. B* 20 (**2001**), pp. 537–541

27 AUSLOOS, M., IVANOVA, K., in *New Directions in Statistical Physics - Econophysics, Bioinformatics, and Pattern Recognition*, (Ed. L.T. Wille), Springer Verlag, Berlin, **2004** pp. 93–114

28 VANDEWALLE, N., AUSLOOS, M., *Int. J. Comput. Anticipat. Syst.*, 1 (**1998**), pp. 342–349

29 ALESIO, E., CARBONE, A., CASTELLI, G., FRAPPIETRO, V., *Eur. J. Phys. B* 27 (**2002**), pp. 197–200

30 VANDEWALLE, N., AUSLOOS, M., BOVEROUX, PH., *Physica A* 269 (**1999**), pp. 170–176

31 AUSLOOS, M., IVANOVA, K., *Comp. Phys. Commun.* 147 (**2002**), pp. 582–585

32 IVANOVA, K., AUSLOOS, M. *Eur. Phys. J. B* 8 (**1999**), pp. 665–669; Err. 12 (**1999**), 613

33 VANDEWALLE, N., AUSLOOS, M. *Eur. J. Phys. B* 4 (**1998**), pp. 257–261

34 LUX, T., AUSLOOS, M., in *The Science of Disaster : Scaling Laws Governing Weather, Body, Stock-Market Dynamics*, (Eds. A. Bunde, J. Kropp, H.-J. Schellnhuber), Springer Verlag, Berlin, **2002**, pp. 377–413

35 ZIPF, G. K., *Human Behavior and the Principle of Least Effort*, Addison Wesley, Cambridge MA, **1949**

36 VANDEWALLE, N., AUSLOOS, M., *Physica A* 268 (**1999**), pp. 240–249

37 AUSLOOS, M., IVANOVA, K., *Physica A* 270 (**1999**), pp. 526–542

38 BRONLET, PH., AUSLOOS, M., *Int. J. Mod. Phys. C* 14 (**2003**) pp. 351–365

39 AUSLOOS, M., BRONLET, PH., *Physica A* 324 (**2003**) pp. 30–37

40 LALOUX, L., CIZEAU, P., BOUCHAUD, J.-P., POTTERS, M., *Phys. Rev. Lett.* 83 (**1999**), pp. 1467–1470

41 LALOUX, L., POTTERS, M., CONT, R., AGUILAR, J.-P., BOUCHAUD, J.-P., *Europhys. Lett.* 45 (**1999**), pp. 1–5

42 PLEROU, V., AMARAL, L. A. N., GOPIKRISHNAN, P., MEYER, M., STANLEY, H. E., *Nature* 400 (**1999**), pp. 433–437

43 PLEROU, V., GOPIKRISHNAN, P., ROSENOW, B., AMARAL, L. A. N., STANLEY, H. E., *Phys. Rev. Lett.* 83 (**1999**), pp. 1471–1474

44 MANTEGNA, R. N., *Eur. Phys. J. B* 11 (**1999**), pp. 193–197

45 BONANNO, G., VANDEWALLE, N., MANTEGNA, R. N., *Phys. Rev. E* 62 (**2000**), pp.R7615-R7618

46 VANDEWALLE, N., BOVEROUX, P., BRISBOIS, F., *Eur. Phys. J. B* 15 (**2000**), pp. 547–549

47 FABRETTI, A., AUSLOOS, M., *Int. J. Mod. Phys. C* 16 (**2005**), pp. 671–706

48 ECKMANN, J. P., KAMPHORST, S. O., RUELLE, D., *Europhys Lett.* 4 (**1987**), pp. 973–977

49 ZBILUT, J. P., WEBBER, C. L., *Phys Lett. A* 171 (**1992**), pp. 199–203

50 ZBILUT, J. P., WEBBER, C. L., GIULIANI, A., *Phys Lett. A* 246 (**1998**), pp. 122–128

51 LAMBERTZ, M., VANDENHOUTEN, R., GREBE, R., LANGHORST, P., *Journal of the Autonomic Nervous System* 78 (**2000**), pp. 141–157

52 ENGLE, R. F., *Econometrica* 50 (**1982**), pp. 987–1008

53 BOLLERSLEV, T., *J. Econometrics* 31 (**1986**), pp. 307–327

54 ENGLE, R. F., BOLLERSLEV, T., *Econometric Reviews* 5 (**1986**), pp. 1–50

55 Nelson, D. B., *Econometrica* 59 (**1991**), pp. 347–370

56 Baillie, R. T., Bollerslev, T., Mikkelsen, H. O., *J. of Econometrics* 74 (**1996**), pp. 3–30

57 Mantegna, R. N., Stanley, H. E., *Nature* 376 (**1995**), pp. 46–49

58 Mantegna, R. N., Stanley, H. E., *Physica A* 254 (**1998**), pp. 77–84

59 Gopikrishnan, P., Plerou, V., Amaral, L. A. N., Meyer, M., Stanley, H. E., *Phys. Rev. E* 60 (**1999**), pp. 5305–5316

60 Gopikrishnan, P., Meyer, M., Amaral, L. A. N., Stanley, H. E., *Eur. Phys. J. B* 3 (**1998**), pp. 139–140

61 Lux, T., *Applied Financial Economics* 6 (1996), pp. 463–475

62 Plerou, V., Gopikrishnan, P., Amaral, L. A. N., Gabaix, X., Stanley, H. E., *Phys. Rev. E* 62 (**2000**), R3023-R3026

63 Scalas, E., *Physica A* 253 (**1998**), pp. 394–402

64 Michael, F., Johnson, M. D., *Physica A* 320 (**2003**), pp. 525–534

65 Ausloos, M., Ivanova, K., *Phys. Rev. E* 68 (**2003**), p. 046122

66 Tsallis, C., Anteneodo, C., Borland, L., Osorio, R., *Physica A* 324 (**2003**), pp. 89–100

67 Johansen, A., Sornette, D., *Eur. Phys. J. B* 1 (**1998**), pp. 141–143

68 Feigenbaum, J. A., Freund, P. G. O., *Int. J. Mod. Phys B* 10 (**1996**), pp. 3737–3745

69 Sornette, D., Johansen, A., Bouchaud, J.-P., *J. Phys. I* 6 (**1996**), pp. 167–175

70 Vandewalle, N., Boveroux, P., Minguet, A., Ausloos, M., *Physica A* 255 (**1998**), pp. 201–210

71 Vandewalle, N., Ausloos, M., Boveroux, P., Minguet, A., *Eur. Phys. J. B* 4 (**1998**), p. 139–141

72 Vandewalle, N., Ausloos, M., Boveroux, P., Minguet, A., *Eur. J. Phys. B* 9 (**1999**), pp. 355–359

73 Drozdz, S., Ruf, F., Speth, J., Wójcik, M., *Eur. Phys. J. B* 10 (**1999**), pp. 589–593

74 Dupuis, H, *Trends/Tendances* 22 (38) (**1997**), p. 26; ibid. 22 (44) (**1997**), p. 11

75 Legrand, G., *Cash* 4 (38) (**1997**), p. 3

76 Sornette, D., *Phys. Rep.* 297 (**1998**), pp. 239–270

77 David, B., *Transaction Survey 1994* Stock Exchange Quaterly, London Stock Exchange, October-December **1994**

78 Eliezer, D., Kogan, I. I., cond-mat/9808240 (**1998**)

79 Cohen, K., Maier, S., Schwartz, R., Whitcomb, W., *J. Pol. Econ.* 89 (**1981**), pp. 287–305

80 Maslov, S., *Physica A* 278 (**2000**), pp. 571–578

81 Slanina, F., *Physica A* 286 (**2000**), pp. 367–376

82 Bak, P., Tang, C., Wiesenfeld, K., *Phys. Rev. A* 38 (**1988**), pp. 364–374

83 Bak, P., *How Nature Works* Copernicus, New York, **1996**

84 Ausloos, M., Ivanova, K., Vandewalle, N., in *Empirical sciences of financial fluctuations. The advent of econophysics*, Tokyo, Japan, Nov. 15-17, 2000 Proc.; (Ed. H. Takayasu) Springer Verlag, Berlin, **2002** pp. 62–76

85 Stauffer, D., Aharony, A., *Introduction to Percolation Theory*, 2nd Ed. Taylor & Francis, London, **1991**

86 Sahimi, M, *Applications of Percolation Theory*, Taylor & Francis, London, **1994**

87 Cont, R., Bouchaud, J.-P., *Macroeconomics Dynamics* 4 (**2000**), pp. 170–196

88 Kirman, A., *Economics Letters* 12 (**1983**), pp. 101–108

89 Eguíluz, V. M., Zimmermann, M. G., *Phys. Rev. Lett.* 85 (**2000**), p. 5659–5662

90 D'Hulst, R., Rodgers, G. J., *Int. J. Theo. App. Finance* 3 (**2000**), p. 609; also in cond-mat/9908481

91 Stauffer, D., Sornette, D., *Physica A* 252 (**1998**), pp. 271–277

92 Bouchaud, J.-P., Cont, R., *Eur. Phys. J. B* 6 (**1998**), pp. 543–550

93 Lux, T., Marchesi, M., *Nature* 397 (**1999**), pp. 498–500

94 Chowdhury, D., Stauffer, D., *Eur. Phys. J. B* 8 (**1999**), pp. 477–482

10
A Thermodynamic Formulation of Social Science
Juergen Mimkes

The thermodynamic formulation of social laws is based on the law of statistics under constraints, the Lagrange principle. This law is called the "free energy principle" in physics and has been applied successfully to all fields of natural science. In social sciences it may be called the "principle of maximum happiness" of societies. The principle may be applied to collective and individual behavior of social, political or religious groups. Opinion formation like US elections may be simulated, as well as the division of the world into hierarchic and democratic countries. The principle applies to data of segregation of Blacks and Whites in the US, to citizen and foreigners in Germany, to political and religious separation in Northern Ireland and Bosnia.

10.1
Introduction

Social models have been presented by many authors like Th. Schelling, W. Weidlich, R. Axelrod, S. Solomon, D. Stauffer, S. Galam and many others. In this paper the Lagrange principle is presented as a very general approach to social science. Socio-economics deals with large systems, the people of a town, a state or a continent. In large crowds people do not know each other. A few symbols like face, skin, hair, clothing or language indicate race, gender, ethnicity, country or religion. The reactions and interactions in large groups depend only on these little bits of information. And since there is often not enough time, most people will quickly generalize and attribute properties to large groups: American travelers are rich; Japanese workers are busy; Jewish business men are smart; Italian men are charming. These generalizations are the properties of groups in the heads of other people, whether they are true or not. This is an "atomization" of social interaction. Accordingly, the elements of socio-economic systems are not people, but agents with few obvious properties. The behavior of agents is determined by probability under constraints. The Lagrange principle is the basis for the understanding and simulation of socio-economic interactions.

10.2
Probability

Calculations of probability have their roots in games of chance and gambling. Figure 10.1 shows a simple nail board, a predecessor of modern flipper machines. Each time a ball hits a nail, the ball has an even chance to fall to the left or to the right side. Each ball has its own course, but at the end all balls fall into different boxes and form a curve shaped like a bell, the normal distribution.

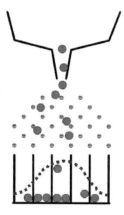

Fig. 10.1 Galton's nail board. A number of balls pass through a nail board and either fall to the left or right side. The course of each single ball is unpredictable and chaotic, but the final distribution in the five boxes leads to a bell-shaped normal distribution.

This normal distribution can be predicted more prescisely the larger the number of balls used. This is a simple but surprising result. The course of each single ball cannot be predicted by mechanics due to the six bifurcations, which lead to a chaotic movement of the balls. But still the sum of all courses is a well-defined normal distribution. This bell-shaped distribution demonstrates that there is order in chaos. The same may be expected for a society of millions of chaotic individuals, who cannot be predicted individually but only as a society.

10.2.1
Normal Distribution

The distribution of N cars parked on the two sides of a street is calculated from the laws of combinations: the probability of N_l cars parking on the left side and N_r cars on the right side of the street is given by

$$P(N_l; N_r) = \frac{N!}{N_l! N_r!} q^{N_l} (1-q)^{N-N_l} \tag{10.1}$$

Fig. 10.2 The distribution of one-half of the cars on each side of the street is most probable. According to Eq. (10.1) the probability for $N_l = 2$ and $N_r = 2$ is given by $P(2;2) = 6:16$ or 37.5%.

Fig. 10.3 The distribution of all cars on one side and none on the other side has always the least probability. For $N_l = 0$ and $N_r = 4$ we find the probability according to Eq. (10.1): $P(0;4) = 1:16$ or 6.25%.

In Fig. 10.2 the cars are evenly parked on both sides of the street. The probability of this even distribution has always the highest probability. According to Eq. (10.1) we find $P(2;2) = 6:2^4 = 6/16$ or 37.5%.

In Fig. 10.3 four cars are all parked on one side of the street and none on the other. The distribution on one side has always the lowest probability. For equal space on both sides, $q = 1/2$ in Eq. (10.1) we find $P(0;4) = 1:2^4 = 1/16$ or 6.25%.

10.2.2 Constraints

In Fig. 10.4 the "no parking" sign on the left side forces the cars to park on the right side, only. The "no parking" sign is a constraint, that enforces the least probable distribution of cars, all on one side. A constraint or law generally enforces a very improbable distribution, which would otherwise not be observed.

In Fig. 10.5 we find one individual driver ignoring the collective "no parking" sign. What is the probability of this unlawful distribution? Figure 10.4 and 10.5 are closely connected to the problem of behavior of socio-economic agents. People often act individually against collectively accepted rules, trends, opinions or laws. What is the probability of this distribution of cars on the road? The problem may be solved by looking at the laws of probability

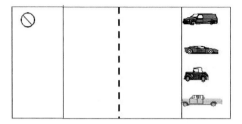

Fig. 10.4 The "no parking" sign enforces the least probable distribution of cars on one side of the street.

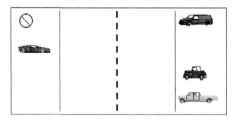

Fig. 10.5 One driver is ignoring the "no parking" sign.

with constraints.

10.2.3
Probability with Constraints (Lagrange Principle)

Probability with constraints may be calculated by a principle introduced by Joseph de Lagrange (1736–1813),

$$L = E + T \ln P \rightarrow \text{maximum!} \tag{10.2}$$

L is the Lagrange function of a system of N interacting elements. P is the combinatorial probability of the distribution of the elements, Eq. (10.1), and $\ln P$ is called entropy. The function E stands for a constraint of the system. In Fig. 10.5 the constraint is given by the "no parking" sign ⊘ on the left side of the road and the corresponding fine for violation. T is the Lagrange parameter. The maximum of L is called the equilibrium of the system. The meaning of the functions L, E, T and $\ln P$ varies with the system. The dimension of the functions depends on the constraint E, $\ln P$ is without dimension:

1. In atomic systems the constraint E is the energy, T is the mean kinetic energy or temperature, $(-L)$ is the free energy. $\ln P$ is the entropy. The atomic system is stable at the maximum of the free energy.

2. Social systems are governed by collective laws (E), $\ln P$ represents the individual behavior. T is the tolerance (of individual behavior), L is the

common happiness of the agents about the behavior in the society. The social system is stable at the maximum of mutual happiness.

3. In political systems emotions (E) between different groups are important constraints, $\ln P$ represents the difference of groups. T is the tolerance (of other groups). L is the happiness of the groups about the structure of the society. The political system is stable at the maximum of mutual happiness.

4. In economic systems the constraint E is the capital, $\ln P$ represents the individual chance. T is the mean capital or standard of living, L is the sum of everybody's self-interest. The economic system is stable, if everybody's self-interest is at a maximum.

The functions L, E, T, P in the Lagrange principle may be interpreted in very general terms. The function E is a law, opinion or decision that acts as a constraint and requires a specific order of the system. The entropy function $\ln P$, on the other hand, is a measure of disorder for the law (E). The parameter T is something like the tolerance of disorder of the law (E). A typical example of order and disorder is a child's room. The order of toys is usually defined by the parents, but due to the high probability of disorder a child's room will be chaotic, if the parents tolerate just a little bit of disorder. The Lagrange parameter T acts as a switch between order and disorder.

The Lagrange principle (10.2) may be interpreted in many different ways: see Table 10.1.

Apparently, the Lagrange equation is a principle that goes far beyond physics or socio-economics. It is one of the most general principles in all aspects of life.

10.3
Elements of Societies

Many people believe that human behavior cannot be calculated by any model, especially if the model has been adopted from atoms and molecules. Human individuals are much more complex than atoms or molecules!

Indeed, no mathematical equation can model the behavior of a single person. But a statistical approach to human nature never applies to single persons, only to large crowds. In a large crowd individual properties are lost and groups may be labelled by certain features:. In a football stadium the competing parties wear different shirt colors. In the streets foreigners are recognized by their accent. Dress, uniform, language, sex, skin, hair are the most obvious labels for unknown people. All other properties remain undiscovered.

10 A Thermodynamic Formulation of Social Science

Tab. 10.1 Interpretations of the Lagrange principle.

Lagrange:	$L = E$	+	$T \ln P$	→maximum!
Probability:	L = improbable	+	T probable	→maximum!
Physics:	L = energy	+	T entropy	→maximum!
Law:	L = order	+	T disorder	→maximum!
Structure:	L = simple	+	T complex	→maximum!
Behavior:	L = collective	+	T individual	→maximum!
Appearance:	L = uniform	+	T variety	→maximum!
Organization:	L = organized	+	T disorganized	→maximum!
Behavior:	L = planned	+	T spontaneous	→maximum!
Welfare:	L = fair	+	T unfair	→maximum!
Ethics:	L = good	+	T bad	→maximum!
Law:	L = right	+	T wrong	→maximum!
Art:	L = beautiful	+	T ugly	→maximum!
Health:	L = healthy	+	T sick	→maximum!
Biology:	L = living	+	T dead	→maximum!
Engineering:	L = correct	+	T defect	→maximum!
Chemistry:	L = cohesion	+	T entropy	→maximum!
Society:	L = bonds	+	T freedom	→maximum!
Economics:	L = ratio	+	T chance	→maximum!

10.3.1
Agents

Agents are the elements of social systems. Agents are representative of persons, as they appear in the mind of other people. Agents are defined by their properties:

Agent: $[A; B; G; H; .....]$

By adding an infinite number of features it would be possible to model a real individual person. But in large crowds people will only notice very few properties of an individual and accordingly, all individuals may be categorized by few properties. These properties may be scalable like tall, fast, strong, beautiful, smart and rich. Scalable properties lead to classes. Unscalable properties are female, black, Muslim, Indian. They lead to groups.

10.3.2
Groups

A group is a crowd of agents with a common unscalable group property. The group property is often linked to a label (A). Women wear different clothes from men; policemen or nurses have their specific uniforms; foreigners have a specific accent; black people stand out in a white society. The group label

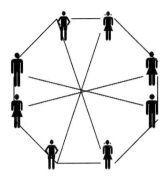

Fig. 10.6 shows a one-dimensional (cyclic) arrangement of equivalent agents in a group. All agents have equivalent nearest-neighbor bonds. The center (hub) is the nucleus of the group (president of a club, teacher of a school class, policeman in traffic).

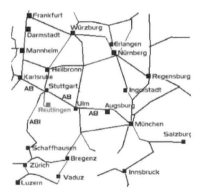

Fig. 10.7 The train network in southern Germany is a two-dimensional lattice of equivalent cities. Nearly all cities have equivalent nearest neighbor connections.

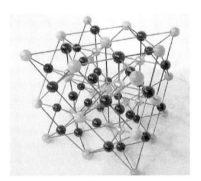

Fig. 10.8 A crystal is a three-dimensional lattice of equivalent atoms. All atoms have equivalent nearest (and second nearest) neighbor bonds. (For a color version please see color plate on page 597.)

may be an idea, an idol or a real person, a president of a club or company, the priest of a parish, the leader of a party, the nucleus of a crystal, the α-animal of a herd. A policeman arrests in the name of the State, a priest marries a couple in the name of the Lord, and a judge makes a decision in the name of the People. In this way it is possible to find a single name for large groups: Europeans, Catholics, Buddhists, British, French, Yankees, Green party, merchants, scientists, workers and tourists. The most obvious labels are: complexion, sex, heritage, nationality, faith, education, profession, etc. Within a group all agents are equivalent. Agents of different groups are also equivalent. There is no natural order for British, French or American citizen, for Moslems, Christians or Jews, for blue, yellow or red shirts of a football team. The loss of individuality leads to the "atomization" of societies. Groups in societies correspond to components in atomic systems. Agents may have several properties at the same time. Bur at one instant an agent will generally show only one specific group property:

Example: A policemen may be married, British and wearing a yellow shirt. But within one group only one property is valid, all other properties are turned off: in his job he is a policeman, at home he is a husband, as a voter he is British and on the football field he wears the yellow shirt of his team.

The size of a group is determined by the number (N_A) of agents with the same group property (A). In social models the size of a group should be larger than one hundred in order to apply the probability calculations with reasonable accuracy.

10.3.3
Interactions

With our subjective impressions of people we develop certain emotions as a response to specific group labels. The uniform of a policeman in a dark street makes us feel secure, the white dress of a doctor raises hope for the sick, the robe of a judge indicates justice, the suit of businessmen, honesty. People of one country will be bound to their heritage, people of one faith will gather in a common church service, scientists will meet at conferences to talk about their field. The group interaction (E_A) is positive, agents of one group are attracted to each other. The solidarity of a group will grow with the number of common labels of the agents. A famous example ist the WASP population (White, Anglo-Saxon, Protestant) in the USA, which feels closely related and tries to maintain power in the country.

The attractive force leads to a specific structure of groups. Figure 10.6 shows a cyclic linear chain and represents a group of equivalent agents with positive nearest-neighbor bonds. All agents are connected to the centre (hub), the representative (king, president, head) of the group. Figure 10.7 shows the train system in southern Germany as a two-dimensional network of equivalent cities. There are only nearest-neighbor connections. (The hub is not shown in the track system, as the head office is connected to the stations by telephone.) A similar network is the system of main highways in most countries, connecting equivalent big cities. All cities again are hubs of local road systems.

In Fig. 10.8 the crystal model shows the three-dimensional network of equivalent atoms. Only nearest neighbors are shown, but second and higher-neighbor interactions are also present. (The hub of the crystal is the electromagnetic field, which is again not visible). The stable structure of a group of agents in socio-economic systems is determined by the maximum of the Lagrange function (10.2).

10.3.4
Classes

A class is a crowd of agents with a common scalable group property. The class property is linked to the Lagrange parameter (T) of the Lagrange function (10.2):

1. In atomic systems the constraints are cohesive energies (E). The Lagrange parameter is a mean energy per atom or temperature, $T = E/N$. This leads to the scalable property "warm".

2. In social systems the constraints are given by emotions (E). Accordingly, the Lagrange parameter tolerance of disorder (T) also has an emotional dimension and leads to the scalable property "tolerant".

3. In economic systems capital is a constraint (E), the Lagrange parameter (T) is a market index, a mean capital per capita or standard of living, $T = E/N$. This leads to the scalable property "rich".

Each system has a specific constraint (E), the Lagrange parameter $T = E/N$, and the corresponding scalable property. Other examples for class properties are tall, fast, strong or beautiful. Agents with the same scalable property are part of the same class. Within a class all agents are nearly equivalent. Members of different classes are not equivalent and may be lined up in a "vertical way". Figure 10.9 shows the vertical structure of classes in different societies: pope, priests, laymen in the Catholic church; king, nobility, peasants in medieval times; general, officers, soldiers in the army; director, managers, workers of a company; professor, assistants, students in a university; adults, youth, children in the family.

The class properties are the constraints (E) of society. Agents in the same class form a group with nearly the same value of (T). Agents in different

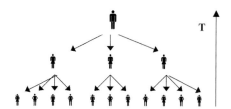

Fig. 10.9 Vertical hierarchic structure of classes in different societies: pope, priests, laymen in the Catholic church; king, nobility, peasants in medieval times; general, officers, soldiers in the army; director, managers, workers of a company. Agents in the same class form a group with the same value of (T). Agents in different classes are at different values of (T). The communications and orders are only one-directional, from the boss to the subordinates. In addition to the arrows of orders the hub has ties of information to all subordinates.

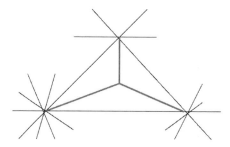

Fig. 10.10 Top view of a hierarchic structure according to Fig. 10.9: three local hubs are connected to the centre hub (top). This network corresponds to international telephone networks with national hubs, international airline cooperation, national computer networks, organization network of companies, schools, churches, etc.

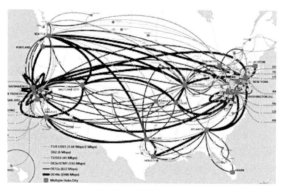

Fig. 10.11 Internet structure of UUNET, one of the world's leading Internet networks. © 2000 - UUNET, an MCI WorldCom Company. (For a color version please see color plate on page 598.)

classes are at different values of (T), this leads to a hierarchic structure. Agents of one class are linked to a person of a higher class, as in Fig. 10.6, indicating that the hub of a group is from a different class. Figure 10.10 shows the hierarchic structure of Fig. 10.9 in a view from the top. The hub has connections to all smaller centres. Figure 10.11 shows the structure of a real network, the Internet (UUNET).

10.3.5
States: Collective vs Individual

The Lagrange equation (10.2) contains two different states: 1. constraints (E) represents the collective, ordered state; and 2. entropy ($S = \ln P$) stands for the individual, chaotic state.

The state of a system depends on the Lagrange parameter (T). This may be observed in atomic systems as well as in socio-economic states.

At low values of T there is little entropy, atoms are in the ordered, collective, solid state, Fig. 10.12. In social systems, at work in the army, agents face a low tolerance (T) of individual behavior and have only one choice (P) like "yes, sir" or "amen". These agents are in a well-ordered, collective and hierarchic state, Fig. 10.14. They cannot decide for themselves.

Fig. 10.12 At low temperature (T) atomic systems are in the ordered, single-crystal state (model).

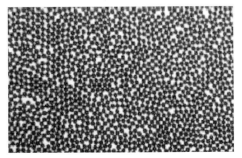

Fig. 10.13 At high temperature (T) atomic systems generally turn into a disordered liquid state (model).

At high values of T there is much entropy, atoms are in the liquid state, Fig. 10.13. After work, the tolerance for personal decisions is high, agents have many choices (P) like "yes" or "no" or "perhaps". These agents are in the individual, disordered state, Fig. 10.15.

Fig. 10.14 At a parade (low tolerance (T) of disorder) soldiers march in perfect order. (With kind permission of the Bundeswehr.)

Fig. 10.15 After work (high tolerance (T) of disorder) the soldiers relax in disorder. (With kind permission of the Bundeswehr.)

States and classes depend on the parameter T. But different classes only show a gradual change with T within the collective state. There are no classes in the individual, disordered state.

Collective and individual states can be observed in all aspects of social and economic life. In economic systems poor people work in production lines sometimes even in collective movement. Most people with low income work in companies at low positions with no opportunity for personal decisions. Only with a higher income will employees have the chance for more individual actions.

In religious systems, people in rich countries promote the right of free, individual opinion. In poor countries people react collectively to individual provocations, as in the cartoon crisis between Denmark and the Moslem world.

In normal human development, children still have a low ability (T). They live collectively in the family hierarchy, they have to obey their parents, they cannot cross the street at their own will. After puberty young people may already begin things on their own, but they still need the family bonds. Only after graduation from school by acquiring their first job will people be "considered grown up" and live their individual lives.

10.4
Homogenious Societies

In a homogenious society all agents have the same label, all belong to the same group, like Blacks, French or Catholics.

10.4.1
The Three States of Homogeneous Societies

Figure 10.16 shows the phase diagram of homogeneous (socio-economic or atomic) societies. At low values of the Lagrange parameter (T) a system will be in the ordered state: socio-economic systems will be in the collective state, political systems in the hierarchic state and atomic systems in the ordered solid state. Above a critical value of T_c the systems will change into the disordered state: socio-economic systems will change into the individual state, political systems into the democratic state and atomic systems in the liquid state. At very high values of T the systems turn into another chaotic state: social and political systems become global and atomic states a gas. This is shown in Fig. 10.16.

10.4.1.1 Atomic Systems: H_2O

Ordered state (solid). At low temperatures (T) the solid is well ordered and inflexible. Atoms have strong bonds and a low mobility. The solubility or tendency to mix with other atoms is low, we usually find segregation.

Chaotic state (liquid). At higher temperatures (T) the liquid is disordered and flexible. Atoms have weak bonds and a higher mobility. The solubility or tendency to mix with other atoms is higher, we find less segregation.

Global state (gas). At very high temperatures (T) the gas is disordered and flexible. Atoms have no bonds and a very high mobility. The tendency to mix with other atoms is very high, we always have full solubility.

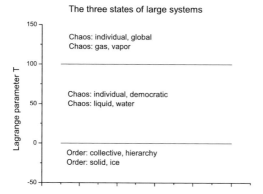

Fig. 10.16 Three states of systems: at low values of the Lagrange parameter (T) the ordered state is stable. Social systems are collective, political systems hierarchic, atomic systems solid. At higher values of T the system turns into the chaotic state: social systems become individual, political systems become democratic, atomic systems become liquid. At very high values of T the system turns into another chaotic state: social and political systems become global and atomic states a gas.

10.4.1.2 Social Systems: Guided Tours

Ordered (collective) state. At low knowledge (T) of the foreign country the tourists follow the guide in a well-ordered and collective manner. The rules for exceptions are inflexible. Tourists have strong bonds to their guide and a low mobility of their own. The tendency to mix with people of the foreign country is low, the tourists usually segregate.

Chaotic state (liquid). At better knowledge (T) of the foreign country the tourists follow the guide in a less ordered and less collective manner. The rules for exceptions are flexible. Tourists have weak bonds to their guide and a higher mobility of their own. The tendency to mix with people of the foreign country is higher, we have less segregation.

Global state (gas). At very good knowledge (T) of the foreign country the tourists have no bonds to a guide and do the sightseeing individually, they have a very high mobility of their own. The tendency to mix with people of the foreign country is very high, they are fully integrated.

10.4.1.3 Economic Systems: Companies

Workers. At low mean income (T) employees often are workers in a well ordered and collective group. The rules of work are inflexible. Workers are strongly bound to their boss and make few individual decisions. They do not communicate with many coworkers.

Higher employee. At higher mean income (T) employees work in a group with more individual initiative. The working rules are more flexible. Workers are less bound to their boss and make more individual decisions. Higher employees have to communicate with many workers.

Employer. At very high mean income (T) people are employers (directors) and work only with their own individual initiative. There are no working rules. Employers have to be very flexible and communicate with all workers. The hierarchy corresponds to Fig. 10.9. The communication is one-directional from the director to the worker.

10.4.1.4 Political Systems: Countries

Figure 10.17 shows the standard of living $T = \text{GNP/capita}$ in 1995 US\$ for all countries as a function of population N (after [2]).

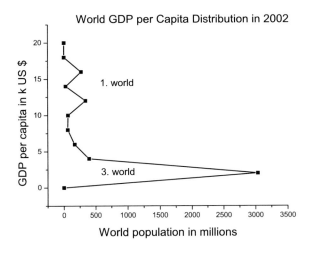

Fig. 10.17 The distribution of the world Gross National Product (GNP) in US\$ per person (1995) is quite uneven. N. America, W. Europe, Japan are at the top, democratic and capitalistic. C. Africa, S. E. Asia are very poor and in a hierarchic, non-capitalistic state (after Barro, 1995). The borderline between democratic/capitalistic and non-democratic/non-capitalistic countries lies somewhere between 2.5 and 5 kUS\$.

1. Countries in North America, Western Europe, Japan and Australia with a high living standard 10 000 US\$ per person are capitalistic economies, the people live in a stable democratic state. Democracies are rich, activities are carried out by private individuals and not ordered by the government, the structures of democracy are flexible, a president will stay

for one or two elective periods. The mobility of the population is high. Family bonds are weak, grandparents do not generally live under one roof with their grandchildren, the average number of children is below two, $f < 2$.

2. Countries with a standard of living less than 2 500 US$ (1995) have non-capitalistic economies, and people mostly live in non-democratic hierarchic structures. Hierarchies are poor, the collective order is high, the structures are generally inflexible, a monarch or dictator is ruling for a lifetime. The mobility of people is low. Families are often large and live all beneath one roof, the fertility is high, $f > 3$.

Obviously, the standard of living acts as an ordering parameter (T) between social states and corresponds to the temperature in materials. At low GNP per capita or standard of living (T) the political state of countries is a collective hierarchy. At high values of (T) states are free, individual democracies.

This result compares to the collective solid state at low temperatures (T) and the individual liquid state at high temperatures. These results lead to an important result:

The standard of living (T) or the gross national product (GNP) per person determines whether or not a state may form a stable democracy!

10.4.2
Change of State, Crisis, Revolution

Different states in Fig. 10.16 are separated by a phase-transition, a crisis, where the old system breaks down. In religious systems the transition is called a reformation, in political systems the transition is called a revolution. In atomic systems this phase transition is called melting, the ordered solid changes abruptly into the disordered liquid.

10.4.3
Hierarchy, Democracy and Fertility

In Fig. 10.18 the standard of living (T) as GNP/person is plotted over fertility (f), the average number of children per woman in a country.

Figure 10.18 indicates that there is no common transition point from hierarchy to democracy. But we may draw a straight line between the diamonds of democratic countries and the squares of hierarchic countries. In this way we may divide the world population into two classes:

1. In hierarchic countries with a low standard of living the father is head of the family and works outside of the house. Many countries have no pension plans, and only a larger number of children can guarantee a

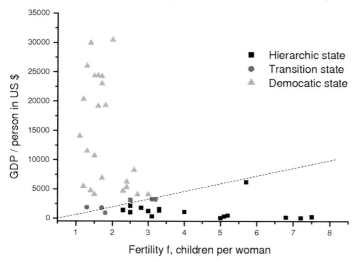

Fig. 10.18 shows the relationship between GNP per person and fertility for 90 states. Countries marked by diamonds are democratic and capitalistic; countries with squares are in a hierarchic state. Countries marked by triangles are in a transition state. The dashed line between democracies and hierarchies represents the line of transition. DSW Weltbevölkerungsbericht (2004).

reliable future for the parents. The mother stays at home to care for the children as well as for the old parents. The family bonds are strong and very important, as in solids.

2. In democratic countries with a high standard of living men and women work hard to obtain the high standard of living. The future depends on their work and the pension plans, not on children. In order to obtain a high living standard, people in democratic, capitalistic countries cannot afford to have many children. The mean number of children per family is less than two, $f < 2$. Families also have little time to care personally for their old parents; seniors will live by themselves or in senior homes. Due to high mobility members of a family often live far apart and family ties are no longer very close, as in liquids.

The democratic change cannot be forced upon a country, unless the standard of living is raised and the way of life in families has changed. Again we find a close relationship between social and atomic systems.

10.5
Heterogeneous Societies

A society with many different groups is called heterogeneous. We will restrict the calculations to binary societies; countries with two ethnic groups like Blacks and Whites in the USA; societies with two religions like Catholics and Protestants in Germany or Northern Ireland; markets with buyers and sellers. The calculations may be extended to societies with large numbers of different groups.

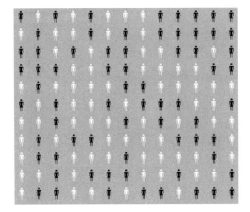

Fig. 10.19 A binary society of women and men or black and white people has four different interactions: AA, AB, BA and BB.

10.5.1
The Six States of Binary Societies

In binary systems we have two groups A and B and four different emotional interactions (constraints) between the two groups: e_{AA}, e_{BB}, e_{AB} and e_{BA}. The probability of the combinations $P(A, B)$ is given by Eq. (10.1). N_A is the number of agents in group A, $w_A = N_A/N$ is the probability of being a neighbor of group A; e_{AA} is the interaction AA between members of group A. The same is valid for interactions BB, AB and BA. The Lagrange function (10.2) is given by

$$L(T, N, N_B) = N_A w_A e_{AA} + N_A w_B e_{AB} + N_B w_A e_{BA} + N_B w_B e_{BB}$$
$$+ T \ln\{(N_A + N_B)!/N_A! N_B!\} \to \text{maximum!} \qquad (10.3)$$

$$L(T, N, x) = N\{e_{AA} + x(e_{BB} - e_{AA}) + \epsilon x(1-x)$$
$$- T[x \ln x + (1-x) \ln(1-x)] + 2\ln 2\} \to \text{maximum!} \qquad (10.4)$$

$$w_B = x = N_B/N; \quad w_A = N_A/N = (1-x) \qquad (10.5)$$

$$\epsilon = (e_{AB} + e_{BA}) - (e_{AA} + e_{BB}). \qquad (10.6)$$

The Lagrange principle (10.4) is the fundamental equation of binary systems that will be applied in all further calculations and simulations of $A\,B$ societies. The Lagrange function L represents the mutual happiness of a binary society about the random distribution of A and B agents. The variable $x = w_B$ stands for the relative size of the minority group (B) and $(1-x)$ the majority group (A). The emotional interactions $e_{AA}, e_{BB}, e_{AB}, e_{BA}$ may be positive or negative.

Equation (10.4) is the Bragg–Williams model of regular solutions in physical chemistry and corresponds to the Ising model of magnetic interactions [3] as well as Schelling's model of social interactions [12].

A surprising result of the calculations for binary societies is the parameter ϵ in Eq. (10.6). There are four different interactions, but in binary systems only one parameter, the difference $\epsilon = (e_{AB} + e_{BA}) - (e_{AA} + e_{BB})$ enters the calculations. This parameter ϵ determines the state and the structure of societies.

A detailed discussion of the parameter $\epsilon = (e_{AB} + e_{BA}) - (e_{AA} + e_{BB})$ in Eq. (10.6) leads to at least six different states in binary societies:

1. Hierarchy: $\epsilon > 0$ and $e_{AB} \neq e_{BA}$
2. Partnership: $\epsilon > 0$ and $e_{AB} = e_{BA}$ $\quad x = 1/2$
3. Segregation: $\epsilon < 0$ and $e_{AB}, e_{BA} > 0$
4. Aggression: $\epsilon < 0$ and $e_{AB}, e_{BA} < 0$
5. Integration: $\epsilon = 0$ and $e_{AB} + e_{BA} = e_{AA} + e_{BB} > 0$
6. Global state: $\epsilon = 0$ and $e_{AB} + e_{BA} = e_{AA} + e_{BB} = 0$

In economic systems (markets) these states correspond to

1. Hierarchy: $\epsilon > 0$ and $e_{AB} \neq e_{BA}$
2. Cooperation: $\epsilon > 0$ and $e_{AB} = e_{BA}$ $\quad x = 1/2$
3. Competition: $\epsilon < 0$ and $e_{AB}, e_{BA} > 0$
4. Aggression: $\epsilon < 0$ and $e_{AB}, e_{BA} < 0$
5. Social market: $\epsilon = 0$ and $e_{AB} + e_{BA} = e_{AA} + e_{BB} > 0$
6. Capitalism: $\epsilon = 0$ and $e_{AB} + e_{BA} = e_{AA} + e_{BB} = 0$

In binary societies we again find hierarchy, democracy and the global state, as in homogeneous societies. But we also find additional states; partnership, segregation and aggression. These additional states are possible only by mixing different kinds of agents. The model function of human relations in binary societies, Eq.(3.1) depends on six parameters: $L(T, x, e_{AB}, e_{BA}, e_{AA}, e_{BB})$: there are two variables: the standard of living T and relative size of the minority x; and four interaction parameters of sympathy or antipathy between the two groups: $e_{AB}, e_{BA}, e_{AA}, e_{BB}$.

All parameters may change gradually with time. But they will be regarded as nearly constant in the simulations. Only T will be assured to change with time, due to productivity. A change in T will again change the states of the society and lead to transitions of state and to the dynamics of societies.

10.5.2
Partnership

Fig. 10.20 A folk dance group of boys and girls show the $AB\ AB$ structure of partnership. Girls are more attracted to boys and boys to girls.

If the attraction to agents of the other group is stronger than to agents of the same group,

$$\epsilon = (e_{AB} + e_{BA}) - (e_{AA} + e_{BB}) > 0 \tag{10.7}$$

Figure 10.20 shows the $AB\ AB$ structure of a folk dance group of boys and girls. In Fig. 10.21 the function $L(x)$ – Eq. (10.4), represents the happiness of a group of boys and girls at a square dance. The relative number of girls is x and the relative number of boys is $(1-x)$. If no girls show up, the happiness $L(x=0)$ of the group will be low, if only girls attend, the happiness $L(x=1)$ will also be low. The maximal happiness of the group will be obtained, if the number of boys and girls are equal, $L(x=1/2)$, when everyone has a partner.

Partnership looks very much like hierarchy, in fact there is no way to detect the difference between partnership and hierarchy from the outside. A married couple may live in true partnership or in (male) hierarchy, the checkerboard structure of the systems are alike. A couple may be dancing in the crowd of dancers, but we do not know whether they are at an equal level or whether he leads her or vice versa. The difference between partnership and hierarchy can only be detected by additional inside information.

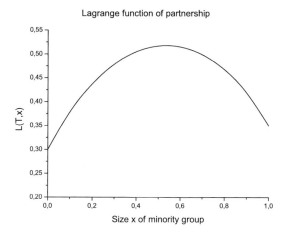

Fig. 10.21 The Lagrange function $L(x)$ – Eq. (10.4), corresponds to the happiness of girls and boys at a square dance; x is the relative number of girls. If no girls show up, the happiness $L(x=0)$ of the boys is low. If there are only girls, the happiness $L(x=1)$ is also low. The maximum happiness is obtained for equal numbers of girls and boys at $L(x=1/2)$, when everyone has a partner.

10.5.3 Integration

If the attraction to the other group is the same as for the same group,

$$\epsilon = (e_{AB} + e_{BA}) - (e_{AA} + e_{BB}) = 0. \tag{10.8}$$

Fig. 10.22 An integrated class of young people from all over the world. The kids are friends and the happiness of each person in the class is independent of the kind of neighbor.

Figure 10.22 shows an integrated class of young people.

The Lagrange function in Fig. 10.23 represents the happiness of a group of boys and girls in the lines for soft drinks after dancing. For the thirsty girls or

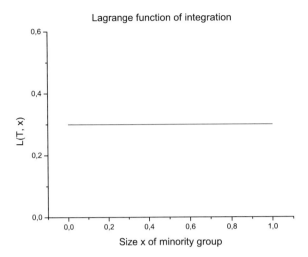

Fig. 10.23 For $\epsilon = 0$ the Lagrange function $L(x)$, Eq. (10.4), corresponds to the happiness of a group of girls and boys in front of the refreshments as a function of the relative number x of girls. The happiness L in the refreshment lines is constant and independent of the number of boys or girls in the lines.

boys in the lines it does not matter who is in front or behind, it only matters to obtain a soft drink, quickly. The happiness of thirsty dancers is constant, Fig. 10.23, independent of the number of girls in the line,

10.5.4
Segregation

If the attraction to the same group is stronger than to the other group,

$$\epsilon = (e_{AB} + e_{BA}) - (e_{AA} + e_{BB}) < 0 \tag{10.9}$$

Bosnia in Fig. 10.24 is a mixture of Moslems and Christian people. Brass in Fig. 10.25 is a mixture of copper and zinc atoms. The structural similarities in both figures is obvious and corresponds to segregation in binary systems. The simulation for $\epsilon < 0$ is given in Figs 10.26 and 10.27.

The Lagrange function or happiness in Fig. 10.26 has two maxima, one for the minority of Moslems in Christian areas, the other for the Christian minority in the Moslem areas. For nearly equal number of Moslem and Christian people, $x = 1/2$, the maximum of the segregated system is given by the diamond on the common tangent. This diamond value is higher than the function $L(x = 1/2)$ for random distribution of agents, as shown by the solid line.

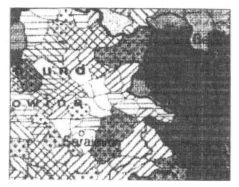

Fig. 10.24 Map of Bosnia 1991. Bosnian society is segregated into a mainly Moslem population (bright areas) and mainly non-Moslem population (dark areas). (With kind permission, Westermann Schulbuch Verlag, Braunschweig.)

Fig. 10.25 Surface of a brass probe after etching. The brass alloy is segregated into areas with mainly zinc (bright regions) and areas with mainly copper (dark regions).

Figure 10.27 is obtained by Monte Carlo simulation of the Lagrange function (10.4).

10.6 Dynamics of Societies

10.6.1 Hierarchy and Opinion Formation

In elections (opinion growth) the winning party leader determines the political direction of more than 100 million people, Fig. 10.28.

This corresponds to the crystallization of liquids. In crystal growth one nucleus determines the crystal direction of 10^{23} molecules, Fig. 10.29. Crystal growth and opinion formation may both be modeled from the Lagrange func-

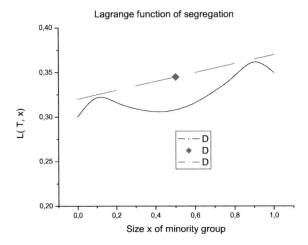

Fig. 10.26 Lagrange function (10.4) of a binary society with preference for the same groups, $\epsilon < 0$. The function shows two maxima, for the minorities in each section.

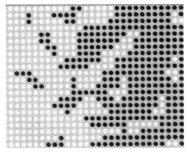

Fig. 10.27 Monte Carlo simulation of the Lagrange function $L(x)$ of the distribution of Moslems and Christians in Bosnia in Fig. 10.23.

Fig. 10.28 Elections are a good example of opinion formation. The winning leader determines the political direction of more than 100 million people.

10.6 Dynamics of Societies

Fig. 10.29 A rock crystal is a well known example of crystal growth. The winning nucleus determines the crystal direction of more than 10^{23} molecules.

tion (10.2), $L = E + T \ln P \to$ maximum:

$$L(T, x_i, x_k) = N\{\Sigma e_{ik} x_i x_k - T \Sigma x_i \ln x_i\} \to \text{maximum!} \quad (10.10)$$

$$e_{ik}(\circ\bullet) = e_{ik}(\bullet\circ) > 0 \text{ (attractive)} \quad (10.11)$$

$$e_{ik}(\bullet\bullet) = e_{ik}(\circ\circ) < 0 \text{ (repulsive)} \quad (10.12)$$

An ideal checkerboard structure can start with $\circ\bullet$ = Bush or with $\bullet\circ$ = Kerry.

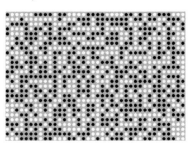

Fig. 10.30 Chaotic distribution of light and dark balls. The distribution corresponds to a disordered binary liquid of GaAs or an undecided crowd before the elctions (yes-no; Bush-Kerry): $\circ\bullet$ = Bush; $\bullet\circ$ = Kerry; $\circ\circ = \bullet\bullet$ = undecided.

Figure 10.30 corresponds to the situation before the election, there is no checkerboard structure, no preference for Bush or Kerry, most people are still undecided. Figure 10.31 and 10.32 are created by a Monte Carlo method. A random agent is asked if he wants to switch positions with his left (right, lower, upper) neighbor. If the Lagrange function Eq. (10.10) is raised by this local movement, the neighbors will switch. The parameter T and the concentrations x_i ($\circ$) and x_k ($\bullet$) remain constant. After some time the pattern of Fig. 10.31 will emerge. The local preference for Bush or Kerry will grow, until both areas meet and are separated by undecided agents. The lines of opposite opinions are not always stable. In a curved line there are more agents

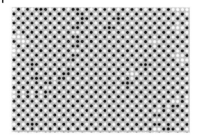

Fig. 10.31 Disordered checkerboard distribution of light and dark balls. The pro Kerry opinion ○● – see upper left corner – and the pro Bush opinion ●○ in the center are separated by a curved line of undecided. Both areas also contain dissidents = ● ○ ○ ○ ● or ○ ● ● ● ○. The pattern simulates the 2004 US elections or crystal growth of a gallium arsenide with grain boundaries and antisites.

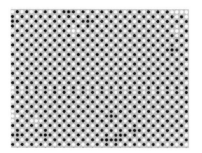

Fig. 10.32 A stable twin formation ○● (see upper corner) and ●○ (lower part). The border line is straight. Both opinions, parties, armies contain local antisites or dissidents. The straight line has equal numbers of opponents and leads to a twin crystal, a stable two-party system or an undecided battle front.

with one opinion than inside with the opposite opinion. With enough time the curved line is pushed out and only one opinion will dominate. A straight line with equal numbers of opponents, Fig. 10.32, will be stable for long times and produce a stable two-party system.

10.6.2
Simulation of Segregation

10.6.2.1 Phase Diagrams

Phase transitions in homogeneous materials are called melting or vaporization. These are often first-order transitions, which are accompanied by a heat of transition and a sudden change of order.

Phase transitions may be discussed on the basis of the Lagrange principle, Eq. (10.4). At equilibrium or maximum the first derivative of L with respect to

x will be equal to zero, if $e_{BB} = e_{AA}$,

$$\epsilon(1 - 2x) - T[\ln x - \ln(1 - x)] = 0 \tag{10.13}$$

Equation (10.13) may be solved for T

$$T = +\frac{\epsilon(1 - 2x)}{\ln x - \ln(1 - x)} \tag{10.14}$$

Equation (10.14) has been plotted in the phase diagram, Fig. 10.33. The plot shows the temperature T that is needed to dissolve a quantity x of platinum in gold.

Mixing gold and platinum is very much like mixing tea and sugar. In cold tea only a little bit of sugar may be dissolved, the rest will segregate at the bottom of the cup. In hot tea much more sugar may be dissolved.

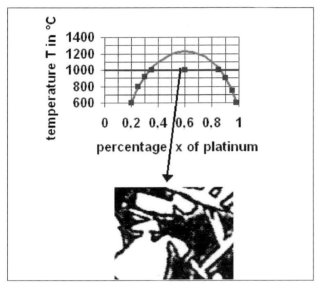

Fig. 10.33 Phase diagram of gold platinum alloys (Hansen, 1958 [7]): The curve $T(x)$ corresponds to the temperature that is needed to dissolve the percentage x of platinum. At $600°C$ gold may solve only 20% platinum. At $1000°C$ the "solubility limit" is $x = 0.35$ or 35$ platinum. Any higher percentage will lead to segregation, as shown by the bright areas with mainly gold, and dark parts with mainly platinum.

10.6.2.2 Intermarriage

Intermarriage data are the phase diagrams in social systems. The diagrams tell us about the degree of integration in a society, they are derived from:

$$P(x) = 2x(1 - x) \tag{10.15}$$

Equation (10.15) of intermarriage is a parabola and corresponds to Mendel's law of mixing white and red peas. Figure 10.34 shows the parabola of intermarriage for Catholics and non-Catholics in 10 different states of Germany and in Switzerland in 1991. Data points on the parabola correspond to states, where Catholics and non-Catholics are integrated. This is observed for states below $x = 0.2$ or 20% Catholics. The value $x = 0.2$ represents an "integration limit" of Catholics in the German society. The constant rate of intermarriage at 33% corresponds to the equilibrium temperature in Fig. 10.33 and indicates the equilibrium in the mutual religious tolerance T of different neighbors. In states with a higher percentage of Catholics we find segregation into mainly Catholic and mainly non-Catholic areas, as shown on the map of the state of Westphalia with 40% Catholics. White areas have $x = 0.2$ or 20% Catholics and about 80% non-Catholics, and dark areas 80% Catholics and 20% non-Catholics. So everybody is most happy to live in the right neighborhood [9]. The intersections of the parabola at $x = 0.2$ and $x = 0.4$ corresponds to the maxima in the Lagrange function, Fig. 10.25.

Mixing tea and sugar or platinum and gold in Fig. 10.33 corresponds to the mixing Catholics and non-Catholics, Fig. 10.34. The maximum rate of intermarriage at the "integration limit" x is equivalent to the solubility limit of sugar in tea or platinum in gold, Fig. 10.33.

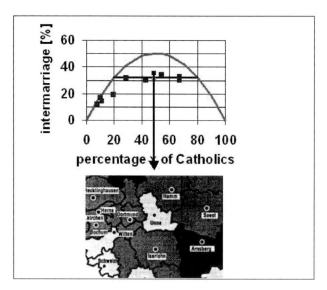

Fig. 10.34 Intermarriage between Catholics and non-Catholics in 10 different states of Germany in 1991 and in Switzerland 1992. Intermarriage is ideal up to the "integration limit" of $x = 0.2\%$ or 20% Catholics in any state of Germany. For states with higher percentage we find segregation. In Westphalia with 40% Catholics, Westphalia segregates into black areas with mainly Catholics and white areas with mainly non-Catholics [11].

Equation (10.15) is equivalent to Schelling's model [12]. However, the present model does not only give a detailed explanation for Schelling's constant c, but also leads to solutions of segregation problems.

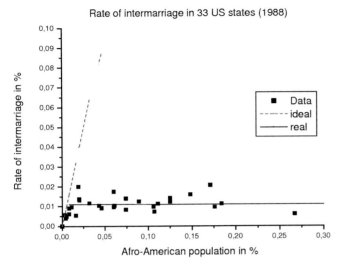

Fig. 10.35 Intermarriage between African and non-African Americans in 33 States of the US in 1988. Intermarriage is ideal up to the "solubility limit" of $x = 0.55\%$ African Americans in each state. For states with a higher percentage we find a constant portion of intermarriage, $I = 1.1\%$, indicating the "social equilibrium temperature" between different states of the US. The "solubility limit" leads again to segregation into predominant white and predominant African American areas. (US Bureau of the Census, 1990.)

The intermarriage diagrams of African and non-African Americans in 33 different states of the US (1988) is given in Fig. 10.35. Again we find an "equilibrium temperature" for mixing African and non-African Americans at a portion $P = 1.1\%$ of intermarriage. The value of P_{max} is independent of the number of African Americans in the particular states and proves again the existence of the Lagrange parameter or mutual racial tolerance T.

10.6.3
Simulation of Aggression

Figure 10.36 shows the intermarriage diagram between 40% Catholics and 60% non-Catholics in Switzerland, Germany and Northern Ireland in 1991. The mutual rates of tolerance are 32% in Switzerland and Germany and 2.3% in Northern Ireland! This low value indicates strong negative emotions and the danger of aggression. In all highly segregated societies with low tolerance like Ireland, Israel or Bosnia, civil war can erupt at any time.

The process of segregation may be simulated by the Lagrange equation Eq. (10.4). This way we can observe how the process of segregation is reflected by the decreasing rate of intermarriage. Starting from a completely

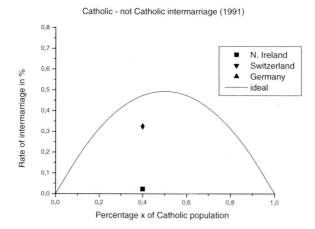

Fig. 10.36 The three countries Germany, Switzerland and N. Ireland had, in 1991, the same percentage of 40% Catholics. But the rate of intermarriage between Catholics and Protestants in Northern Ireland is $I = 2.3\%$, compared to $I = 32\%$ in Germany and Switzerland and reflects the low tolerance, danger of aggression and civil war in each country [17].

integrated minority, intermarriage and tolerance drops from the maximum to a minimum value due to segregation or aggression. This special simulation gives a proper interpretation of the danger of aggression and civil war due to changing emotions within the population.

If emotions towards other groups become negative, such as: hate, envy, distrust or aggression, we have the condition for aggression,

$$\epsilon > 0 \text{ and } (e_{AB} + e_{BA}) < 0 < (e_{AA} + e_{BB}) \tag{10.16}$$

Negative emotions to others will lead to total separation of the groups, and vice versa, a total separation of groups will generally lead to aggression. This is presently observed in many places with binary populations like Bosnia, Northern Ireland or Israel.

The close relation between separation and negative emotions is also observed in the popular prisoner dilemma of game theory [11]. The separation of prisoners generally leads to distrust, $\epsilon < 0$ and defection.

10.7
Conclusion

In the thermodynamic formulation of social science the laws of societies have been derived from statistical laws under constraints. No further assumptions have been made. The calculated functions are supported by data, which all seem to agree very well. The results indicate that natural science may very easily be applied to social science. The thermodynamic approach, which has been most successful in all natural sciences like physics, chemistry, metallurgy, meteorology, biology and engineering, is also a successful basis in the fields of socio-economic sciences. The approach to social science is very general and the data seem to fit well. In this view the Lagrange statistics seems to be a valuable addition to the standard statistical theory of social science.

References

1 AXELROD, R., *The Evolution of Cooperation*, Basic Books, NY, **1984**

2 BARRO, R. J., MARTIN, X.S., *Economic Growth*, McGraw Hill, **1995**

3 BECKER, R., *Theorie der Wärme*, Springer Verlag Heidelberg, **1966**, p. 154

4 Bundesamt für Statistik der Schweiz, Verlag Neue Zürcher Zeitung, Zürich, **(1992)**

5 DSW Weltbevölkerungsbericht **(2004)**

6 GALAM, S., *Physica A* 238 **(1997)**, p. 66–80

7 Hansen, M., Constitution of binary alloys, McGraw Hill, New York, 1958

8 MIMKES, J., *J. Thermal Analysis* 43 **(1995)**, p. 521–537

9 MIMKES, J., *J. Thermal Analysis* 60 **(2000)**, p. 1055–1069

10 MIMKES, J., *Die familiale Integration von Zuwanderern und Konfessionsgruppen – zur Bedeutung von Toleranz und Heiratsmarkt in Partnerwahl und Heiratsmuster*, Klein, Th., Editor, Verlag Leske und Budrich, Leverkusen, **2000**

11 Statistisches Jahrbuch für das vereinte Deutschland, Statistisches Bundesamt, Wiesbaden **(1991)** and **(2000)**

12 SCHELLING, TH., *Journal of Mathematical Sociology* 1 **(1971)**, p. 143–186

13 STAUFFER, D., *J. Artifial Societies and Social Simulation* 5 **(2002)**

14 SOLOMON, S., WEISBUCH, G., DE ARCANGELIS, L. J. N., STAUFFER, D., *Physica A* 277 **(2000)**, p. 239

15 Statistisches Jahrbuch der Schweiz, Verlag Neue Zürcher Zeitung, Zürich **(1992)**

16 U. S. Bureau of the Census **(1990)**

17 WEIDLICH, W., *Collective Phenomena* 1 **(1972)**, p. 51

18 WEIDLICH, W., *Sociodynamics*, Harwood Acad. Publ., Amsterdam, **2000**

11
Computer Simulation of Language Competition by Physicists
Christian Schulze and Dietrich Stauffer

Computer simulation of languages is an old subject, but since the paper of Abrams and Strogatz [10] several physics groups have independently taken up this field. We briefly review their work and give more details of our own simulations.

11.1
Introduction

Human languages dististinguish us from other animals, but birds and ants also have systems of communication. Also, humans have invented alphabets and other formalized forms of writing. In principle, the methods to be described here could also be applied to these other forms of communication, but mostly we are interested here in the, currently, about 10^4 different human languages on this planet [1]. We leave it to linguists to distinguish languages from dialects or language families; when we mention "language" readers may read dialect or family instead.

Everyday language contains thousands of words for different aspects of life, and including the special words of science, medicine, etc., we have even more. For the same concept of everyday life, each different language in general has a different word, and thus the number of possible languages is enormous and difficult to simulate. Things become easier if we look only at grammar; do we order (subject, object, verb) or (subject, verb, object) or ...? Briscoe [2] mentioned about 30 independent binary grammatical choices, which leads to a manageable $2^{30} \simeq 10^9$ possible languages, which can be symbolized by a string of $\ell = 30$ bits. Thus many of the simulations described here use bit-strings with $\ell = 8, 16 \ldots 64$.

The present situation is not in equilibrium; about every ten days a human language dies out, and in Brazil already more than half of the indigenous languages have vanished as a result of the European conquest. On the other hand, Latin has split in the last two millennia into several languages, from Portuguese to Romanian, and many experts believe that Latin and the other

Econophysics and Sociophysics: Trends and Perspectives.
Bikas K. Chakrabarti, Anirban Chakraborti, Arnab Chatterjee (Eds.)
Copyright © 2006 WILEY-VCH Verlag GmbH & Co. KGaA, Weinheim
ISBN: 3-527-40670-0

Indo-European languages spoken 600 years ago from Iceland to Bengal (and now also in the Americas, Australia, Africa) originated from the people who invented agriculture in the Konya plane of Turkey, 10^4 years ago. Thus similar to biology, languages also can become extinct or speciate into several daughter languages.

In contrast to biology, humans do not eat humans of other languages as regular food, and thus one does not have a complex ecosystem of predators eating prey as in biology. Instead, languages are meant for communication, and thus there is a tendency for only one language to dominate in one region, like German in Germany etc. Will globalization lead to all of us speaking one language in the distant future? For physics research, that situation arrived many years ago. If we follow the Bible, then at the beginning Adam and Eve spoke one language only, and only with the destruction of the Tower of Babel did different languages originate.

Thus in the history of mankind we may have had first a rise, and later a decay, in the number of different languages spoken. In Papua New Guinea [3] there are now 10^3 languages, each spoken by about 10^3 people; can this situation survive if television and mobile phones become more widespread there?

While we cannot answer these questions, we can at least simulate such "survival of the fittest" among languages, in a way similar but not identical to biology. We will not emphasize here the longer history of computer simulations of how children learn a language [4], see also [5], or how mankind developed the very first language out of ape sounds [6]. Instead, we talk about the competition between different languages for adults. And we will emphasize the "agent-based" models simulating individuals, analogous to Monte Carlo and Molecular Dynamics for spins and molecules.

The second section deals with differential equations (a method we regard as outdated), followed by agent-based simulations with few languages in Section 11.3 and with many languages in Section 11.4. Further results from the two many-language models are given in the Appendix.

11.2
Differential Equations

Nettle [7] has already suggested a very simple differential equation to see how the number L of languages changes with time:

$$dL/dt = 70/t - L/20$$

Here the time unit is a thousand years. (Actually L is the number of different language groups, and time is discrete.) The second term on the RHS means a loss of five percent per millennium; the first term indicates the formation of new languages which became more difficult when the population increased since then the higher demand for communication reduced the chances of new languages developing. The aim was to explain why the recently populated Americas have a higher language diversity than Africa and Eurasia with their older human population. For long times, this differential equation means that L decays exponentially towards zero.

Nowak et al. [4] use

$$dx_j/dt = \left(\sum_i f_i Q_{ij} x_i\right) - \phi x_j$$

for the fraction x_j of a population speaking language $j = 1, 2, 3 \ldots L$. (Actually they apply this equation to the learning of languages or grammars by children; the interpretation for competition between adult languages is ours.) Here the fitness $f_i = \sum_j F_{ij} x_j$ of language i is determined by the degree F_{ij} to which a speaker of language i is understood by people speaking language j. The average fitness is $\phi = \sum_i f_i x_i$ and is subtracted to keep the sum over all fractions x_j independent of time. The probability that children from i-speaking parents later speak language j is Q_{ij}.

For a large number L of languages, there are numerous free parameters in the matrices Q_{ij} and F_{ij}. With most of them the same one finds a sharp phase transition [4] as a function of mutation rates Q_{ij}. If one starts with only one language, then at low mutation rates most of the people still speak this language and only a minority has switched to other languages. For increasing mutation rates, suddenly the system jumps from this dominance of one language to a fragmentation state where all languages are spoken equally often. If, in turn, we start from such a fragmented state then it stays fragmented at high mutation rates. With decreasing mutation rates it suddenly jumps to the dominance of one language (numerically, one then has to give this one language a very slight advantage). The two jumps do not occur at the same mutation rate but show hysteresis. Starting with dominance and increasing the mutation rate allows dominance for higher mutation rates than when we start with fragmentation and decrease the mutation rate.

Qualitatively, these properties remain if the many matrix elements are selected randomly instead of being the same [8] except that the hysteresis has become very small. Figure 11.1(a) shows the case where we start with dominance and Fig. 11.1(b) the case where we start with fragmentation. The time development for two of the 30 simulated languages is shown for two slightly different mutation rates, and we see how, for the lower mutation rate but not for the higher rate, one of the two languages starts to dominate, at the expense of the other.

These 30 languages are more mathematical exercises, but Fig. 11.2 applies these methods to up to $L = 8000$ languages, using two 8000×8000 random matrices F and Q. We show the size distribution of languages, where the size

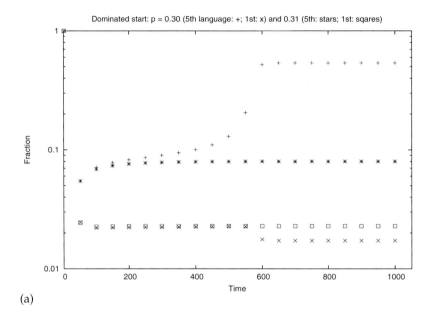

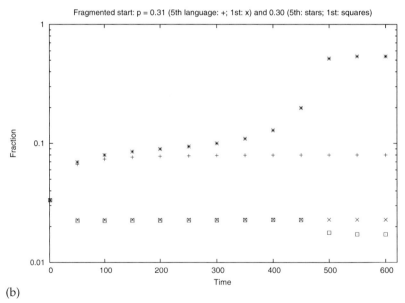

Fig. 11.1 Development of one dominating language, or lack of such dominance, for the model of Nowak et al. [4], with random matrix elements. In (a) we start from the dominance of another language, in (b) part all 30 languages start with the same number of speakers. The different symbols correspond to two suitably selected languages and two slightly different mutation rates $p \simeq 0.3$. From [8].

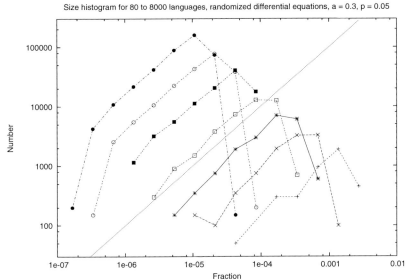

Fig. 11.2 Size histogram, ignoring the dominating language, for the model of Nowak et al. [4] with random matrix elements. The number of simulated languages varies from 80 on the right to its real value 8000 on the left. The straight line has slope 1 in this log-log plot. From [8].

is the fraction of people speaking this language. On this log-log plot we see roughly, parabolas, shifting to the left with increasing number L of languages. These parabolas roughly correspond to log-normal distributions, as observed empirically in Fig. 11.3.

(Similar to Komarova [4] we assume the average F to be 0.3 except for $F_{ii} = 1$ and the average Q to be $p/(L-1)$ except $Q_{ii} = 1-p$; the actual values are selected randomly between zero and twice their average.)

There are two problems in this comparison of Figures 11.2 and 11.3. In these simulations, the (logarithmic) range over which the language sizes vary is quite small and does not change with increasing L. And the real distribution is unsymmetric, having higher values for small languages [9] than the lognormal distribution; this enhancement is missing in the simulation of Fig. 11.2. Finally, we have cheated. Figure 11.2 was taken in the dominance regime and the dominating language was ignored in the statistics.

Much more attractive for physicists was the one-page paper of Abrams and Strogatz [10] which was, within weeks, followed by a poster of Patriarca and Leppanen [11]. This pair of papers then triggered apparently independent research in Spain [12], Greece [13], Germany [14], Argentina [15] and at two different places in Brazil [16, 17], all on language competition.

The Abrams–Strogatz differential equation for the competition of a language Y with higher social status $1-s$ against another language X with lower

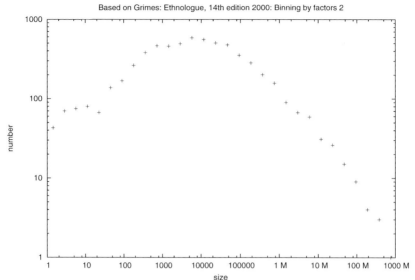

Fig. 11.3 Size histogram for human languages, from [1]. We bin language sizes by factors of two, just as in Fig. 11.2. Thus, the leftmost point corresponds to size = 1, the second sums sizes 2 and 3, the third sums sizes 4 to 7, etc.

social status s is

$$\mathrm{d}x/\mathrm{d}t = (1-x)x^a s - x(1-x)^a(1-s)$$

where $a \simeq 1.3$ and $0 < s \leq 1/2$. Here x is the fraction in the population speaking language X with lower social status s while the fraction $1 - x$ speaks language Y. Figure 11.4 with no status difference, $s = 1/2$, shows as intended that the language which is initially in the majority wins; the other language dies out. For $x(t=0) < 1/2$ the language Y wins and for $x(t=0) > 1/2$ the language X wins. This is highly plausible. If we would immigrate to Brazil where most of the people speak Portuguese, then we would also have to learn Portuguese, not because of status but because of numbers. If the initial minority language has the higher status, as happened 500 years ago when Portuguese ships landed in Brazil, then it may win in the end, thanks to guns, writing, and other status aspects, as is the case in Brazil. Figures for unequal status are published in [8] from where also Fig. 11.4 is taken.

The Finnish group [11] generalized this simple differential equation to a square lattice, where it became a partial differential equation including a Laplacian $\nabla^2 x(\mathbf{r})$. Then having in one part of the lattice a higher status for X compared to Y, and in the other part the opposite status relation, they showed that the languages X and Y can coexist beside each other, with a narrow interface in between. This reminds us of Canal Street in New Orleans which

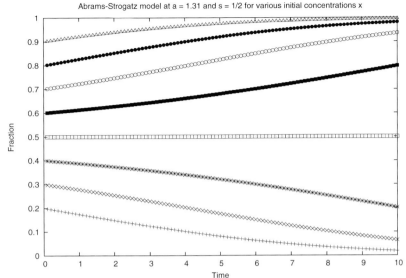

Fig. 11.4 The initial majority wins in an Abrams–Strogatz competition between two languages. The initial concentration of the plotted language is 0.9, 0.8,..., 0.1 from top to bottom; $a = 1.31$, $s = 1/2$. From [8].

in earlier times separated the French quarter from the English-speaking part. We will later return to such geographical coexistence also without status differences, arising merely from an initial separation into an X part and a Y part [14, 18].

If we set $a = 1$ the simple logistic Verhulst equation results [19],

$$dx/dt = (2s - 1)(1 - x)x$$

which had already been applied to languages [20] before Abrams and Strogatz. This case was generalized to two Verhulst equations [15] describing the two populations of people speaking languages X and Y. Now as in Lotka–Volterra equations for predators eating prey, both populations can coexist with each other in some parameter range, which in the usual Abrams–Strogatz model is possible only for $x(t = 0) = s = 1/2$.

The competition of two languages is changed if some people become bilingual, that means they learn to speak the other language which was not their mother tongue [12]. This was applied to Gallego versus Castellano in Spain ; of course, some may regard Castellano, spoken in Madrid, as the proper Spanish, and Gallego as its dialect spoken in Galicia. As citizens of the Prussian occupied Westbank of the Rhine River, we know that publicly going into such details before liberation may be dangerous. A language is a dialect with an army and a navy behind it.

Of course, all these differential equations are dangerous approximations, just as mean-field theory for critical phenomena in statistical physics is dangerous. We have known for 80 years that the one-dimensional Ising model has a positive Curie temperature T_c in the mean-field approximation, while in reality $T_c = 0$. So do the Abrams–Strogatz results remain correct if we are dealing with individuals who randomly change from one language to the other, with probabilities corresponding to the original differential equation?

In general, the answer is yes [21]. As long as both s and the initial concentration $x(t = 0)$ are not 1/2, one language still dies out, and it does so exponentially. This holds for the case of everybody influencing everybody, Fig. 11.5, as well as for a square lattice where everybody is influenced only by their four lattice neighbors, Fig. 11.6. The line in Fig. 11.5 is the solution of the differential equation and agrees qualitatively with the Monte Carlo results represented by the separate symbols for various total constant populations N. Only for a completely symmetric start, $s = x(t = 0) = 1/2$, when the differential equation gives an equilibrium (stable for $a < 1$ and unstable for $a > 1$), does the microscopic Monte Carlo simulation give one or the other language dying out, while the differential equation then predicts both always to comprize half of the population each. More details are given in [21].

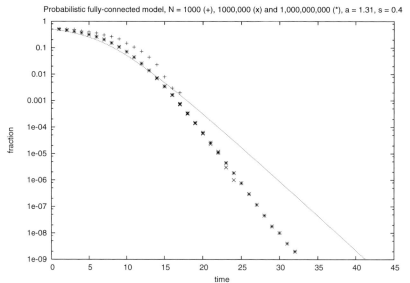

Fig. 11.5 Exponential decay for the language with lower status, consisting initially of half the population. The symbols give Monte Carlo simulations where each individual is influenced by the whole population N, while the line is the result of the differential equation of Abrams and Strogatz. $a = 1.31$, $s = 0.4$, $N = 10^3, 10^6, 10^9$.

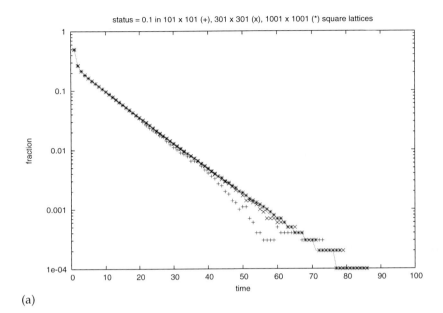

(a)

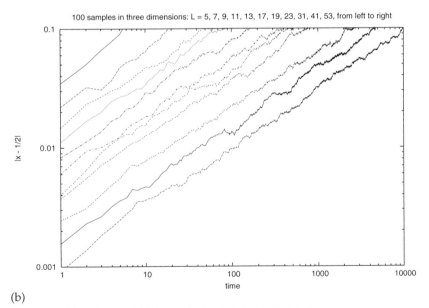

(b)

Fig. 11.6 (a) As for Fig. 11.5 but on 101×101, 301×301, 1001×1001 square lattices with $a = 1.31$, $s = 0.1$. Part b: Three-dimensional lattices, at the symmetry point $x(t = 0) = s = 1/2$, $a = 1$. After a long time, the concentration moves towards zero or one.

Finally we mention the model of [3] which does not deal with individuals either but avoids differential equations.

11.3
Microscopic Models

Here we deal with the more modern methods of language simulation, based on individuals instead of on overall concentrations. Such methods have been applied in physics for half a century and are called agent-based in some fields outside physics. First we review two models for only two (or a few) languages, then, in much greater detail, the two models for many languages.

11.3.1
Few Languages

The model of Kosmidis et al. [13] for mixing two languages X and Y uses bit-strings of length 20; each bit can be 0 (representing a word or grammatical aspect which is not learned) or 1 (an element which this individual has learned). If someone speaks language X perfectly and language Y not at all, the bit-string for this person is 11111111110000000000 while 00000000001111111111 corresponds to a perfect Y-speaker. People can become perfectly bilingual, having all 20 bits at 1, but this is rare. This model is particularly useful for explaining the generation of a mixture language Z out of the two original languages X and Y. One merely has to take about ten bits equal to one and distribute them randomly among the 20 bit positions. This may then correspond to the creation of Shakespeare's English out of the Germanic language spoken by the Anglo-Saxons and the French spoken by the Normannic conquerors of the year 1066.

Biological ageing was included in the model of Schwammle [16], using the well-established Penna model [8, 19, 22] of mutation accumulation. Two languages X and Y are modeled. Individuals learn to speak from father and mother (and thus may become bilingual) and move on a square lattice in search of emptier regions. Bit-strings are also used here, but only for the ageing part to store genetic diseases; the two languages have no internal structure here. A bilingual person surrounded by neighbors speaking mostly language X forgets with some probability the language Y, and vice versa. The model allows for the coexistence of the two languages, each in a different region of the lattice, as in [11] but without giving one language a higher status than the other.

In his later model [16], that author allows for up to 16 languages. Again the structure of languages is ignored. Only young people can learn languages from others, and sometimes they learn a new language by themselves. As a function of the "mutation" probability to learn a new language independently, the model gives dominance of one language for small mutation rates, and fragmentation of the population into many languages for high mutation rates, with a sharp phase transition separating these two possibilities, e.g., at a mutation rate near 1/4. This phase transition is similar to that found by [4] as reviewed above.

11.3.2
Many Languages

To explain the existence of the 10^4 present human languages, we need different models [14, 17] which we review now.

11.3.2.1 Colonization

After the first human beings came to the American continent by crossing the Bering strait several ten thousand years ago, presumably at first they all spoke one language. Then they moved southward from Alaska and separated into different tribes which slowly evolved different languages. This first colonization was modeled by Viviane de Oliveira and collaborators [17] by what we call the Viviane model.

Languages have no internal structure but are labelled by integers $1, 2, 3, \ldots$. Human population starts at the centre site of a square lattice with language 1, and from then on humans move to empty neighbor sites of already populated areas. Each site can carry a population of up to about 10^2 people, selected randomly. The size or fitness of a language is the number of people speaking it. On every new site, the population selects as its own language that of a populated neighbor site, with a probability proportional to the fitness of the neighboring language. In addition, the language can mutate into a new language with a probability α/size. To prevent this mutation rate from becoming too small, this denominator is replaced in their later simulations by some maximum $\simeq 10^3$, if the actual language size is larger than this cut-off value. The simulation stops when all lattice sites have been populated. A complete Fortran program is listed in [14](d).

Figure 11.7(a) shows, for a mutation coefficient $\alpha = 0.256$, that the simulated language sizes in the Viviane model can reach the thousand millions of Chinese, but the shape differs from Fig. 11.3 and corresponds more to two power laws than to one roughly log-normal parabola. In contrast to other models [4, 14, 16] there is no sharp phase transition between the dominance of one language and the fragmentation into numerous languages; Fig. 11.7(b) instead shows a smooth decrease, with increasing mutation coefficient α, of the fraction of the population speaking the largest language.

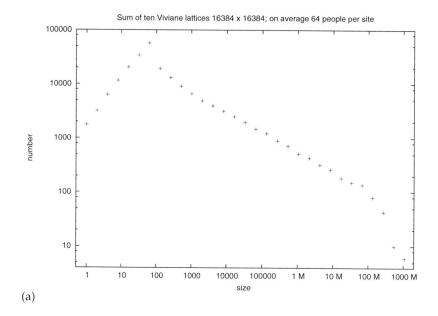

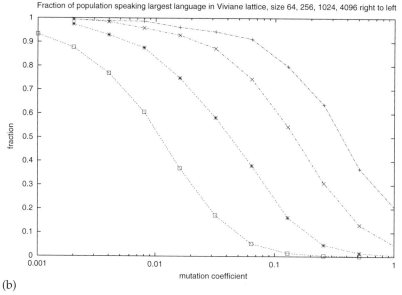

Fig. 11.7 (a) Language size distribution in the Viviane model with 10^{10} people and sizes reaching 10^9 speakers, as in reality for Mandarin Chinese. (Sum over ten lattices 16384×16384.) (b) Lack of sharp phase transition in Viviane model as a function of α with linear lattice dimension 64, 256, 1024 and 4096 from right to left. From [14](d).

This Brazilian group [23] had earlier claimed that the language size distribution follows two power laws, both indicating a decay of the number of languages with increasing language size. This fit, however, applies only to the large-size tail and not for small sizes where the power law would indicate an unrealistic divergence. Figure 11.3 in contrast, shows a very small number, with less than 10^2 of the 10^4 languages spoken by only one person [9]. The cumulative number of languages spoken by at least s people thus should be quite flat for small s instead of diverging with a power law for $s \to 0$, as fitted in [23]. A log-normal distribution gives a much better overall fit and is for large sizes not necessarily worse than the two power laws of [23].

Further results from the Viviane model are given in our Appendix.

11.3.2.2 Bit-string Model

Our own model uses bit-strings as in [13, 16] but for different purposes. Each different bit-string represents a different language though one may also define slightly different bit-strings as representing different dialects of the same language. Lengths ℓ of 8 to 64 bits have been simulated, and the results for 16 bits differed little from those of longer strings, while 8 bits behaved differently.

We used three different probabilities p, q, r though most properties can be also obtained from the special cases $q = 0$, $r = 1$. When a new individual is born its language is mutated with probability p compared to that of the mother. One of the ℓ bits is selected randomly and reverted, which means a zero bit becomes one and a one bit becomes zero. This p is the mutation probability per bit-string; the probability per bit is therefore p/ℓ.

When q is not zero, then the above mutation process is modified. With probability $1 - q$ it happens randomly as above, and with probability q the new value of the bit is obtained, not by reverting it, but by taking over the corresponding bit value of a randomly selected individual from the whole population. This transfer probability q thus describes the effect that one language can learn concepts from other languages. Many words of higher civilization in the German language came from French, while French beers sometimes have German names.

Thus far the simulations are similar to biology with vertical (p) and horizontal (q) gene transfer. Specific human thinking enters into the third probability $(1 - x^2)r$ (also $(1 - x)^2$ instead of $1 - x^2$ was used) to give up their own language and to switch to the language of another randomly selected person. Here x is the fraction of people speaking the old language, and thus this probability to abandon the old language is particularly high for small languages. The new language is selected by a random process, but since it is that of a randomly selected person and not a randomly selected language, most likely the new language is one of the major languages in the population. In this way we simulate the same trend towards dominating language which was already modeled by Abrams and Strogatz, as described above in the example of our emigration to Brazil. This flight from small to large languages, through the parameter r, distinguishes the language competition from biological competition between species in an ecosystem, and takes into account human consideration of the utility of the language.

The population size is kept from going to infinity by a Verhulst death probability proportional to the actual population size. Thus if we start with one person, the population will grow until it reaches the carrying capacity given by the reciprocal proportionality factor. More practical is an initial population which is already about equal to the final equilibrium population. With the latter choice one can start with either everybody talking the same language, or everybody talking a randomly selected language. A complete Fortran program for the simple case $q = 0$, $r = 1$ is listed in [8].

Compared to the Viviane model explained above, our model is more complicated since it has three probabilities p, q, r instead of only one coefficient α. However, one can set $q = 0$, $r = 1$ in our model and then one has the same number of free parameters. The Viviane model simulates the flight from small to large languages by a mutation probability inversely proportional to the size of the languages while we separate the mutations (independent of language size) from the flight probability $(1 - x^2)r$. Moreover, we simulate a continuous competition of languages while the Viviane model simulates the unique historical event of a human population spreading over a continent where no humans lived before.

The results of our model are reported in [8,14]. Most important is the sharp phase transition, for increasing mutation rate p at fixed q and r, between dominance at small and fragmentation at large p. For dominance, at least three-quarters of the population speak one language, and most of the others speak a variant differing by only one bit from that language. For fragmentation, on the other hand, the population spreads over all possible languages. If we start with dominance, the phase transition to fragmentation was already described in the biblical story of the Tower of Babel. If we start with fragmentation, we get dominance for long enough times and small enough mutation rates, if we use $(1 - x)^2$ instead of $1 - x^2$ for the flight probability. Figure 11.8 shows the phase diagram for $\ell = 8$ and 16 if we start from fragmentation. In Fig. 11.9, particularly long simulations for $\ell = 64$ and one million people show how an initial dominance decays into fragmentation.

Tesileanu and Meyer-Ortmanns [18] introduced into this model the Hamming distance as a measure of dissimilarity between languages. This Hamming distance counts the number of different bits in a position-by-position comparison of two bit-strings. Thus the $\ell = 4$ strings 0101 and 1010 have a Hamming distance of four. This distance can be normalized to lie between zero and one, through division by ℓ. Figure 11.10 shows this normalized Ham-

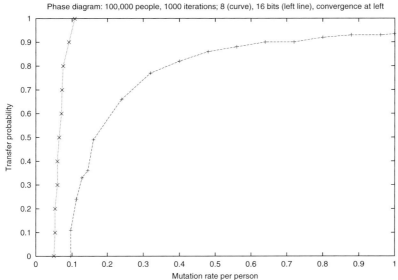

Fig. 11.8 Phase diagram for dominance in the upper left part and fragmentation in the lower right part. The higher the mutation rate p and the lower the transfer rate q the more fragmented is the population into many different languages. We start with an equilibrium distribution of 100 000 languages. each speaking a randomly selected language. The curve corresponds to $\ell = 8$ bits, the nearly straight line to $\ell = 16$; $r = 1$ in both cases. From [14].

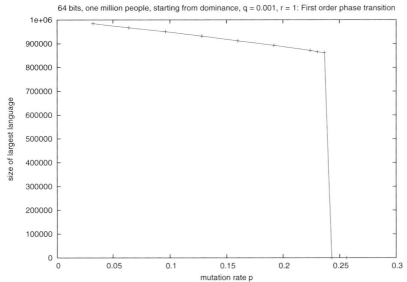

Fig. 11.9 Phase transition from dominance to fragmentation for one million people and 64 bits, i.e., many more possible languages than people. We show the size of the most-often spoken language after 300 iterations; it jumps from 10^6 to 10^2.

ming distance for both the two largest languages and the average over all possible pairs. Not much difference is seen except that the one for the single pair fluctuates much more strongly than the average over all pairs. And for dominance the difference is very small, while for fragmentation it is nearly 1/2. Thus for fragmented populations, the various languages are nearly uncorrelated, and half their bits agree accidentally while the other half disagree. For dominance, the minor languages are mostly one-bit mutants of the dominating language. Figure 11.10, like Fig. 11.9 before, shows a clear first-order phase transition, which means a sharp jump. Thus far we were not able to modify this model such that it gives a second-order transition where the fraction of people speaking the largest language goes continuously to zero at a sharp critical point. Such a modification might give a more realistic distribution of language sizes.

The time dependence of the size of the largest cluster, if we start with fragmentation, suggests a complicated nucleation process. Originally all languages are about equal in size, and then due to random fluctuations one language happens to be somewhat more equal than the others. This language then wins over, first slowly, then rapidly, Fig. 11.11. The time needed for one language to win increases about logarithmically when the population increases from 10^3 to 10^8 in Fig. 11.11. Thus, for an infinite population, as simulated by deterministic differential equations of the Nowak et al. style [4], the emergence of dominance out of a fragmented population might never happen in our model.

First-order phase transitions like those in Figs 11.9 and 11.10 are usually accompanied by hysteresis, such as when undercooled liquid water crystallizes into ice. Thus, we should get different positions of the effective transition (for fixed population size and fixed observation time) depending on whether we start from dominance or fragmentation. This is shown in Fig. 11.12, using in both cases $(1-x)^2$ for the flight probability.

The language size distribution in Fig. 11.13 shows the desired shape of a slightly asymmetric parabola on the log-log plot (log-normal distribution) but the actual language sizes are far too small compared with reality. This is not due to lack of computer power but comes from the sharp first-order transition, Figs 11.9 and 11.10. Either one language dominates as if 80 percent of the world speaks Chinese. Or all 2^ℓ languages are equivalent apart from fluctuations and thus each is spoken only by a small population. If the first-order transition were changed into a second-order one, the results for mutation rates slightly below the critical point might be better.

An alternative was suggested by linguist Wichmann [24]. The present language distribution is not in equilibrium. If we assume that parts of the world are on one side and parts on the other side of the phase transition from dominance to fragmentation (or from fragmentation to dominance), then the above

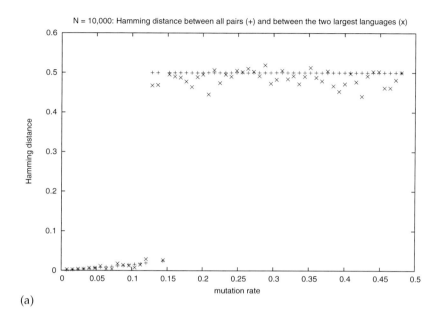

(a)

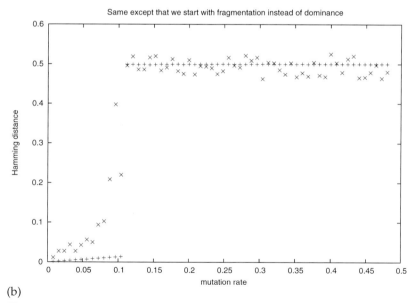

(b)

Fig. 11.10 Difference between languages, as measured by the normalized Hamming distance = fraction of different bits. We show both the average distance between all pairs and that between the two largest languages, for 10 000 people and $q = 0$. (a) starts with dominance, (b) with fragmentation. From [27]

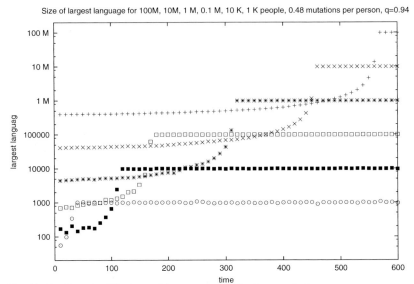

Fig. 11.11 Growth of the largest language out of a fragmented population for $\ell = 8$. The larger the population is the longer we have to wait.

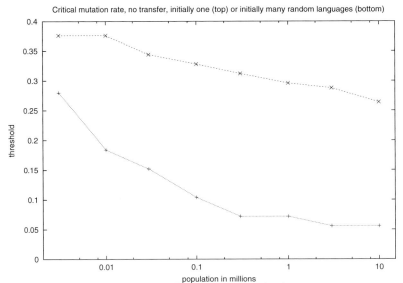

Fig. 11.12 Hysteresis for position of phase transition; for the upper curve we start from dominance, for the lower curve from fragmentation. Below the curve, one language dominates at the end, above the curve we end up with fragmentation.

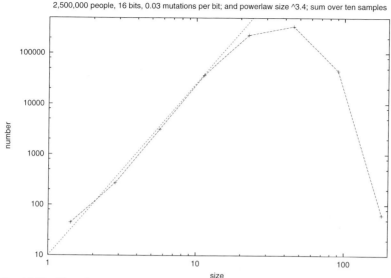

Fig. 11.13 Size distribution in equilibrium, showing a realistic shape but not a realistic size scale, from [14].

equilibrium results are not good. Instead, we show, in Fig. 11.14, two runs for a nonequilibrium situation of about 5000 iterations at very low mutation rate, starting from fragmentation. The results are averaged over the second half of the simulation with the time adjusted such that the phase transition of Fig. 11.11 happened during that second half. Now the language sizes vary over five orders of magnitude, much better than in Fig. 11.13. (If we start from dominance the size distribution is similar but more symmetric [24].)

11.4 Conclusion

The last few years have seen the development of a variety of different approaches to simulating the competition between existing languages of adult humans. Each model has its advantages and disadvantages.

If we follow the tradition of physics, that theories should explain precise experimental data, then the size histogram of the 10^4 human languages, Fig. 11.3, seems to be the best candidate. Empirically it is based on Grimes [1] and was analyzed, e.g., by [1,3,9,23]. In order to simulate this language size distribution, we need models for 10^4 different languages, and only two of them have been published thus far, the Viviane model and our model [14,17].

Future work with these models could look at the similarities and differences between the languages (bit-strings), as started in Fig. 11.10 and [18], or the geography of languages and their dialects [25], as started in Fig. 11.15 and [17].

We thank our coauthors [21,24] for collaboration.

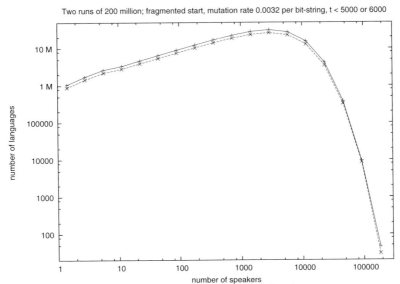

Fig. 11.14 Size distribution far from equilibrium during the phase transiton from fragmentation to dominance, $\ell = 16$, $q = 0$. Additional smoothening by random multiplications was applied [24].

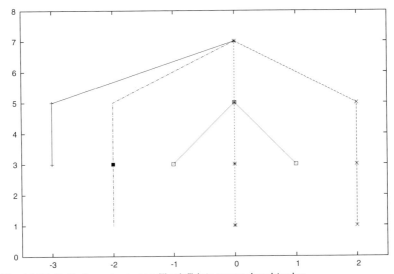

Fig. 11.15 Split of one language ("Latin") into many daughter languages ("Spanish") which then split into further languages ("Catalan") and dialects ("Mallorquin"). Many dead-ends are omitted for clarity. Time increases from top to bottom, but in the Viviane model no language dies out.

11.5
Appendix

This Appendix gives some more results for many languages, first on the Viviane model [17] and then on our model [14].

11.5.1
Viviane Colonization Model

For the model of [17], one can look at the history of how languages split from mother languages, and later produce more daughter languages. In contrast to linguistic field research, which looks only at the last few thousand years, computer simulations can store and analyze the whole of history since the beginning. Figure 11.15 shows for a small 64×64 lattice from [14](d), how one language split into daughters, etc., very similar to biological speciation trees. For clarity we omitted numerous languages which had no "children". For larger lattices we found that, even for many thousand languages, a few steps suffice on average, to reach the oldest ancestor language on the top of the tree from any of the languages in this tree. Other tree simulations were published in [26].

Often a conquering population imposes its language on the native population. Perhaps in Europe, before the arrival of Indo-European farmers, the Cro Magnon people spoke a language family of which the Basque language is the only present survivor. Better documented, though not necessarily any more true, is the story of the single Gallic village in today's France which resisted the Roman conquest two millennia ago, thanks to the efforts of Asterix and Obelix (helped by doping). In the Viviane model, where people may adopt the language of their neighbors, such a single resistance center can influence many other sites during the later spread of languages. Figure 11.16 shows that indeed a rather large fraction of the total population is influenced by Asterix, particularly for large mutation rates [14](d). In physics, such simulations of the influence of a single "error" are called "damage spreading".

11.5.2
Our Bit-string Model

While the Viviane model always happens on a lattice, for our model the lattice is optional. If we want to study the geographical coexistence of two languages in adjacent regions, then of course a lattice is needed [14](b). Now on every lattice site live many people. Figure 11.17 shows how, without any difference in status, as opposed to [11], on one side one language dominates and on the other side the other language dominates, if initially each region was occupied only by speakers of its own language. Also, in the transition region, the other $2^\ell - 2$ languages play no major role. The situation in this figure may

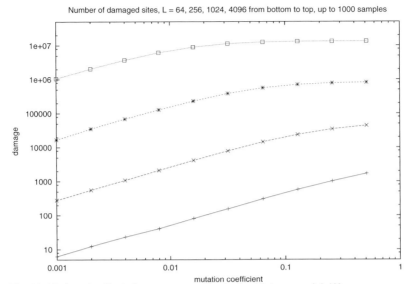

Fig. 11.16 Asterix effect. Damage spreading in the Viviane model. We plot the number of sites that are influenced by one stubborn site which refuses to obey the rules and sticks to its own language, versus the mutation coefficient.

correspond to New Orleans a long time ago, where Canal Street separated the French quarter from the newer English settlement. These methods could be applied to dialectometry, as documented for France by Goebl [25].

Bit-strings allow only $Q = 2$ choices per position, but the lattice model was also generalized to $Q = 3$ and 5 choices. Surprisingly, the phase transition curve, Fig. 11.18, between dominance for low and fragmentation for high mutation rates was independent of this number Q of choices. Only when ℓ was changed were the different symbols in Fig. 11.18 obtained.

If we want to apply the lattice model to geography we want compact geographical language regions to emerge from a fragmented start. Then it is not only the transfer of language elements but also the flight to another language which needs to be restricted to lattice neighbors, i.e., people learn new elements or a new language only from one of the four nearest neighbors, randomly selected. Figure 11.19 shows how one language, accidentally dominating at intermediate times, grows until is covers nearly the whole lattice.

One may look, without a lattice, on the history of people speaking one randomly selected language in an initially fragmented population. Because of mutations, after a long enough time everybody has moved at least once to another language. But since the number $L = 2^\ell$ of possible languages is finite, some people move back to their original language, like emigrants whose offspring later return to their old country. Thus after 50 iterations to give an

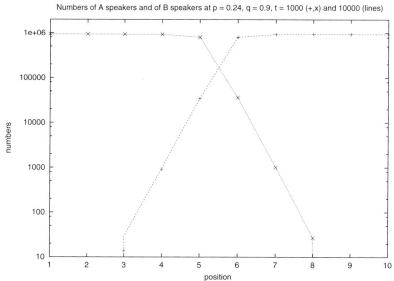

Fig. 11.17 Interface between two languages spoken on a square lattice; from [14](b).

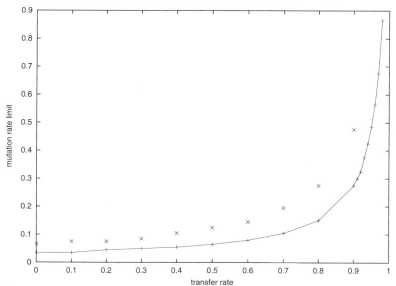

Fig. 11.18 Phase diagram. Dominance in the lower right and fragmentation in the upper left parts. One sample 301×301, $\ell = 8$, $t = 300$. The thresholds are about the same for $Q = 2, 3, 5$ and are shown by plus signs connected with lines. The single $\times$ signs refer to $Q = 10$, $\ell = 4$ and agree with those for $Q = 3$, $\ell = 4$. Here we start from dominance; if we start from fragmentation, the transition lines shifts to lower mutation rates.

334 | *11 Computer Simulation of Language Competition by Physicists*

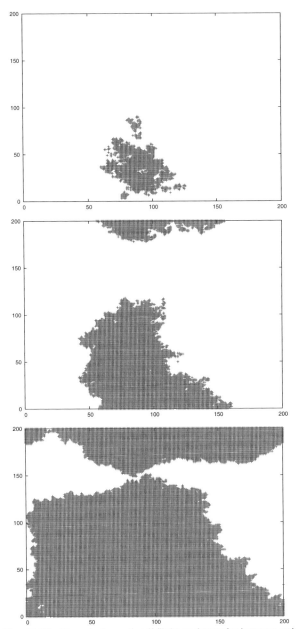

Fig. 11.19 Domain formation if flight and transfer happen only to/from a language learned from a lattice neighbor. We mark the sites where the largest language is spoken, after 240, 330, and 450 iterations. For $t \geq 514$ nearly everybody speaks this language. ($L = 200$, $p = 0.016$, $q = 0.9$, $r = 1$, $\ell = 16$, $Q = 2$, periodic boundary conditions).

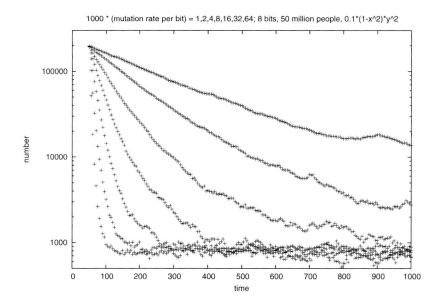

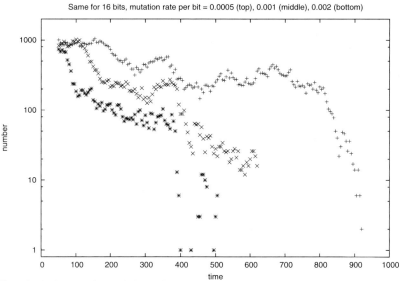

Fig. 11.20 Decay of population originally speaking one particular language for $\ell = 8$ (a) and 16 (b).

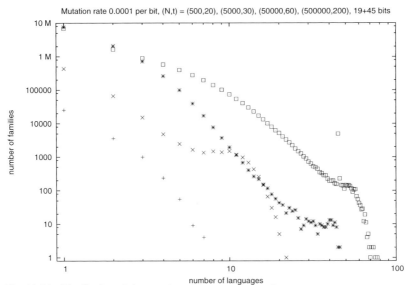

Fig. 11.21 Distribution of the number of languages in a language family, from sums over 100 or 10 independent simulations at various population sizes, $p = 0.0064$. The observation time increases slightly with increasing population size. From [27]

equilibrium, we mark all those speaking language zero. Their offspring carry that mark also, even if they mutate their language, and we count at each time step the number of marked people speaking language zero.

Then we see a rapid decrease in that number; to slow down the decay we modified the flight probability to $0.1(1 - x^2)y^2$ where y is the fraction for the language which the individual from fraction x decides to switch to. Then a slower decay as in Fig. 11.20 results, faster for higher mutation rates. For $\ell = 8$ bits we see clearly the random background of less than a thousand people (from 50 million) who returned to the language zero of their ancestors; for $\ell = 16$ both the initial and the final number of zero speakers are much smaller since the 50 million can now distribute among 65 536 instead of only 256 possible languages.

Human languages can be grouped into families, like the Indo-European family of 10^2 different languages. To simulate language families we need a criterion for which bit-strings belong in one family. Thus we worked with $\ell = 64$ bits and assumed, following Wichmann, that the leading 19 bits determine the family and the remaining 45 bits the different languages within one family. The numbers 2^{19} and 2^{45} of possible families and languages are so large that our computer simulations, with less than a million people, do not notice their finite size. Indeed, the results in Fig. 11.21 for 500, 5000, 50 000 and 500 000 people are roughly independent of population size and show a mostly monotonically decaying probability distribution function for the number of languages within one family. Empirical observations have been published, e.g., by Wichmann [1].

References

1 GRIMES, B. F., *Ethnologue: languages of the world* (14th edn. 2000). Dallas, TX: Summer Institute of Linguistics; WICHMANN, S., *J. Linguistics* 41 (**2005**), 117

2 BRISCOE, E. J., *Language* 76 (**2000**), 245

3 NOVOTNY V., DROZD, P., *Proc. Roy. Soc. London* B267 (**2000**), 947

4 NOWAK, M. A., KOMAROVA, N. L., NIYOGI, P., *Nature* 417 (**2002**), 611; KOMAROVA, N. L., *J. Theor. Biology* 230 (**2004**), 227

5 BARONCHELLI, A., FELICI, M., CAGLIOTI, E., LORETO, V., STEELS, L., arXiv:physics/0509075, arXiv:physics/0511201, arXiv:physics:0512045

6 BRIGHTON, H., SMITH K., KIRBY, S., *Physics of Life Reviews* 2 (**2005**) 177; CULICOVER, P, NOWAK, A., *Dynamical Grammar*, Oxford University Press, Oxford, **2003**; *Simulating the Evolution of Language*, Eds. A. Cangelosi, D. Parisi, Springer, New York, **2002**

7 NETTLE, D., *Proc. Natl. Acad. Sci. USA* 96 (**1999**), 3325

8 STAUFFER, D., MOSS DE OLIVEIRA, S., DE OLIVEIRA, P. M. C., SA MARTINS, J. S., *Biology, Sociology, Geology by Computational Physicists*, Elsevier, Amsterdam, **2006**

9 SUTHERLAND, W. J., *Nature* 423 (**2003**), 276

10 ABRAMS, D. M., STROGATZ, S. H., *Nature* 424 (**2003**) 900

11 PATRIARCA, M., LEPPANEN, T., *Physica A* 338 (**2004**) 296

12 MIRA, J., PAREDES, A., *Europhys. Lett.* 69 (**2005**), 1031

13 KOSMIDIS, K., HALLEY, J. M., ARGYRAKIS, P., *Physica A*, 353 (**2005**), 595; KOSMIDIS, K., KALAMPOKIS A., ARGYRAKIS, P., *Physica A* in press, physics/0510019

14 SCHULZE, C., STAUFFER, D., *Int. J. Mod. Phys. C* 16 (**2005**), 781; *Physics of Life Reviews* 2 (**2005**), 89; *AIP Conference Proceedings* 779 (**2005**), 49; *Comput. Sci. Engin.* 8 (May/June **2006**), 86

15 PINASCO, J. P., ROMANELLI, L., *Physica A* 361 (**2006**), 355

16 SCHWAMMLE, V., *Int. J. Mod. Phys. C* 16 (**2005**), 1519; *ibidem* 17 (**2006**), p. 103

17 DE OLIVEIRA, V. M., GOMES, M. A. F., TSANG, I. R., *Physica A* 361 (**2006**), 361; DE OLIVEIRA, V. M., CAMPOS, P. R. A., GOMES M. A. F., TSANG, I. R., *Physica A* e-print physics/0510249

18 TESILEANU, T., MEYER-ORTMANNS, H., *Int. J. Mod. Phys. C* 17 (**2006**), p. 259

19 *The Logistic Map and the Route to Chaos*, Eds. M. Ausloos, M. Dirickx, Springer, Berlin, **2006**

20 WANG, W. S. Y., KE, J., MINETT, J. W., in: *Computational linguistics and beyond*, Eds. C. R. Huang, W. Lenders, Academica Sinica : Institute of Linguistics, Taipei, **2004**; www.ee.cuhk.edu.hk/~wsywang

21 STAUFFER, D., COSTELLO, X., EGUÍLUZ, V. M., SAN MIGUEL, M., *Physica A* in press; arXiv:physics/0603042

22 PENNA, T. J. P., *J. Stat. Phys.* 78 (**1995**), 1629; TICONA BUSTILLOS, A., DE OLIVEIRA, P. M. C., *Phys. Rev. E* 69 (**2004**), 021903

23 GOMES, M. A. F., VASCONCELOS, G. L., TSANG, I. S., TSANG, I. R., *Physica A* 271 (**1999**), 489

24 STAUFFER, D., SCHULZE, C., LIMA, F. W. S., WICHMANN, S., SOLOMON, S., arXiv:physics/0601160, *Physica A*, in press

25 GOEBL, H., *Mitt.Österr. Geogr. Ges.* 146 (**2004**), 247 (in German)

26 WANG, W. S. Y., MINETT, J. W., *Trans. Philological Soc.* 103 (**2005**), pp. 121

27 WICHMANN, S., STAUFFER, D., LIMA, F. W. S., SCHULZE, C., arXiv:physics/0604146.

12
Social Opinion Dynamics
Gérard Weisbuch

12.1
Introduction

First of all, we will here discuss very simplified models of opinion dynamics. At this level of description, the same set of models are used concerning the diffusion of opinions and the resulting decisions, although this distinction might be of importance in many circumstances in real life.

Opinion dynamics is a central topic in sociology, especially in sociodynamics. Decision dynamics under social influence have been under scientific scrutiny for many decades (see for instance the book *Diffusion of Innovations* by Everett Rogers [1]). In fact, the issue of how opinions are made and decisions adopted by crowds is mentioned even by historians from the Antiquities: Livius, for instance, wonders what were the processes that made the Plebs retire on the Aventine hill.

Opinion and decision dynamics under social influence occur in economic, political, or even in more personal contexts. An important field of study with socio-political implications is the diffusion of working practices. The earlier empirical studies concerned the choice of corn varieties in the US (by Bryce Ryan and Neal C. Gross, rural sociologists at Iowa State University [2]) and the diffusion of antibiotic prescriptions by doctors (by James S. Coleman, Elihu Katz, and Herbert Menzel [3]). Later literature discusses the empirical and normative aspect of the diffusion of innovations in agriculture (the "Green revolution" or environmental friendly practices, [4]) or in the organization of the economy (the study of the 90s transition in Eastern Europe [5]).

The basic process that we here hypothetise is Gabriel Tarde's [6] imitation principle "do as the others do". The rationale for this imitation behavior are of two kinds.

- Bounded rationality: in a situation when agents lack objective information, such as choosing a new movie or an unknown restaurant, they might suppose that the other agents whom they observe have made a documented choice.

- Externalities (an expression from economics): in some cases doing as the majority might bring some advantage, such as sharing knowledge and getting a better service in the case of technological equipment, for instance.

Among the important simplifications that we introduce by only considering social processes (influence, imitation, etc.) we neglect the importance of the following

- Knowledge: Some objective knowledge might exist which would bias opinions.
- Individual cognitive processes which would for instance modify opinions according to direct personal experience.

All the dynamic processes that we discuss here privilege social interactions as the main vector for opinion change. The general framework is then as follows.

- At any given time a distribution of opinions exists in the population.
- At each time step some group of individuals interact and as a result, one or several opinions are shifted, generally towards some consensus among the group.
- As a result of these interactions, some form of clustering of opinions is observed.

In the spirit of complex system studies, we focus on the attractor of the dynamics and their spatio-temporal structure because the characterization of attractors are the generic properties of the dynamics [7, 8]. The models that we discuss here are based on such strong simplifications of the processes occurring in the real world, that the eventual conclusions that may come out of these models are only the generic properties of the dynamics, to be compared with what economists call stylized facts.

We will, in general, be interested in:

- the nature of the attractors, consensus or diversity of opinon;
- their number;
- possible patterns on social networks
- regime transitions, and so on ...

The specifics of the different variants further described concern the following.

- The interaction space: are all interactions *a priori* possible (the "well mixed case") or do interactions only occur via a neighborhood structure such as a "social network"?

- The "dimension" of the opinion; the most standard studies referred to binary opinions or binary choices: good/bad, buy/not buy, for/against. The generalization to continuous opinions, how much is this worth?, or to vector opinions as in some models of cultural dynamics, yields surprisingly different results.

- The detailed results might depend upon the specifics of the iteration modes, parallel/sequential, the heterogeneity of the agents or the reversibility of the decision.

Not surprisingly, the role of the dimensions of the interaction space and of the opinion variable reminds us of the universality classes of the renormalization-group approach [9].

The present contribution first recalls the main results obtained with the binary choice models. The next section is devoted to continuous opinion dynamics. We then discuss vector opinion models. The exposition in each section starts a with well-mixed case and continues with generalizations to social networks. We examine, in the conclusion section, the possible connections with empirical data and the experimental approach of social psychology, and some "normative" aspects: how to use some understanding of opinion dynamics to plan a campaign to convince customers, firms or fellow citizens. We have chosen to present simulations rather than the formalisms which sometimes yield exact results, but the bibliographical references to formal methods are given. We also tried to give some indication about connections with the Social Sciences literature.

12.2
Binary Opinions

Binary opinions and binary choices are the most standard approach to opinion dynamics. The basic model is that, at each time step, one agent samples a small group of other agents and chooses the majority opinion. In the simplest version, closely parallel to epidemiology (Susceptible-Infected models), one starts from a large majority of agents with opinion 0. Whenever they encounter agents with opinion 1, they are irreversibly converted to opinion 1.

When all binary encounters are *a priori* possible, a differential equation approach is possible when the number of agents tends towards infinity; it yields exact results.

12.2.1
Full Mixing

Let us call N the total number of agents, and n the number of agents with opinion 1. When N is large we can postulate that the probability of encountering agents with any opinion 0 or 1 is equal to their proportion in the population. If we suppose that only agents in state 0 can change their opinion, the dynamics of n can be written:

$$\frac{dn}{dt} = a \cdot (N - n) \cdot n \tag{12.1}$$

where a is a kinetic coefficient. This equation is readily integrated and gives the S-shaped solution represented in Fig. 12.1:

$$n = \frac{N}{1 + \alpha \exp(\beta t)} \tag{12.2}$$

where α and β relate respectively to initial conditions and to the kinetic coefficient. The S curve, and the integrated bell curve, are standard material in

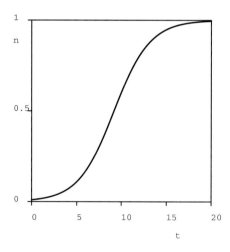

Fig. 12.1 Time evolution of the fraction of agents with opinion 1 in a full-mixing topology.

textbooks on marketing, representing the sales curves of new products. The S curve is also the rate adoption curve of new technologies. This simple model when applied to the adoption of new products or techniques, supposes that the transition from choice 0 to choice one is always beneficial: each transition is thus asymmetrical. The status of the initial adopters supposes that they had some special reason to adopt as opposed to the later adopters who adopted because of social influence.

The precise set-up that we used, namely an infinitely large number of agents and asymmetry of opinions is important. Other set-ups can give a different outcome.

For instance, the two-restaurants problem [10], where newcomers would visit restaurant A or B with a probability proportional to the number of agents already inside each restaurant, is a standard problem among economists discussing "herd behavior". The set-up is symmetrical with respect to either choice, irreversibility of choice, but a finite number of customers. A first customer facing the two empty restaurants would choose randomly among them. New customers would choose randomly with a probability proportional to the number of customers present in each restaurant. The first steps in the series of customer choices are crucial in determining the ultimate choices of later customers. It can be shown that attractors of the dynamics of the fraction of customers in one restaurant (say A) are distributed in the interval [0,1] depending upon the initial customer's choice: the problem is equivalent to Polya's urn. The term "information cascade" is often used by economists to describe these dynamics [11, 12].

12.2.2
Lattices as Surrogate Social Nets

Even the earliest empirical studies revealed that social influence occurs on some social network. Human agents are not influenced by unknown individuals (as implied by a full mixing topology); they rather trust some selected individuals with whom they have stronger connections. According to the problem they have to solve, they might use family connections, seek professional advice, or check the opinion of their peers. The topic of social network has always been very active and has received a lot of attention in the last ten years. For instance, small-worlds [13] and scale-free networks [14] were proposed as topological structures more relevant to social dynamics than lattices or random networks.

We nevertheless base our exposition on periodic lattices because :

- they are easy to visualise;

- they display a high degree of "betweenness", a property shared with real social networks.

(Betweenness is the proportion of your neighbors which are themselves neighbors; the property is important in the models that we will use). Most of the concepts that we develop here apply to structures more disordered than lattices.

12.2.3
Cellular Automata

The lattice version of binary dynamics is expressed by cellular automata "counter" rules (so-called because the new state of the automaton is obtained by counting how many neighbors are in state 1). The name "voter rules" applies when a majority rule is applied. Each cell i of the network represents an agent and its binary state S_i represents an opinion. Cells are connected to some neighborhood of z "neighbors". Cells can be in state 0 or 1. At any time step, a cell i updates its state S_i taking into account the number of its neighbors S_j in state 1 according to:

$$\text{if} \quad \sum_{j=1}^{z} S_j > \theta_i \quad \text{then} \quad S_i = 1 \tag{12.3}$$

$$\text{else} \quad S_i = 0 \tag{12.4}$$

where θ_i is a fixed integer threshold.

These models are formally equivalent to the Ising model of ferromagnetism at 0 temperature with an external field $h_i = z/2 - \theta_i$, where z is the number of neighbors of cell i.

In the "socio-economic" interpretation of counter automata, the S variable represents the binary decision, for instance, buy/not buy. The threshold θ represents the (economic) utility of choice S, including eventually the price as a negative component. Alternatively, in a buy brand A or buy brand B interpretation, the threshold θ represents the difference in utility of the two brands A and B. The sum term represent the social influence of the neighborhood on the choice of agent i.

This simple set-up can result in many different dynamics according to updating rules, choice of thresholds, probabilistic versus deterministic transition rules etc.

Let us first discuss the most standard case of the cellular automata rules.

- Parallel iteration: the state of all cells is updated simultaneously.
- Homogeneity: all thresholds θ_i are identical.
- We use a square lattice with von Neumann neighborhood: each cell is connected to four neighbors (see Fig. 12.2).

12.2.3.1 Growth

The necessary specifications of "counter" automata rules are completed by the choice of neighborhood and threshold. Several neighborhoods have been pro-

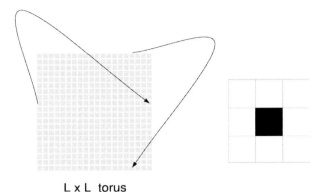

L x L torus

Fig. 12.2 Interaction geometry. Periodic boundary connections are displayed on the left. On the right, depending upon the neighborhood choice, the black cell in the center is influenced by four or eight neighbors (the white cells).

posed on square lattices, the von Neumann neighborhood which include the four neighbors N, S, E and W; and the Moore neighborhood with 8 neighbors, N, NW, W, WS, S, SE, E and NE. In fact the i site itself may also be included; the standard choices respecting the symmetry of the square lattice are 4, 5, 8 or 9 neighbors.

Let us describe possible dynamics for the five neighbors case; the results are easily generalized to other neighborhoods. One classifies "counter" automata by the attractors that are reached when the initial configuration consists of a few automata in state 1 surrounded by a "sea" of automata in state 0. Small thresholds favor the growth of contiguous regions of cells in state 1, while larger thresholds ($\theta_i > 3$) favour the growth of contiguous regions of cells in state 0. But as discussed below, there are limits to growth and some "local" conditions have to be fulfiled to allow growth [15].

When θ_i is 0 or less, growth of a 1's region accross the whole lattice occurs under any initial condition in one time step.

For a threshold of 1, any initial configuration with at least one cell in state 1 generates clusters of 1's which grow to fill the entire lattice. A possible interpretation of the case of threshold 1 is that state 1 corresponds to the situation when the agent is "informed" while agents in state 0 are not yet "informed". One important variant concerns the case when not all lattice sites are occupied by an automaton: this is the percolation problem (more later on this issue).

The next couple of figures display the possible evolution when the threshold is two neighbors. When the density of ones is low, figure 12.3 display the evolution of isolated small clusters of 1's.

Larger initial clusters evolve towards their convex hull after possible trimming of corners as seen in Fig. 12.4.

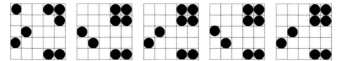

Fig. 12.3 Evolution in five time steps from left to right of small initial clusters of 1's for a threshold of 2 neighbors out of 5. Black circles represent cells in state 1, cells in state 0 are empty. Isolated 1 disappear, horizontal (and vertical) pairs are stable, oblique pairs oscillate.

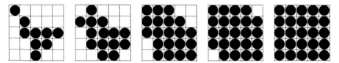

Fig. 12.4 Growth from an initial cluster (left pattern) to its convex hull (right pattern) in 5 time steps for a threshold of 2 neighbors out of 5. Growth proceeds along cluster edges and stops when all are completed.

The evolution of the system thus depends on the initial conditions; low densities of initial 1's end up in isolated clusters of 1's; larger densities might result in filling up the lattice with 1's with the exception of a few clusters of 0's. In any case coarsening of the clusters is obtained: small clusters disappear, and boundaries between clusters are vertical or horizontal lines.

When the threshold for growth is three neighbors, coarsening occurs for initial densities of close to 0.5 and large domains can be observed. But unbalanced initial densities bias the evolution of the system towards larger densities of 0's or 1's.

Larger thresholds from 4 to 6 yield symmetrical behavior with respect to the 0's and 1's regions.

The description given above for threshold automata with a neighborhood of 5 is easily generalized to other neighborhoods (4, 8 or 9 neighbors). The important conclusions still hold. Except for the limit cases of extreme thresholds which favor homogeneous conditions, the following conditions hold.

- The dynamics favor coarsening of the initial domains.

- Attractor configurations depend upon initial conditions.

- Growth necessitates "seed" configurations to occur; it occurs at the boundaries of clusters.

12.2.4
INCA

The term INCA INhomogeneous Cellular Automata has been proposed by H. Hyman and G. Vichniac [15] and applies to a variety of models. The idea is that all automata on the lattice obey the same "kind" of rules, but with different parameters. INCAs belong to a large class of systems described by theoretical physicists as disordered systems. The "philosophy" of disordered systems is to replace structural disorder observed in natural systems, such as glasses or diluted magnetic alloys, by interaction disorder on regular lattices.

A general expression can be written:

$$\text{if} \quad \sum_j J_{ij} S_j > \theta_i \quad \text{then} \quad S_i = 1 \quad (12.5)$$

$$\text{else} \quad S_i = 0 \quad (12.6)$$

where interaction strength J_{ij} and threshold θ_i are local variables. These models are formally equivalent to Ising spin glasses at 0 temperature, and to neural nets.

Let us briefly discuss a few examples.

- The percolation problem [16] can be interpreted as a cellular automaton with homogeneous threshold 1 when a fraction of the cells are empty. Percolation models display abrupt transitions in the diameter of the largest cluster (at equilibrium) of cells at state one as a function of the fraction of empty cells.

- A number of models of social influence proposed by the "Polish school" [17]–[21] are based on INCA's. A simple version proposed by Kacperski and Holyst [17] discusses the role of a strong leader with a large influence on his neighborhood in the case of a homogeneous threshold. Most of these papers were written to describe the "transition" in Eastern Europe from a socialist to a free market economy: state 0 corresponds to the choice of socialism by economic actors and state 1 to free market choice. The threshold is interpreted as some external "global" signal applied uniformly to all actors. These systems display first-order transitions and hysteresis when the threshold is varied[1].

More generally, networks of threshold automata are also called neural nets, and are widely use as models in cognitive science [7].

One application of this formalism to coalition formation in social and political sciences has been proposed by R. Axelrod [22]. In this model the J_{ij} are computed according to common traits, to social actors, in a version quite close to Hebb's rule in cognitive science. J_{ij} are constructed starting from 0 and adding +1 for each common trait and −1 for each different trait. J_{ij} can then be negative. The signal S_i corresponds to which coalition is chosen: one example used by Axelrod is WWII with two coalitions Allies and Axis represented,

---

1) First-order transitions and hysteresis are not specific of INCAs and also happen with homogeneous cellular automata when the threshold is varied.

for instance, by $S_i = 1$ and $S_i = -1$. No spatial structure is involved and all interactions are, a priori, possible. The attractors of the dynamics correspond to stable coalitions. Axelrod applied his model of coalition formation from the twentieth century political situation to the adoption of different UNIX standards among computer firms.

Analytical solutions derived from analogies with disordered physical systems have been proposed for both the percolation model (by the Renormalization-group approach) [16] and the inhomogeneous threshold automata model (by the Mean-field formalism) [17].

12.2.5
Probabilistic Dynamics

In the above models, the disorder is quenched: it only concerns the initial conditions, plus eventually the sequence of which cells are chosen to be updated when sequential updating is used. Most of these models have also been studied in their probabilistic (or finite temperature version). The expressions are easier written with variables $S = 1$ or $S = -1$:

$$\Delta E_i = -2 \sum_j J_{ij} S_j - \theta_i \quad (12.7)$$

$$S_i = 1 \quad \text{with probability} \quad \frac{1}{1 + \exp(\beta \Delta E_i)} \quad (12.8)$$

$$\text{else} \quad S_i = -1 \quad (12.9)$$

where β is the inverse temperature coeeficent, and ΔE_i the difference in "energy" between states 1 and -1. The J_{ij} coefficients represent the intensity of the influence of agents j on agent i. Larger values of β yield behavior similar to those observed for the deterministic case described above. Low values of β yield disordered behavior with cells randomly changing states. A second-order phase transition is observed for some intermediate β_c value depending upon interaction and threshold distribution. The analogy with magnetic systems at finite temperature and Glauber dynamics is obvious.

In the probabilistic versions of cellular automata with large values of β, the limits to growth and the necessity of having favorable initial conditions lose their compulsory character, but simply appear as rate-limiting steps to growth. For instance, in the case of threshold equal to four out of eight neighbors, convex hulls of ones do not limit growth for ever: they rather slow down the growth of "faceted macro crystals". The same is true for the seed configurations allowing growth for lower thresholds: one observes metastable configurations which evolve towards uniform configurations of ones after the long-awaited apparition of seeds. These delay phenomena are typical of first-order phase transitions.

Finally, let us note that this transcription from Boltzmann statistics to opinion dynamics is already familiar to economists [23] since the 70s: they use utility functions where physicists use energy function, although with a minus sign. The Boltzmann factor in expression 12.8 is often called the logit function by economists [24]. Various interpretations of this factor have been proposed from a noise term to optimization between exploitation vs exploration compromise [25]

12.2.6
Group Processes

In the above discussion, we only considered processes when one agent would survey other agent(s) (voters models), and choose a new opinion accordingly. Some authors (Sznajd-Weron et al. [26], Galam [27]) considered the possibility of elementary processes involving more than one agent. Agents are engaged in small discussion groups and eventually changing opinions, accordingly. Both groups of authors had in mind modeling votes in a political context. The Galam model [27], is discussed in the present volume by its author.

The idea of Sznajd-Weron et al. [26] is based on the formula: "United we stand divided we fall", hence the proposed acronym for their model USDF. USDF is a model for "propaganda", where pairs of sites in the same state convince their neighbors to share their views. On the contrary, pairs of opposite views propagate opposite views to their neighbors. The model was originally proposed with binary states on a 1D line, and gives attractors which are either homogeneous with all nodes at state 0 or 1 (equivalent to ferromagnetism) or alternation of zero and 1 sites (equivalent to anti-ferromagnetism). Later variants by Stauffer [28] and his "Brazilian connection" have generalized to larger dimensions and integer opinions. They can be compared to the models described in the next sections.

12.3
Continuous Opinion Dynamics

The rationale for binary versus continuous opinions might be related to the kind of information used by agents to validate their own choice:

- the actual choice of the other agents, a situation common in economic choice of brands: "do as the others do";

- the actual opinion of the other agents, about the "value" of a choice: "establish one's opinion according to what the others think or at least according to what they say".

One often encounters situations when opinions concern quantities rather than two options.

- How much is this worth?
- How should we share?

etc.

On the empirical side, there exist well-documented studies about social norms concerning sharing between partners. Henrich et al. [29] compared through experiments, shares accepted in the ultimatum game and showed that people agree upon what a "fair" share should be, which can of course vary across different cultures. Young and Burke [30] report empirical data about crop-sharing contracts, whereby a landlord leases his farm to a tenant laborer in return for a fixed share of the crops. In Illinois as well as in India, crop sharing distributions are strongly peaked upon "simple values" such as 1/2–1/2 or 1/3–2/3. The clustering of opinions about "fair shares" is the kind of stylized fact that continuous opinion models try to reproduce.

The model discussed here was introduced [31] to interpret decisions made by farmers about accepting or refusing grants to change their agricultural practices in favour of environmentally friendly practice. According to the survey results, they first evaluate the economic outcome of new practices and compare them with their old practice in order to make their decision.

Modeling of continuous opinions dynamics was earlier started by applied mathematicians and focused on the conditions under which a panel of experts would reach a consensus, ([32–35], further referred to as the "consensus" literature).

The purpose of this section is to present results concerning continuous opinion dynamics subject to the constraint that convergent opinion adjustment only proceeds when opinion difference is below a given threshold. The rationale for the threshold condition is that agents only interact when their opinions are already close enough; otherwise they do not even bother with discussion. The reason for refusing discussion might be, for instance, lack of understanding, conflicting interest, or social pressure. The threshold would then correspond to some openness character. Another interpretation is that the threshold corresponds to uncertainty: the agents have some initial views with some degree of uncertainty and would not care about other views outside their uncertainty range.

12.3.1
The Basic Case: Complete Mixing and one Fixed Threshold

Let us consider a population of N agents i with continuous opinion x_i. We start from an initial distribution of opinions, most often taken as uniform on

[0,1] in the computer simulations. At each time step any two randomly chosen agents meet: they re-adjust their opinion when their difference of opinion is smaller in magnitude than a threshold d. Suppose that the two agents have opinion x and x'. Iff $|x - x'| < d$ opinions are adjusted according to:

$$x = x + \mu \cdot (x' - x) \qquad (12.10)$$
$$x' = x' + \mu \cdot (x - x') \qquad (12.11)$$

where μ is the convergence parameter whose values may range from 0 to 0.5.

In the basic model, the threshold d is taken as constant in time and across the whole population. Note that we here apply a complete mixing hypothesis plus a random serial iteration mode[2].

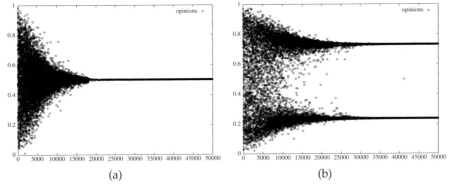

Fig. 12.5 Time charts of opinions for different interaction thresholds (a) $d = 0.3$, (b) $d = 0.2$. Each dot correspond to a sampled opinion at a time given by the x-axis postion. One time unit corresponds to sampling a pair of agents. In both cases $\mu = 0.5$; $N = 2000$ for (b) and $N = 1000$ for (a).

- For large threshold values ($d > 0.3$) only one cluster is observed at the average initial opinion as seen Fig. 12.5(a), which represents the time evolution of opinions starting from a uniform distribution of opinions.

- For lower threshold values, several clusters can be observed (see Fig. 12.5(b)). Consensus is then NOT achieved when thresholds are low enough.

Obtaining clusters of different opinions does not surprise an observer of human societies, but this result was not obvious since we started from an initial configuration where transitivity of opinion propagation was possible through

---

2) The "consensus" literature most often uses the parallel iteration mode when they suppose that agents average, at each time step, the opinions of their neighborhood. Their implicit rationale for parallel iteration is that they model successive meetings among experts.

the entire population: any two agents, however different in their opinions, could have been related through a chain of agents with closer opinions. The dynamics that we describe resulted in gathering opinions in clusters on the one hand, but also in separating the clusters in such a way that agents in different clusters no longer exchange.

The number of clusters varies as the integer part of $1/2d$: this is to be further referred to as the "$1/2d$ rule" (see Fig. 12.6).

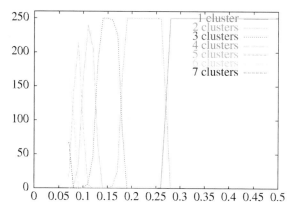

Fig. 12.6 Statistics of the number of opinion clusters as a function of d on the x axis for 250 samples ($\mu = 0.5$, $N = 1000$). (For a color version please see color plate on page 598.)

An alternative method to directly runing simulations is to solve the master equations describing the time evolution of populations of bins of opinions. This approach is used for instance by Redner and co-authors [36] to finely describe the structure of the attractors as a function of the interaction threshold d.

Rather than working on continuous opinions, Stauffer and other authors [37] have proposed to use integer opinions: in that case convergence is rapidly obtained and easily tested.

12.3.2
Social Networks

The literature on social influence and social choice also considers the case when interactions occur along social connections between agents [23] rather than randomly across the whole population.

Apart from the similarity condition, we now add to our model a condition on proximity, i.e., agents only interact if they are directly connected through a pre-existing social relation.

We then started from a two dimensional network of connected agents on a square grid. Any agent can only interact with his four connected neighbors

(N, S, E and W). We used the same initial random sampling of opinions from 0 to 1 and the same basic interaction process between agents as in the previous sections. At each time step a pair is randomly selected among *connected agents* and opinions are updated according to Eqs (12.1) and (12.2) provided of course that their distance is less than d.

At equilibrium, clusters of homogeneous opinions appear. The opinions themselves are not very different from those observed with nonlocal opinion mixing as described in the previous section,

For the larger values of d (e.g., $d = 0.3$ as seen in Fig. 12.7(a)), the lattice is filled with a large majority of agents who have reached consensus around $x = 0.5$ while a few isolated agents have "extremists" opinions closer to 0 or 1. The importance of extremists is the most noticeable difference with the full mixing case described in the previous section.

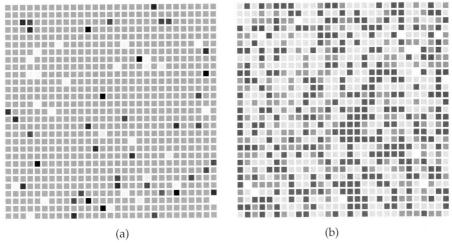

(a)   (b)

Fig. 12.7 Display of final opinions of agents connected on a square lattice of size 29x29. Opinions between 0 and 1 are coded by gray level (0 is black and 1 is white). The pattern (a) is obtained for a larger interaction threshold, $d = 0.3$. Note the percolation of the large cluster of homogeneous intermediate opinion and the presence of isolated "extremists". The pattern (b) is obtained for a lower interaction threshold, $d = 0.15$. It displays many small clusters.

Interesting differences are noticeable for the smaller values of $d < 0.3$ as observed in Fig. 12.7(b) several opinion clusters are observed and none percolates across the lattice. Similar opinions, but not identical, are shared across several clusters. The difference of opinions between groups of clusters relate to d, but the actual values inside a group of clusters fluctuate from cluster to cluster because homogenization occurred independently among the different clusters: the resulting opinion depends on fluctuations of initial opinions and histories from one cluster to the other. The same increase in fluctuations compared to the full mixing case is observed from sample to sample with the same parameter values.

The clustering effects observed with full mixing and lattice topologies, and the observation of outliers in lattice are maintained for disordered topologies such as scale-free networks of the Barabasi–Albert type [14]. Simulations were run by D. Stauffer and H. Meyer-Ortmanns [38], and by Weisbuch [39]. To summarize some of the results obtained on scale-free networks:

- One does observe clustering effects, and the opinions in the main clusters do not differ much from what they are for equivalent tolerance thresholds in lattices.

- As a result of the scale-free distribution of node connectivity, well-connected nodes are influenced by other nodes and are themselves influential. Most of them belong to the big cluster(s) after the convergence process.

- Even more outlier nodes are observed, and most of them are among the less connected nodes.

12.3.3
Extremism

The above results where obtained when all agents have the same invariant threshold. Several models were tried with distributions of thresholds and eventually thresholds which vary according to the agents' experience [31]. One of the most interesting variants concerns modeling populations with large numbers of open-minded "centrist" agents and a small proportion of extremists.

The issue of whether extremist or moderate opinions are adopted in committees is thoroughly discussed by social psychologists, see for instance Moscovici and Doise [40]. A connection with statistical physics and the Ising model was soon proposed, for instance, by Galam [41] who collaborated with Moscovici. They considered binary opinions as in the vast majority of the literature on binary social choice.

Fascinating results were obtained in the "extremism" model of Deffuant et al. [42]. When interaction thresholds are unevenly distributed, and in particular when agents with extreme opinions are supposed to have a very low threshold for interaction, extremism can prevail, even when the initially extremist agents are in a very small proportion. The so-called "extremist model" can be applied to political extremism, and a lot of the heat of the discussion generated by these models relates to our everyday concerns about extremism. But we can think of many other situations where some "inflexible" agents are more sure about their own opinion than others. Inflexibility can arise for instance:

- because of knowledge. Some agents might know the answer while others only have opinions; think of scientific knowledge and the diffusion of new theories;

- Some agents might have vested interests which are different from others.

Although the model has some potential for many other applications, we will here use the original vocabulary of extremism.

The model[3] for extremism introduce by Deffuant et al. [42] is based on two more assumptions than the original continuous-opinion model.

- A few extremists with extreme opinions at the ends of the opinion spectrum and with very low threshold for interaction are introduced.

- Whenever the threshold allows interaction, both opinions and threshold are readjusted according to similar expressions.

Iff $|x - x'| < d$

$$x = x + \mu \cdot (x' - x) \quad (12.12)$$
$$d = d + \mu \cdot (d' - d) \quad (12.13)$$

A symmetrical condition and equations apply to the other agent of the pair with opinion x' and tolerance d' but when thresholds are different the influence can be asymmetric: the more "tolerant" agent (with larger d) can be influenced by the less tolerant (with smaller d) while the less tolerant agent is not. This "effective" asymmetry is responsible for the outcome of "extremist" attractors.

Let us demonstrate the issue for the very simple case of a single extremist in the presence of a large majority of centrist agents [43]. We check opinion and tolerance dynamics by time plots of single simulations.

The time plots display different dynamical regimes according to the eventual predominance of the extremists: sometimes they remain isolated and most agents cluster as if there were no extremist (e.g., as represented in Fig. 12.8(a)); otherwise extremism prevails and most agents cluster in the neighborhood of one (e.g., as represented in Fig. 12.8(b)) or both extremes.

Still, in both cases, a phase of convergence towards an average opinion of most initially centrist agents is observed (for roughly 10 updatings per agent). The initial convergence towards the center is due to the much larger number of centrists as compared to extremists. After this preliminary phase, the center clustered agents can either slowly evolve towards extremism if they still

---

3) In fact we here give a simplified version (called bounded confidence) of the model actually used by Deffuant et al. in their original model (relative agreement model).

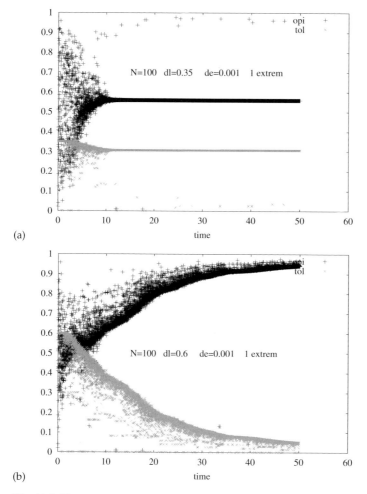

Fig. 12.8 Time plots of opinion (black '+') and tolerance (grey 'x') dynamics exhibiting a "centrism" attractor (a) and a single extremist cluster (b). Time is given in the average number of updating per agent. The number of agents is $N = 100$, extremists' tolerance is $de = 0.001$ and two opposite extremists are initially present. Any pair of agents can, a priori, interact. The centrism attractor was obtained when the initial centrist tolerance was $dl = 0.35$ and the extremist attractor when $dl = 0.6$

feel their influence, in other words when their tolerance is larger than their distance to extremists ($dl > 0.5$); otherwise (when $dl < 0.5$) they are no longer under extremist influence and remain at the center. Due to the random character of the initial opinion distribution and pair sampling, the $dl = 0.5$ is not a sharp boundary, but rather indicates a dynamical crossover.

The convergence characteristic time then differs: convergence is fast for the centrism attractor and slow for the extremism attractor. The ratio in convergence time is approximately the initial fraction of centrists and extremists.

More generally, which attractor is reached depends mainly upon the parameters of the simulation (number and initial tolerance of extremists, and the initial tolerance dl of the other agents). A sketchy conclusion is that some kind of extremism prevails for larger values of the tolerance of initially nonextremist agents when $dl > 0.5$, and centrism when $dl < 0.5$. In other words, the outcome of the dynamics is largely determined by the tolerance of the nonextremists agents. But systematic studies show *co-existence* parameter regions where several attractors can be reached depending upon the specific initial distribution of opinions and upon the specific choice of updated pairs.

12.4
Diffusion of Culture

Axelrod [44] proposed in 1997 a model for the dissemination of cultures based on the idea that cultures were sets of cultural traits which could exist under different specifications. For instance, cultural traits would be described by how major challenges are answered; how houses are built, cooking, language, hunting habits and so on. Axelrod's model inspired by the analogy between memes (cultural traits) and genes which exist as different alleles, goes beyond more classical models of cultural dynamics based on population genetics, because the dynamics of cultural traits are not taken as independent from each other. The Axelrod model involves space modeled by a lattice. We here start with a simpler model based on binary traits and full mixing.

12.4.1
Binary Traits

Usually people have opinions on different subjects, which can be represented by vectors of opinions. A typical case is political opinion: citizens can have positive or negative opinions on m topics such as social security, free trade or protectionism, pro or anti-nuclear energy, abortion, etc. In some sense a traditional view about politics such as the standard view in France about single dimension position on a right/left axis would corresponds to continuous opinions, while the more pragmatic Anglo-Saxon view of political platforms would correspond to binary opinion vectors.

In accordance with our previous hypotheses, we suppose [31] that agents interact according to their distance in opinions. In order to simplify the model, we use binary opinions. An agent is characterized by a vector of m binary opinions about the complete set of m topics. We use the notion of Hamming distance between binary opinion vectors (the Hamming distance between two binary opinion vectors is the number of different bits between the two vectors). We only treat the case of complete mixing; any pair of agents might interact and adjust opinions according to how many opinions they share.

The adjustment process only occurs when agents agree on at least $m - d$ subjects (i.e. they disagree on $d - 1$ or fewer subjects). The rules for adjustment are as follows: among the binary traits which are different for the two agents, one is selected and the two agents take the same opinion (randomly selected among the two possible) with probability μ. Obviously this model has connections with population genetics in the presence of sexual recombination when reproduction only occurs if the genome distance is smaller than a given threshold. Such a dynamics results in the emergence of species (see [45]). We are again interested in the clustering of opinion vectors. In fact, clusters of opinions here play the same role as biological species in evolution.

12.4.2
Results

μ and N only modify convergence times towards equilibrium; the most influential factors are the threshold d and m the number of topics under discussion. Most simulations were done for $m = 13$. For $N = 1000$, convergence times are of the order of 10 million pair iterations. For $m = 13$ we find the following.

- When $d > 7$, the radius of the hypercube, convergence towards a single opinion occurs (the radius of the hypercube is half its diameter which is equal to 13, the maximum distance in the hypercube).

- Between $d = 7$ and $d = 4$ a similar convergence is observed for more than 99.5 per cent of the agents with the exception of a few clustered or isolated opinions distant from the main peak by roughly 7.

- For $d = 3$, one observes from 2 to 7 significant peaks (with a population larger than 1 per cent) plus some isolated opinions.

- For $d = 2$ a large number (around 500) of small clusters is observed.

The same type of results are obtained with larger values of m: two regimes, uniformity of opinions for larger d values and extreme diversity for smaller d values, are separated by one critical d_c value for which a small number of clusters is observed (e.g., for $m = 21$, $d_c = 5$, d_c seems to scale in proportion with m).

Figure 12.9 represents the populations of the different clusters at equilibrium (iteration time was 12 000 000) on a log-log plot according to their rank-order (Zipf plots). No scaling law is obvious from these plots, but we observe the strong qualitative difference in decay rates for various thresholds d.

The main difference between continuous opinion and vector opinion dynamics is in the sharpness of the transition and the scaling of the number of clusters. Let us recall that for continuous opinions, the number of clusters follow a staircase: every time the quantity $int(1/2d)$ increases, this number

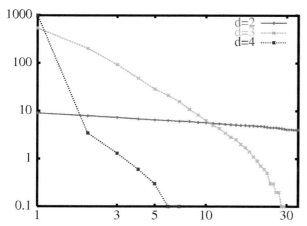

Fig. 12.9 Log-log plot of average populations of clusters of opinions arranged by decreasing order for $N = 1000$ agents ($\mu = 1$).

increases by one. For the low threshold values, in the case of vector opinion dynamics the number of clusters scale increases exponentially in m in the limit of large N. A finite fraction of the sites of the hypercube $\{0,1\}^m$ is occupied by clusters. The transition from consensus to an exponential number of clusters is abrupt.

12.4.3
Axelrod Model of Cultural Diffusion

In the Axelrod model [44], cultural features in some small number (d) can take different integer values (from 1 to q). Agents are situated at the nodes of a two-dimensional lattice. Interactions only occur among neighbors; each agent has four nearest neighbors. At each time step, one pair of neighbors is selected and they interact with a probability proportional to their overlap, i.e., their relative number of common features. As a result of the interaction, one trait, among those which are different, is copied from one agent to the other.

When starting from random initial conditions, one observes a sharp transition between cultural homogeneity and the persistence of clusters with many different cultures depending upon the values of q and d. Cultural homogeneity is observed for smaller q values, and cultural diversity when q is above a threshold which increases with d and the network size.

Castellano et al. [46] derived master equations describing the evolution of the population of links P_m between sites which share m common traits. Numerical integration of the master equation yields the same transition in behavior as was directly observed on the original system. Several variants were described by Klemm and co-authors [47], including a probabilistic version.

12.5
Conclusions

12.5.1
Range and Limits of Opinion Dynamics Models

All the models that we have presented, whether binary, continuous or vector opinions were based on simple hypotheses about imitation processes.

Unfortunately, we still lack a solid experimental and empirical basis to validate these hypotheses. Too few experiments[4] have been conducted in social psychology which could validate a choice among rationality, seen as individuals only taking into account knowledge or direct experience, versus complete conformism, not to mention, in the case of conformism, a possible distinction between bounded confidence or total openness.

The above skepticism is not shared in the agent-based models community; computer scientists often take as given the "psychological" hypotheses that describe their agents individual behavior and envisage a possible validation against empirical data. The sociophysics approach is more prone to admit the shortcomings of the simple agent description and search for generic properties rather than for a "Popperian" validation.

One can still hope, that modeling will provoke experiments, in the same way as the General Equilibrium Model inspired the empirical work of Tversky and Kahneman [48] to check its assumptions.

Many empirical studies have been done on social networks: who influences whom under what conditions, see for instance [49]. Empirical evidence about elections [50] have been proposed to validate the results of some variant of the Sznajd model. Some empirical data might assist the predictions of binary opinions models: consensus of either type, everyone choosing one solution or the other. For instance, one observation of the adoption of environmentally friendly contracts by farmers in northern Italy in two neighboring regions, showed that the adoption rate was nearly zero in Piemonte and maximum in Lombardia [51]. A book, *How Hits Happen* [52] is entirely devoted to the occurence of these very contrasting success rates in the media and toy industry, for instance.

12.5.2
How to Convince

Understanding opinion dynamics, or just believing that you understand it, leads to strategies to convince: producers, sellers, political leaders, government agencies are all faced with the issue.

- Producers and designers have to decide the specifics of a product, including price, in view of the buyers expectations, including the accep-

---

[4] References to experimental work include [18, 20, 40].

tation dynamics of the product. Externalities and social factors are of prime importance in computers and information technology, but also in the media industry.

- Political science often discusses politicians strategies in terms of game theory; but choosing a political platform, including those topics on which to campaign, also relates to the approaches that we have discussed.

- The importance of social dynamics for the acceptance of reforms by the stakeholders has long been discussed by the academics working on policy design [4]; how much of it is used used by international institutions such as IMF or the World Bank is another story.

Marketing specialists could, of course, dream of designing a campaign with a full knowledge of the social network and of the local processes which would convince most individuals, at the lowest cost. We are far from such knowledge and this might be a good thing!

But we can still observe sellers' tricks using networking knowledge to convince prospects to change their cars or to buy insurance policies. Such tricks are also used by politicians to persuade voters or public agencies and to convince citizens to adopt new practices more favorable to the environment.

The multiplicity of attractors leading to strong differences in market shares or uptake rates of proposals is sometimes perceived as a source of indeterminacy by scientists. But it is an asset for marketing people: it implies that they do have openings to interfere and modify the outcome of the dynamics. This would not be the case if there were only one attractor.

12.5.3
How to Make Business

Several papers were written on coupled seller/buyers dynamics: how one seller should adjust his prices taking into account opinion dynamics; or what would happen to a seller who would re-adjust her prices as a function of the volume of sales, ignoring opinion dynamics. The hypotheses made by sociophysicists are often in contradiction with the perfect knowledge hypothesis of the General Equilibrium Theory and, not surprisingly, they give rise to totally different predictions.

When using the formalism of networks of "counter" automata, the S variable represents the binary decision buy/not buy or eventually buy brand A or buy brand B. The threshold θ represents the (economic) utility of choice S, including eventually the price as a negative component. A rule such as Eq. 12.1 is interpreted as buy when social influence (the sum term) is larger than differences in utility (the threshold).

We have seen in binary choice models that, because of interactions, the fraction of buyers undergoes sharp transitions as a function of the threshold distribution. In terms of economics, the aggregated demand as a function of prices is not convex as hypothetized by economists, and hence the standard equilibrium analysis does not apply.

Let us take the Solomon et al. [53] social percolation model as an example of the socio-physics approach. It starts from a lattice or a random graph, with sites occupied by agents i susceptible to go and see a given movie. The quality q of the movie is only known to agents who viewed it. It would determine whether an agent knowing about it would go and see it.

Each of the agents i initially informed about the movie decides to go and see it, if and only if the quality q of the movie is larger than his/her personal preference p_i, i.e., if $q > p_i$.

However, the initial agents who decided to go and see the movie become themselves sources of information about the movie: their first neighbors j are informed about the movie and decide (according to the $q > p_j$ criterion) whether to see it or not. One continues until the procedure stops by itself, i.e., until all the neighbors of all the agents which went to the movie up to now, either have already been to the movie or have already decided not to go.

Since the agent preferences p_i are frozen, the present model is a classical percolation problem. For instance, if the personal preferences p_i on various sites i take independent random values distributed uniformly between 0 and 1, then the average probability for an agent to go to the movie, once one of its neighbors went, is the movie quality q. Consequently, if a movie happens to have a quality q lower than the percolation threshold p_c, i.e., $q < p_c$, then after a certain short time its diffusion among the public will stop. The movie will then be a flop and will not reach any significant percentage of its potential public (which is a fraction q of the lattice).

On the other hand, when the movie quality q is larger than p_c (and not too close to p_c), the movie will reach most of its potential viewers as the islands of interest will percolate. The movie will be viewed by roughly a fraction q of the entire viewing population. Up to here the model predicts the existence of a percolation transition regime for some values of the quality q and preferences p_i.

When the possible moves of the provider of the goods, in the above case the movie producer, are taken into account, the resulting dynamics can be very different from the supply/demand equilibrium predicted by economists.

- Let first suppose a "myopic" behavior of the producer, who adjusts her price according to a "Walrasian tatonnement": she increases the price when she has too many customers, and decreases it when she has too few. In the presence of a sharp discontinuity in the demand/price curve, she would never be able to adjust the price to a desired production level.

Depending upon the adjustment rate of the producer versus the propagation of information among customers several dynamical regimes can be observed including self organised criticality or limit cycles (see [53] for details).

- Gordon, Nadal and co-workers [54] have studied the case of a more "strategic" producer facing a crowd of buyers who imitate each other. The preferences of the buyers, the θ_i in our formalism, are distributed according to the derivative of a logit distribution (see Eq. (12.8)) with coefficient β. All are biased by the selling price set by the producer. Buyers imitate each other with the same influence coefficient J. For some values of β and J coefficients, around a first-order transition in the market share versus price, there are two possible attractors of the opinion dynamics. The seller is interested in selling more. On the other hand, he might make more profit by selling less at a higher price. He is then facing two possible strategies in the transition region as seen in Fig. 12.10.

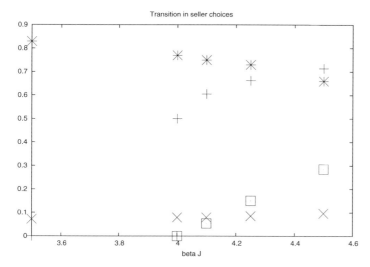

Fig. 12.10 Variation of prices and market shares of a producer facing a crowd of buyers who imitate each other, as a function of the intensity of imitation βJ. Around the transition ($\beta J = 4.18$) two solutions coexist: a low price (□), large market share (+) solution and a high price (*), small market share (x) solution. Further from the transition only one solution is stable: for $\beta J < 4$ the high price, small market share solution and for $\beta J > 5$ the low price, large market share solution. (By courtesy of J-P Nadal.)

12.5.4
Final Conclusions

The issues which we addressed using physics concepts and methods are fundamental issues in the social sciences:

> "If people tend to become more alike in their beliefs, attitudes and behavior when they interact, why do not all differences eventually disappear?"

Axelrod (1997) [44]

Although physicists' models are too simple to give any accurate prediction concerning social systems, they are good enough to provide an appropriate framework to formulate questions. Notions such as dynamics, consensus versus diversity attractors, the importance of seeds and edges in opinion dynamics, scaling laws, etc., shed a new light on old issues in Social Sciences.

Acknowledgment

My whole contribution to modeling in Social Sciences not surprisingly involved collaboration! It is a pleasure to acknowledge the help of many friends and colleagues, Frédéric Amblard, Gérard Boudjema, Guillaume Deffuant, Jean-Pierre Nadal, Sorin Solomon, Dietrich Stauffer, the Santa Fe Institute, and European collaborations: IMAGES (FAIR 3 2092), COSIN (IST-2001-33555) and EXYSTENCE (IST-2001-32802).

References

1 ROGERS, E. M., *Diffusion of Innovations*. The Free Press, New York, **1962**

2 RYAN B., GROSS, N. C., *Rural Sociology* 8 (**1943**), 15

3 COLEMAN, J. S., KATZ, E., MENZEL, H., *Sociometry*, 20, (**1957**), 253-270

4 ROLING, N., *Extension Science: Information Systems in Agricultural Development*. Cambridge, England: Cambridge University Press, **1989**

5 *Restructuring Networks in Post-Socialism* (Ed. Grabher, G., Stark, D.), Oxford University Press, **1997**

6 TARDE, G., *Les lois de l'imitation, Etude sociologique*, Paris : Alcan, **1890**

7 WEISBUCH, G., *Complex Systems Dynamics*, Santa-Fe Institute Studies in the Sciences of Complexity, Addison-Wesley, Redwood City, CA, USA, **1990**

8 ROKHSAR, D. S., ANDERSON, P.W., STEIN, D. L., *J. Mol. Evol.* V.23. No.2. (**1986**), p. 119

9 TOULOUSE, G., PFEUTY, P., *Introduction to the Renormalization Group and to Critical Phenomena*, Wiley, **1977**

10 BANERJEE, A. V., *Quarterly Journal of Economics*, 107, (**1992**), p. 797-818

11 ORLÉAN, A., *Journal of Economic Behavior and Organization* 28, (**1995**), p. 257–274

12 CHAMLEY, C., *Rational Herds : Economic Model of Social Learning*, Cambridge University Press, **2003**

13 WATTS, D. J., STROGATZ, S. H., *Nature* 393 (**1998**) p. 440–442

14 ALBERT, R., BARABÁSI, A.-L., *Rev. Mod. Phys.* 74 (**2002**), p. 47

15 VICHNIAC, G. Y. and HYMAN H., VICHNIAC, G. Y. in *Disordered Systems and Biological Organization*, Eds. Bienenstock, E., Fogelman-Soulié, F., Weisbuch, G., Springer, **1986**

16 STAUFFER, D., AHARONY, A., *Introduction to Percolation Theory*, Taylor and Francis, London, **1994**

17 HOLYST, J. A., KACPERSKI, K., SCHWEITZER, F., *Annual Review of Comput. Phys.* 9, 253-273 ((**2001**), KACPERSKI, K., HOLYST, J. A., *Physica A* 269 (**1999**) 511-526

18 LATANÉ, B., *Am. Psychologist* 36 (**1981**) 343

19 LATANÉ, B., *J. Commun* 46 (**1996**), 13

20 VALLACHER, R. R., NOWAK, A., in VALLACHER, R. R., NOWAK, A., (Eds.), *Dynamical Systems in Social Psychology*, San Diego: Academic Press, **1994**

21 NOWAK, A., SZAMREJ, J., LATANÉ, B., *Psych. Rev.* 97 (**1990**) 362

22 AXELROD, R., *J. Conflict Resolut.* 41 (**1997**), 203

23 FÖLLMER, H., *Journal of Mathematical Economics* 1/1, (**1974**), 51-62

24 ANDERSON, S., DE PALMA, A., THISSE, J, *Discrete Choice Theory of Product Differentiation*, MIT Press, Cambridge, **1992**

25 WEISBUCH, G., KIRMAN, A., HERREINER, D., "Market Organisation and Trading Relationships" *Economic Journal*, Vol. 110, no 463, (**2005**), 411-436

26 SZNAJD-WERON, K., SZNAJD, J., *Int. J. Mod. Phys.* C 11 (**2000**), 1157

27 GALAM, S., *J. Stat. Phys.* 61 (**1990**), 943-951

28 SZNAJD-WERON, K., SZNAJD, J., *Int. J. Mod. Phys.* C 11, (**2000**), 1157; STAUFFER, D., SOUSA, A. O., MOSS DE OLIVEIRA, S., *Int. J. Mod. Phys.* C 11 (**2000**), 1239; STAUFFER, D., *Journal of Artificial Societies and Social Simulation* 5, issue 1, paper 4 (jasss.soc.surrey.ac.uk) (**2002**)

29 HENRICH, J., BOYD, R., BOWLES, S., CAMERER, C., FEHR, E., GINTIS, H., MCELREATH, R., *Am. Econ. Rev.* 91, 2 (**2001**), pp. 73-78

30 YOUNG, H. P., BURKE, M. A., *American Economic Review*, 91 (**2001**), 559- 573

31 WEISBUCH, G., DEFFUANT, G., AMBLARD, F., NADAL, J.-P., *Complexity* 7(3) (**2002**), 55

32 STONE, M., *Ann. of Math. Stat.* 32 (**1961**), 1339-1342

33 CHATTERJEE, S., SENETA, E., *J. Appl. Prob.* 14 (**1977**), 89-97

34 COHEN, J. E., HAJNAL, J., NEWMAN, C. M., *Stochastic Processes and their Applications* 22 (**1986**), 315-322.

35 HEGSELMANN, R., KRAUSE, U., Journal of Artificial Societies and Social Simulation vol. 5, no. 3 (**2002**) http://jasss.soc.surrey.ac.uk/5/3/2.html

36 VAZQUEZ, F., KRAPIVSKY, P. L., REDNER, S., *J. Phys. A* 36 (**2003**, L61; (cond-mat/0209445). BEN-NAIM, E., KRAPIVSKY, P. L., REDNER, S., *Physica D* 183 (**2003**), 190; (cond-mat/0212313).

37 STAUFFER, D., *Computing in Science and Engineering* 5 (May/June2003), 71; and in *The Monte Carlo Method on the Physical Sciences*, Ed. J. E. Gubernatis, AIP Conference Proceedings, in press = cond-mat/0307133.

38 STAUFFER, D., MEYER-ORTMANNS, H., *Int. J. Mod. Phys.* C 15, issue 2 (**2004**)

39 WEISBUCH, G., *Eur. Phys. J. B* 38 (**2004**), 339-343

40 DOISE, W., MOSCOVICI, S., *Dissensions et consensus* PUF, Paris, **1992**

41 GALAM, S., MOSCOVICI S., *European Journal of Social Psychology*, 21 (**1991**), 49-74

42 DEFFUANT, G., AMBLARD, F., WEISBUCH, G., FAURE, T., *Journal of Artificial Societies and Social Simulation*, 5, (**2002**) (4), 1, http://jasss.soc.surrey.ac.uk/5/4/1.html

43 WEISBUCH, G., DEFFUANT, G. ET AMBLARD, F., *Physica A*, 353 (**2005**), p. 555–575

44 AXELROD, R., in *The complexity of cooperation*, Princeton University Press, **1997** p. 145–177

45 HIGGS, P. G., DERRIDA, B., *J. Phys. A: Math. Gen.* 24 (**1991**), 985-991

46 CASTELLANO, C., MARSILI, M., VESPIGNANI, A., *Phys. Rev. Lett.* 85 (**2000**), 3536-3539

47 Klemm, K., Eguíluz, V. M., Toral, R., San Miguel, M., *J. Econ. Dyn. Control* 29 (**2005**), 321-334; Klemm, K., Eguíluz, V. M., Toral, R., San Miguel, M., *Phys. Rev. E* 67 (**2003**), 026120; Klemm, K., Eguíluz, V. M., Toral, R., San Miguel, M., *Phys. Rev. E* 67 (**2003**), 045101R

48 Tversky, A., Kahneman, D., *Science* 185 (**1974**), 1124-1131

49 Lazega, E., *The Collegial Phenomenon: The Social Mechanisms of Cooperation Among Peers in a Corporate Law Partnership*, Oxford, Oxford University Press, **2001**

50 Bernardes, A. T., Stauffer, D., Kertesz, J., *Eur. Phys. J . B* 25 (**2002**), 123

51 Deffuant, G., *Final report of project FAIR 3 CT 2092. Improving Agri-environmental Policies : A Simulation Approach to the Cognitive Properties of Farmers and Institutions*, **2001** http://wwwlisc.clermont.cemagref.fr/ImagesProject/freport.pdf

52 Farrell, W., *How Hits Happen*, Harper-Collins, New York, **1998**

53 Solomon, S., Weisbuch, G., de Arcangelis, L., Jan, N., Stauffer, D., *Physica A* 277 (**2000**), 239

54 Gordon, M. B., Nadal, J.-P., Phan, D., Vannimenus, J., *Physica A*, 356 **2005**, 628-640; Nadal, J.-P., Phan, D., Gordon, M. B., Vannimenus, J., *Quantitative Finance* 5 (**2005**), 557-568

13
Opinion Dynamics, Minority Spreading and Heterogeneous Beliefs
Serge Galam

The connection between contradictory public opinions, heterogeneous beliefs and the emergence of democratic or dictatorial extremism is studied, extending our former two-state dynamic opinion model. Agents are attached to a social-cultural class. At each step they are distributed randomly in different groups within their respective class to evolve locally by majority rule. In the case of a tie the group adopts either one opinion with respective probabilities k and $(1-k)$. The value of k accounts for the average of individual biases driven by the existence of heterogeneous beliefs within the corresponding class. It may vary from class to class. The process leads to extremism with a full polarization of each class along either opinion. For homogeneous classes the extremism can be along the initial minority, making it dictatorial. As a contrast, heterogeneous classes exhibit a more balanced dynamics which results in a democratic extremism. Segregation among subclasses may produce a coexistence of opinions at the class level thus averting global extremism. The existence of contradictory public opinions in similar social-cultural neighborhoods is seen in a new light. To conclude, it shows how the model can be used to make solid predictions with respect to nationwide voting concerning a given issue like the 2005 French referendum about the European Constitution project.

13.1
The Interplay of Rational Choices and Beliefs

In recent years the study of opinion dynamics has become a main area of research in physics [1–15]. It was initiated a long time ago [16–18] and is part of sociophysics [19].

Outside physics, research has concentrated on analyzing the complicated psycho-sociological mechanisms involved in the process of opinion-forming. In particular, focusing on those in which a huge majority of people give up to an initial minority view [20, 21]. The main ingredient being, for instance, in the case of a reform proposal, that the prospect of losing definite advantages is much more energizing than the corresponding gains which, by nature, are hypothetical.

Such an approach is certainly realistic in view of the very active nature of minorities involved in a large spectrum of social situations. However, we have shown [9,14,15] that even in the case of nonactive minorities, public opinion obeys some internal threshold dynamics which breaks its democratic character. Although each agent does have an opinion, they may find themselves in a local unstable doubting collective state while discussing with others in small groups. In such a case, we postulate that all group members adopt the same opinion, the one which is consistent with the common beliefs of the group.

Examples of such common beliefs may be substantiated by sayings such as *"There is no smoke without fire"* or *"When in doubt, leave well alone"*. Such a possibility of occurrence of local doubts results in highly unbalanced conditions for the competition of opinions within a given population even when each opinion has an identical convincing weight [5,9].

In the present chapter we introduce a spectrum of social-cultural classes to include all agents which may meet together to discuss an eventual public issue. Each class is characterized by some common beliefs that result from the class average of all heterogeneous individual biases. Accordingly, in the case of a tie in a local group, the resulting choice is one opinion with respective probabilities k and $(1-k)$ where k accounts for the corresponding common belief of the social-cultural class. The value of k is constant within each class with $0 \leq k \leq 1$ and may vary from class to class.

Considering a class with a common belief k and groups of size 4, O denoting an opponent to the issue at stake and S a supporter, our update rules can be written as:

- $SSSS$ and $OSSS \rightarrow SSSS$
- $OOOO$ and $OOOS \rightarrow OOOO$
- $OOSS \rightarrow \begin{cases} OOOO \text{ with probabiliy } k \\ SSSS \text{ with probabiliy} (1-k) \end{cases}$

where all permutations are allowed. In our earlier works $k=1$ or $k=0$ except in [15] and in an application to cancerous tumor growth [22]. For $0 \leq k \leq 1$ the opinion flow dynamics still converges towards a full opinion polarization but the separator location is now a function of k and the group size distribution. It may vary from 0% to 100% as shown in Figs 13.1 and 13.2.

The flow is fast and monotonous. When the initial majority make the final decision, the associated extremism is democratic, otherwise it is more or less dictatorial depending on the value of the initial minority which eventually spreads over the whole class. Various extreme social situations can thus be

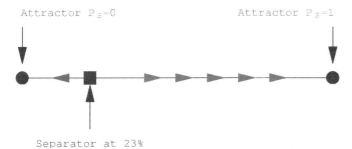

Fig. 13.1 Opinion flow diagram with the two attractors at 0 and 1 and the separator at a nonsymmetric value 77%. To survive a public debate the opinion, which does not conform to common beliefs, must start at an initial support of more than 77%. The debate is anti-democratic, while weakening an initial huge majority. It will eventually disappear.

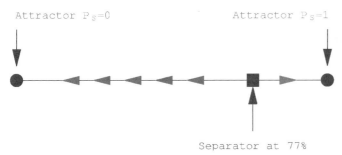

Fig. 13.2 Opinion flow diagram with the two attractors at 0 and 1 and the separator at a nonsymmetric value 23%. To spread via a public debate the opinion, which conforms to common beliefs, needs to start at an initial support at above only 23%t to invade the whole population. The resulting associated extremism is dictatorial.

described. In the case of an homogeneous high risk aversion sharing ($k = 0$) it is found that a reform proposal may need an initial support of more than 90% of the population to survive a public debate [9]. On the contrary, a shared high prejudice against some ethnic or religious group ($k = 1$) can make an initial false rumor, shared by only a few percent of the population, spread over the whole population [5]. Moreover, very small fluctuations in the initial conditions may lead to the opposite extremism.

In contrast, heterogeneous populations ($k = \frac{1}{2}$) are found to exhibit a more balanced dynamics (see Fig. 13.3). Extremism becomes democratic. However, segregation among subclasses leads to the opposite extremism within each subclass, which in turn stabilizes the associated overall class opinion at a permanent coexistence of both opinions, thus averting extremism.

Earlier versions of a threshold dynamical process can be found in the study of voting in democratic hierarchical systems [17, 23]. There, groups of agents

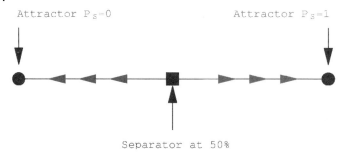

Fig. 13.3 Opinion flow diagram with the two attractors at 0 and 1 and the separator at a symmetric value 50%. To survive a public debate one opinion must start at an initial support of just a bit more than fifteen percent. The initial minority support tends to weaken before it eventually disappear. The associated extremism is democratic.

vote for a representative to the higher level using a local majority rule. Going up the hierarchy turns out to be exactly identical to an opinion-forming process in terms of equations and dynamics. Instead of voting, agents update their opinions. The probability of electing a representative at some hierarchy level n is equal to the proportion of opinions which are the same after n updates [15, 17]. However, within all these earlier studies the value of k is always taken equal to one except in [15] where it was a function of the rational majority/minority within some finite-sized neighborhood. The existence of threshold dynamics in social phenomena was advocated a long time ago in qualitative studies of some social phenomena [24, 25].

The rest of the paper is organized as follows. The model is defined in the next section. It is then solved exactly within the simple case of groups of size three in Section 13.3. The counter-intuitive case of groups of size 4 is studied in Section 13.4. In Section 13.5 small fluctuations are shown to sometimes lead to contradictory public opinions. It sheds a new light on the fact that very similar areas, in terms of respective beliefs, can hold an opposite view, for instance about the feeling of safety. Segregation effects are studied in Section 13.6 and we find that they can drive either democratic extremism or coexistence of opinions, thus avoiding global extremism. The inclusion of a size distribution for local groups is presented in Section 13.7 and the last section contains some further discussion.

13.2
Rumors and Collective Opinions in a Perfect World

We start partitioning a given population within different social-cultural classes. Then we consider each class independently to study the dynamics of opinion-forming among its N individual members facing some issue.

It may be a reform proposal, a behavior change like stopping smoking, a foreign policy decision or a belief in some rumor. The process is held separately within each class of agents.

We discriminate between two levels in the process of formation of the global opinion, an external level and an internal one. The first one is the net result from the global information available to every one, the private information some persons may have and the influence of mass media. The second level concerns the internal dynamics driven by people discussing freely among themselves. Both levels are interpenetrating but here we decouple them to specifically study the laws governing the internal dynamics.

Accordingly, choosing a class of agents at a time t prior to the public debate the issue at stake is given a support by $N_s(t)$ individuals (denoted S) and an opposition from $N_o(t)$ agents (denoted O). Each person is supposed to have an opinion with $N_s(t) + N_o(t) = N$. Associated individual probabilities of being in favor or against the proposal at time t are

$$p_{s,o}(t) \equiv \frac{N_{s,o}(t)}{N} \tag{13.1}$$

with

$$p_s(t) + p_o(t) = 1 \tag{13.2}$$

From this initial configuration, people start discussing the project. However, they don't meet all the time and all together at once. Gatherings are shaped by the geometry of social life within physical spaces like offices, houses, bars, restaurants and others. This geometry determines the number of people which meet at a given place. Usually it is of the order of just a few. Groups may be larger but in these cases spontaneous splitting always occurs with people discussing in smaller subgroups.

To emphasize the mechanism at work in the dynamics which arises from local interactions, no advantage is given to the minority with neither lobbying nor organized strategy. People discuss in small groups. To implement the psychological process of a collective mind driven update, a local majority rule is used within each group. Moreover, an identical individual persuasive power is assumed for both sides with the principle "one person – one argument". On this basis all members of a group adopt the opinion which had the initial majority. In case there exists no majority, i.e., at a tie in a group of even size, all members yet adopt the same opinion, but now either one with respective probabilities k and $(1-k)$ where k is a function of the class.

13.3
Arguing by Groups of Size Three

We first consider the case of update groups with the same size, three. Accordingly to our local majority rule groups with either 3 S or 2 S end up with 3 S. Otherwise it is 3 O. The probability of finding one supporter S after n successive updates is

$$p_s(t+n) = p_s(t+n-1)^3 + 3p_s(t+n-1)^2(1 - p_s(t+n-1)) \quad (13.3)$$

where $p_s(t+n-1)$ is the proportion of supporters S at a distance of $(n-1)$ updates from the initial time t.

Equation (13.3) exhibits three fixed points $p_{s,0} = 0$, $p_{s,1} = 1$ and $p_{c,3} = \frac{1}{2}$. The first two correspond to a total opinion polarization along, respectively, a total opposition to the issue with zero supporters left and total support. Both are attractors of the dynamics. The last point with a perfectly balanced opinion-splitting is the separator of the dynamics as shown in Figs 13.2 and 13.3.

To reach the attractor, the dynamics requires a sufficient number of updates. In solid terms each update means some real time measured in numbers of days whose evaluation is outside the scope of the present work. An illustration is given starting from $p_s(t) = 0.45$. We get successively $p_s(t+1) = 0.42$, $p_s(t+2) = 0.39$, $p_s(t+3) = 0.34$, $p_s(t+4) = 0.26$, $p_s(t+5) = 0.17$, $p_s(t+6) = 0.08$ down to $p_s(t+7) = 0.02$ and $p_s(t+8) = 0.00$. Within eight successive updates, 45% of the agents who were supporting the issue have shifted against it. The process has preserved and reinforced the initial majority making the resulting extremism democratic. The dynamics is perfectly symmetric with respect to both opinions as seen from Figs 13.2 and 13.3. It is worth stressing that here the social-cultural character of the class is not activated since a local majority is always found.

13.4
Arguing by Groups of Size Four

It is only when dealing with even groups that the common-beliefs driven bias can be analyzed. Here, the "one person – one argument" rule allows for the possibility of a tie with no local majority. In such a case participants are in a nondecisional state. They are doubting collectively, both opinions being supported by an equal number of arguments. Here we invoke a common belief "inertia principle" to remove the doubt. We state that at a tie the group eventually adopts the opinion O with a probability k and the opinion S with the probability $(1-k)$ where k accounts for the collective bias produced by the

13.4 Arguing by Groups of Size Four

common beliefs of the group members [26]. Some specific situations are considered below.

To illustrate the model we consider groups of size four. The probability of finding one supporter S after n successive updates becomes,

$$p_s(t+n) = p_s(t+n-1)^4 + 4p_s(t+n-1)^3\{1 - p_s(t+n-1)\} \\ + 6(1-k)p_s(t+n-1)^2\{1 - p_s(t+n-1)\}^2 \quad (13.4)$$

where $p_s(t+n-1)$ is the proportion of supporters at a distance of $(n-1)$ updates from initial time t. The last term includes the tie case contribution (2S-2O) weighted with the probability k.

From Eq. (13.4) both attractors $p_{s,0} = 0$ and $p_{s,1} = 1$ are recovered. However, the unstable fixed point $p_{c,4}$ has now departed from the symmetric value $\frac{1}{2}$ to the nonsymmetric value,

$$p_{c,4} = \frac{(6k-5) + \sqrt{13 - 36k + 36k^2}}{6(2k-1)}, \quad (13.5)$$

except at $k = \frac{1}{2}$ where $p_{c,4} = \frac{1}{2}$. Figure 13.4 shows the variation of $p_{c,4}$ as a function of k. The effect of the common beliefs of the class in the formation of the associated public opinion is seen explicitly. For instance, an initial support of $p_s = 0.40$ leads to a an extremism in favor of the issue at stake for the whole range of bias $0 \leq k \leq 0.36$. In contrast the extremism is against it when $0.36 \leq k \leq 1$.

For instance, when the issue relates to some reform proposal and the class shares a high risk aversion, a tie supports the status quo, i.e., $k = 1$ which

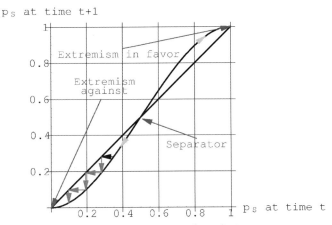

Fig. 13.4 Variation of the proportion $p_s(t+1)$ of supporters as a function of $p_s(t)$ for groups of size three. Arrows show the direction of the flow for an initial support $p_s(t) < p_{c,3} = \frac{1}{2}$. The extremism is democratic.

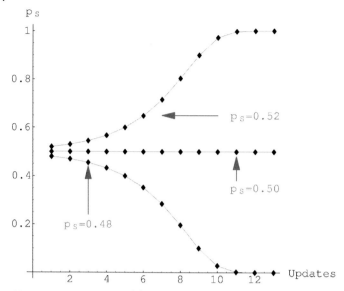

Fig. 13.5 Variation of $p_s(t)$ for groups of size three as a function of repeated updates with three initial supports $p_s(t) = 0.48, 0.50, 0.52$. The resulting extremism is democratic since it is along the initial majority.

yields $p_{c,4} \approx 0.77$. The initial support for the reform to make the final public opinion has thus to start with more than 77% as shown in Figs 13.5 and 13.6. When it does, the corresponding extremism is democratic.

On the other hand, an issue within a context where novelty is preferred drives a nondecisional state to adopt the opinion S making $k = 0$ with, in turn, $p_{c,4} \approx 0.23$. An initial support of more than 23% is now sufficient to invade the whole population. If it happens, the resulting extremism is dictatorial since it is along an initial minority view. Associated flow dynamics are shown on the right side of Figs 13.7.

In the case of groups of size four the number of updates to reach a full polarization is smaller than is the case for size three as shown in Fig. 13.7 at $k = 1$ and $k = 0$. The number of required updates to have an extremism completed can be evaluated as

$$n \simeq \frac{1}{\ln[\lambda]} \ln\left[\frac{p_c - p_S}{p_c - p_+(t)}\right] \tag{13.6}$$

where λ is the first derivative of $p_s(t+1)$ with respect to $p_s(t)$ taken at $p_s(t) = p_c$. Moreover, $p_S = 0$ if $p_+(t) < p_c$ while $p_S = 1$ when $p_+(t) > p_c$. The number of updates being an integer, its value is obtained from Eq. (13.6) rounding to an integer. The number of updates diverges at p_c. The situation is

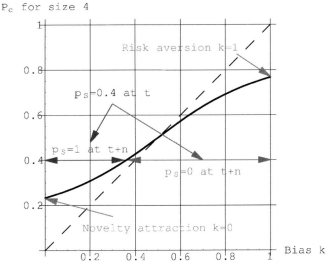

Fig. 13.6 Variation of $p_{c,4}$ as a function of k. For a collective risk aversion ($k = 1$), $p_{c,4} \approx 0.77$ while for a collective novelty attraction ($k = 0$), $p_{c,4} \approx 0.23$. In the case of no collective bias ($k = \frac{1}{2}$), $p_{c,4} = \frac{1}{2}$. An initial support of $p_s = 0.40$ is shown to lead to a an extremism in favor ($p_S = 1$) of the range of bias $0 \leq k \leq 0.36$. In contrast the extremism is against ($p_S = 0$) for the whole range $0.36 \leq k \leq 1$.

symmetric with respect to $k = 0$ and $k = 1$ with the divergence at, respectively, $p_c = 0.23$ and $p_c = 0.77$. It occurs at $p_c = 0.50$ for $k = 3$.

For instance, starting as above with groups of size three from $p_s(t) = 0.45$ we get with $k = 1$ the series $p_s(t+1) = 0.24$, $p_s(t+2) = 0.05$ and $p_s(t+3) = 0.00$. Within three successive updates 45% support has shifted to the opposition. Even an initial support above 50% with $p_s(t) = 0.70$ yields $p_s(t+1) = 0.66$, $p_s(t+2) = 0.57$, $p_s(t+3) = 0.42$, $p_s(t+4) = 0.20$, $p_s(t+5) = 0.03$, and $p_s(t+6) = 0.00$. Only six updates are enough to have 70% of supporters shift their opinion.

13.5
Contradictory Public Opinions in Similar Areas

We now discuss the effect of small differences in shared beliefs in the making of public opinion of neighboring groups. For instance, we consider a city area and its suburb as shown in Fig. 13.8. One class covers the city with $k = 0.49$ and another one the suburb with $= 0.47$. Such a minor difference is not explicitly felt while crossing from one area to another. Both areas are perceived as identical. However, the study of the dynamics of opinion starting from the

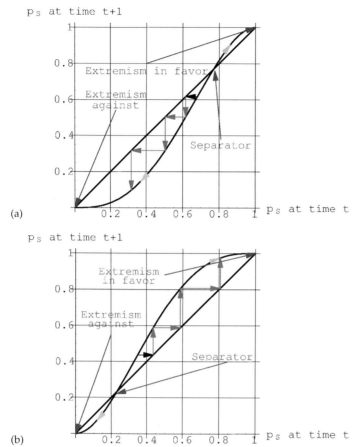

Fig. 13.7 Variation of $p_s(t+1)$ as a function of $p_s(t)$ for groups of size four. (a) $k = 1$ making $p_{c,3} \approx 0.77$. Arrows show the direction of the flow for an initial support $p_s(t) < 0.77$. (b) $k = 0$ making $p_{c,3} \approx 0.23$. Arrows show the direction of the flow for an initial support $p_s(t) > 0.23$.

same initial conditions within each area shows that sometimes huge differences can be driven by minor differences in the initial conditions.

To illustrate our statement we consider two very similar initial conditions with an issue at stake having, respectively, 49% and 51% support among both city and suburb populations. We then follow the dynamics of the corresponding public opinion flows.

In the first case, 51% of the support for the issue results in both populations fully supporting the issue, making both geographical areas identical as seen in Fig. 13.9. The process is completed within an estimate of ten updates. However, a tiny decrease of 2% in the initial support down at 49% splits the two neighboring similar areas. The city is now fully opposed to the issue while the suburb stays unchanged with full support for it (see Fig. 13.9).

13.5 Contradictory Public Opinions in Similar Areas

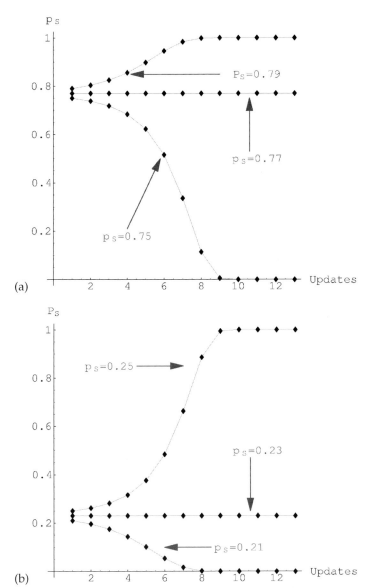

Fig. 13.8 Variation of $p_s(t)$ for groups of size four as a function of repeated updates. (a): $k = 1$ with three initial supports $p_s(t) = 0.75, 0.77, 0.79$. To make the extremism, the initial support has to be larger than 77%. When it happens it is democratic. (b): $k = 0$ with three initial supports $p_s(t) = 0.21, 0.23, 0.25$. The resulting extremism is dictatorial when it corresponds to an initial minority.

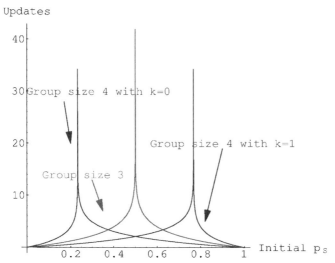

Fig. 13.9 Variation in the number of required updates to reach full extremism for groups of size three and four at $k = 0$.

The above case may shed new light on situations in which contradictory feelings or opinions are sustained in areas which are nevertheless very similar, such as, for instance, the feeling of safety. It shows how an insignificant change in either the initial support or the bias driven by the common beliefs, may yield drastic differences in the outcome of public opinion. Figure 13.10 shows the variations in the number of updates needed to reach extremism for several values of the bias k. Here too, the same initial support is shown to lead to totally different outcomes as a function of k. A value $p_S = 0.30$ leads to $p_S = 0$ for both $k = 0.50$ and $k = 0.70$ while it yields $p_S = 1$ for $k = 0.10$. For $p_S = 0.70$ all three cases lead to $p_S = 1$.

13.6
Segregation, Democratic Extremism and Coexistence

Up to now we have considered as a tie, an average local bias k which results from a distribution of heterogeneous beliefs within a population. It means that all members of that population do mix together during the local group updates whatever the individual respective beliefs. At this stage it is worth noting that different situations may arise in the distribution of the k_i.

We discuss two cases for which either all k_i are equal, i.e., an homogeneous population, or they are all distributed among two extreme values, for instance 0 and 1. There the existence of subclasses as a result of individual segregation may become instrumental in producing drastic changes in the final global public opinion of the corresponding class.

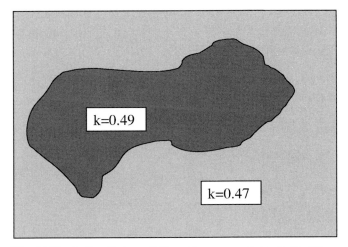

Fig. 13.10 A city with $k = 0.49$ and its surrounding suburb with $k = 0.47$.

Consider first two different homogeneous classes A and B in two different areas with, respectively, $k_i = 0$ for all i in A and $k_j = 1$ for all j in B. From Eqs (13.5) and (13.6) an initial support $p_s = 0.25$ yields $p_S = 1$ for A and $p_S = 0$ for B as seen in Fig. 13.11. For A the extremism in support of the issue is dictatorial, since along the initial minority $p_s = 0.25$. When odd, in B the extremism is against the issue and democratic since along the initial majority of $1 - p_s = 0.75$ is against it.

Now consider the above classes A and B but as subclasses of the same class within one unique area. Two situations may occur, as illustrated in Fig. 13.12 where A individuals are represented by circles and B individuals by squares. They are white when in favor and black if against. In the first situation (higher part of the figure) people from each subclass do not mix together while updating their individual opinions. They are segregating from each other yet sharing the same class within the same geographical area. As a result, two opposite extremisms for each subclass are obtained as above. However, the novelty with respect to distinct geographical areas is that here the resulting public opinion of the global class, which includes the whole population, is no longer exhibiting any extremism. The dynamics of segregated updates of opinion has produced a stable coexistence of both initial opinions with thus a balance collective stand. A poll over the bias would reveal the average value $k = 1/2$.

In the second situation (lower part of the figure) people from each subclass do mix together while updating their individual opinions. As a result, at a tie with mixed individuals, A people adopt the opinion in favor while B people go against. With two sub-populations having more or less the same (large

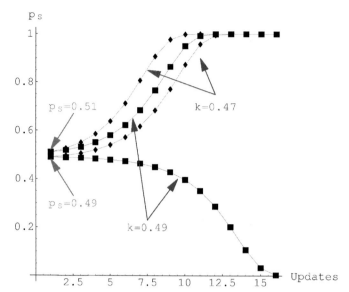

Fig. 13.11 Evolution of $p_s(t)$ as a function of the updates for groups of size four with $k = 0.49$ and $k = 0.47$. Two initial supports are considered. For $p_s = 0.51$ both $k = 0.49$ and $k = 0.47$ leads to $p_S = 1$. But for $p_s = 0.49$ only $k = 0.47$ leads to $p_S = 1$, while $k = 0.49$ leads to $p_S = 0$.

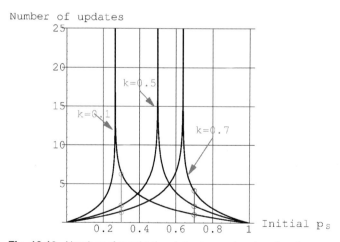

Fig. 13.12 Number of required updates to reach extremism for several values of the local bias with $k = 0.10$ $k = 0.50$ and $k = 0.70$. Associated values are shown for an initial support of respectively 30% and 70%.

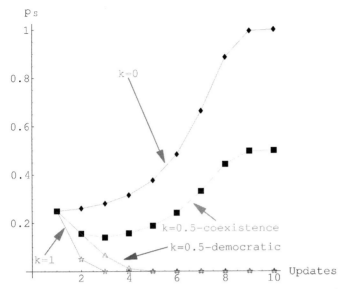

Fig. 13.13 The dynamic of public opinions for two populations on different areas with respectively $k = 0$ and $k = 1$ at an initial support of $p_s = 0.25$ is shown on the upper and lower curves. In between the two populations are in the same area as subclasses of one unique class. The upper one has segregation and yields a coexistence of opinions. The lower one has mixing and reveals a democratic extremism.

enough) size this process is equivalent on average to a bias of $k = 1/2$. The resulting extremism is democratic since it is along the initial global majority among the whole population. Mixing or segregation within the same situation may thus lead to drastically different public opinions as illustrated in Fig. 13.12.

13.7
Arguing in Groups of Various Sizes

Most of our analysis deals with groups of size four. However, in real life, people meet and discuss in groups of different sizes. Usually it is of the order of just a few persons from two up to five or six. Groups may be larger but it is rare. Indeed, most of the time when a large group of people meet to discuss, smaller groups form spontaneously [9]. These different sized gatherings are shaped by the geometry of social life within physical spaces like offices, houses, bars and restaurants.

This geometry determines on average the number of people who meet at a given place. Accordingly, a given social life yields a random local geometry landscape characterized by a probability distribution for gathering sizes $\{a_i\}$,

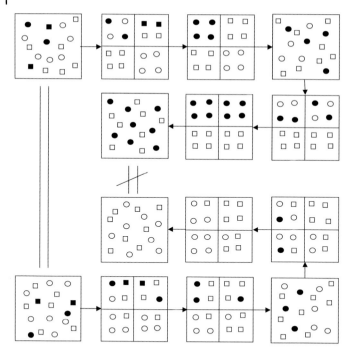

Fig. 13.14 A population composed of two subclasses sharing opposite beliefs. The circles have $k = 0$ while the squares have $k = 1$. The initial proportion in favor (white) is identical for both of them at $p_s = 0.25$. The black color expresses an opinion against. In the upper series, people segregate while updating their opinions. The result is a perfect balance of the global public opinion with all circles against (blacks) and all squares in favor (whites). In contrast, in the lower series, circles and squares do mix together while updating. The result is a democratic extremism with all circles and squares in favor (white).

which satisfy the constraint,

$$\sum_{i=1}^{L} a_i = 1 \tag{13.7}$$

where $i = 1, 2, \ldots, L$ stands for respective sizes $1, \ldots, L$ with L being the larger group.

On this basis we can extend Eq. (13.4) to the general update equation

$$p_s(t+n) = \sum_{i=1}^{L} a_i \left\{ \sum_{j=N[\frac{i}{2}+1]}^{i} C_j^i p_s(t+n-1)^j p_o(t+n-1)^{(i-j)} \right.$$
$$\left. + (1-k_i) V(i) C_{\frac{i}{2}}^i p_s(t+n-1)^{\frac{i}{2}} p_o(t+n-1)^{\frac{i}{2}} \right\} \tag{13.8}$$

where $C_j^i \equiv \frac{i!}{(i-j)!j!}$, $N[\frac{i}{2}+1] \equiv$ the integer part of $(\frac{i}{2}+1)$, $p_s(t+n-1)$ is the proportion of supporters after $(n-1)$ updates and $V(i) \equiv N[\frac{i}{2}] - N[\frac{i-1}{2}]$. It gives $V(i) = 1$ for i even and $V(i) = 0$ for i odd. We also have introduced the possibility of having the local bias k_i be a function of the size i of the group.

Clearly an infinite size distribution $\{a_i\}$ is possible combined with a whole spectrum of various values of k_i. More sub-cases can also be considered depending on the mixing of difference sub-classes. However, whatever configuration is chosen, the existence of a local collective nondecisional state monitored by the occurrence of local ties in groups of even size will always occur. Such a feature, whatever its amplitude, produces an asymmetry in the polarization dynamics towards either one of the two competing opinions, thus preserving the main result of the simple version of the model presented here.

To emphasize the effect of the group size distribution in a simple frame, we thus restrict the study of the effect of size on the opinion dynamics in the extreme homogeneous case $k_i = 1$. Within this frame Eq. (13.8) reduces to

$$p_s(t+n) = \sum_{i=1}^{L} a_i \sum_{j=N[\frac{i}{2}+1]}^{i} C_j^i p_s(t+n-1)^j p_0(t+n-1)^{(i-j)} \qquad (13.9)$$

In the course of time, at each new encounter a given agent will find itself discussing locally with a varying number of other agents. A simple case is illustrated in Fig. 13.15, 13.16, 13.17 and 13.18. It shows how a population composed of 33 agents with 13 against the issue and 20 supporting it changes to 19 against and 14 supporters after one cycle of discussion among four groups of one agent, two groups of two, two groups of three, two groups of four, one group of five and one group of six.

Figure 13.19 shows the variation of $p_S(t+1)$ as a function of $p_S(t)$ for two particular sets of the $\{a_i\}$. The first one is $a_1 = a_2 = a_3 = a_4 = 0.2$ and $a_5 = a_6 = 0.1$ where $L = 6$. There $P_S = 0.74$ which puts the required initial support to an issue success at a very high value of more than 74%. Simultaneously, an initial minority above 26% is enough to produce a final total refusal. The second set is $a_1 = 0, a_2 = 0.1$ and $a_3 = 0.9$ with $L = 3$ and $P_S = 0.56$. There the situation is much milder but also unrealistic since pair discussions are always much more numerous than just 10%.

To make a quantitative illustration of the dynamics leading to minority democratic spreading, let us consider the above first situation with an initial $p_S(t) = 0.70$ at time t. The associated series in time is $p_S(t+1) = 0.68$, $p_S(t+2) = 0.66$, $p_S(t+3) = 0.63$, $p_S(t+4) = 0.58$, $p_S(t+5) = 0.51$, $p_S(t+6) = 0.41$, $p_S(t+7) = 0.27$, $p_S(t+8) = 0.14$, $p_S(t+9) = 0.05$, $p_S(t+10) = 0.01$ and eventually $p_S(t+11) = 0.00$. Eleven cycles of discussion make all 70% of supporters for an issue turn against it by merging with the initial 30% of their opponents.

In order to get some quantitative feeling about the associated time, we might consider that, on average, an effective discussion, i.e., a discussion which could produce an agent change of opinion, takes place once a week.

Fig. 13.15 A population composed of 33 agents of which 13 are against the issue (dark squares) and 20 are supporting it (light squares).

Within such a time frame, three months of debate are required to reach a total crystallization of the opposition against the current issue. Moreover, a majority against the issue is obtained already within six weeks (see Fig. 13.19). Depending on the intensity of the public debate this time connexion should be re-assessed.

Changing the parameters slightly with $a_1 = 0.2$, $a_2 = 0.3$, $a_3 = 0.2$, $a_4 = 0.2$, $a_5 = 0.1$ and $a_6 = 0$ gives $P_S = 0.85$. A higher value of the required initial support which makes any reform proposal quite impossible. Indeed, a realistic reform project could never start with more than 85% support in the population.

Starting from $p_S(t) = 0.70$ yields successively $p_S(t+1) = 0.66$, $p_S(t+2) = 0.60$, $p_S(t+3) = 0.52$, $p_S(t+4) = 0.41$, $p_S(t+5) = 0.28$, $p_S(t+6) = 0.15$, $p_S(t+7) = 0.05$, $p_S(t+8) = 0.01$ before $p_S(t+9) = 0.00$. The number of local meetings has shrunk from 11 to 9. Within two months the whole population stands against a reform proposal which started with 70% support. The initial 30% of opponents grow to more than 50% in a month or so (see Fig. 13.3).

We have thus found that, while the two stable fixed points are unchanged at respectively zero and one, the separator, which is an unstable fixed point, varies with both size and the a_i distribution. To determine the specific con-

13.7 Arguing in Groups of Various Sizes | 385

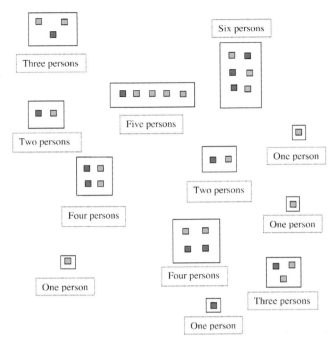

Fig. 13.16 First distribution of agents among various groups of size 1, 2, 3, 4, 5 and 6.

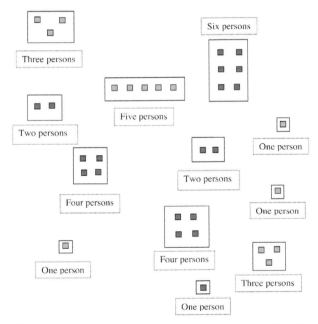

Fig. 13.17 One update has been performed within each group which is now polarized along one single opinion. Here the new distribution of opinions is 19 opponents (plus 6) and 14 supporters (minus 6).

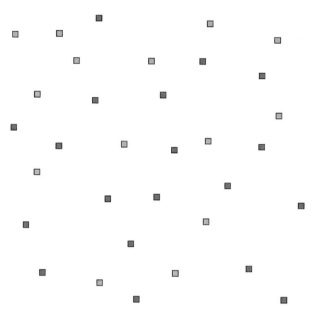

Fig. 13.18 The same population composed of 33 agents of which 19 are now against the issue (dark squares) and 14 are supporting it (light squares). Another distribution of agents within the same group size configuration as before can now be carried out.

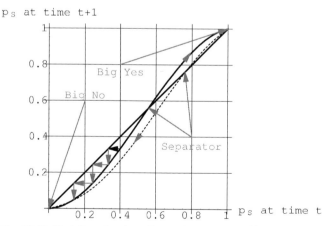

Fig. 13.19 Variation of $p_S(t+1)$ as a function of $p_S(t)$. The dashed line is for the set $a_1 = a_2 = a_3 = a_4 = 0.2$, $a_5 = a_6 = 0.1$, $L = 6$ and $P_S = 0.74$. The full line is for the set $a_1 = 0$, $a_2 = 0.1$ and $a_3 = 0.9$ with $L = 3$ and $P_S = 0.56$. Arrows show the direction of the flow.

tribution of each gathering size to the aggregation effect leading to the actual value of the separator, we now evaluate the value of the unstable fixed point for groups from size two up to six. Values are shown in Table 13.1. The flow landscape is identical for all odd sizes with an unstable fixed point at $\frac{1}{2}$. On the other hand, for even sizes, the unstable fixed point starts at one for size two and decreases to 0.65 at size six, via 0.77 at size four.

Tab. 13.1 Values of the various fixed points for each group size from two to six. SFP ≡ Stable Fixed Point and UFP ≡ Unstable Fixed Point.

Group Size	SFP Total refusal P_S	UFP Separator P_S	SFP Total support P_S
2	0	1	none
3	0	$\frac{1}{2}$	1
4	0	$\frac{1+\sqrt{13}}{6} \approx 0.77$	1
5	0	$\frac{1}{2}$	1
6	0	≈ 0.65	1

To illustrate the interplay dynamics between even and odd sizes, let us look at more details in the hypothetical case of discussion groups restricted to only two or three persons. Putting $a_1 = a_4 = .. = a_L = 0$ and $a_3 = 1 - a_2$, Eq. (13.9) reduces to

$$p_S(t+1) = a_2 P_+(t)^2 + (1-a_2)\{P_+(t)^3 + 3P_+(t)^2(1-P_+(t))\} \quad (13.10)$$

whose stable fixed points are still 0 and 1 with the unstable one located at,

$$P_S = \frac{1}{2(1-a_2)} \quad (13.11)$$

For $a_2 = 0$ (only three size groups) we recover $P_S = \frac{1}{2}$ while it reaches $P_S = 1$ already at $a_2 = \frac{1}{2}$. This shows that the existence of pair discussion has a drastic effect on creating doubt that, in turn, produces a massive spreading of the opposition. Few weeks are now enough to get a total issue rejection.

Keeping an initial $p_S(t) = 0.70$ with the above set of $\{a_i\}$, the time series become $p_S(t+1) = 0.64$, $p_S(t+2) = 0.55$, $p_S(t+3) = 0.44$, $p_S(t+4) = 0.30$, $p_S(t+5) = 0.16$, $p_S(t+6) = 0.05$ before $p_S(t+7) = 0.00$. The issue supporters falling off is extremely sharp as shown in Fig. 13.20. Within a bit more than two weeks a majority of the people is already standing against the issue which had started with 70% support.

Seven weeks later the issue at stake is completely out of reach with not one single supporter left. Considering instead only 30% of pair discussion groups, the falling off is weakened but still within sixteen days we have $p_S(t+16) = P_N 0.00$ (see Fig. 13.20).

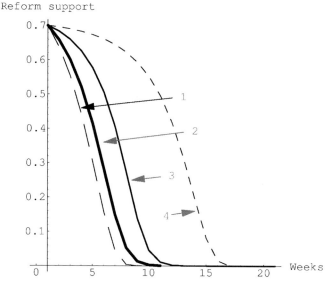

Fig. 13.20 Variation $p_S(t)$ as a function of successive weeks with $L = 6$. The initial value at $t = 1$ is $p_S(1) = 0.70$. Long dashed line (1): $a_1 = 0$, $a_2 = \frac{1}{2}$, $a_3 = \frac{1}{2}$, $a_4 = a_5 = a_6 = 0$ with $P_S = 1$. Heavy thick line (2): $a_1 = 0.2$, $a_2 = 0.3$, $a_3 = 0.2$, $a_4 = 0.2$, $a_5 = 0.1$ and $a_6 = 0$ with $P_S = 0.85$. Other line (3): $a_1 = a_2 = a_3 = a_4 = 0.2$, $a_5 = a_6 = 0.1$ with $P_S = 0.74$. Dashed line (4): $a_1 = 0$, $a_2 = 0.3$, $a_3 = 0.7$, $a_4 = a_5 = a_6 = 0$ with $P_S = 0.71$.

Whatever combination of the $\{a_i\}$ is chosen, the existence of temporary local doubts will always produce a strong polarization towards social refusal. At this stage it is worth stressing that, in real-life situations, not every person is open to a mind change. Some fractions of the population will keep their opinion whatever happens. Including this effect in the model will not qualitatively change the results. It will mean that the polarization process is not total with the two stable fixed points shifted towards larger and smaller values than zero and one, respectively.

13.8
The Model is Capable of Predictions: the Example of the 2005 French Referendum against the European Constitution Project

To give some real-life illustrations of our model, we can cite events related to the European Union which all came as a surprise. From the beginning of its construction there has been never a large public debate in most of the countries involved. The whole process came through government decisions although most people always seemed to have agreed on this arrangement.

At the same time, European opponents have been systematically lobbying for public debate. Such a demand sounds absurd knowing a majority of people favor the European union. But even so most European governments have been reluctant to hold a referendum on the issue.

Surprisingly, several years ago French president Mitterand decided to run a referendum on the acceptance of the Maastricht agreement [27]. While the great success of the "Yes" was taken for granted, it achieved only a little more than the required fifty percent. The more people discussed, the less support there was for the proposal. It is even possible to conjuncture that an additional two weeks extension of the public debate would have caused the "Nos" to win. More recently an Irish "No" [28] came as a blow to all analysts. The difference in the French case was certainly the weaker initial support.

But, if explaining past events is possible from a quantitative approach, predicting a forthcoming event is a much more solid decisive criterion with which to validate a model. And indeed, the minority spreading model combined with the inclusion of heterogeneous beliefs and group-size distribution was able to make a very improbable counterintuitive prediction in early 2006. The event is related to the first proposition of a European constitution which had to be validated by a large number of EC members in order to be implemented. Accordingly each EC country had to decide to adopt it using either parliamentary procedure or a national referendum.

In France, president Jacques Chirac decided to hold a referendum. His decision was made public several months ahead of the voting date. At that time, the "No" had only a few percent of support while the "Yes" to the constitution proposal had a huge majority. Everyone, including most of the scaremongers of the "No", were convinced the "Yes" would win definitively.

Then, against all the odds, including my own conviction, I came to predict the "No" would eventually win the vote. This prediction was the result of applying my opinion dynamics model [9, 26] to the current French situation. Besides oral announcements at some seminars, my prediction went out on print in an interview published by the French daily newspaper Le Monde [29]. From most of the reactions I got, it was then looked upon as ridiculous.

After a long and fierce campaign from the supporters of the constitution proposal, on May 29, 2005 the "No" won with a score of 55%, thus validating the "ridiculous" prediction. Of course, one single prediction is far from being enough to establish the validity of a model. Nevertheless, it gives a solid credibility to the model. In particular it proves that the model is worthy of more study and investigation. In addition, it legitimizes the sociophysical approach as a feasible new tool to tackle social and political problems or at least to view them in a new counterintuitive light.

13.9
Sociophysics is a Promising Field

To conclude, we have presented a simple model which is able to reproduce some complex and counterintuitive dynamics of social reality. In particular it suggests that holding a democratic debate to decide on an issue may indeed turn a free public debate onto a dictatorial machine to propagate the opinion of a tiny minority against the initial opinion of an overwhelming majority.

The dynamics of opinion-forming articulates around successive local opinion updates combining majority rule rationality and the possible occurrence of local doubt. In that case, it is the collective beliefs which are activated to make a choice. The emergence of a stable collective opinion is found to obey a threshold dynamic. It is found that the degree of heterogeneity in the distribution of common beliefs determines the current value of the threshold, which may well vary from 10% to 90%, depending on the issue at stake.

Our study demonstrates the inherent polarization effect associated with the holding of democratic debates towards social immobility. Moreover, this process was shown to be often *de facto* anti-democratic, since even in a perfect world it causes an initial minority refusal to spread almost systematically very quickly convincing the whole population. The existence of natural random local temporary doubts, which have been perceived as an additional feature of a democratic debate, is indeed instrumental in this dictatorship phenomenon, driven by the geometry of social life.

The model may generalized to a large spectrum of social, economical and political phenomena that involve propagation effects. In particular could it shed a new light on both processes of fear propagation and rumor spreading, like the French hoax about September eleven. It enlightens other issues including national voting, behavior changes, like smoking versus nonsmoking, support or opposition for a military action like the war in Iraq, and the impossibility of getting reform proposals adopted.

At this stage we have not addressed the difficult question of how to remedy this reversal of opinion phenomenon, which leads to the natural establishment of dictatorial extremism. The first hint might be in avoiding the activation of a common general background in the social representation of reality by the agents.

However, direct and immediate votes could be also rather misleading. Holding an immediate vote without a debate as soon as a new issue arises has other drawbacks. At this stage, the collaboration with psychosociologists as well as political scientists would be welcome.

Last, but not least, in 2005, for the first time, a highly improbable political voting outcome was predicted using our model of opinion dynamics. Moreover, the prediction was made several months ahead of the actual vote against all predicted outcomes by polls and analyses. It is the first demonstration of the heuristic power of sociophysics. Clearly, extensions [30] and limits of the approach should always be discussed before making any prediction.

References

1 TESSONE, C. J., TORAL, R., AMENGUAL, P., WIO, H. S., SAN MIGUEL, M., *Eur. Phys. J. B* 39 (**2004**), p. 535

2 WU, F., HUBERMAN, B. A., arXiv: cond-mat/0407252

3 GONZ'ALEZ, M. C., SOUSA, A. O., HERRMANN, H. J., *Int. J. Mod. Phys. C* 15, issue 1 (2004), to appear

4 SLANINA, F., LAVICKA, H., cond-mat/0305102

5 GALAM, S., *Physica A* 320, (**2003**), p. 571

6 STAUFFER, D., KULAKOWSKI, K., *Task Quarterly* 2 (**2003**), p. 257

7 MOBILIA, M., REDNER, S., *Phys. Rev. E* 68 (**2003**), p. 046106

8 STAUFFER, D., MEYER-ORTMANNS, H., cond-mat/0308231

9 GALAM, S., *Eur. Phys. J. B* 25 Rapid Note (**2002**), p. 403

10 SCHWEITZER, F., ZIMMERMANN, J., MÜHLENBEIN, H., *Physica A* 303(1-2) (**2002**), p. 189

11 SZNAJD-WERON, K., SZNAJD, J., *Int. J. Mod. Phys. C* 11 (**2002**), p. 1157

12 SOLOMON, S., WEISBUCH, G., DE ARCANGELIS, L., JAN, N., STAUFFER, D., *Physica A* 277 (**2000**), p. 239

13 DEFFUANT, G., NEAU, D., AMBLARD, F., WEISBUCH, G., *Advances in Complex Systems* 3 (**2000**), p. 87

14 GALAM, S., *Le Monde*/ 28 mars/ **2000**, p. 18

15 GALAM, S., CHOPARD, B., MASSELOT, A., DROZ, M., *Eur. Phys. J. B* 4 (**1998**), p. 529

16 GALAM, S., GEFEN, Y., SHAPIR, Y., *Math. J. Sociol.* 9 (**1982**), p. 1

17 GALAM, S., *Jour. Math. Psychology* 30 (**1986**), p. 426, *J. Stat. Phys.* 61 (**1990**), p. 943

18 GALAM, S., MOSCOVICI, S., *Euro. J. of Social Psy.* 21 (**1991**), p. 49

19 GALAM, S., *Physica A* 336 (**2004**), p. 49

20 FRIEDMAN, R. D., FRIEDMAN, M., *The Tyranny of the Status Quo*, Harcourt Brace Company **1984**

21 MOSCOVICI, S., *Silent majorities and loud minorities*, Communication Yearbook, 14. J. A. Anderson. Sage Publications Inc. **1990**

22 GALAM, S., RADOMSKI, J. P., *Phys. Rev. E* 63 (**2001**), p. 51907

23 GALAM, S., *Physica A* 285 (**2000**), p. 66

24 SCHELLING, T. C., *Micromotives and Macrobehavior*, New York, Norton and Co. **1978**

25 GRANOVETTER, M., *American J. Socio.* 83 (**1978**), p. 1420

26 GALAM, S., *Phys. Rev. E* 71 (**2005**), p. 046123

27 FRANKLIN, M., MARSH, M., MCLAREN, L., *Uncorking the Bottle: Popular Opposition to European Unification in the Wake of Mastricht*, Annual meeting of the Midwest Political Science Association (**1993**)

28 http://news.bbc.co.uk/hi/english/world/europe/ Prodi tackles Irish sceptics (2001)

29 LE HIR, P., *LeMonde* (Samedi 26 Fevrier) **2005**, p. 23

30 GALAM, S., *Physica A* 333 (**2004**), p. 453

14
Global Terrorism versus Social Permeability to Underground Activities
Serge Galam

Terrorism designates the use of random violence against civilians, the main purpose being to kill them. The phenomenon has existed for centuries. Up to September 11, 2001, it has always been geographically localized. In modern times, Basque, Irish and Corsican terrorism are emblematic of this "classical terrorism" acting in western countries. In each case, only one single flag leads the fight requiring the independence of a given territory to which most of the terrorist actions are confined. It is a one-dimensional terrorism.

On the other hand, the 2001 attack on the US has revealed an unprecedented worldwide range by destruction of a terrorist group. Later, the following attacks on Bali (2002), Madrid (2004) and London (2005) have confirmed this novel unbounded geographical status. Moreover, this "global terrorism" does not restrict its fight to the liberation of one single territory. Insted, it articulates around several simultaneous different flags. It is a global multi-dimensional terrorism.

It is worth noting that methods used to fight classical terrorism in democratic countries including Spain, England and France have been unable to eradicate it to date. It seems that coordinated efforts to thwart global terrorism have also failed so far. At the same time, each specific terrorism has been prevented from extending to a level which would eventually lead to its victory.

On this basis, it could happen that some new point of view is necessary to embed some missing mechanisms by which terrorism operates. Accordingly, we suggest studying the effect of the existence of individual passive supporters on the operational capabilities of a terrorist organization. In others words, we aim to quantify the extensions of social spaces which are open to terrorist activities. To complete this goal we use the physical concept of percolation in a multi-dimensional social space. Inded, social dimensions are singled out and added to the usual geographical ones to determine the geometry of the social space. The results are surprising and shed a new light on terrorism. It may provide novel skills in order to curb it. The model applies to all underground activities.

14.1
Terrorism and Social Permeability

The terrorist attack of 2001 September 11 on the United States came as a tremendous blow to all the experts in terrorism and intelligence services. Besides the horror of the attack, there was the question of how it could happen so far away from the associated terrorist sanctuary. Overall, it is possible to state that the phenomenon of terrorism is still not understood.

Indeed, while fighting terrorism is extremely complex, complicated and difficult, a different view, from the physics aspects, could be useful to clarify in a radically new way some part of the problem which seems to be resistant to traditional anti-terrorist protocols. Most studies focus on the terrorist groups themselves [1,2], The primary focus of anti-terrorist frameworks is on the terrorist network itself. It concentrates, on the one hand, on an active military side with the destruction of its organization and the blocking of its finances. On the other hand, it develops a passive side which has to figure out the potential targets and to design an efficient defense. Strategic vulnerable sites must also be protected against potential attacks.

However, we take here a radically different viewpoint based on the very little attention paid to the social environment in which the terrorist action proceeds. It considers the total society which includes, simultaneously, the terrorists, the potential terrorists and all the people living in the society. On this basis we investigate the permeability of the social space to terrorist members [3]. Permeability here means, to the terrorists, the possibility of living, moving and performing terrorist actions.

This social background seems essential in the delimitation of the space accessible to the terrorists, since it is within this space that terrorist damage can take place. To establish our new approach of the problem we study how, independently of a terrorist group's internal mechanisms, the range of its action results directly from the degree of permeability of the social background in which its members evolve and move.

The permeability results from the aggregation of a passive adhesion from some individuals to the claim of a particular terrorist cause. No concrete engagement nor any public statement with the terrorist group itself is required. This permeability constitutes a strategic stake in a possible victorious fight against terrorism [4]. It results from a particular collective effect by the series of individuals who agree passively, at least partly, with the terrorist cause. It often happens that the majority of these people even reject the terrorist action in itself. However, they support the corresponding flag. We will show how the degree of permeability of the social background determines the limits of social space open to terrorist action.

The physical concept of percolation is particularly well adapted to treat this kind of collective passive aggregated problem. Using this concept gives a coherent framework to the comprehension of the geographical space of the terrorist action. On this basis, this work does not aim to provide an exhaustive description of the complexity of terrorism. It only attempts to clarify the geometrical connection between an individual passive attitude and the perpetuation of a terrorist act far away from the associated passive supporters [3–5]. Such a study of terrorism can also be seen from the logics of complex adaptive systems [6]. It is part of sociophysics [7,8].

14.2
A Short Introduction to Percolation

Percolation deals with the study of the properties of the connection of objects distributed randomly on a surface [9]. Let us take, for instance, the classical case of stones thrown into a lake, each staying above the water surface. Two stones located at less than fifty centimeters can thus be reached without getting wet. They are said to be connected. For a given density of random coverage of the lake with the stones, some connected paths appear with an average size. However, it is not possible to move from one little island made up of a cluster of stones to another one, since they are not connected.

Throwing in more stones results in an increase in the average size of the unconnected islands but no qualitative change occurs. One still cannot cross the lake without getting wet. If someone is put on one of the existing islands of connected stones, they can move on it but remain stuck on it. One illustration is shown in Fig. 14.1.

Nevertheless, at some precise value of the cover rate a qualitative change suddenly occurs. A connected path appears, at once connecting both opposite sides of the lake. One can now cross it without getting wet. At this value of coverage, the stones are said to percolate (see Fig. 14.2).

What is remarkable in this phenomenon of percolation is that if one repeats the experiment a great number of times, the value of the cover rate to which the stones percolate is always the same in the limit of an infinite lake with a fixed given geometry. This value is an invariant of the problem denoted by the percolation threshold p_c. It changes only if the rules of connection are modified or the dimension of the space in which the process is taking place. In the case of the lake it is a space of two dimensions. The connecting cluster is denoted as the infinite cluster.

This phenomenon was first studied by mathematicians and then by physicists. The theory of percolation shows that the percolation threshold p_c for the appearance of an infinite cluster is almost always lower or equal to 50%. Depending on the geometry it can be calculated precisely either analytically for a small number of regular lattices or numerically for most lattices. Then as soon as this threshold is reached, within the framework of a completely ran-

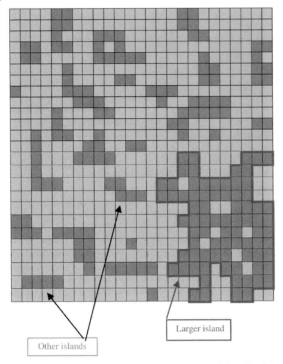

Fig. 14.1 The whole lake is dotted with little islands of more or less the same width. The dark color indicates the stones. The little islands are disjointed and so the lake cannot be crossed on foot.

dom cover, there always exists, in a clear direction, a path to go from one end to the other of the surface concerned, which may also be a hypervolume in an hyperspace.

Using this lake illustration which is not simply fortuitous, the application of the physical phenomenon of percolation to our problem of social permeability to terrorism, can be implemented in a straightforward manner as will be seen below. The curtains at the windows replace the stones and the space open to the terrorist movement, replaces the stone paths allowing movement on the lake. Accordingly, one can build a coherent and unified framework for the geographical deployment of terrorist action.

14.3
Modeling a Complex Problem as Physicists do

Dealing with terrorism is an extremely delicate matter. Too much sorrow, passion, hate and death are at stake. But all that should not prevent us from trying to model the issue using a scientific standpoint, equations and a simplistic

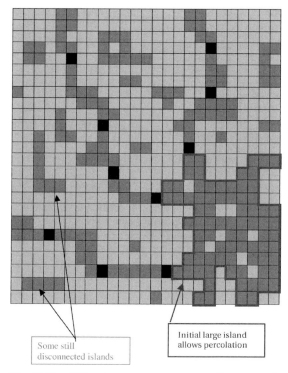

Fig. 14.2 Additional stones have produced the percolation phenomenon. From now on the whole lake can be crossed without getting wet. Small islands are still separated from the percolating island.

point of view. On the other hand, any new insights which could eventually help in reducing terrorism are urgent and to be welcomed.

But modeling does not mean discovering one unique truth. It only aims to provide a different viewpoint which, in turn, may lead to useful steps against terrorism. Initially, we proceed to a modeling of the problem, as physicists do, without keeping account of its full complexity. The maximum number of simplifications of reality are made, keeping only some minimum number of ingredients which can still highlight the essential mechanisms of the real problem. Here this single ingredient is the role and effect of passive supporters to a given terrorist cause.

Of course the terrorist net itself is of central importance. Clearly, without it, there will be no terrorism. However in order to act, a terrorist must move to reach their target. In order to move they need to find a path within the social space which is safe. Therefore, to determine the terrorist range of destruction it is not the terrorist net itself which matters, but the social permeability of its members. It is worth remembering Mao Zed Dong stating "A revolutionary activist must be like a fish in water while in the popular masses".

Applied to terrorists, this metaphor means they need to find the water if any, where they can swim freely. The water is made of people who are passive supporters to the flag for which the network is fighting.

Throughout this paper we are dealing with passive supporters who have a positive attitude towards the terrorist organization, but it could also be a passive support driven by fear, indifference or profit, as in some cases of underground activities or illicit practices. This means the model can apply to a large spectrum of clandestine activities including guerilla warfare as well as tax evasion, corruption, illegal gambling, illegal prostitution and black markets.

Although this work does not aim at an exact description of the complexity of terrorism; thanks to its simplicity based on many rough approximations, it may lead to the discovery of an essential feature of terrorist activity. In particular, it connects the capacity for destruction to the attitude of the surrounding population.

14.4
The World Social Grid

To start the modeling we regard the world as a surface constituted by a "social grid" composed of all the individuals on the planet. On average, an individual square of the grid corresponds to the earth's surface divided by the corresponding total population of several billion individuals. Each square of the grid is occupied by one person.

Of course, in the real world most spaces are not populated, like the seas, the mountains, fields, etc. But to keep the presentation clear and simple we ignore these empty spaces and consider a totally occupied grid. The corresponding social grid is formed by the juxtaposition of all the individual spaces of the world's billion human beings. We call an individual space a box. Each box of the grid is occupied by one person as shown in Fig. 14.3 for a small area.

Now we make the assumption that, for an individual to move on the grid, it must pass from one box to another one which is contiguous to it, and so on. Thus, departing from a given box, to reach another precise box, the agent must follow a continuous path crossing a series of adjacent boxes connecting these two extreme boxes, departing from "D" and arriving at "A".

As already stated above, between the two boxes D and A, a large number of paths are potentially possible. In theory, there is even an infinity number. It is thus impossible to predict which is actually followed by the person moving from D to A. Each individual space is open to other movements. Figure 14.4 illustrates one case.

Note that, to cross a box, the agreement of the person who is there is necessary, at least in principle. But as the space of the box is sufficiently large to

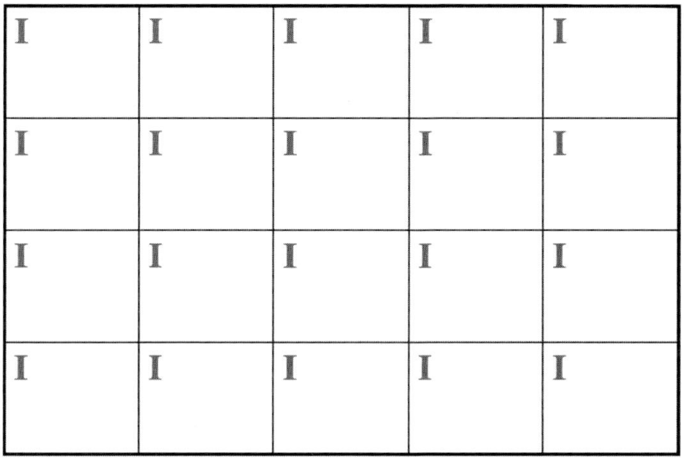

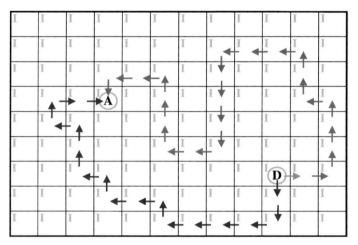

Fig. 14.3 A portion of the earth's surface: each little square is occupied by one individual I. All spaces are open.

Fig. 14.4 From a departure point "D" to reach an arrival point "A", several different paths may be taken. Two are shown by the series of arrows. The one to be used is unpredictable.

contain at least two people, the agreement of the occupant of the box is reduced in fact to nonopposition. In other words, any box is open in the passing of each one provided that its occupant does not close it physically, which is a very rare event. All the boxes are thus *a priori* open to the movement of anyone. They are all permeable. The grid shown in Fig. 14.4 is totally permeable.

14.5
Passive Supporters and Open Spaces to Terrorists

Now let us consider a terrorist wanting to move on the grid from their base to another location. Passing through a box is now an active move which can endanger their freedom since each occupant of the box becomes passed *de facto* one possible threat by noticing the passing and then reporting it to the appropriate institution. Being passive, while noticing a terrorist move, has the same effect whatever the motivation, either support or indifference.

An occupant of a box which is hostile to terrorism in general, or to the terrorist cause defended by the terrorists acting in its neighborhood, will block the passage of its box by just showing it is aware something suspicious is going on. A terrorist will not use force. They need to stay unnoticed to achieve their deadly goal.

Each box is thus opened or closed to the public passing through it according to the will of its occupant. The passive attitude means the box is open. It is necessary to act to close the box, even if this action is reduced to nothing very important such as watching locally. Metaphorically we say that a passive supporter closes their curtains while an active opponent to terrorism will keeps the curtain wide open as shown in Fig. 14.5.

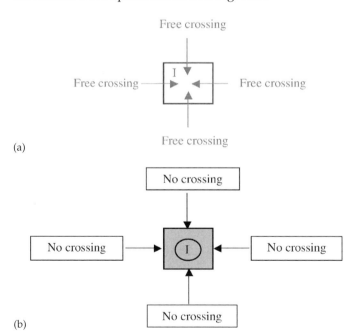

Fig. 14.5 (a) is a box in the passive position, "closed curtains", a terrorist can cross the box starting from any one of its four sides. (b) is a box in the active position, "opened curtains", a terrorist cannot cross the box safely from any side, under scrutiny.

More precisely, passive supporters are normal people who share some sympathy with the terrorist cause, but without any involvement in it. They just don't oppose it when they could. They are normal people, you, me, others who, facing a terrorist cause, may support the principle of the terrorist's goal although they are against the killing of civilians and are not involved at all with the terrorist organization.

The main feature of a passive supporter is that it is someone who would not oppose a terrorist move if it could, since it shares the terrorist goal, at least in part. Or it could also be because the supporter opposes the political power which is fighting terrorism. It is of importance to emphasize that most passive supporters are never confronted by any terrorist move. It is more of a silent and dormant attitude and is almost never activated. It might possibly be perceived as a neutral tolerant attitude towards the political content of a terrorist trend.

Passive supporters are obviously not the only ones who constitute the "terrrorist water" relating to Mao Zed Dong's statement, but they definitely constitute a substantial part of it. Here we focus on them and their effect on the capability of destruction of the corresponding terrorist organization.

From this arises the question of how passive supporters become connected with terrorist movements? To answer this, it is necessary to specify the very important fact that the terrorist's need do nothing to pass a given box. They only need to be sure that the corresponding curtains are closed.

Any movement always comes from a box with the intention of moving to another one. Given the starting point, to proceed with the plan, they must find a pathway of contiguous and open boxes which enables them to reach their point of arrival where their target is located. Several ways may be possible, and they will follow one of them, each time trying to go in the direction of the target. But it may well happen that the open path is not the shortest, as seen in Fig. 14.6. It may also be that no open path exists to reach the selected target.

Therefore, starting from the base, any infrastructure, to be regarded as a potential target, must be connected by at least a continuous path of closed curtain windows. Since each window state is independent of the terrorist, it is in fact all of the existing possible paths starting from the terrorist base, which determine the social space open to terrorist action, the Active Open Space (AOS). It is worth noting that, in general, other Open Spaces (OS) always exist simultaneously but are not connected to the terrorist base, so they are not accessible to terrorist action (Fig. 14.6). They are safe, but potentially vulnerable as soon as a terrorist base is established within the overlapping Open Space.

All the boxes being adjacent to each other, a given box is characterized by its four sides, each one being either open or closed according to the active or passive position of the corresponding occupant. Thus, for a passage to be open between two boxes, it is necessary that its two occupants, on both sides,

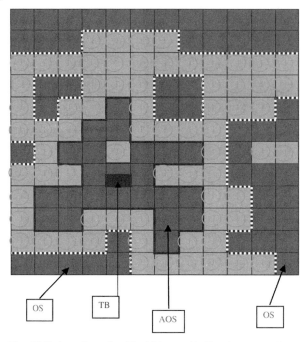

Fig. 14.6 A portion of a 12×12 box grid. Passive supporters occupy dark boxes, their curtains are closed. On the other hand, clear boxes are closed to terrorist movement, their curtains are open. The juxtaposition of dark boxes produces Open Spaces (OS) to terrorist action. But only the one including the terrorist base (TB) is active (AOS) under actual terrorist threat. The other OS are inaccessible to terrorist action.

are in the passive position. As soon as one or both is in the active position, then the passage is closed as seen in Fig. 14.7.

It is remarkable to note that it is enough for an occupant to put himself in an active position (open curtains) in order to block the four sides of the box, independently of what the occupants of the nearby boxes may do. The passive and active states are thus asymmetric with the latter being much stronger as indicated on Fig. 14.8.

It also should be specified that the state of a given box is *a priori* unknown. The state of a box is dormant. It is only when the terrorist arrives to pass by a box that they activate the dormant state. Consequently, a terrorist moves on the grid looking out for a path of connecting contiguous open sites to reach the target from the initial starting point. A terrorist can also be obliged to go back if a way is not found to continue towards the target. It can also hit the target.

It is a series of these individual free open paths which are used by any individual who departs from a point and want to reach an arrival point. To

14.5 Passive Supporters and Open Spaces to Terrorists

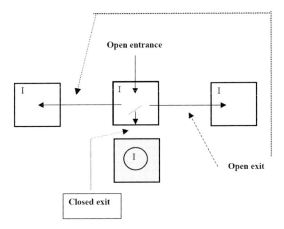

Fig. 14.7 A terrorist arriving from the top to cross the middle box which is open. They can then proceeds to the right side where the next box is open as well as to the left. However, they cannot go downwards since the box is closed.

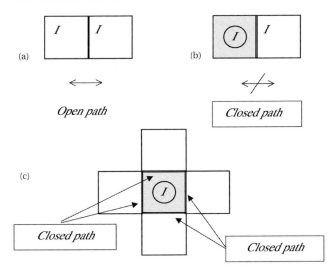

Fig. 14.8 (a) the path is open while it is closed in (b). (c) One closed box is shown to block four paths from open contiguous boxes.

join the departure and arrival points, several different paths may be available, forcing a choice. The one to be used is unpredictable. It depends on the walkers choice and may reach a dead-end thus failing to reach the target. A few people may simultaneously use one individual space, but not many.

Accordingly, the model clarifies the conditions of existence of an open way between two points, the source and the target. Social permeability allows the calculation of the different possibilities by using a probabilistic approach.

14.6
The Geometry of Terrorism is Volatile

We are now in a position to apply the model of the box presented above to describe some features and characteristics of terrorist activity. It turns out to be a geometrical problem or at least we will treat it from a geometrical viewpoint making *de facto* percolation theory totally adequate to solve the problem.

More precisely, we initially study the permeability of the social background of a given terrorist movement. Then we associate the characteristics of this permeability to the potential range of the destructive capacity of this same terrorism. We will see why and how until now terrorist organizations have always evolved and moved in a regional geographical context.

The recent appearance of a form of terrorism with international significance will be explained in the following section. We start by considering a given terrorist group. It is always attached to a geographical sector specific with a local cause. Very often it is based on a territorial claim, either of independence, or of autonomy. Usually, a substantial part of the corresponding population have some sympathy with the terrorist cause in question, even if most people oppose the choice of violence employed by the terrorist group. But this opposition is only formal. Nothing is never done against the terrorist group and its members. In particular, nobody denounces active nationalists involved in the terrorist network.

There can be a passive identification with the terrorist cause without any real involment in it. These passive supporters are completely normal people, well under any reported level. Not only do they not participate in the terrorist activity, but they do not even need to explicitly express their support or identification. It is a dormant attitude, just an individual opinion. The passive supporters are thus very often completely invisible.

Their essential characteristic is that they will not oppose the passing of a terrorist by their box, if that should occur. And it is very important to stress that, in the vast majority of cases, passive support never occurs. Moreover, passive supporters do not need to communicate between themselves. They simply independently share the same opinion in favor in the correctness of the terrorist cause. They are distributed within the whole population but find themselves naturally concentrated in the main sector of the terrorist group territory claim.

Thus, some passive supporters are localized side by side on adjacent open boxes, while others will be isolated, surrounded by active opponents to the terrorist group activities. From this distribution emerge all the potentially open paths to terrorist displacement. As noted earlier, they are nevertheless virtual and invisible. Let us call them permeable clusters.

These are distributed on the territory claimed by the terrorist group and their sizes can vary from one cluster to another, but they are all of a finite size, i.e. a limited geographical extension. They are disconnected from one another. On the other hand, their number, like their average size, depends on the total number of people passively agreeing to the terrorist cause.

Each one of its clusters creates a permeable social space in which a terrorist can strike any target located in the geographical field covered by the cluster in question, provided it includes its base. We call such a space an "Open Space" to the terrorist action (OS). Any terrorism which starts to spread from such a OS can reach any target within the OS. The problem for the terrorist is to reach this OS which, according to our model, it can only do if its base is localized within the given OS.

It thus implies that, while several OS may exist simultaneously, terrorist acts can be perpetrated in only one of them. It is the one which physically includes the base of the terrorist network (TB). This OS is dangerous and is denoted "Dangerous Open Space" to the terrorist action (DOS). All the other OS whatever their number may be, are out of reach of actual terrorist action. This safety state is not the result of a strategic choice by the terrorist group, it is rather the physical result of the absence of accessibility to them from the terrorist base. An illustration is shown in Fig. 14.9.

It appears that, for most cases, when moving away from the claimed geographical sector, the number of people passively agreeing with the terrorist flag decreases quickly towards zero. Far from it, people are often unaware of the cause of terrorism. So the density of OS falls drastically to zero as soon as the frontiers of the terrorist claimed area are crossed.

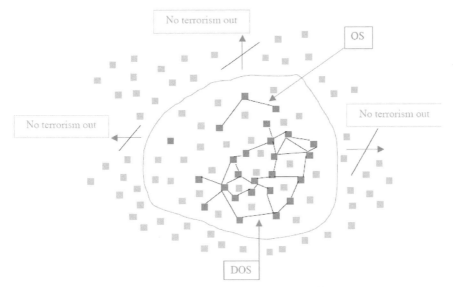

Fig. 14.9 Schematic view of an area with two Open Spaces (OS) of which only one is dangerous (DOS).

It should be noted again that, as the various OS are inaccessible to the terrorist action, the potential targets which are localized there are out of danger. On the other hand, all those which are inside the space covered by the DOS are directly threatened. But it is necessary to keep in mind that the situation is dynamic and fragile. The DOS is volatile.

Indeed, it can be enough that one or two people change state, passing suddenly from the active position to the passive position. Closing the curtains of even one box may immediately connect an OS to the DOS making its size increase at once, to cover a much larger area. Many up to then safe targets become vulnerable, although in fact not much has changed at the global level regarding support for terrorism or even with respect to the terrorist infrastructure itself.

The same type of individual change can also have the opposite effect and reduce the DOS while cutting it in two unequal parts, a remainder staying as the DOS, the other becoming an OS. This last is not inevitably the smallest cluster. It is the one which does not contain the Terrorist Base. One illustration is exhibited in Figs 14.10–14.12.

In Fig. 14.10, the first of the series, an area is shown where there exists one Dangerous Open Space (DOS) where a base of the terrorist group (BT) is located and six Open Spaces (OS), all in grey.

In Fig. 14.11 one person has turned to a passive supporter state indicated by a smaller square within the light colored square. By doing so the former Dangerous Open Space (DOS) has merged with a former Open Space increasing its size. There are now one DOS and five Open Spaces (OS).

Figure 14.12 shows Fig. 14.10 but where now one passive supporter has become an active opponent, opening the curtains. This is indicated by a circle with a square. The change has reduced the DOS by making one part of it an Open Space. The Dangerous Open Space has shrunk and there are now seven Open Spaces.

14.7
From the Model to Some Real Facts of Terrorism

The model we have developed to this point, although extremely simple, enables us to understand a certain number of characteristics of the various forms of known terrorism. In particular, it explains two essential characteristics of terrorist activity.

The first is its geographical anchoring. The model provides an explanation as to why the selected targets are in the immediate vicinity of its base. It is not the result of a choice, either ideological or strategic, it is quite simply because of the physical constraint of access to the terrorists. This limitation seems clear with respect to classical terrorism like those active in Corsica, Ireland and Euskadi.

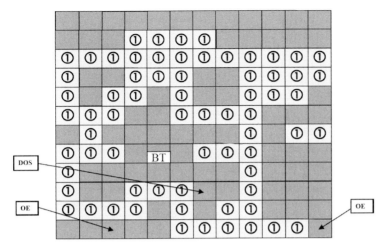

Fig. 14.10 A view of an area including one Dangerous Open Space (DOS) where a base of the terrorist group (BT) is located and six Open Spaces (OS), all in grey.

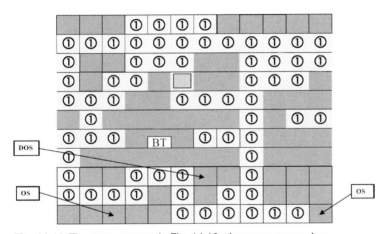

Fig. 14.11 The same area as in Fig. 14.10 where one person has turned to a passive supporter state indicated by a smaller square within the light colored square. By doing so the former Dangerous Open Space (DOS) has merged with a former Open Space increasing its size. There are now one DOS and five Open Spaces (OS).

The second characteristic is the impossibility of precisely determining the space which is accessible to terrorist actions since this space is simultaneously multiple, invisible, dormant and volatile as shown explicitly in the series of figures 14.10–14.12.

These two features may provide an explanation to the ongoing impotence of police and anti-terrorist services to curb the continuous and constant actions of these terrorist groups which have been active for many years now. On the

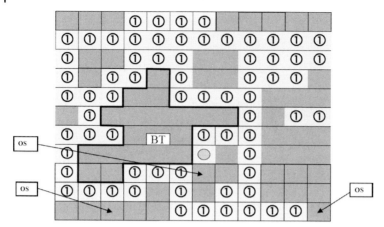

Fig. 14.12 The same area as in Fig. 14.10 where one person has turned from passive supporter to an active opponent, opening its curtains. It is indicated by a circle with a square. This individual change has reduced the DOS by making one part of it an Open Space. The Dangerous Open Space has shrunk and there are now seven Open Spaces.

other hand, an overall assessment of the current support existing for a terrorist cause can make it possible to evaluate the number, the average size and the distribution of the OS within a well-defined geographical area.

The fact that those people passively agreeing to the terrorist cause are distributed in a random way and are dormant but concentrated in a given area makes the theory of percolation perfectly fitted to the evaluation of these quantities, which then can provide a clear picture of the real situation. In particular the application of the physical theory of percolation will even allow some quantitative evaluation of the situation on the ground.

At this stage of our modeling, we also see why the forms of terrorism that one could qualify as traditional or classical ones are confined geographically. They lack sufficient passive support outside their respective area of anchorage. For each terrorist group, there is a DOS concentrated at the heart of the disputed area, and some OS which are dispersed around the DOS in a perimeter of varying width. But beyond that territory, it may happen that some tiny Open Space exists but most of the people do not share the terrorist cause. Then it is a total vacuum for the terrorist capability of moving beyond its natural territory.

Figure 14.9 shows such an example of regional terrorism. Active terrorism takes place within the claimed zone and is locked within it. The only possibility of extension for the terrorist group is the connection of the DOS to the other OS of the area. All the other parts of the world are completely prohibited to its action.

According to our reading of the situation we come to the conclusion that the recent current global terrorism which was able to strike in New York, Bali, Madrid and London must have used a Dangerous Open Space covering the whole world. This statement implies within our frame, the existence of passive supporters spread throughout the world. And in addition these passive supporters were so distributed as to become connected to one another in the way we have described earlier. It underlines the existence of a huge number of passive supporters.

Accordingly an essential feature of the current global terrorism has been, and still is, its capacity to create a very great number of people who support its cause throughout the whole world. During the years before the September 11, 2001 attack on the US, no world Dangerous Open Space existed. But going backwards we could say that, before September 11, many Open Spaces did exist all over the world but went unnoticed since they were disconnected from the DOS based in the west. Therefore they remained dormant.

In this context, the continuous dynamics of future global terrorism in convincing more and more people of the correctness of its ideology, during recent years has caused neither concern nor worry of any kind. After all, this expansion was reduced to a question of opinion, which is *a priori* perfectly legitimate in the western standards of free societies. Although an increasing number of new OS were seen all over the world, they did not generate any particular problem. The reason can be perfectly understood within the framework of our approach.

More and more permeable spaces were created for the terrorist action, but remained inaccessible to the terrorists themselves. So this basic change of the situation did not in any way modify the level of safety implemented in the western world. For several years in a row the Open Spaces must have grown gradually bigger due to the great indifference of the world intelligence services. An illustration is shown in Fig. 14.13

However, at a certain time, a huge and brutal phenomenon occurred with respect to the connectivity of all widespread Open Spaces. Well-known in physics, this phenomenon is called a phase transition of second order. In the geometrical case which interests us here, it is called percolation. But what does this mean exactly?

14.8
When Regional Terrorism Turns Global

For several years the dynamics of expansion of the OS had remained without effect on the extension of the current geographical area touched directly by terrorist action. And that was in spite of the substantial increase in the passive number of supporters. But at a certain level of passive supporters, the swing

14 Global Terrorism versus Social Permeability to Underground Activities

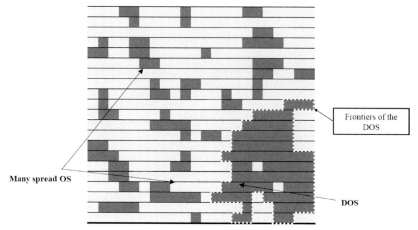

Fig. 14.13 One area is highly permeable to the terrorist actions but yet is cut from the rest of the surface. Although there exists many Open Spaces outside the frontiers of the DOS, terrorism does not reach them. *De facto* outside the frontiers of the DOS terrorism does not exist.

of only a few additional people from their active position to a passive position has produced a massive and immediate connection of many OS randomly spread all over the world to the regional DOS of the terrorist group.

This simultaneous fusion of several OS to the DOS immediately produced a total permeability of all the territory concerned, which happens to be the whole planet. In physics the DOS is said to have percolated DOS, i.e., it joins one side of the surface of the Earth to the other side. One illustration of the change is shown in Fig. 14.14.

We suggest that such a phenomenon, which is unique in the history of humanity, did occur very recently sometime before September 11, 2001. It is the first phenomenon of percolation in the history of terrorism. The passive character of the event results from the fact that no interaction is required between the passive supporters. This geometrical result does not require exchanges among its components. It is only a question of the connectivity of the individuals. All of a sudden and at once the level associated with the terrorist threat became infinite and spread all over the world. Figure 14.15 shows a case where the passive supporter for a terrorist group has percolated to the level of a country.

Since the territory of the whole planet has now become easily accessible, the number of possible targets jumps to high figures. The attacks of September 11 against the United States were the first immediate sign of this social permeability which percolated from one end of the planet to the other. The following attack on Bali, Madrid and London have confirmed this hypothesis showing that the 2001 attack was not an isolated unlucky case. It is useful to note that,

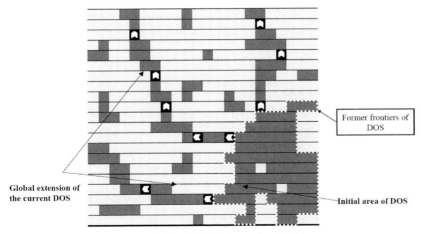

Fig. 14.14 The same area as in Fig. 14.13 but now the DOS has percolated throughout the whole surface. While terrorism did exist outside the former frontiers of the DOS terrorism, it has emerged suddenly everywhere.

from such a percolating situation, the majority of the potential targets are not only accessible but are also reachable in several different ways.

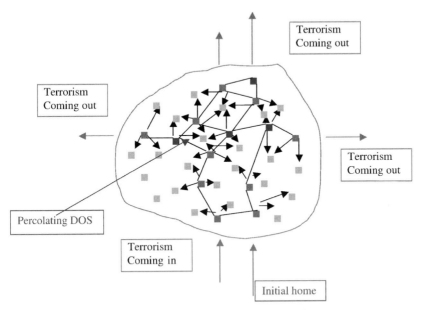

Fig. 14.15 One area is highly permeable to terrorist action. Moreover the enclosed DOS has percolated up to the frontiers of the country concerned. The associated terrorism can now extend outside the frontiers of the DOS without any particular internal change.

14.9
The Situation Seems Hopeless

At this stage, we can conclude that for a given territory the distribution and the size of aggregate spaces, yielded by its passive supporters, directly determines the range of terrorist action. It is the relative value of the passive support p of the population compared with the value of the critical point p_c of the corresponding space, which determines the effective amplitude of the terrorist threat.

If the passive support is larger than the percolation threshold with $p > p_c$, all the territory falls under the terrorist threat. However, while in a physical system the size is infinite, here the delimitation of the size of the territory considered is essential data. One can have a density of passive supporters p which satisfies the condition $p > p_c$ for a given geographical surface, and at the same time for a larger territory, including the first one another density p satisfying $p' < p_c$. Indeed, the number of passive supporters always strongly decreases far from the home area of terrorism. That is the case for traditional terrorism mentioned previously. For example, in this diagram related to Corsica terrorism one can say that it does percolate at the level of Corsica but not at the level of France and even less at the world level as shown in Fig. 14.9.

Terrorist deployment thus obeys a universal scheme of activity with two phases: a percolating and a nonpercolating one. The only difference between one terrorism and another being the scale on which the passive supporters are spread and the geographical area in which percolation may take place. Obviously, if the change of scale does not change the nature of the terrorist phenomenon, it substantially modifies the number of threatened people, which clearly is not a negligible difference.

September 11, while revealing for the first time the existence of a world percolation also showed that, from now on, the whole world population is in potential danger.

In this context, for a territory under active terrorist threat, the lifting of this threat requires the suppression of the percolation of the passive supporters. Indeed if any destruction of a terrorist cell has obvious immediate advantages, it is without effect on the range and severity of the threat. As soon as a new terrorist group is formed, it can strike again immediately in all the space which remains accessible to its members. The strategic challenge is thus to bring the condition $p > p_c$ back to $p' < p_c$ by lowering the density of passive supporters from p to p. Such a reduction drastically shrinks all the territory accessible to terrorist action and reduces its open space to a narrow area. The only problem is that the completion of such a program on the military level is simply terrifying and unacceptable.

Indeed, a reduction of even only a few percent in the density of passive supporters of any population would require the neutralization of, at a minimum, several tens of thousands of people in the case of a not too widespread terrorism. The figure can grow up to tens of millions of people for current world terrorism. For instance, to lower 15% of passive supporters down to 10% for a population of one billion requires neutralizing 5% of 10^9 yielding 5×10^7 (fifty million).

However, these people being dispersed at random in a ratio of one to six, gives a figure of 300 million people to be neutralized. Moreover, the large majority of these people, here 250 million, are opposed to the terrorist cause, making such an option totally unthinkable.

Therefore, an efficient military solution is completely unacceptable for ethical and humanitarian reasons as well as from the point of view of morality and justice. It would lead to the destruction of a good part of the planet, although existing weapons of mass destruction would allow it.

At the same time, any partial military solution would result in sheer waste being without any powerful effect on the level and broadness of the terrorist threat. The conclusion from our study thus seems completely depressing, without any hope. Accordingly, international terrorism would seem to be a fatality against which one can do nothing; the anti-terrorist fight being limited to specific actions against the networks themselves. The current danger would thus remain unchanged with a world social permeability available on request.

On this basis we can conclude that no military solution can solve a terrorist problem which has strong enough passive support from the relevant population.

14.10
Reversing the Strategy from Military to Political

However, facing the above conceptual and strategic dead-end, an alternative strategy to tackle the problem exists. It consists in reversing the step used in physics to remove a percolation. Let us come back to the lake illustration. It is possible to cross it while keeping dry because the density of stone p satisfies the condition of percolation $p > p_c$. Accordingly, to increase the possibility of a dry passage, it is necessary to suppress the percolation of the stones, which is done quite naturally by removing some of them. The number of removed stones just need to be sufficient to make the new density p pass below the critical point value satifying $p < p_c$. As soon as that condition is met there is no island which extends from one side of the lake to the other.

But let us imagine that it is possible, at least in principle, to restructure the lake topology in order to modify its percolation threshold without modifying the current density of stones. In other words, the goal becomes to turn the initial lake topology into a new one which has a new threshold value p_c which is larger than its former value p_c.

Consequently, with an unchanged density of stone, one could satisfy the condition $p < p_c$. Accordingly the percolation would be removed at once and simultaneously and the possibility of a dry crossing of the lake would no longer exist.

Unfortunately, such a scheme is not feasible in a real lake. But for the case of a virtual lake on a computer screen, things are different. And jumping from the computer screen to our application to terrorist phenomena, the temptation is strong since it would open the way for the possible treatment of the major difficulty with the terrorism problem, which is to act quantitatively and in a drastic way on the terrorist capacities for action without physically touching its passive supporters. At this stage we have, at least in theory, a clear procedure to implement. But how to put it into practice?

Before answering the ambitious and challenging question of modifying the value of the percolation threshold of a given problem, we should first come back to the identification of the parameters which determine that very percolation threshold. From mathematical studies and numerical simulations the percolation threshold depends primarily on two independent parameters of the network on which the the phenomenon takes place. These two parameters are respectively the connectivity of the network q (the average number of immediate neighbors of a site) and the dimension d of the space in which the network is imbedded.

For example, a checkerboard is a square network of connectivity $q = 4$ and dimension $d = 2$. Its threshold of site percolation is $p_c = 0.59$. On the other hand, for its cubic extension to $d = 3$ dimensions, the connectivity is $q = 6$ and the threshold is smaller at $p_c = 0.31$. For the hypercube at $d = 5$ dimensions the connectivity is $q = 8$ and the threshold decreases to $p_c = 0.20$.

One understands intuitively that increasing the dimension d and or, the connectivity q creates more options of finding ways to connect from one site to another site further away. Such an increase of possible paths leads automatically to a reduction in the value of the percolation threshold.

For a social application of percolation, the corresponding connectivity could be of the order of 15 for a dimension $d = 2$ (the surface of the earth), which would correspond to an unknown network in physics. Thus the corresponding percolation threshold is unknown. However, it turns out that just a few years ago, studying some problems related to disordered magnetic systems a universal formula for the percolation threshold was discovered [10]. For site percolation it is written as

$$p_c = a[(d-1)(q-1)]^{-b} \qquad (14.1)$$

with $a = 1.2868$ and $b = 0.6160$. Plugging in $d = 2$ and $q = 15$ yields $p_c = 0.25$. Our model would thus imply a density of 25% of passive supporters all over the world. Even if the real number of world passive supporters is unknown, such a figure seems totally out of context.

Accordingly, either our approach does not make sense or something is wrong in the estimation of q or d.

14.11
Conclusion and Some Hints for the Future

Taking $q = 15$ seems reasonable but even considering larger values like $q = 30$ or $q = 50$ yields too large thresholds with $p_c = 0.16$ and $p_c = 0.12$. However, keeping $q = 15$ but increasing the dimension to $d = 10$ gives $p_c = 0.06$ which could be close to the real value.

On this basis we are led to consider the hypothesis that, in a human application of percolation, there exist additional dimensions on top of the direct geometrical ones. These additional dimensions could be produced by social paradigms in which individuals may position themselves in a similar way to that in which they do on the earth.

Thus, in addition to geographical space, it is necessary to consider the social space of terrorism. Besides the two dimensions of the terrestrial surface on which we move, there exist "flags" by which a terrorist group exhibits its fight. To each one of these flags, people may identify with more or less support thus producing a dimension in social space.

Typically, for most terrorist groups, the first flag is a claim to territorial independence. This flag constitutes a social dimension independent of those from the geography. Also, as soon as a terrorist group engages in action, it induces a more or less hard repression against itself, which in turn determines a new additional flag, since people may disapprove of the intensity of repression. This gives us four dimensions. Thus, any terrorist social dimension seems to be at least of value four.

With regard to the current new global terrorism, the situation seems qualitatively different. It appears to have clearly succeeded in widening the spectrum of independent flags on which it deploys its claims. On top of the traditional territorial claim, it has a religious dimension, an ethnic dimension, a bipolarizing dimension of partitioning the world, a social dimension, a regional dimension and an historic dimension. As seen above, that brings its social dimension to at least $d = 10$ which lowers its percolation threshold down to 6% which is a realistic estimate for world support.

Such an identification opens new ways to decrease current global terrorism by fighting on the level of social representations to reduce its number of flags. Such a strategy will cost no life and would be quite efficient.

Nevertheless the determination of the means appropriate to such a political fight to neutralize these social representation flags cannot be due to physicists alone. It requires joint research with experts from the corresponding social sciences. Unfortunately, an effective deathless fight against international terrorism requires an interdisciplinary research program which is perhaps a more challenging task than is the classical fight against terrorism.

References

1 SANDLER, T., TSCHIRHART, J. T., CAULEY, J., *American Political Science Review* 77:1 (**1983**), p. 36

2 FRANCART, L., DUFOUR, I., *Strategies and decisions: "The crisis of September 11"*, Economica, Paris **2002**

3 GALAM, S., *Eur. Phys. J B* 26, Rapid Notes (**2002**), p. 269

4 GALAM, S., MAUGER, A., *Physica A* 323 (**2003**), p. 695

5 GALAM, S., *Physica A* 330 (**2003**), p. 139

6 AHMED, E., ELGAZZAR, A. S., HEGAZI, A. S., *Phys. Lett. A* 337 (**2005**), p. 127

7 GALAM, S., GEFEN, Y., SHAPIR, Y., *Math. J. Sociol.* 9 (**1982**), p. 1

8 GALAM, S., *Physica A* 336 (**2004**), p. 49

9 STAUFFER, D., AHARONY, A., *Introduction to Percolation Theory*, Taylor and Francis, London, **1994**

10 GALAM, S., MAUGER, A., *Phys. Rev. E* 53 (**1996**), p. 2177

15
How a "Hit" is Born: The Emergence of Popularity from the Dynamics of Collective Choice

Sitabhra Sinha and Raj Kumar Pan

> **hit** (*noun*) a person or thing that is successful
>
> **popular** (*adj.*), from Latin *popularis*, from *populus*: the people, a people
> 1: of or relating to the general public,
> 2: suitable to the majority: as (**a**) adapted to or indicative of the understanding and taste of the majority, (**b**) suited to the means of the majority: inexpensive,
> 3: frequently encountered or widely accepted,
> 4: commonly liked or approved.
>
> Merriam-Webster Online Dictionary [1].

15.1
Introduction

In a pioneering study of how apparently rational people can behave irrationally as part of a crowd, Charles MacKay [1] gave several illustrations of certain phenomena becoming wildly popular without any discernible reason. In fact, he had focussed specifically on examples where the individuals were behaving clearly contrary to their self-interest or that of society as a whole, as for example, the habit of duelling or the practise of witch-hunting. MacKay termed these episodes "moral epidemics", long before the formal introduction of the concept of social contagion [2] and the use of biological epidemic models to study such phenomena, ascribing their origin to the nature of men to imitate the behavior of their neighbors.

However, such herding behavior [2] is not limited to the examples given in MacKay's book, nor do the outcomes of such behavior need to be so dramatic in their impact as, say, financial market crashes or publicly sanctioned geno-

---

1) http://www.m-w.com/dictionary/
2) MacKay referred to such behavior as "gregarious", in its original sense of "to flock".

Econophysics and Sociophysics: Trends and Perspectives.
Bikas K. Chakrabarti, Anirban Chakraborti, Arnab Chatterjee (Eds.)
Copyright © 2006 WILEY-VCH Verlag GmbH & Co. KGaA, Weinheim
ISBN: 3-527-40670-0

cides. In fact, the sudden emergence of a popular product or idea, that is otherwise indistinguishable in quality from its competitors, is a more common example of the same process at work. These events occur so often that we take such phenomena for granted; however, the question of why certain products or ideas become much more popular than what their intrinsic quality would warrant remains a fascinating and unanswered problem in the social sciences. Watts [3] points this out when he says "... for every *Harry Potter* and *Blair Witch Project* that explodes out of nowhere to capture the public's attention, there are thousands of books, movies, authors and actors who live their entire inconspicuous lives beneath the featureless sea of noise that is modern popular culture."

It may be worth mentioning that such popularity may be of different kinds, one being runaway popularity immediately upon release, and, another being modest initial popularity followed by ever-increasing popularity in subsequent periods. The former is thought to be driven by the advertising blitz preceding the release or launch of the product while the latter has sometimes been explained in terms of self-reinforcing effects, where a slight relative edge in terms of initial popularity results in more consumers being inclined towards the slightly more popular product, thereby increasing its popularity even further and so on, driving up its popularity through positive feedback.

As physicists, we are naturally interested to see whether there are general trends that can be observed in popularity phenomena across a large range of contexts in which they are observed. An allied question is whether this popularity can be related to any of the intrinsic properties of the products or ideas, or whether this is entirely an outcome of a sequence of chance events. The fact that often popular products are seen to be not all that qualitatively different from their competitors, or in some cases, actually somewhat inferior, seems to weigh against the former possibility. However, we would like to see whether the empirically observed popularity distributions also suggest the latter alternative. We also need to see whether pre-release advertising does indeed play a role in creating a high initial burst of popularity.

In this article, we first approach the problem empirically, looking at previous work done on measuring popularity distributions, as well as presenting some of our recent analysis of the popularity phenomena occurring in a variety of different contexts. One remarkable universality we find is that most popularity distributions we examine seem to have long tails, and can be fitted either by a log-normal or a power-law probability distribution function, the exponent of the latter often being quite close to -2. Another interesting feature observed for some distributions is their bimodal character, with the majority of instances occurring at extreme ends of the distribution, while the center of the distribution is remarkably under-represented. Both of these features indicate a significant departure from the Gaussian distribution that may have been naively expected. Next, we survey possible theoretical models for explaining the above features of the empirical distributions. In particular, we discuss how log-normal distributions can arise through several agents making independent decisions in choosing from a range of products with randomly distributed qualities. We also present a model of agent-agent interaction that shows a transition from unimodal to bimodal distribution of the collective choice, when agents are allowed to learn from their previous experience. We conclude with a short discussion on how log-normal and power-law tail distributions can be generated from the same theoretical framework, the former occurring when agents choose independently of other agents (basing their decisions on individual perceptions of quality) and the latter emerging when agent-agent interactions are crucial in deciding the desirability of a product.

15.2
Empirical Popularity Distributions

In studying the popularity distribution of products, the first question one needs to resolve is how to measure popularity. While in some cases this may seem rather obvious, e.g., the number of people buying a particular book, in other cases it may be difficult to identify a unique measure that will satisfy everyone. For example, the popularity of movies can be measured either in terms of an average over critics' opinions published in major periodicals, web-based voting in movie-related online communities, the income generated when a movie is running in theaters, or the cumulative sales and rentals from DVD stores. In most cases, we have let the quality of the available data decide our choice of which popularity measure to use.

An equally important question one needs to answer is the nature of the statistical distribution with which to fit the data. In almost all cases reported below, we observe distributions that deviate significantly from the Gaussian distribution in having extremely long tails. The occurrence of such fat-tailed distributions in so many instances is very exciting, as it indicates that the process of emergence of popular products is more than just N agents independently making *single* binary (i.e., *yes* or *no*) decisions to adopt a particular choice. However, to go beyond this conclusion and to identify the possible process involved, one needs to ascertain accurately the true nature of the distribution. This brings up the question of how to obtain the probability density function (PDF) from the empirical data. The method generally used is to arrange the data into a suitable number of bins to obtain a histogram, which in an appropriate limit will provide the PDF. This works fine when the underlying distribution is Gaussian with sharply decaying tails; however, for long-tailed distributions, it is exactly the extreme ends one is interested in, which have the least representation in the data. As a result, the PDF is extremely noisy at the tails, and hence it is often hard to conclude the nature of the distribution. Often, one can remove some of the noise by using the PDF to generate the cumulative distribution function (CDF), which is essentially the probability that an event is larger than a given size [3]. As larger quantities of data points are now accumulated in each of the bins, the tail becomes smoother in the CDF plot. However, the data-binning process is susceptible to noise, that can significantly change the shape of the distribution, depending on the size and boundary values of each bin. This can lead to serious errors, e.g., wrongly identifying the tail of the distribution to be following a power law. Even if the distribution indeed has a power-law tail, one may obtain a quantitatively erroneous value for the power-law exponent by using graphical methods based on linear least-squares fitting on a double logarithmic scale [4].

A better way to examine the nature of the tail of a distribution is to avoid binning altogether and to switch to a rank-ordered plot of the data, which allows one to focus on the upper tail of the distribution containing the data points of largest magnitude. These plots are often referred to as *Zipf plots*, after the Harvard linguist, G. K. Zipf, who used such rank-frequency plots of the occurrence of the most common words in the English language to establish a scaling relation for written natural languages [5,6]. In this procedure, the data points are ranked or arranged in decreasing order of their magnitude. Note that the CDF can be obtained from the rank-ordered plot by simply exchanging the abscissae and the ordinate, and suitably scaling the axes. Thus, by avoiding binning one can make a better judgement of the nature of the distribution. To quantitatively determine the parameters of the distribution, one of the most robust methods is the maximum likelihood estimation (MLE) [7]. For example, if the underlying distribution $P_c(x)$ has a power-law tail, then the CDF exponent can be obtained from the MLE method by using the formula

$$\alpha = n \sum_{i=1}^{n} \left[\ln \frac{x_i}{x_{min}} \right]^{-1} \tag{15.1}$$

where, x_{min} corresponds to the minimum value of x for which the power-law behavior holds. Similarly, one can obtain maximum likelihood estimates of the parameters for log-normal and other distributions.

It is, of course, obvious that the results from the three different plots, namely, the PDF, the CDF and the rank-ordered, should be related to each other. So, for example, if the CDF of an empirically obtained distribution is found to exhibit a power-law tail which can be expressed as,

$$P_c(x) \sim x^{-\alpha} \tag{15.2}$$

---

[3] The CDF, $P_c(x)$, of a given process is obtained by integrating the corresponding PDF, $P(x)$, i.e., $P_c(x) = \int_x^{\infty} P(x')dx'$.

with the characteristic exponent [4] α, it is easy to show that the PDF and the rank-ordered plots will also exhibit power-law behavior [9]. Moreover, the exponents of the power law seen in these two cases will be related to the characteristic exponent of the CDF, α, as follows: the PDF will follow the relation

$$P(x) \sim x^{-(\alpha+1)} \tag{15.3}$$

while, the rank-ordered plot will exhibit the relation

$$x_k \sim k^{-1/\alpha} \tag{15.4}$$

where x_k denotes the kth ranked data point. The above examples are all given for the case when the underlying distribution has a power-law tail; similar relations can be derived for other underlying distributions, e.g., log-normal.

15.2.1
Examples

In the following paragraphs we have briefly surveyed previous empirical work on popularity distribution, as well as, presenting some of our own recent analysis of popularity data from a broad variety of contexts. In most cases, we have characterized the empirical CDF with a log-normal fit over the entire distribution. However, in those cases where the data is available only for the upper tail of the distribution, such a procedure is not possible. In these cases, we have presented a rank-ordered plot of the data and have tried to fit a power-law characterized by the CDF exponent, α. In this context, we note that most previous observations of popularity distributions had focussed on the upper tail, and fitted a power law onto this. However, we find that the entire distribution is very often a much better fit to the log-normal distribution [10]. We conclude with a brief discussion as of why data that fit log-normal much better have often been reported in the literature to follow a power-law tail.

15.2.1.1 City Size

Possibly the first ever empirical observation of a long-tailed popularity distribution is that of cities, as measured by their population, which was first proposed in 1913 by Auerbach [11]. Later, this basic idea was refined by many others, most notably Zipf [6]. In fact, the last mentioned work has become so well known that often the term *Zipf's law* is used to refer to the idea that city sizes follow a cumulative probability distribution having a power-law tail [12] with exponent $\alpha = 1$. Over the years, several empirical studies have

---

4) This exponent α is often referred to as the *Pareto exponent*, after the Italian economist, V. Pareto, who was the first to report power-law tails for the CDF of income distribution across several European countries [8].

been published in support of the validity of Zipf's law [13]. However, other empirical studies have found significant deviations from the exact form given by Zipf [14]. In a recent review, the combined estimate of the exponent α from 29 different studies is found to be significantly larger than 1 suggesting a less extended tail than is implied by a strict interpretation of Zipf's law [15]. All these studies have focussed on the upper tail (i.e., larger cities) of the distribution. If one also considers the smaller cities, the whole distribution often fits a double-Pareto log-normal, i.e., a distribution which is log-normal in the bulk but has long tails at the two ends [16]. Even the power-law fit of the tail has itself been called into question by a study of the size distribution of US cities over the period 1900–1990 [17]. These results are of special significance to our study, as it shows that the fat-tailed distribution of the popularity of cities need not be a power law but could be explained by other distributions.

15.2.1.2 Company Size

Almost of similar vintage to the city-size literature is the work on company size, measured in terms of sales or employees. Note that both of these are measures of the popularity of the company, the former measuring its popularity among the consumers of its products, while the latter measures its popularity in the labor market. In 1932, Gibrat formulated the *law of proportional growth*, essentially a multiplicative stochastic process for explaining company growth, which predicts that the distribution of firm size would follow a log-normal distribution [18, 19]. While this has indeed been reported from empirical data [20, 21], there have also been reports of a power-law tail [22]. In particular, Axtell [23] has looked at the size of US companies (listed in the US Census Bureau database) in terms of the number of employees which yields a CDF with power-law tail whose exponent $\alpha \sim 1$. When the size was expressed in terms of receipts (in dollars) this also yielded a power-law CDF with $\alpha \sim 0.99$.

15.2.1.3 Scientists and Scientific Papers

The study of popularity in the field of science has a rich and colorful history [24]. One of the earliest such studies is that on the visibility of scientists, as measured by subjective opinions elicited from a sample of the scientific community [25]. The skewed nature of the visibility because of misallocation of credit in the field of science, where an already famous scientist gets more credit than is due compared to less well-known colleagues, has been termed as the *Mathew effect* [26]. This is quite similar to the unequal degree of popularity seen in show-business professions, e.g., among movie actors and singers. A more objective measure for the popularity of scientists is the total number of citations to their papers [27].

15.2 Empirical Popularity Distributions

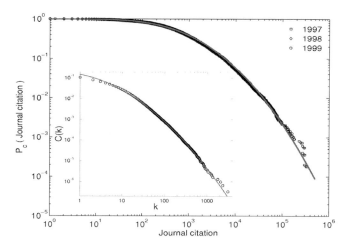

Fig. 15.1 The cumulative distribution function for the total number of citations to a journal in a given year, for all journals ($\sim$ 5500) listed in ISI Journal Citation Report (Science edition) for the years 1997–1999, fitted by a log-normal curve. The inset (from [31]) shows the cumulative probability distribution of citations, $C(k)$, against the number of citations, k, to all papers published from July 1893 through June 2003 in the *Physical Review* journals, fitted by a log-normal curve.

The popularity of individual scientific papers can also be analyzed in terms of citations to them [28]. Price [29] had tried to give a theoretical model based on *cumulative advantage* along with supporting evidence showing that the distribution of citations to papers follows a power-law tail. More recently, in a study [30] analyzing papers in the Institute for Scientific Information (ISI) database, as well as papers published in *Physical Review D*, Redner concluded that the probability distribution of citations follow a power-law tail with an exponent close to -3. However, in a later work looking at all papers published in *Physical Review* journals over the past 110 years, this distribution was found to be fit better by a log-normal [31] (Fig. 15.1, inset).

In addition to the popularity of individual papers measured by the number of their citations, one can also define the popularity of the journals in which these papers are published by considering the total number of citations to all articles published in a journal. In Fig. 15.1, we have plotted the cumulative distribution of the total citations in 1997–99 to all papers ever published in a journal. The data has been fitted with a log-normal distribution; maximum likelihood estimates of parameters for the corresponding distribution are $\mu = 6.37$ and $\sigma = 1.75$.

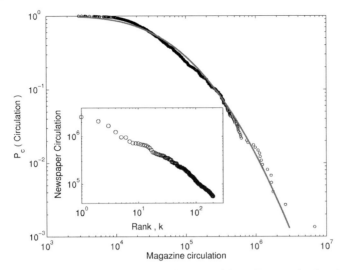

Fig. 15.2 Cumulative distribution function of the 740 most circulated magazines in the UK, fitted by a log-normal curve. The inset shows the rank-ordered plot of the top 200 newspapers in the USA according to circulation.

15.2.1.4 Newspaper and Magazines

The popularity of scientific journals naturally leads us to wonder about the popularity distribution for general interest magazines as well as newspapers. An obvious measure of popularity in this case is the circulation figure. Figure 15.2 shows the CDF of the top 740 magazines according to average net circulation per issue in the United Kingdom [5] in 2005. The figure shows an approximately log-normal fit; the maximum likelihood estimates of parameters for the corresponding distribution are $\mu = 10.79$ and $\sigma = 1.18$. Next, we analyzed the circulation figures for the top 200 newspapers in the USA for the year 2005 according to their circulation [6]. Figure 15.2(inset) shows the corresponding rank-ordered plot with an approximate power-law fit over a decade yielding Zipf's law, which is supported by the maximum likelihood estimate of the exponent for the cumulative probability density function, $\alpha \sim 1.12$.

15.2.1.5 Movies

Movie popularity can be measured in a variety of ways, e.g., by looking at the votes given by users of various movie-related online forums. One of the largest of such forums is the Internet Movie Database (IMDb) [7] that allows registered users to rate films (and television shows) in the range 1–10 (with 1

---

5) http://www.abc.org.uk
6) http://www.accessabc.com/reader/top150.htm
7) www.imdb.com

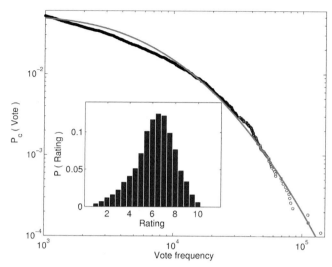

Fig. 15.3 Cumulative probability distribution of the number of votes given by registered users of IMDb to movies and TV series released or shown between the years 2000 and 2004, fit by a log-normal curve. (Inset) The probability distribution of the IMDb rating of a movie, averaged over all the votes received.

corresponding to "awful" and 10 as "excellent"). We looked at the cumulative distribution of all votes received by movies or TV series shown between 2000 and 2004 (Fig. 15.3). The tail of the distribution approximately fits a log-normal distribution, with maximum likelihood estimates of the corresponding parameters, $\mu = 8.60$ and $\sigma = 1.09$. Next, we look at the distribution of average rating given to these items. As the minimum and maximum ratings that an item can receive are 1 and 10, respectively, this distribution is necessarily bounded. The skewed probability distribution of the average rating resulting from our analysis is shown in Fig. 15.3 (inset).

The measures used above have many drawbacks as indicators of movie popularity, particularly so when they are aggregated to produce average values. For example, users may judge different movies according to very different information, with so-called classic movies faring very differently from recently released movies about which very little information is available. Also, it does not cost anything to vote for a movie, so that the vital element of competition among movies to become popular is missing in this measure. In contrast, looking at the gross income distribution of movies that are being shown at theaters gives a sense of the relative popularity of movies that have roughly equal amounts of information available about them. Also, this kind of "voting with one's wallet" is a truer indicator of the viewer's movie preferences. The freely available datasets about weekly earnings of most movies released

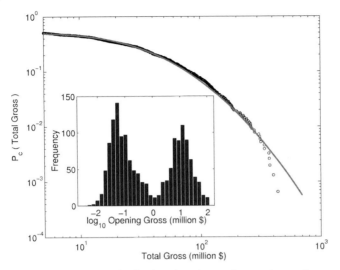

Fig. 15.4 Cumulative distribution of total gross income for movies released across theaters in the USA during 2000–2004, fitted by a log-normal curve. The inset shows the distribution of movie income according to the opening weekend gross.

across theaters in the USA makes this a practical exercise. For our study we have concentrated on data from *The Movie Times* [8] and *The Numbers* [9] websites for the period 2000–2004. Although total gross may be a better measure of movie popularity, the opening gross is often thought to signal the success of a particular movie. This is supported by the observation that about 65–70 % of all movies earn their maximum box-office revenue in the first week of release [32]. The rank-ordered distribution for the opening, as well as the total gross, show an approximate power law with an exponent $1/\alpha \sim -1/2$ in the region where the top grossing movies are located [33]. However, when the data are aggregated together we find that the distribution (Fig. 15.4) is better fitted by a log-normal [10] (similar to the observation of Redner vis-a-vis citations) [34]. The maximum likelihood estimates of the log-normal distribution parameters yield $\mu = 3.49$ and $\sigma = 1.00$. Further, we observe that the total gross distribution is just a scaled version of the opening distribution, which essentially implies that the popularity distribution of movies is decided at the opening itself. An additional feature of interest is that both the opening and the total gross distributions are bimodal (Fig. 15.4, inset), implying that most movies either do very well or very badly at the box office.

8) http://www.the-movie-times.com
9) http://www.the-numbers.com/
10) We have also verified this for the income distribution of Indian movies.

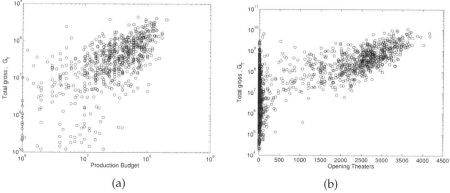

Fig. 15.5 (a) The total gross (G_T, in dollars) of a movie vs its production budget (in dollars). (b) The total gross (G_T, in dollars) of a movie vs the number of theaters at which it is released on the opening weekend.

We have tried to see whether the popularity of individual movies correlates with its production quality (as measured by production budget). Figure 15.5(a) shows a plot of the total gross vs production budget for a large number of movies released between 2000 and 2004 whose budget exceeded 10^6 $. As is clear from the figure, although in general, movies with higher production budget tend to earn more, there is no significant correlation (the correlation coefficient is only 0.62). One can also argue that the determination of success of a movie on its opening implies the key role of pre-release advertising. Although the data for an advertising budget is often unavailable, we can use as a surrogate, the data about the number of theaters at which a movie is initially released, since the advertising cost will scale with this quantity. As is obvious from Fig. 15.5(b), the correlation here is worse, indicating that advertising has often a very little role to play in deciding the success or otherwise of a movie in becoming popular. In this context, one may note that De Vany and Walls [32] have looked at the distribution of movie earnings and profit as a function of a variety of variables, such as, genre, ratings, presence of stars, etc., and have not found any of these to be significant determinants [35].

To make a quantitative analysis of the relative performance of movies, we have defined the persistence time τ of a movie as the time (measured in number of weekends) up to which it is being shown at theaters. We observe that most movies run for up to about 10 weekends, after which there is a steep drop in their survival probability. The empirical data seem to fit a Weibull distribution quite well.

15.2.1.6 Websites and Blogs

Zipf's law for the distribution of requests for pages from the web was first reported by Glassman [36]. By tracing web accesses from DEC's Palo Alto facilities, 10^5 HTTP requests were gathered and the rank-ordered distribution of pages was shown to have an exponent ~ -1. This was supported by a popular article [37] which observed Zipf's law when analyzing the incoming page requests to a single site (www.sun.com). However, subsequent investigation of the page-request distribution seen by web proxy caches using traces from a variety of sources, found the rank-order exponent to vary between 0.64 and 0.83 [38]. The deviation from the earlier result (showing an exact Zipf's law) was ascribed to the fact that web accesses at a web server and those at a web proxy are different, because the former includes requests from all users on the Internet while the latter includes only those users from a fixed group. Access statistics for web pages have also been analyzed by Adamic and Huberman from the access logs of about 60 000 individual usage logs from America Online [39]. The resulting cumulative distribution of website popularity, according to the number of unique visits to a website by users, showed a power-law fit with α very close to 1.

Another obvious measure of webpage popularity is the number of links to it from another webpage. Distribution of incoming links to a webpage (i.e., URLs pointing to a certain HTML document) for the nd.edu domain, have been shown to obey a power law with exponent $\simeq -2.1$ [40]. This power law was quantitatively confirmed (i.e., the same exponent value of 2.1 was reported) over a much larger data set involving a web-crawl on the entire WWW with 2×10^8 webpages and 1.5×10^9 links [41]. While the power-law distribution of popularity of websites according to the number of incoming links has been well established as a power law, among webpages of the same type (e.g., the set of US newspaper homepages) the bulk of the distribution of incoming links deviates strongly from a power law, exhibiting a roughly log-normal shape [42].

The finding that the micro-structure of popularity within a group is closer to a log-normal distribution has created some controversy among researchers involved in measuring the popularity distribution of blogs [11] which have, over the past few years, picked up a large following all over the web. Shirky [43] had arranged 433 weblogs in rank order according to number of incoming links from other blogs and had claimed an approximate power law distribution. In contrast to this, Drezner and Farrell [44] conducted a study of

---

11) A blog or weblog has been defined as a web page with minimal to no external editing, providing on-line commentary, periodically updated and presented in reverse chronological order, with hyperlinks to other online sources [44]. Blogs can function as personal diaries, technical advice columns, sports chat, celebrity gossip, political commentary, or all of the above.

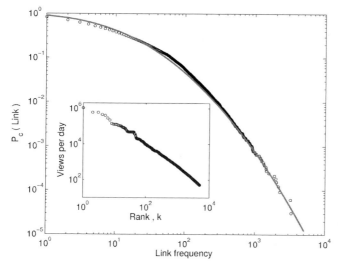

Fig. 15.6 Cumulative distribution function for blog popularity measured by the number of incoming links a blog receives from other weblogs listed in the TTLB Blogosphere ecosystem within the past 7–10 days. The curve is the best log-normal fit to the data. The inset shows the rank-ordered plot of blog popularity according to the number of visits to a blog in a single day, in the TTLB ecosystem.

the incoming link distribution of over 4000 blogs dealing almost exclusively with political topics, and found the distribution to be much better fitted by a log-normal than a power law. Other studies have made contradictory claims about whether the popularity of blogs is better fitted by a log-normal or a power-law tailed distribution [45, 46].

We have also analyzed the popularity distribution of blogs according to citations in other blogs, using three different blogosphere ecologies, i.e., directories of blog listings. Such ecologies scan all blogs registered with them for (i) the number of links they receive from other blogs in their list, as well as (ii) the number of visits to that blog. These two measures of popularity complement each other, as the former looks at who is getting the most links from other bloggers, while the latter shows which blogs are actually receiving the most readers. The most extensive data that we have analyzed comes from the TTLB Blogosphere ecosystem [12] that lists 52048 blogs. In Fig. 15.6 we show the CDF for the popularity of blogs from this ecology, measured from the number of links to that blog seen in the "front page" of other member blogs within the past 7–10 days. This can be considered a rolling snapshot of the relative popularity of different blogs at a particular instant of time. For comparison, we also looked at data from two other ecologies, namely, the Tech-

12) http://truthlaidbear.com/

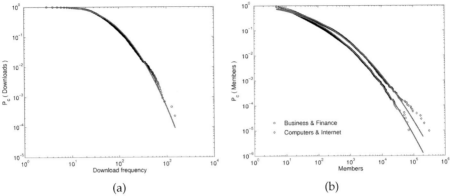

Fig. 15.7 (a) Cumulative distribution function for the number of downloads of different files in 1 month during early 2006 from the MATLAB file exchange site. (b) Cumulative distribution function for the number of members in different Yahoo groups under the Business & Finance (squares) and Computers & Internet (diamonds) categories. Groups with less then five members are not considered. For both figures, the curves are the best log-normal fits to the data.

norati [13] and the Blogstreet [14] ecosystems, and observed qualitatively almost identical behavior. The CDF (Fig. 15.6) shows an approximately log-normal fit; maximum likelihood estimates of parameters for the corresponding distribution are $\mu = 1.98$ and $\sigma = 1.51$. We have also analyzed the popularity of blogs listed in the TTLB ecosystem according to traffic, i.e., views per day (Fig. 15.6, inset), which shows a power law over almost two decades for the rank-ordered plot. The maximum likelihood estimate of the corresponding exponent for the cumulative probability density yields $\alpha \sim 0.67$.

15.2.1.7 File Downloads

Another web-related measure of popularity is that of file downloads. There are numerous file repositories in the net which allow visitors to download files either freely or for a fee. We focussed on files stored in the MATLAB Central File Exchange [15], which are computer programs. We looked at the number of downloads of all files over a period of one month during early 2006. The CDF (Fig. 15.7(a)) shows an approximately log-normal fit; maximum likelihood estimates of parameters for the corresponding distribution are $\mu = 3.76$ and $\sigma = 0.89$.

13) http://www.technorati.com/
14) http://www.blogstreet.com/
15) http://www.mathworks.com/matlabcentral/fileexchange/

15.2.1.8 Groups

A fertile area for observing the distribution of popularity is in the arena of social groups. While the membership of clubs, gangs, co-operatives, secret societies, etc., are difficult to come by, with the rising popularity of the internet it is easy to obtain data for online communities such as those in Yahoo [16] or Orkut [17]. By observing the memberships of each of the groups in the community that a user can join, one can have a quantitative measure of the popularity of these groups. An analysis of the Yahoo groups resulted in a fat-tailed cumulative distribution of the group size [47]. Even though the distribution has a significant curvature over the entire range, the tail fits a power law for slightly more than a decade, with exponent $\alpha = 1.8$.

We have recently carried out a smaller-scale study of the popularity of Yahoo groups [18]. As in the earlier study, the popularity of the groups in each category has been estimated by the number of group members. Fig. 15.7(b) looks at the cumulative distributions of the group size for two categories, namely Business & Finance and Computer & Internet, which comprize 182 086 and 172 731 groups, respectively. However, unlike the power law reported in the earlier study, we found both the distributions to approximately fit a log-normal form, with the parameters for the corresponding distributions being $\mu = 2.80$, $\sigma = 2.00$ and $\mu = 3.10$, $\sigma = 2.05$, respectively.

One can also look at the popularity of individual members of an online group, which has been analyzed for a different type of community in the web: that formed by the users of the *Pretty-Good-Privacy* (PGP) encryption algorithm. To ensure that identities are not forged, users certify one another by "signing" the other person's public encryption key. In this manner, a directed network (the "web of trust") is created where the vertices are users and links are the user certifications. A measure of popularity in this case will be the number of certifications received by a user from other users, i.e., the number of incoming links for a vertex in the "web of trust". The in-degree cumulative distribution has been reported to be a power law with the exponent $\alpha \simeq 1.8$ [48].

15.2.1.9 Elections

Political elections are processes that can be viewed as contests of popularity between individual candidates, as well as parties. The fraction of votes received by candidates is a direct measure of their popularity, regardless of whether the electoral system uses a majority voting rule (where the candidate with the largest number of votes wins) or a proportional representation

---

[16] http://groups.yahoo.com
[17] http://www.orkut.com
[18] The entire Yahoo groups community is divided into 16 categories, each of which are then further divided into subcategories.

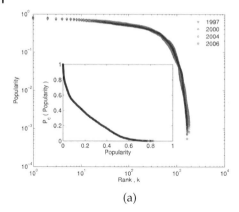 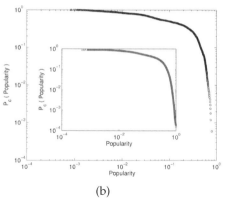

Fig. 15.8 Canadian elections: (a) The rank-ordered plot of candidate popularity measured by the fraction of votes received by him or her, for four successive general elections. The inset shows the cumulative frequency distribution function for this popularity measure. Note the region of linear decay in the middle of the curve. (b) Cumulative probability distribution function for the fraction of votes received by a candidate for all constituencies in the 2000 general election. The inset shows the cumulative distribution function of the vote fraction for candidates for all polling booths at each constituency in the above election. Note that a constituency can have hundreds of polling booths.

(parties getting representation at the legislative house proportional to their fraction of the popular vote). Such studies have been carried out for, e.g., the 1998 Brazilian general elections [49], which looked at the fraction of votes received by candidates for the positions of state deputies. The resulting frequency distribution was fitted by a power law with exponent very close to -1. The cumulative distribution, however, revealed that about 90% of the candidates' votes followed a log-normal distribution, with a large dispersion that resulted in the apparent power law.

We have carried out an analysis of the distribution of votes for a number of general elections in Canada and India. The data about votes for individual candidates in Canada was obtained from the website Elections Canada On-line [19] for the general elections held in 1997, 2000, 2004 and 2006. The total number of candidates in each election varied between 1600 and 1800, there were over ~ 300 electoral constituencies and the total number of votes cast varied around 13 million. Each constituency was divided into hundreds of polling stations, thereby allowing us to obtain a micro-level picture of the popularity of the candidates at a particular constituency across the different polling stations. Figure 15.8(a) shows the results of our analysis, indicating an exponential decay of the tail of the popularity distribution for all the elections being considered. The results do not change even if we consider the number of votes, rather than the vote fraction. Figure 15.8(b) shows that the distribution of popularity across polling stations has almost an identical distribution

19) http://www.elections.ca/

15.2 Empirical Popularity Distributions

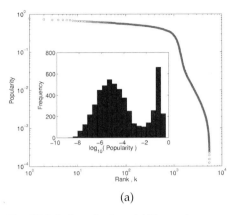
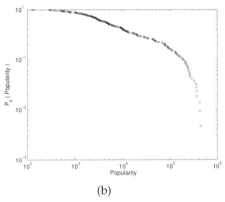

(a)   (b)

Fig. 15.9 Indian election: (a) The rank-ordered plot of candidate popularity measured by the fraction of votes received, for the 2004 Lok Sabha election. The inset shows the frequency distribution of the vote fraction, clearly indicating a bimodal nature with candidates receiving either most of the votes cast or very few. (b) Cumulative probability distribution function of party popularity for the 2004 election, measured by the fraction of votes received by candidates from that party, over all the constituencies it contested.

to that seen over the larger scale of electoral constituencies. Note that we did not observe the popularity of parties for Canada, as the total number of parties was only about 10.

Next, we looked at the corresponding data for the 2004 general elections in India obtained from the website of the Election Commission of India [20]. The total number of candidates is 5435, about half of whom belonged to 230 registered parties, who contested from a total of 543 electoral constituencies, while the total number of votes cast was about 400 million. Figure 15.9(a) shows that the rank-ordered popularity (measured by the vote fraction) distribution for candidates in an Indian general election is qualitatively similar to that of Canada, except for the presence of a kink indicative of the bimodal nature of the distribution. This implies that candidates either receive most of the votes cast by electors in that constituency or very few votes. It may be due to the very large number of independent candidates (i.e., without affiliation to any recognized party) in Indian elections compared to Canada. This is supported by our analysis of the popularity of recognized political parties (Fig. 15.9(b)) that shows an exponential decay at the tail. Note that the popularity of a party is measured by the total votes received by a party divided by the number of constituencies in which it contested. This is the same (up to a scaling constant) as the percentage of votes received by candidates belonging to a party, averaged over all the constituencies in which the party had fielded candidates.

20) http://www.eci.gov.in/

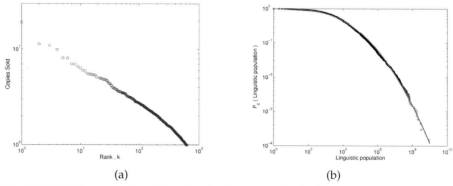

Fig. 15.10 (a) The rank-ordered plot of best-selling books (that sold 2 million copies or more) according to the number of copies sold in the USA between 1895 and 1965. Adapted from [7], data provided by M. E. J. Newman. (b) Cumulative distribution function for the size of the population of first-language speakers for over 6650 languages. The data was obtained from *Ethnologue*. The curve indicates the best log-normal fit to the data.

15.2.1.10 Books

An obvious popularity distribution based on product sales is that of books, especially in view of the record-breaking sales in recent times of the *Harry Potter* series of books. However, the lack of freely available data about exact sales figures has so far prevented detailed analysis of book popularity. It was reported in a recent paper [50], that the cumulative distribution of book sales from the online bookseller Amazon [21] has a power-law tail with $\alpha \sim 2$. However, one should note that Amazon does not reveal exact sales figures, but rather only the rank according to sales; therefore, this distribution was actually based on an heuristic relation between rank and sales proposed by Rosenthal [51]. Needless to say, this is at best a very rough guide to the exact sales figures (e.g., although the sale of *Harry Potter and the Half-Blood Prince* fluctuated a lot during the few weeks following its publication, it remained steady as the top ranked book in Amazon) and is likely to yield a misleading distribution of sales. A more reliable dataset, if somewhat old, has been compiled by Hackett [52] for the total number of copies sold in the USA of the top 633 best-selling books between 1895 and 1965. Newman [7] has reported the maximum likelihood estimate for the exponent of the power-law fit to this data as $\alpha \sim 2.51$. Figure 15.10(a) shows the rank-ordered plot of this data, indicating an approximate power-law fit for slightly more than a decade, with an exponent of -0.4.

21) www.amazon.com

15.2.1.11 Language

Figure 15.10(b) shows the cumulative distribution of the first-language speaker population for different languages around the world. The data has been obtained from *Ethnologue* [22] which provides the number of first-language speakers (over all countries in the world) wherever possible. Out of a total of 7299 languages listed in its 15th edition, we have considered above 6650 languages for which information about the number of speakers is available. The figure shows a long tail with an approximately log-normal fit; the maximum likelihood estimates of parameters for the corresponding distribution are $\mu = 8.78$ and $\sigma = 3.17$. Note that this kind of popularity distribution is different from the others we have discussed so far as the speakers are not really free to choose their first language; rather this is connected to the population growth rate of a particular linguistic community. A similar kind of popularity distribution is that for family names, which has been analyzed by Miyazima et al. [53] for Japanese family names and Newman [7] for American family names, both reporting cumulative distribution functions with power-law tails having α close to 1. However, for Korean family names [54] the distribution was reported to be exponentially decaying.

Other Popularity Distributions Unlike the distribution of family names discussed above, the frequency of occurrence of given names (or first names) are indeed subject to waves of popularity, with certain names appearing to be very common at a particular period. A recent study [55] has looked at the distribution of most popular given names in England and Wales over the past millennium, and has claimed a long-tailed distribution for the same.

Another popularity distribution is that of tourist destinations, as measured by the number of tourist arrivals over a time period. A study [56] that has ranked 89 countries, focussing on the period 1980–1990, has found evidence for a log-normal distribution as the best fit to the data.

The occurrence of superstars (i.e., extremely successful performers) in popular music has led to a relatively large amount of literature by economists on the occurrence of popularity [57–60]. Chung and Cox have used the number of gold-records by performers as the measure of their artistic success, and found the tail of this popularity distribution to approximately follow a power law [61]. Another study [62] looked at the longevity of music bands in the list of Top 75 best-selling recordings, and observed a stretched exponential distribution [23]. However, a more recent study [63] has shown the survival probability of a music recording on the *Billboard* Hot 100 chart to be better fitted by the log-logistic distribution.

22) http://www.ethnologue.com/
23) While the term "stretched exponential distribution" is quite common in the physics literature, we observe that in other scientific fields it is more commonly referred to as a Weibull distribution.

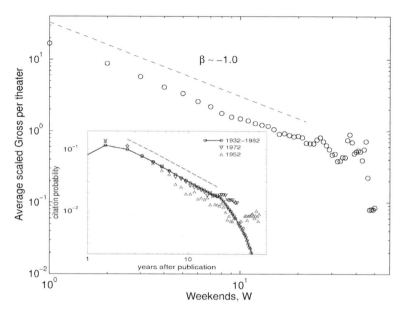

Fig. 15.11 Weekend gross per theater for a movie (scaled by the average weekend gross over its theatrical lifespan), after it has run for W weekends, averaged over the number of movies that ran for that long. The initial decline follows a power-law with exponent $\beta \simeq -1$ (the fit is shown by the broken line). The inset (from [65]) shows the probability that a paper will be cited t years after publication in a *Physical Review* journal, in the years 1952 and 1972, as well as over the period 1932–1982. Over the range of $2-20$ years the integrated data is consistent with a power law decay having an exponent -0.94 (broken line).

15.2.2
Time-evolution of Popularity

Here we look briefly at how popularity evolves over time. For movies, we look at the gross income per theater over time (Fig. 15.11). This is a better measure of the dynamics of movie popularity than the time-evolution of the weekly overall gross income, because a movie that is being shown in a large number of theaters has a bigger income simply on account of higher accessibility for the potential audience. Unlike the overall gross that decays exponentially with time, the gross per theater shows a power-law decay in time with exponent $\beta \simeq -1$ [64]. This has a striking similarity with the time-evolution of popularity for scientific papers in terms of citations. It has been reported that the citation probability to a paper published t years ago, decays approximately as $1/t$ [65] (Fig. 15.11 (inset)). Note that Price [29] had also noted a similar behavior for the decay of citations to papers listed in the Science Citation Index. In a very different context, namely, the decay in the popularity of a website (as measured by the rate of download of papers from the site) over time t has also been reported to follow an inverse power-law, but with a different exponent [66].

15.2.3 Discussion

The selection of (mostly) long-tailed empirical popularity distributions presented above underlines the following broad features of such distributions:

1. The entire distribution seem to be fitted by a log-normal curve (in the few cases where the entire distribution is not available, the upper tail seems to fit a power law with characteristic exponent α which is often close to 1, corresponding to the exact form of Zipf's law).

2. In some cases the distribution shows a bimodal character, with most of the instances occurring at the two ends of the distribution.

3. The decay of popularity in some cases seem to show a simple power law decay, declining inversely with time elapsed since release.

4. The persistence time at high levels of popularity shows a Weibull distribution in many instances.

The first of these features may come somewhat as a surprise, because for many popularity distributions, power-law tails have been reported with various exponents, often significantly different from 1. However, we observe that very often log-normal distributions have been mistakenly identified as having power-law tails. In fact this is a very common error, especially if the variance of the log-normal distribution is sufficiently large. To see this, note that the log-normal distribution

$$P(x) = \frac{1}{x\sigma\sqrt{2\pi}} e^{-(\ln x - \mu)^2 / 2\sigma^2} \tag{15.5}$$

can be written as (on taking logarithm of both sides)

$$\ln P(x) = -(\ln x)^2 / 2\sigma^2 + (\mu/\sigma^2 - 1) - \ln \sqrt{2\pi}\sigma - \mu^2 / 2\sigma^2 \tag{15.6}$$

which is a quadratic curve in a doubly logarithmic plot. However, a sufficiently small part of the curve will appear as a straight line, with the slope depending on which segment of the curve one is focussing attention [7,67]. This is the origin of most of the power-law tails with exponent $\alpha \neq 1$ that have been reported in the literature on popularity distributions.

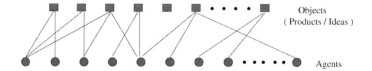

Fig. 15.12 A schematic diagram of the emergence of popularity as a relation between agents and objects (products or ideas).

15.3 Models of Popularity Distribution

From the perspective of physics, popularity can be viewed as an emergent outcome of the collective decision process in a society of individual agents exercising their free will (as reflected in their individual preferences) to choose between alternative products or ideas (Fig. 15.12). In a system without authoritarian control, agents differ in their personal preferences which are determined by the information available to the agent about the possible alternatives. However, in any real-life scenario with uneven access to information, a seemingly well-informed agent may influence the choice of several other agents [68]. Thus, the emergence of a popular product is a result of the self-organized coordination of choices made by heterogeneous entities.

The simplest model of collective choice is one where the agents decide independently of each other and select alternatives at random with a one-step decision process. It is easy to see that the possible alternatives will not be significantly different from each other in terms of popularity. In particular, the popularity distribution arising from such a process will not have long tails. There are two possible alternative modifications of this simple model that will allow it to generate distributions similar to those seen empirically. The first option is to allow interactions between agents where the choice of one agent can influence that of another. While this is often true in real life, we also observe long-tailed distributions much before the interaction among agents (and the resulting dissemination of information) has had a chance to influence popularity. For example, the long-tailed distribution of movie popularity, in terms of gross earning is seen at the opening weekend itself, long before potential movie viewers have had a chance to be influenced by other moviegoers. The second option for generating realistic popularity distribution gets around this problem: here we replace the single-step decision process by one comprizing multiple sub-decisions (as there may be many factors involved in making a particular decision), each of which contributes to the overall decision to purchase a particular product. Therefore, the probability of any particular entity achieving a particular degree of popularity can be expressed as the product of probabilities of each of the underlying factors satisfying the required condition to make an agent opt for that entity. As is easily seen, the resultant distribution arising from such a multiplicative stochastic process has a log-normal form, agreeing with many of the empirically observed distributions [24].

While the bulk of the popularity distributions, showing a log-normal nature, can therefore be plausibly explained as the product of the multiplicative stochastic structure underlying even apparently simple decision processes, this would still leave unanswered the reason for the wide occurrence of Zipf's law in other instances. We now turn to the first option for extending the simple model outlined above, i.e., investigating the influence of an agent's choice behavior on other agents. It turns out that there have been many proposed mechanisms to explain the ubiquity of power-law tailed distributions employing interactions. However, from the point of view of the present paper, the most relevant (and general) model seems to be the Yule process [69], as modified by Simon [70]. This is essentially a cumulative advantage process by which the relatively more popular entities become even more popular by virtue of being more well known.

The Yule–Simon process can be described as follows. Suppose initially there are n agents, each of whom are free to choose one of a number of products. Subsequently, the number of agents is augmented by unity at each time step. At any point in time, when the total number of agents is m, the number of distinct products, each of which have been chosen by k agents is denoted by $f(k,m)$. Then, given that, (i) there is a constant probability, γ, that an agent chooses a completely new product (i.e., one that has not been chosen before by any of the agents) and (ii) the probability of choosing a product that has already been chosen by k agents is proportional to $kf(k,n)$, one obtains an asymptotic popularity distribution that has a power-law tail [25] with exponent $\alpha = \frac{1}{1-\gamma}$. If the appearance of a new product is relatively infrequent, i.e., γ is extremely small, then the exponent $\alpha \simeq 1$ (i.e., Zipf's law).

Another feature of popularity distributions that has been mentioned earlier is that, in some cases, they appear to have a bimodal nature. We now present a simple agent-based model [73] that shows how bimodal and unimodal distributions of popularity can arise very simply through agents interacting with each other, and reacting to information about what the majority are choosing in the previous time step.

24) One can argue that the probability distribution of collective choice may also reflect the distribution of quality amongst various competing entities; however, in this case the popularity distribution would be essentially identical to the quality distribution, which *a priori* can follow any arbitrary distribution. The universality of long-tailed popularity distributions and the seeming absence of any correlation between popularity and quality (when it can be measured in any well-defined manner) would argue against this hypothesis.

25) Note that, the models of Price [29], Barabasi-Albert [71] and Redner [72] are all special cases of this general mechanism.

15.3.1
A Model for Bimodal Distribution of Collective Choice

We have already discussed the simplest model of collective choice in which individual agents make completely independent decisions. For binary choice (i.e., each agent can only choose between two options) the emergence of collective choice is equivalent to a one-dimensional random walk with the number of steps equal to the number of agents. Therefore, the outcome will be normally distributed, with the most probable outcome being an equal number of agents choosing each alternative. While such unimodal distributions of popularity are indeed observed in some situations, as mentioned earlier in this article, many real-life examples show the occurrence of bimodal distributions indicative of highly polarized choice behavior among agents, resulting in the emergence of a highly popular product. This polarization suggests that agents not only opt for certain choices based on their personal preferences, but are also influenced by other agents in their social neighborhood. Also, the personal preferences may themselves change over time as a result of the outcome of previous choices, e.g., whether or not their choice agreed with that of the majority. This latter effect is an example of the global feedback process that we think is crucial in the occurrence of bimodal behavior.

We now present a general model of collective decision that shows how polarization in the presence of individual choice volatility can be achieved with an adaptation and learning dynamics of the personal preference. In this model, the choice of individual agents are not only affected by those of their neighbors, but, in addition, their preference is modified by their previous choice as well as information about how successful their previous choice behavior was in coordinating with that of the majority. Here it is assumed that information about the intrinsic quality of the alternative products is inaccessible to the agent, who takes the cue from what the majority is choosing, to decide which one is the "better choice". Examples of such limited global information about the majority's preference available to an agent are the results of consumer surveys and publicity campaigns disseminated through the mass media.

The simplest binary choice version of our model is defined as follows. Consider a population of N agents, each of whom can be in one of two choice states $S = \pm 1$ (e.g., to buy or not to buy a certain product, to vote Party A or Party B, etc.). In addition, each agent has an individual preference, θ, that is chosen from a uniform random distribution initially. At each time step, every agent considers the average choice of its neighbors at the previous instant, and if this exceeds its personal preference, makes the same choice; otherwise, it makes the opposite choice. Then, for the ith agent, the choice dynamics is described by:

$$S_i^{t+1} = \text{sign}\left(\sum_{j \in \mathcal{N}} J_{ij} S_j^t - \theta_i^t\right) \qquad (15.7)$$

where sign $(x) = +1$, if $x > 0$, and $= -1$, otherwise. The coupling coefficient among agents, J_{ij}, is assumed to be a constant ($= 1$) for simplicity and normalized by z ($= |\mathcal{N}|$), the number of neighbors. In a lattice, $\mathcal{N}$ is the set of spatial nearest neighbors and z is the coordination number, while in the mean-field approximation, $\mathcal{N}$ is the set of all other agents in the system and $z = N - 1$.

The individual preference, θ, evolves over time as:

$$\begin{aligned}\theta_i^{t+1} &= \theta_i^t + \mu S_i^{t+1} + \lambda S_i^t &&\text{if } S_i^t \neq \text{sign}(M^t) \\ &= \theta_i^t + \mu S_i^{t+1} &&\text{otherwise}\end{aligned} \qquad (15.8)$$

where $M^t = (1/N) \sum_j S_j^t$ is the collective decision of the entire community at time t. Adjustment to previous choice is governed by the adaptation rate μ in the second term on the right-hand side of Eq. (15.8), while the third term, governed by the learning rate λ, represents the correction when the individual choice does not agree with that of the majority at the previous instant. The desirability of a particular choice is assumed to be related to the fraction of the community choosing it; hence, at any given time, every agent is trying to coordinate its choice with that of the majority. Note that, for $\mu = 0, \lambda = 0$, the model reduces to the well-known zero-temperature, random-field Ising model (RFIM).

Random-neighbor and mean-field model. For mathematical convenience, we choose the z neighbors of an agent at random from the $N - 1$ other agents in the system. We also assume this randomness to be "annealed", i.e., the next time the same agent interacts with z other agents, they are chosen at random anew. Thus, by ignoring spatial correlations, a mean-field approximation is achieved.

For $z = N - 1$, i.e., when every agent has the information about the entire system, it is easy to see that, in the absence of learning ($\lambda = 0$), the collective decision M follows the evolution equation rule: $M^{t+1} = \text{sign}[(1 - \mu)M^t - \mu \sum_{\tau=1}^{t-1} M^\tau]$. For $0 < \mu < 1$, the system alternates between the ordered states $M = \pm 1$ with a period $\sim 4/\mu$. The residence time at any one state ($\sim 2/\mu$) diverges with decreasing μ, and for $\mu = 0$, the system remains fixed at one of the ordered states corresponding to $M = \pm 1$, as expected from RFIM results. At $\mu = 1$, the system remains in the disordered state, so that $M = 0$. Therefore, we see a transition from a bimodal distribution of the collective decision, M, with peaks at nonzero values, to a unimodal distribution of M centered about 0, at $\mu_c = 1$. When we introduce learning, so that $\lambda > 0$, the agents try to coordinate with each other and at the limit $\lambda \to \infty$ it is easy to see that $S_i = \text{sign}(M)$ for all i, so that all the agents make an identical choice. In the simulations, we note that the bimodal distribution is recovered for $\mu = 1$ when $\lambda \geq 1$.

For finite values of z, the population is no longer "well mixed" and the mean-field approximation becomes less accurate the lower z is. For $z \ll N$, the critical value of μ at which the transition from a bimodal to a unimodal distribution occurs in the absence of learning, $\mu_c < 1$. For example, $\mu_c = 0$ for $z = 2$, while it is 3/4 for $z = 4$. As z increases μ_c quickly converges to the mean-field value, $\mu_c = 1$. On introducing learning ($\lambda > 0$) for $\mu > \mu_c$, we again notice a transition to an ordered state, with more and more agents coordinating their choice.

Lattice. To implement the model when the neighbors are spatially related, we consider d-dimensional lattices ($d = 1, 2, 3$) and study the dynamics numerically. We report results obtained in systems with absorbing boundary conditions; using periodic boundary conditions leads to minor changes but the overall qualitative results remain the same. It is worth noting that the adaptation term disrupts the ordering expected from results of the RFIM for $d = 3$, so that for any nonzero μ the system is in a disordered state when $\lambda = 0$.

In the absence of learning ($\lambda = 0$), starting from a initial random distribution of choices and personal preferences, we observe only very small clusters of similar choice behavior (Fig. 15.13(a)) and the average choice M fluctuates around 0. In other words, at any given time an equal number (on average) of agents have opposite choice preferences. Introduction of learning in the model ($\lambda > 0$) gives rise to significant clustering as well as a nonzero value for the collective choice M. We find that the probability distribution of M (Fig. 15.14(b)) evolves from a single peak at 0, to a bimodal distribution as λ increases from 0. This is similar to a second-order phase transition in systems undergoing qualitative changes at a critical threshold. The collective decision M switches periodically from a positive value to a negative value having an average residence time which diverges with λ and with N. For $\mu > \lambda > 0$, large clusters of agents with identical choice are observed to form and dissipate throughout the lattice (Fig. 15.13(b)). After sufficiently long times, we observe the emergence of structured patterns having the symmetry of the underlying lattice, with the behavior of agents which belong to a particular structure being highly correlated. Note that these patterns are dynamic, being essentially concentric waves that emerge at the center and travel to the boundary of the region, which continually expands until it meets another such pattern. Where two patterns meet their progress is arrested and their common boundary resembles a dislocation line. In the asymptotic limit, several such patterns fill up the entire system. These patterns indicate the growth of clusters with strictly correlated choice behavior. The central site in these clusters act as the "opinion leader" for the entire group. This can be seen as analogous to the

15.3 Models of Popularity Distribution

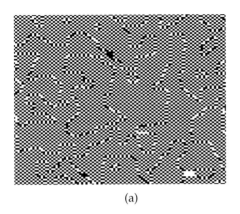
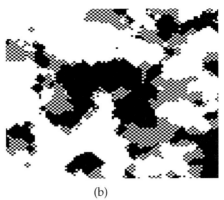

Fig. 15.13 (a) The spatial pattern of choice (S) in the absence of learning ($\lambda = 0$) in a two-dimensional square lattice of 1000×1000 agents after 500 iterations starting from a random configuration. The figure is a magnified view of the the central 100×100 region showing the absence of long-range correlation among the agents. (b) The spatial pattern of choice (S) with learning ($\lambda = 0.05$) in the same system, with a majority of agents now in the choice state $S = +1$. The magnified view of the central 100×100 region shows coarsening of regions having agents aligned in the same choice state.

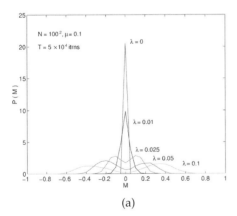
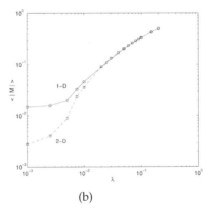

Fig. 15.14 (a) The probability distribution of the collective decision M in a two-dimensional square lattice of 100×100 agents. The adaptation rate $\mu = 0.1$, and the learning rate λ is increased from 0 to 0.1 to show the transition from unimodal to bimodal behavior. The system was simulated for 5×10^4 iterations to obtain the distribution. (b) The order parameter $< |M| >$ for one and two-dimensional lattices. The adaptation rate is $\mu = 0.1$, while λ is increased gradually to show the transition to an ordered state. Note that for higher values of μ the two curves are virtually identical. There is very little system size-dependence of the curves.

formation of "cultural groups" with shared preferences [74]. It is of interest to note that distributing λ from a random distribution among the agents disrupts the symmetry of the patterns, but we still observe patterns of correlated choice behavior. It is the global feedback ($\lambda \neq 0$) which determines the formation of large connected regions of agents having similar choice behavior. This is reflected in the order parameter, $< |M| >$, where $<>$ indicates time averaging. Figure 15.14(b) shows the order parameter increasing with λ in both one and two-dimensional lattices, signifying the transition from a disordered state to an ordered state, where neighboring agents have coordinated their choices.

Our model seems to provide an explanation for the observed bimodality in a large number of social or economic phenomena, e.g., in the distribution of the gross income for movies released in theaters across the USA during the period 1997–2003 [33]. Bimodality in this context implies that movies either achieve enormous success or are dismal box-office failures. Based on the model presented here we conclude that, in such a situation, the moviegoers' choice depends not only on their neighbors' choice, but also on how well previous action based on such neighborhood information agreed with media reports and reviews of movies indicating the overall or community choice. Hence, the case of $\lambda > 0$, indicating the reliance of an individual agent on the aggregate information, imposes correlation among agent choice across the community which leads to a bimodal gross distribution.

Based on a study of the rank distribution of movie earnings according to their ratings [75], we further speculate that movies made for children (rated G) have a significantly different popularity mechanism than those made for older audiences (PG, PG-13 and R). The former show striking similarity with the rank distribution curve obtained for $\lambda = 0$, while the latter are closer to the curves corresponding to $\lambda > 0$. This agrees with the intuitive notion that children are more likely to base their choices about movies (or other products, such as toys) on the choice of their friends or classmates, while adults are more likely to be swayed by reports in mass media about the popular appeal of a movie. This suggests that one can tailor marketing strategies to different segments of the population depending on the role that global feedback plays in their decisions. Products whose target market has $\lambda = 0$ can be better disseminated by distributing free samples in neighborhoods; while for $\lambda > 0$, a mass-media campaign blitz will be more effective.

15.4
Conclusions

In this article we have primarily made an attempt to ascertain the general empirical features inherent in many popularity phenomena. We observe that the distribution of popularity in various contexts often exhibits long tails, the nature of which seem to be either following a log-normal form or a power law with the exponent $\alpha \simeq 1$ (Zipf's law). While the log-normal distribution would arise naturally in any multiplicative stochastic process, in the context of popularity it would be natural to interpret it as a manifestation of the interplay of the multiple factors involved in an agent making a decision to adopt a particular product or idea. Further, there is no necessity for interactions among agents for this particular distribution in popularity to be observed. On the other hand, distributions with power-law tails would seem to necessarily entail inter-agent interactions, e.g., a process whereby agents follow the choice of other agents, with a particular choice becoming more preferable if many more agents opt for it [26]. This is not necessarily an irrational "herding" effect; for example, in the case of the popularity of cities, the larger the population of a city, the more likely it is to attract migrants, owing to the larger variety of employment opportunities. Thus the very fact that more agents have chosen a particular alternative may make that choice more preferable than others. Seen in this light, the popularity distribution should show a lognormal distribution in situations where individual quality preferences play an important role in making a choice, while, in cases where the choice of other agents is a paramount influence in the decision process of an agent, Zipf's law should emerge [27]. In either case, a stochastic process is sufficient to generate the popularity distributions seen in reality. This suggests that the emergence of popularity can be explained entirely as an outcome of a sequence of chance events.

Acknowledgments

We would like to thank S. Redner and M. E. J. Newman for permission to use figures from their papers. SS would like to thank S. Raghavendra for the many discussions during the early phase of the work on popularity distributions. We would also like to thank D. Stauffer, B. K. Chakrabarti, M. Marsili and S. Bowles for helpful comments and suggestions at various stages of this work.

26) In the economics literature, this is referred to as positive externality [76].

27) Montroll and Shlesinger [77] have shown that a simple extension to multiplicative stochastic processes can generate power-law tails from a log-normal distribution. Recently, Bhattacharyya et al. [78] have also proposed a very simple model showing the asymptotic emergence of Zipf's law in the presence of random interaction among agents; it is interesting in the context of our statements here that, if the mean field-theoretic arguments used in the above paper are extended to the case of no interactions amongst agents, they would suggest a log-normal distribution.

Part of this work was supported by the IMSc Complex Systems Project funded by the DAE.

References

1. MACKAY, C., *Memoirs of Extraordinary Popular Delusions And The Madness Of Crowds*, National Illustrated Library, London, **1852**
2. MORRIS, S., *The Review of Economic Studies* 67 (**2000**), pp. 57–78
3. WATTS, D. J., *Six Degrees: The Science of a Connected Age*, Vintage, London, **2003**
4. GOLDSTEIN, M. L., MORRIS, S. A., YEN, G. G., *European Physical Journal B* 41 (**2004**), pp. 255–258
5. ZIPF, G. K., *Selected Studies of the Principle of Relative Frequency in Language*, Cambridge, MA., Harvard University Press, **1932**
6. ZIPF, G. K., *Human Behaviour and the Principle of Least Effort*, Addison Wesley, Cambridge, Massachusetts, **1949**.
7. NEWMAN, M. E. J., *Contemporary Physics* 46 (**2005**), pp. 323–351
8. PARETO, V., *Cours d'Economie Politique*, vol 2. Macmillan, London, **1897**
9. ADAMIC, L. A., HUBERMAN, B. A., *Glottometrics* 3 (**2002**) pp. 143–150.
10. LIMPERT, E., STAHEL, W., ABBT, M., *Bioscience* 51 (**2001**), pp. 341–352
11. AUERBACH, F., *Petermann's Geographische Mitteilungen* 59 (**1913**), pp. 74–76
12. LI, W., *Glottometrics* 5 (**2003**), pp. 14–21
13. GABAIX, X., *Quarterly Journal of Economics* 114 (**1999**), pp. 738–767
14. SOO, K. T., *Regional Science and Urban Economics* 35 (**2005**), pp. 239–263
15. NITSCH, V., *Journal of Urban Economics* 57 (**2005**), pp. 86–100
16. REED, W. J., *Journal of Regional Science* 41 (**2002**), pp. 1–17.
17. BLACK, D., HENDERSON, V., *Journal of Economic Geography* 3 (**2003**), pp. 343–372
18. GIBRAT, R., *Les inégalités économiques*, Recueil Sirey, Paris, **1932**
19. SUTTON, J., *Journal of Economic Literature* 35 (**1997**), pp. 40–59
20. STANLEY, M. H. R., BULDYREV, S. V., HAVLIN, S., MANTEGNA, R. N., SALINGER, M. A., STANLEY, H. E., *Economics Letters* 49 (**1995**), pp. 453–457
21. CABRAL, L. M. B., MATA, J., *American Economic Review* 93 (**2003**), pp. 1075–1090
22. RAMSDEN, J. J., KISS-HAYPÁL, G., *Physica A* 277 (**2000**), pp. 220–227
23. AXTELL, R. L., *Science* 293 (**2001**), pp. 1818–1820
24. HAGSTROM, W. O., *The Scientific Community*, Basic Books, New York, **1965**
25. COLE, S., COLE, J. R., *American Sociological Review* 33 (**1968**) pp. 397–413
26. MERTON R. K., *Science* 159 (**1968**), pp. 56–63
27. LAHERRÈRE. J. SORNETTE, D., *European Physical Journal B* 2 (**1998**), pp. 525–539
28. PRICE, D. J. S., *Science* 149 (**1965**), pp. 510–515
29. PRICE, D. J. S., *Journal of the American Society for Information Science*, 27 (**1976**), pp. 292–306
30. REDNER, S., *European Physical Journal B* 4 (**1998**), pp. 131–134
31. REDNER, S., *Physics Today* 58 (**2005**), pp. 49–54
32. DE VANY, A., WALLS, W. D., *Journal of Cultural Economics* 23 (**1999**), pp. 285–318
33. SINHA, S., RAGHAVENDRA, S., *European Physical Journal B* 42 (**2004**) pp. 293–296
34. PAN, R. K., SINHA, S., *forthcoming* (**2006**)
35. DE VANY, A., *Hollywood Economics*, Routledge, London, **2003**
36. GLASSMAN, S., *Computer Networks and ISDN Systems* 27 (**1994**), pp. 165–173
37. NIELSEN, J., Do Websites Have Increasing Returns?, http://www.useit.com/alertbox/9704b.html (**1997**)
38. BRESLAU, L., CAO, P., FAN, L., PHILLIPS, G., SHENKER, S., *IEEE INFOCOM* 1 (**1999**), pp. 126–134
39. ADAMIC, L. A., HUBERMAN, B, A., *Quarterly Journal of Electronic Commerce* 1 (**2000**), pp. 5–12

40 ALBERT, R., JEONG, H., BARABÁSI, A.-L., *Nature* 401 (**1999**) pp. 130

41 BRODER, A., KUMAR, R., MAGHOUL, F., RAGHAVAN, P., RAJAGOPALAN, S., STATA, R., TOMKINS, A., WIENER, J., *Computer Networks* 33 (**2000**), pp. 309–320

42 PENNOCK, D. M., FLAKE, G. W., LAWRENCE, S., GLOVER, E. J., GILES, C. L., *Proc. Natl. Acad. Sci. USA* 99 (**2002**), pp. 5207–5211

43 SHIRKY, C., Power Laws, Weblogs, and Inequality, http://www.shirky.com/writings/ powerlaw_weblog.html (**2003**)

44 DREZNER, D. W., FARRELL, H., The Power and Politics of Blogs, http://www.henryfarrell.net/ blogpaperapsa.pdf (**2004**)

45 HINDMAN, M., TSIOUTSIOULIKLIS, K., JOHNSON, J. A. "Googlearchy": How a few heavily-linked sites dominate politics on the web, www.princeton.edu/~mhindman/googlearchy–hindman.pdf (**2002**)

46 ADAMIC, L., GLANCE, N., The Political Blogosphere and the 2004 U.S. Election: Divided They Blog, http://www-idl.hpl.hp.com/ blogworkshop2005/adamic.pdf (**2005**)

47 NOH, J. D., JEONG, H. C., AHN, Y. Y., JEONG, H., *Physical Review E* 71 (**2005**), 036131

48 GUARDIOLA, X., GUIMERÀ, R., ARENAS, A., DÍAZ-GUILERA, A., STREIB, D., AMARAL, L. A. N., *Arxiv preprint cond-mat/0206240* (**2002**)

49 FILHO, R. N. C., ALMEIDA, M. P., ANDRADE, J. S., MOREIRA, J. E., *Physical Review E* 60 (**1999**), pp. 1067–1068

50 SORNETTE, D., DESCHATRES, F., GILBERT, T., AGEON, Y., *Physical Review Letters* 93 (**2004**), 228701

51 ROSENTHAL, M., What Amazon Sales Ranks Mean, http://www.fonerbooks.com/surfing.htm (**2004**)

52 HACKETT, A. P., *70 Years of Best Sellers, 1895-1965*, R.R. Bowker Company, New York, **1967**

53 MIYAZIMA, S., LEE, Y., NAGAMINE, T., MIYAJIMA, H., *Journal of the Physical Society of Japan* 68 (**1999**), pp. 3244–3247

54 KIM, B. J., PARK, S. M., *Arxiv preprint cond-mat/0407311* (**2004**)

55 GALBI, D. A., *Arxiv preprint physics/0511021* (**2005**)

56 ULUBASOGLU, M. A., HAZARI, B. R., *Journal of Economic Geography* 4 (**2004**), pp. 459–472

57 ROSEN, S., *American Economic Review* 71 (**1981**), pp. 845–858

58 ADLER, M., *The American Economic Review*, 75 (**1985**), pp. 208–212

59 MACDONALD, G., *Economic Review* 78 (**1988**), pp. 155–166

60 HAMLEN, W., *Economic Inquiry* 32 (**1994**), pp. 395–406

61 CHUNG, K. H., COX, R. A. K., *The Review of Economics and Statistics* 76 (**1994**), pp. 771–775

62 DAVIES, J. A., *European Physical Journal B* 27 (**2002**), pp. 445–448

63 GILES, D. E., *Econometrics Working Paper*, EWP0507 (**2005**)

64 SINHA, S., PAN, R. K., in (Eds. A. Chatterjee, S. Yarlagadda and B. K. Chakrabarti), *Econophysics of Wealth Distributions*, Springer, **2005**, pp. 43–47

65 REDNER, S., *Arxiv preprint physics/0407137* (**2004**)

66 JOHANSEN, A., SORNETTE, D., *Physica A* 276 (**2000**), pp. 338–345

67 MITZENMACHER, M., *Internet Mathematics* 1 (**2003**), pp. 226–251

68 BIKHCHANDANI, S., HIRSHLEIFER, D., WELCH, I., *Journal of Political Economy* 100 (**1992**), pp. 992–1026

69 YULE, G. U., *Phil. Trans. Roy. Soc. B* 213 (**1925**), pp. 21–87

70 SIMON, H. A., *Biometrika* 42 (**1955**), pp. 425–440

71 BARABÁSI, A.-L., ALBERT, R., *Science* 286 (**1999**), pp. 509–512

72 REDNER, S., *Physica A* 306 (**2002**), pp. 402–411

73 SINHA, S., RAGHAVENDRA, S., *SFI Working Paper* 04-09-028 (**2004**)

74 AXELROD, R., *Journal of Conflict Resolution* 41 (**1997**), pp. 203–226

75 DE VANY, A., WALLS, W. D., *Journal of Business* 75 (**2002**), pp. 425–451

76 BRIAN ARTHUR, W., *Economic Journal* 99 (**1989**) pp. 116–131

77 MONTROLL, E. W., SHLESINGER, M. F., *Proc. Natl. Acad. Sci. USA* 79 (**1982**) pp. 3380–3383

78 BHATTACHARYYA, P., CHATTERJEE, A., CHAKRABARTI, B. K., *Arxiv preprint physics/0510038* (**2005**)

16
Crowd Dynamics
Anders Johansson and Dirk Helbing

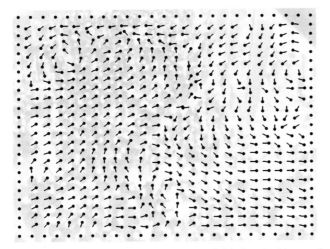

Fig. 16.1 Snapshot of the velocity vector field of pedestrian motion under crowded conditions.

Modeling of pedestrian motion is performed for different reasons: quantitative research for evacuation dynamics, investigation or assessment of the space in public facilities, or simply to gain more knowledge about how pedestrians interact with each other and how the design of the infrastructure affects the flow of pedestrians. For illustration of the potential complexity of pedestrian flows, Fig. 16.1 shows the velocity vector field computed from a video of an extremely dense crowd. Notice that some parts of the crowd possess coherent motion, while other parts shows turbulent or chaotic motion.

16.1
Pedestrian Modeling: A Survey

For modeling pedestrians properly, several levels of abstraction need to be taken into consideration, reaching from the lowest level – crowd dynamics –

Econophysics and Sociophysics: Trends and Perspectives.
Bikas K. Chakrabarti, Anirban Chakraborti, Arnab Chatterjee (Eds.)
Copyright © 2006 WILEY-VCH Verlag GmbH & Co. KGaA, Weinheim
ISBN: 3-527-40670-0

to higher levels – route choice, activity chains, scheduling, and so on. Here, we will focus on the lowest level, i.e., crowd dynamics. Simplifying, the higher levels use the conscious part of the brain for reasoning, where the lowest level, on the other hand, is mainly based on unconscious actions and reactions. This makes it possible to create simple models which describe how a pedestrian will respond to different impulses and environmental impacts. Such models can mimic how pedestrians interact to each other and with the surroundings, i.e., such simple stimulus-response models are able to quite realistically reproduce the complex dynamics of crowds.

A more behavioral aspect of crowd dynamics is how people learn and optimize walking and interaction patterns and therefore how collective crowd dynamics evolve. The publications by Bolay [11] and Helbing et al. [13] (2001) address this aspect of crowd dynamics in more detail.

16.1.1
State-of-the-art of Pedestrian Modeling

A variety of different approaches to modeling of pedestrian movement has been proposed, and here will be given a short overview. For a more complete review, the reader is encouraged to read the PhD thesis of Daamen [2]. The different approaches to pedestrian modeling can be classified according to their abstraction level and detail of description.

- Microscopic models describe each pedestrian as a unique entity with its own properties.

- Macroscopic models, delineate the average or aggregate pedestrian dynamics by densities, flows, and velocities, as a function of space and time.

- Mesoscopic (gas-kinetic) models are in between the two previously mentioned levels, taking into account the velocity distribution.

16.1.1.1 Social-force Model

The *Social-Force Model* [5,14,28,29] is a microscopic model, which is continuous both in space and time. This model is based on driving forces and generalizes Newtonian mechanics to the motion of pedestrians.

The forces consist of repulsive forces with respect to other pedestrians and boundaries, friction forces, attractive forces among group members, and forces related to desired velocities. A superposition of all these forces gives a resultant force and determines the acceleration of the pedestrians.

16.1.1.2 Cellular Automata Models

Another popular approach to pedestrian modeling is based on cellular automata [7, 8, 11]. The exact specification of these models differs, but the common idea is to divide the walkable space into a lattice, where each cell has an area corresponding to the size of a human body, approximately 40 cm × 40 cm. Each cell can either be occupied by *nobody* or *one* pedestrian. The movements of pedestrians are performed by iterating the time in steps of about 0.3 s. At each time step the pedestrians can move to unoccupied neighboring cells. However, even though the basic idea of the cellular automata models is simple, it often becomes complex with up to 30 rules for how the movement can be performed.

Since the cellular automaton models are discrete both in time and space, and due to the fact that they use only local interactions, they are used for simulating large crowds.

One drawback of the cellular automata models is that the accuracy of the dynamics is questionable, especially for high crowd densities, because the pedestrian bodies are not compressible and phenomena like clogging due to high pressures or turbulent high density flow are not well reproduced.

16.1.1.3 Fluid-dynamic Models

When the crowd density is high, the flow of pedestrians resembles fluid flow. Therefore a macroscopic approach to crowd modeling is to use fluid-dynamic models adapted to the case of pedestrians [16, 17]. A potential advantage of fluid-dynamic modeling of pedestrians is that it becomes possible to make analytical evaluations of changes in the infrastructure or changes of boundary conditions.

16.1.1.4 Queueing Models

Queuing models [9, 10] make further simplifications for simulating crowds. They are used to analyze how pedestrians are moving around in a network of modules, where the nodes and links can, for example, be doors and rooms, or intersections and roads. It is important to stress that the dynamics inside each node is usually not explicitly taken into consideration. The idea is rather to grasp how the different modules are interacting with each other, by analyzing the queues in the system where each node has a certain service rate and pedestrians move to the next queue as soon as they have been served.

16.1.1.5 Calibration and Validation

When a model should be used for producing quantitative results, there is, of course, a need to validate and calibrate it correctly. The classical way to do this is to fit the so-called *fundamental diagram*, which is the relation between flow and crowd density. Fitting the fundamental diagram (see, e.g., [1]) gives a rough idea of the accuracy of a pedestrian model. However, it should be stressed that currently there is no empirically verified specification of the fundamental diagram in the high-density domain, i.e., for crowd densities above six pedestrians per m^2. Figure 16.2 shows recent measurements for dense, but forward-moving pedestrian crowds.

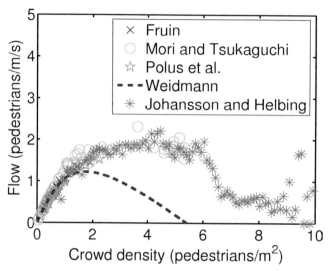

Fig. 16.2 Fundamental diagram gained from measurements of high-density crowds. Note that the local flow tends to increase again at extreme densities and that both maximum flows and maximum densities significantly depend on the body-size distribution of pedestrians (i.e., the average area occupied by them).

As the fundamental diagram does not reflect the complex dynamics of the crowd, a better approach to calibration and validation is to use trajectories of pedestrians, extracted from video recordings. See Fig. 16.3 for an example of such trajectories. Finally, we want to mention some recent studies based on walking experiments ([3,4]).

The reader is refered to Section 16.5.2 for a further explanation of how this calibration is performed.

16.2
Self-organization

Given a set of simple rules of local interactions between pedestrians, complex dynamical patterns can emerge in the crowd. These collective spatio-temporal patterns of motion are refered to as *self-organization* phenomena.

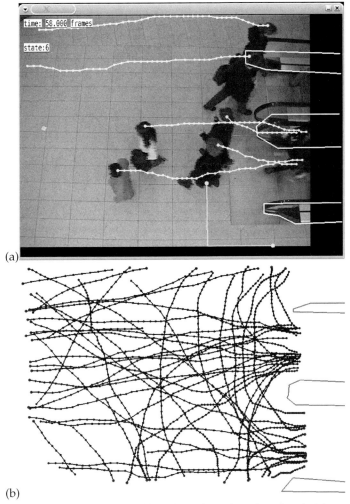

Fig. 16.3 Video tracking used to extract the trajectories of pedestrians from video recordings. (a) Illustration of the tracking of pedestrian heads; (b) Resulting trajectories after being transformed onto the two-dimensional plane. (For a color version please see color plate on page 599.)

16.2.1
Lane Formation

A simple measure of the discomfort of walking is the average change of velocity in time, i.e., $\frac{1}{T}\int_t^{t+T}||\vec{a}(t')||dt'$, where the absolute value of the acceleration $\vec{a}(t)$ is integrated over a certain time window T. If the crowd density around a certain pedestrian is low, a simple straegy can be used to minimize this inefficiency, e.g., to adapt the speed in order to avoid potential encounters with the few persons around. However, when the crowd density is higher and especially when the pedestrians around have different desired directions of walking, a more sophisticated strategy is needed. A special case which we see quite often in nature, where opposite pedestrians have mainly two different desired directions of walking, in opposite directions (bi-directional movement), people tend to follow someone who is walking in roughly the same direction and with the same speed to reduce the number of encounters with oppositely moving people. This causes the self-organization phenomenon of *lane formation*. Lane formation is characterized by the separation of pedestrians into lanes of similar walking direction, where the walkable space is divided into a number of lanes which increases with the width of the walkway.

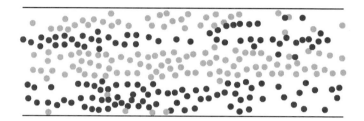

Fig. 16.4 Lane formation. The color coding is used to distinguish movement in the left direction from movement in the right direction.

16.2.2
Strip Formation

If we generalize the principle of lane formation to two pedestrian flows intersecting at an arbitrary angle it turns out that self-organization also occurs in this case, decreasing the inefficiency of walking for all pedestrians. It is more tricky to come up with a strategy for angle of encounter different from 180°, since the two different groups of pedestrians have to cross each others' paths.

It turns out that the resulting dynamics of the pedestrians is similar to the zipper strategy observed in vehicular traffic when two lanes merge into one. The pedestrians cluster together in groups and the walkway is dynamically subdivided into strips made up of groups with the two different directions of motion.

We define the *main direction* as the mean direction of the two directions, and refer to movement along this direction as longitudinal movement, while we call the movement orthogonal to this, transversal movement. With the strategy of strip formation, it is easy to walk in the longitudinal direction, since all groups are moving along this direction. Also the movement along the transversal direction turns out to be easy since all transversal movements occur within the strips, i.e., in the same group, and there is no directional conflict within the groups.

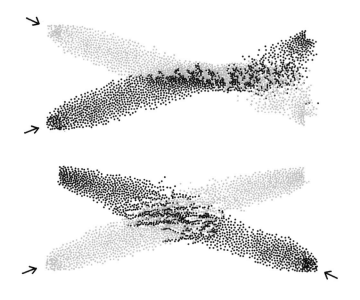

Fig. 16.5 Illustration of strip formation, which allows pedestrians to move through a crossing pedestrian stream without having to stop (after [14]).

16.2.3
Turbulent and Stop-and-go Waves

In dense crowds, especially in large open areas, turbulent and stop-and-go waves may develop and propagate through the crowd. At the front edge of turbulent waves (for an illustration see Fig. 16.1), strong forces may act on the people, which give them a push. This is especially dangerous for the persons next to boundaries, e.g., walls, since these persons may be pushed into the boundary and might get seriously injured. People may also fall which causes a risk of being trampled. It has been measured that the forces in the crowd can reach magnitudes of up to 4500 Newtons per meter, or more. This is even strong enough to bend steel barriers.

An understanding of how turbulent waves are started can be gained by assuming that people within a dense crowd are pushing each other in order to get ahead, and this pressure builds up in the crowd.

However, the pressure in dense crowds is not homogeneous. Instead, force networks build up, which can cause eruptive events of pressure release. During these events, people experience large displacements of up to 10 m or more, which causes an irregular crowd motion in all possible directions.

16.3
Other Collective Crowd Phenomena

16.3.1
Herding

It is well known that people tend to show herding behavior in situations where it is advantageous to follow the majority or where it is not clear what would be the best behavior. One example is the stock market, where there is a tendency to follow other people's advice, as price changes are hard to anticipate. Another example is fashion, where people follow a trend in order to increase the likelihood of a positive feedback to their appearance.

Herding also occurs in crowds, particularly when the crowd is nervous, as in many evacuation situations [5]. As a consequence, people tend to follow others. In an unfamiliar environment (a hotel or stadium), this can help to find the closest exit faster, if someone knows the way and takes the lead. Otherwise, a large number of people may take the same wrong decision. For example, emergency exits are often not used or underutilized, since many people will head back to the entrance. This causes an unbalanced use of evacuation capacities, which can considerably delay evacuation.

16.3.2
Synchronization

Self-organization of crowds also includes spontaneous synchronization, in one way or another. As a first example, Neda et al. [26] discovered that an applauding audience can eventually develop synchronized clapping.

Another synchronization phenomenon is the Mexican wave, *La Ola*, which first appeared at the 1986 World Cup in Mexico. Farkas et al. [25] have shown how the arousal of a critical mass might trigger a wave of excitement propagating around the audience in a sports stadium. The explanation is that every person in the audience has a certain arousal level, and if this level is high enough, the person will jump up to their feet with a probability that depends on the activity level of the persons around him/her, while he/she desires a period of rest before being activated again. In other words, if a certain person is idle and the neighbors of this person are jumping up, the probability that this person will also do so is very high. The result is that a local maximum in the activity level at one position in the audience may spread through the rest of the crowd. The speed of such waves has been measured to be around 22 seats per second.

Synchronization has been also observed in pedestrian crowds, when there is a vibrational feedback as on the Millenium Bridge in London (see Strogatz et al., [27]). This can cause dangerous oscillation of bridges. Therefore, the marching of soldiers or large crowds on light weight bridge constructions are a serious concern.

16.3.3
Traffic Organization in Ants

For any pedestrian model outlined in Section 16.1, some minor modifications can be made so that the model describes ant movement as well. See, e.g., [21] for a cellular automaton ant model and [22] and [23] for a network-based approach for modeling ants.

A simplified explanation of ant motion can be made by a biased random walk with reinforcement. Initially, the ants in the system have no preference for any specific direction of motion and propagate from the nest in arbitrary directions of motion (exploration phase). However, each ant also leaves behind a track of pheromone, which attracts other ants. The result is the reinforcement of a network of pheromone trails, which decay at about 1/40 per minute. Altogether, the movement of each ant is a combination of stochasticity and deterministic following of pheromone trails.

Finally, if an ant finds food, it returns to the nest. This attracts other ants to walk along the same path, which eventually reinforces the pheromone trails towards the food sources.

This simplified description of how ant navigation works, has been found to mimic ant behavior quite realistically.

Recent work [23] has additionally found that an encounter of two ants often results in one of the ants being pushed aside. This occurs more often if part of the system is overcrowded, so that this repulsive interaction mechanism can establish a dynamic load balancing between alternative paths as illustrated in Fig. 16.6.

16.3.4
Pedestrian Trail Formation

Another self-organization phenomenon relevant for pedestrians is trail formation [18–20]. Analogous to the phenomenon of lane formation described in Section 16.2.1, for movement in terrain such as grass or snow, it is easier to follow in the steps of a predecessor, rather than producing new paths. This behavior is similar to the pheromone-based trail following of ants, but this time the role of the attractive chemicals is replaced by the greater comfort of a trail that is used more often. However, if the path of a predecessor deviates too much from the desired path of the pedestrian, he or she will take a new path

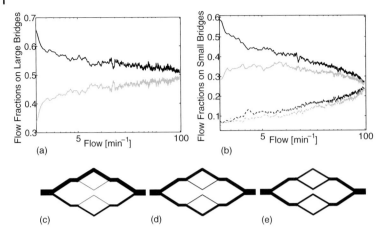

Fig. 16.6 Simulation results for a completely symmetrical bridge between the nest and a food source with two large branches that are again subdivided into two small branches each (after [30]). The overall width of the bridge (i.e., all branches) is the same everywhere. (a) Fraction of ants on the two large branches as a function of the overall ant flow ϕ. (b) Fraction of ants on the four small branches. The lower figures give a schematic representation of the distribution of ant flows over the large and small branches (c) at very low flows, (d) at medium flows and (e) at large flows.

in the terrain. When these two principles are put together with some added stochasticity, permanent paths will eventually evolve in the terrain. Interestingly, the shape of these trails looks different from conventional pedestrian ways created by civil engineers. Especially when it comes to the angle of two paths merging or separating, naturally evolving paths will never have the 90° angle of a "T" shaped intersection. Pedestrians rather prefer Y-shaped intersection designs. which is common for engineered roads.

16.4
Bottlenecks

16.4.1
Uni-directional Bottleneck Flows

Naturally, the most problematic locations during evacuations are bottlenecks, i.e., the locations with the lowest walkable width or other obstructions. The traditional view of a bottleneck is that the flow of pedestrians is proportional to the walkable width and that the flow characteristics are not further modified. This is roughly true for large widths, but for small bottleneck-like exits, things become more complex.

For example, at narrow bottlenecks, it is not possible for pedestrians to overtake their predecessors as it would be under normal, relaxed conditions, and

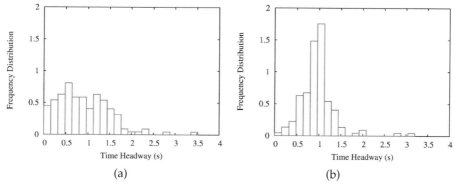

Fig. 16.7 Time headway distributions of pedestrians in a unidirectional stream before a short bottleneck (a) and after it (b). The mean values of the inverse time gaps (the flows) are the same before and after the bottleneck. However, due to the interactions in the bottleneck area, the most probable time gap is reduced, and the maximum is more pronounced after the bottleneck (after [14]).

this removes one degree of freedom, which creates a smoother flow. Experiments have shown that the distribution of time headways is more distinct directly after a narrow bottleneck than before, as Figure 16.7 shows [14]. This indicates a condensation-like kind of ordering transition. See Hoogendoorn and Daamen [24] for a quantitative study of bottlenecks.

16.4.1.1 Analytical Treatment of Evacuation Through an Exit

The evacuation of a large crowd through a small door raises additional questions. As the number of pedestrians is conserved, we may describe the aggregate, two-dimensional particle flow by the continuity equation for the pedestrian density ρ. Due to the assumed semicircular shape of the waiting crowd, it makes sense to write this equation in cylindrical (polar) coordinates:

$$\frac{\partial \rho}{\partial t} + \frac{1}{r}\frac{\partial}{\partial r}(r\rho v) + \frac{1}{r}\frac{\partial}{\partial \theta}(\rho v_\theta) = 0 \tag{16.1}$$

where t denotes the time, r the distance from the opening (exit), $v = v_r$ the velocity component in the radial direction, v_θ the angular velocity, and $v_z = 0$ the velocity in the z-direction. In the following, we will assume for simplicity that there is no dependence of the escape dynamics on the angle, which is approximately true. Then, the continuity equation simplifies to

$$\frac{\partial \rho}{\partial t} + \frac{\partial(\rho v)}{\partial r} = -\frac{\rho v}{r} \tag{16.2}$$

where the term on the right-hand side reflects the merging effect towards the bottleneck. Note that $v \leq 0$, as the pedestrians walk opposite to the radial coordinate r. By use of logarithmic derivatives, the above equation can be rewriten as

$$\frac{\partial \ln \rho(r,t)}{\partial t} = v(r,t)\frac{\partial \ln[r\rho(r,t)v(r,t)]}{\partial r} \tag{16.3}$$

Let us study first the stationary case with $\partial \ln \rho / \partial t = 0$. Then, from $\partial \ln(r\rho v)/\partial r = 0$ we just get that the overall flow $\pi r \rho(r) v(r)$ through any half-circle of radius r is constant in the stationary case:

$$\pi r \rho(r) v(r) = \pi r q(r) =: Q_{\text{in}} = \text{const.} \leq 0 \tag{16.4}$$

Herein, $q(r) = \rho(r)v(r)$ has the meaning of the pedestrian flow through a cross-section of unit length, while $\pi r q(r)$ is the flow through the half-circle of length πr. Q_{in} is the overall inflow into the system.

In the following, we will start with the simplest possible, i.e., linear velocity-density relationship

$$v(r) = V(\rho(r)) = -v_0\left(1 - \frac{\rho}{\rho_{\max}}\right) \tag{16.5}$$

where v_0 has the meaning of the maximum ("free") pedestrian speed and $\rho_{\max}$ the meaning of the maximum pedestrian density (which is a growing function of v_0). With (16.4) and the abbreviation $c = -Q_{\text{in}}/\pi > 0$ we find the quadratic equation $r\rho(r)v_0[1 - \rho(r)/\rho_{\max}] = c$, from which we get the stationary density profile

$$\rho_\pm(r) = \frac{\rho_{\max}}{2} \pm \sqrt{\left(\frac{\rho_{\max}}{2}\right)^2 - \frac{c\rho_{\max}}{v_0 r}} \tag{16.6}$$

For reasons of consistency with the limit $r \to \infty$, one has to apply the minus sign, i.e., $\rho(r) = \rho_-(r)$. $\rho(r) = \rho_-(r)$ is also expected to be the asymptotic solution after a transient time period, if the initial density profile differs.

We have to require that the discriminant of Eq. (16.6) stays non-negative. Interestingly, this is only the case for $r \geq r_{\text{crit}}$ with

$$r_{\text{crit}} = \frac{4c}{v_0 \rho_{\max}} \tag{16.7}$$

At this distance, the density becomes $\rho_{\max}/2$, corresponding to the maximum flow $q_{\max} = -\rho_{\max} v_0 / 4$. For the existence of the above stationary density profile $\rho(r)$, the absolute value $\pi r_{\text{crit}} |q_{\max}|$ of the overall flow through the semi-circle of the critical radius r_{crit} must not exceed the absolute value of the overall outflow $Q_{\text{out}} = 2r_0 q_{\text{out}}$ through the opening of width $2r_0$. This requires $|Q_{\text{in}}| \leq |Q_{\text{out}}|$, i.e., the absolute value of the overall inflow must stay below that of the overall inflow. If this condition is not fulfilled, no stationary solution exists. Instead, we expect the formation of a growing queue of pedestrians.

16.4.1.2 Intermittent Flows and Faster-is-slower Effect

The above analysis seems to work well in cases of continuous outflows, which are found for large enough openings or small enough velocities v_0. If the desired velocity v_0 is high, however, the maximum density $\rho_{\max}$ reaches values characterized by contact forces and frictional interactions. In such situations, intermittent outflows with power-law size distributions have been found. This intermittent behavior will be addressed in the following.

If two pedestrians are walking next to each other and have to pass a narrow bottleneck, they will eventually interfere with each other. After some period of mutual interaction or obstruction, one of them will pass the bottleneck first. This is normally not a problem, but if the utilization of space in front of the bottleneck is high, i.e., in the case of high crowd density, and if the crowd is nervous or excited, the cooperation level will decrease and the bottleneck will not be used in an optimal way. Rather, clogging of individuals may occur, which can happen when the individuals are pushing too much. Then, the counter-intuitive "slower-is-faster effect" sets in [5]. This means that calmly evacuating a building is faster than rushing to the exit and pushing, which often causes chaotic situations and slows down the evacuation due to clogging and mutual obstructions.

16.4.1.3 Quantifying the Obstruction Effect

Apart from the reduced capacity at the location of a bottleneck, a bottleneck has another negative influence on the crowd. If people are moving towards a bottleneck, sooner or later they will start to physically interfere with each other, given that the bottleneck is narrow enough. To determine this effect, we have set up a Monte Carlo simulation, where pedestrians are distributed randomly in front of a bottleneck, with a given crowd density in the interval $\rho \in [0, \rho_{max}]$. ρ_{max} is the maximum density at which they are standing shoulder to shoulder. To quantify the probability of mutual obstruction, we have measured the relative frequency with which two pedestrians would overlap, if they were moving a distance towards the bottleneck corresponding to their step size. Figure 16.10 shows the results of the simulation, at a distance 5m from a virtual bottleneck with zero width, indicating that everyone is moving towards the same focal point.

The quantification of this obstruction effect helps one to understand intermittent outflows and the "faster-is-slower effect". If a crowd is leaving a room through a narrow door, the flow is smooth if the crowd stays relaxed.

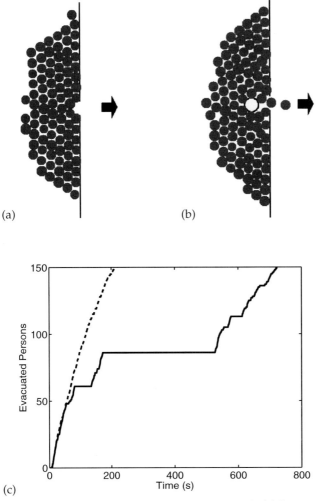

Fig. 16.8 Evacuation of a room through a narrow exit. (a) Conventional scenario. (b) Improved scenario, where a pillar is added in front of the exit to absorb pressure in the crowd. (c) Number of evacuated persons as a function of time. The dashed line shows the scenario with the pillar, and the solid line shows the conventional scenario without a pillar.

However, in an emergency evacuation the flow can be interrupted sometimes, which reduces the throughput. The reason is that the densities in nervous, pushy crowds are higher and their distances smaller, so that mutual obstructions are more likely.

16.4.2
Bi-directional Bottleneck Flows

So far, only uni-directional flows through bottlenecks have been discussed, but for the case of bi-directional flows, another interesting self-organization phenomenon occurs. When people are using a bottleneck in both directions and the bottleneck is narrow enough to allow only one person to pass at a time, it is obvious that the flow can be zero, in either of both possible passing directions. Remembering the discussion of lane formation, it is easier to follow a nearby pedestrian who is walking in the same direction as oneself, than to move against the flow. Therefore, it is easy to see that, once the flow is established in one of both directions, it will possess a forward momentum that sustain the flow for some time. However, the crowd that wants to pass in the other direction will become more impatient and pushy, the longer the time they have to wait, so eventually they will be eager enough to stop the flow, and then change the flow direction, so they can pass the bottleneck. As this scheme is repeated, the resulting flow is oscillatory in time (see Fig. 16.9). This also applies to narrow bottlenecks allowing more than one person to pass at a time, but then the oscillation effects may be less pronounced.

16.5
Optimization

16.5.1
Pedestrian Flow Optimization with a Genetic Algorithm

When a large group of pedestrians is to be evacuated from a building, one of the most dangerous locations is at the doors. In emergency situations when pedestrians panic or fear for their lives, they tend to force their way out, even if the exits are jammed, which creates a clogging phenomenon that is much more pronounced during an evacuation than under normal conditions. It has been pointed out that it is possible to increase the outflow by suitably placing a pillar or some other type of obstacle in front of the exit, which reduces the inter-pedestrian pressure in front of the door, decreases the magnitude of clogging and therefore makes the overall outflow higher and more regular [5, 14]. Figure 16.8 illustrates that a small modification in the infrastructure can, in fact, significantly reduce the evacuation time. To investigate how the architectural infrastructure in the vicinity of a door, or other bottleneck, should be constructed to maximize the pedestrian outflow under evacuation conditions, we present a method based on a Genetic Algorithm.

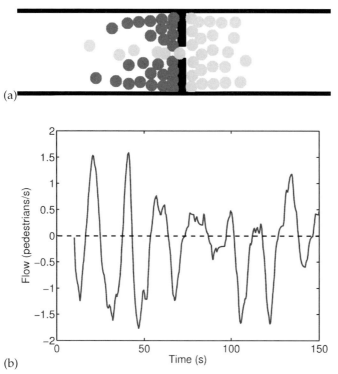

(a)

(b)

Fig. 16.9 Illustration of oscillations in bi-directional flows at a bottleneck. (a) Snapshot of the simulation of a bottleneck with the social force model. (b) Resulting oscillations in the pedestrian flow, from the simulation. The data are averaged over a 5 s time window.

16.5.1.1 Boolean Grid Representation

Previous optimization studies ([11–13]) have involved the comparison and further parameter optimization of alternative designs, but the topologies of the geometrical designs were given. To allow for more freedom in the evolution of the infrastructure, we choose a representation similar to that of Deb and Goel [15], which uses a Boolean grid where 0 means *no obstacle* and 1 means *an elementary obstacle*, similar to cellular automata models. To create an initial population of obstacle grids $X_{i,j}^{\alpha}$, we use the following scheme.

1. Initialization rule:

$$\text{Set } X_{i,j} = \begin{cases} 1 & \text{with probability } p_1 \\ 0 & \text{with probability } 1 - p_1 \end{cases}$$

where $i \in [1, n]$ and $j \in [1, m]$, n is the number of cells in the x-direction and m is the number of cells in the y-direction.

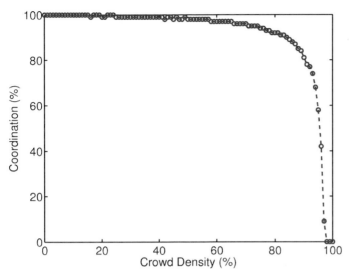

Fig. 16.10 Degree of coordination at a bottleneck. The deviation from 1 reflects the likelihood of mutual obstructions of pedestrians when moving towards a bottleneck.

2. Update rule (clustering):

   Choose i and j at random $n * m * c$ number of times, where c is a clustering constant. Then perform the following nonlinear updating rule, where the constant d denotes the size of the neighborhood taken into consideration:

   $$s_{i,j} = \sum_{k,l} X_{k,l}, \quad \text{for} \quad k \in [i-d, i+d] \quad \text{and} \quad l \in [j-d, j+d]$$

   $$p_1 = \tfrac{1}{2} + \frac{\arctan(k[2s_{i,j} - (2d+1)^2])}{\pi}$$

   Then set $X_{i,j} = 1$ with probability p_1 in the same manner as in the *initialization rule*.

The use of arctan in the *update rule* is the key to forcing the clustering of the matrix elements and create smooth clusters of solid nonfragmented obstacles. The advantage if this method is the generation of a flexible number of objects with variable sizes and shapes. If properly designed and located, such "obstacles" can surprisingly improve the throughput, efficiency and safety of pedestrian flows, while reducing the travel and evacuation times. To avoid unwanted noise (small islands of only a few elements), numerical experiments indicate that the clustering constant should be chosen as $c = 20$.

To reach this, we have implemented a simple Genetic Algorithm to improve the clustered grids for evacuation scenarios, according to the following scheme.

1. Generate a population of N obstacle matrices, $X_{i,j}$ according to the Boolean grid method specified in Section 16.5.1.1, and perform the clustering method.

2. Choose four geometrical setups at random and make a pedestrian simulation for 600 s each, e.g., with the *social force model*.

3. Assign a fitness value to each of the four geometrical setups, e.g., according to the outflow rate of pedestrians. Keep the two setups with the best fitness, $X_{i,j}^{\alpha}$ and $X_{i,j}^{\beta}$, and update the two worst ones, for all i and j, according to the following rule with pseudo random numbers $p_{\alpha} = rnd[0,1]$ equally distributed between 0 and 1:

$$X_{i,j} = \begin{cases} X_{i,j}^{\alpha} & \text{with probability } p_{\alpha} \\ X_{i,j}^{\beta} & \text{otherwise} \end{cases} \quad (16.8)$$

Then smooth the outcome by the *update rule* described in Section 16.5.1.1.

16.5.1.2 Results

We have simulated a rectangular corridor with an exit, where the outflow is around 0.50 pedestrians per second, and determined variants of this corridor with obstacles that reduce the overall cross-section of the corridor. Two of the best outcomes, each from a class with different characteristics, are shown in Fig. 16.11. By implementing the best evolved solution, the outflow may be increased by up to a factor of four. Note, however, that this does not mean that the throughput will always be four times higher by corresponding design changes. It should rather be seen as an upper limit, how much the throughput may be increased under certain conditions, mostly for extreme crowding.

From the various resulting improved solutions, we conclude that the following design elements are potentially favourable.

1. Asymmetry.

2. Funnel shape, i.e., an eventual decrease in the walkable width towards the bottleneck.

3. Zig-zag shape, i.e., a series of turns in different directions.

For example, an asymmetrically placed pillar in front of an exit contains the favourable elements 1 and 2 in the above list. A particular zig-zag-shaped

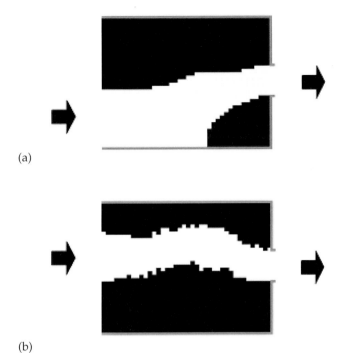

Fig. 16.11 Representatives of two classes of optimized bottleneck designs. (a) Funnel-shape (fitness 1.99). (b) Zig-zag shape (fitness 1.78).

design has been recently suggested to improve the safety of egress routes in sports stadia [14].

The applicability of design solutions obtained with a Genetic Algorithm based on Boolean grids goes beyond outflow maximization. It gives general insights into how interfaces may be optimized between areas with different capacities. Our future work will focus on the further improvement of automated, simulation-based design optimization of pedestrian facilities. In particular, we want to test multi-goal functions considering not only flows, but also other performance measures. This will also help to avoid artefacts like designs with unused walkable areas without exits in the walking direction (see Fig. 16.11).

16.5.2
Optimization of Parameter Values

Over the years, the social force model has been simulated using several different types of force-field function. The outcome is that the model is stable enough to give reasonable results for a variety of functional specifications.

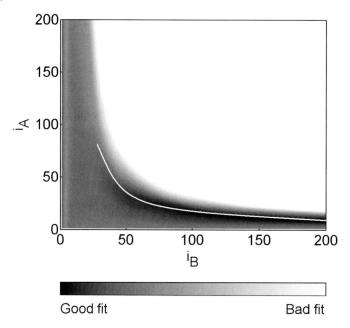

Fig. 16.12 Goodness of fit for different parameter values $A = 10(i_A/200)^2$ and $B = 10(i_B/200)^2$, when pedestrian trajectories are simulated with the social-force model and the force specification 16.10. Below-average fitness values are not displayed. The white line indicates the optimal combination of A and B values.

However, this does not automatically mean that all of these models give good quantitative results outside the scenarios for which they were made or calibrated. To determine the interaction forces empirically we extract the pedestrian trajectories from several crowd videos and evaluate different functional force specifications to see which one corresponds to best real pedestrian movement behavior. Our main focus here is on the social repulsive force between two pedestrians, since this is the strongest influence on crowd dynamics under non-pushing conditions.

We separate the repulsive force field around each human into two parts:

$$F(d, \phi) = F_s(d) F_{angle}(\phi) \quad (16.9)$$

The first part controls how the force changes with the distance from the pedestrian,

$$F_s(d) = A e^{(r_\alpha + r_\beta - d)/B} \quad (16.10)$$

where d is the distance between the centers of pedestrians α and β, and r_α, r_β are the effective radii of the two pedestrians.

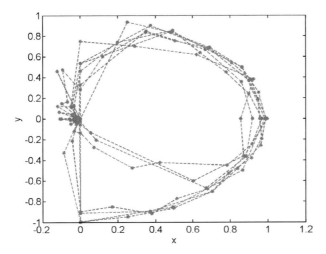

Fig. 16.13 Angular dependence of the influence of other pedestrians. The direction along the positive x axis corresponds to the walking direction of the pedestrian, y, to the perpendicular direction.

The second part depends on the angle ϕ between the velocity vector $\vec{v}_\alpha$ of pedestrian α and the vector $\vec{r}_\beta - \vec{r}_\alpha$ reflecting the direction and distance of pedestrian β. This force could be specified in different ways:

Isotropic [5]:

$$F_{angle}(\phi) = 1 \qquad (16.11)$$

Anisotropic [6]:

$$F_{scale}(\phi) = \lambda + (1-\lambda)\frac{1+\cos(\phi)}{2} \qquad (16.12)$$

Here, we chose a more general approach by using a polygon with n edges at positions $\phi \in 2\pi i/n, i \in [0, n-1]$,

$$F_{angle}(i) \in \{r_0, r_1, r_2, ..., r_{n-1}\} \qquad (16.13)$$

where r_i is the distance from the pedestrian center to the polygon edge, at angle $2\pi i/n$. To obtain some smoothness, linear interpolation is carried out between each of the edges.

Figure 16.12 shows the optimal parameters for the distance dependence, and Fig. 16.13 shows the angular dependence. Apparently, pedestrians are only sensitive to what happens inside an 180° angle in front of them, which roughly corresponds to the visually perceived area.

Fig. 16.14 Simulation of pedestrian streams in the city center of Dresden, Germany, serves to assess the impact of a new theater or shopping mall. (For a color version please see color plate on page 600.)

16.6
Summary and Selected Applications

In this chapter, we have given an overview over the main phenomena of crowd dynamics, including some recent developments in the field. For models and details, however, we must refer the reader to the available review articles and original contributions.

In order to finally show how the pedestrian models discussed in the previous sections can be used to solve real-world problems, we present two examples here.

The simulation represented in Fig. 16.14 is performed to investigate how changes in the infrastructure of an urban area (such as new shopping malls or theaters) will modify the pedestrian flow.

Our second example compares the conventional design of a soccer stadium with a proposed improved design. The improved design has zig-zag shaped downward staircases in order to release the crowd pressure in the case of an incident (see Fig. 16.15). It also uses funnel-shaped egress routes to better cope with the higher demand near the exit area.

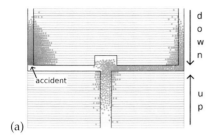
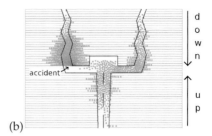
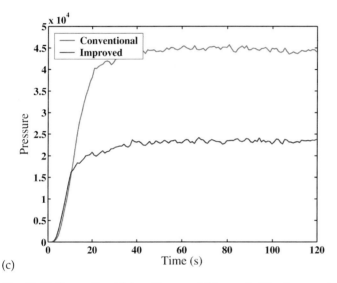

Fig. 16.15 Conventional (a) and improved (b) design of a seating sector of a sports stadium (after [5]). For safety assessment we have assumed an accident (fallen people) at the end of the left downward staircase. (c) Numerical results for the pressure level at the end of the downward staircase, in the case of a blockage of the outflows. The evacuation simulations have been performed with the social-force model. (For a color version please see color plate on page 601.)

Acknowledgments

The authors are grateful to Joachim Mathiesen, Mogens H. Jensen, and Alex Hansen for suggesting the idea of an analytical approach to evacuation through a bottleneck.

Furthermore, the authors are also grateful to Pradyumn Shukla for his cooperation and for introducing the interesting work of Deb and Goel, used for the optimization of bottlenecks with Genetic Algorithms.

Last but not least, we would like to thank the German Research Foundation (DFG project He 2789/7-1) for financial support.

References

1. WEIDMANN, U., *ETH Zuerich, Schriftenreihe IVT-Berichte 90*, Zuerich (In German), **1993**
2. DAAMEN, W., *PhD thesis*, Delft University Press, **2004**
3. HOOGENDOORN, S. P., DAAMEN, W., BOVY, P. H. L., in: *Pedestrian and Evacuation Dynamics*, E. R. Galea (ed.), 89-100, CMS Press, London, **2003**
4. SEYFRIED, A., STEFFEN, B., KLINGSCH, W., BOLTES, M., *J. Stat. Mech.* (**2000**), P10002
5. HELBING, D., FARKAS, I., VICSEK, T. *Nature*, 407 (**2000**), p. 487
6. HELBING, D., *Reviews of Modern Physics* 73 (**2001**), p. 1067
7. MEYER-KÖNIG, T., KLÜPFEL H., SCHRECKENBERG, M., in: *Pedestrian and Evacuation Dynamics*, M. Schreckenberg and S. D. Sharma (eds.), Springer, Berlin, **2002**
8. BLUE, V. J.., ADLER, J. L., *Transportation Research Record*, 1676 (**2000**), p. 135
9. WATTS, J. M., *Fire Safety Journal*, 12 (**1987**), p. 237
10. LÖVÅS, G. G., *Transportation Research*, 28B (**1994**), p. 429
11. BOLAY, K., *Nichtlineare Phänomene in einem fluid-dynamischen Verkehrsmodell*, Diploma thesis, University of Stuttgart, (**1998**)
12. ESCOBAR, R., DE LA ROSA, A., in: *7th European Conference on Artificial Life (ECAL 2003)*, Dortmund, Germany, **2003**
13. HELBING, D., MOLNÁR, P., FARKAS, I., BOLAY, K., *Environment and Planning B*, 28 (**2001**), p. 361
14. HELBING, D., BUZNA, L., JOHANSSON, A., WERNER, T., *Transportation Science* 39(1) (**2005**), p. 1
15. DEB, K., GOEL, T., in: *Proceedings of the First International Conference on Evolutionary Multi-Criterion Optimization (EMO-2001)*, Dortmund, Germany, **2001**, p. 385
16. HELBING, D., *Complex Systems* 6 (**1992**), p. 391
17. HUGHES, R. L., *Annual Review of Fluid Mechanics* 35 (**2003**), p. 169
18. HELBING, D., MOLNÁR, P., SCHWEITZER, F., in: *Evolution of Natural Structures (Sonderforschungsbereich 230)*, Stuttgart, Germany, **1994**, p. 229
19. HELBING, D., KELTSCH, J., MOLNÁR, P., *Nature* 388 (**1997**), p. 47
20. HELBING, D., SCHWEITZER, F., KELTSCH, J., MOLNÁR, P., *Physical Review E* 56 (**1997**), p. 2527
21. CHOWDHURY, D., GUTTAL, V., NISHINARI, K., SCHADSCHNEIDER, A., *J. Phys. A: Math. Gen.* 35 (**2002**), p. L573
22. PETERS, K., JOHANSSON, A., HELBING, D., *Künstliche Intelligenz* 4 (**2005**), p. 11
23. DUSSUTOUR, A., FOURCASSIÉ, V., HELBING, D., DENEUBOURG, J.-L., *Nature* 428 (**2004**), p. 70
24. HOOGENDOORN, S. P., DAAMEN, W., *Transportation Science* 39(2) (**2005**), p. 147
25. FARKAS, I., HELBING, D., VICSEK, T., *Nature* 419 (**2002**), p. 131
26. NEDA, Z., RAVASZ, E., BRECHET, Y., VICSEK, T., BARABÁSI, A.-L., *Nature* 403 (**2000**), p. 849
27. STROGATZ, S. H., ABRAMS, D. M., MCROBIE, A., ECKHARDT, B., OTT, E., *Nature* 438 (**2005**), p. 43
28. HELBING, D., *Behavioral Science* 36 (**1991**), p. 298
29. HELBING, D., MOLNÁR, P., *Physical Review E* 51 (**1995**), p. 4282
30. PETERS, K., JOHANSSON, A., DUSSUTOUR, A., HELBING, D., *Advances in Complex Systems*, (**2006**), submitted

17
Complexities of Social Networks: A Physicist's Perspective
Parongama Sen

17.1
Introduction

A large number of natural phenomena in the universe continue, quite oblivious of the presence of any living being, let alone a society. Fundamental science subjects like physics and chemistry deal with laws which would remain unaltered in the absence of life. In biological science, which deals with living beings, the idea of a society may exist in a basic form, such as food chains, or a struggle for existence, etc., among different species whereas present day human society has many added factors like politics, economics or psychology which to a large extent, dictate its evolution.

A society of human beings can be conceived in different ways as it involves different kinds of contact. Based on a certain kind of interaction, a collection of human beings may be thought of as a network where the individuals are the nodes and the links are formed whenever two of them interact in the defined way [1]. An interesting aspect of several such social networks is that these show a small-world effect, a phenomenon which has been shown to exist in diverse kinds of networks [2]. The Small world effect and other interesting features shared by real-world networks of different nature have triggered off a tremendous activity of research among scientists of different disciplines [3–5]. Physicists' interests in networks lie in the fact that these show interesting phase transitions as far as equilibrium and dynamic properties are concerned. The tools of statistical physics come in handy in the analytical and numerical studies of networks. Also, physics of systems embedded in small world networks raise interesting questions.

In this article we discuss the properties of social networks which have received much attention during recent years. The topic of social networks is vast and covers many aspects. Obviously, all the issues are too difficult to address in a single review and thus we have tried to give an overview of the subject, emphasising a physicist's perspective whenever possible.

Econophysics and Sociophysics: Trends and Perspectives.
Bikas K. Chakrabarti, Anirban Chakraborti, Arnab Chatterjee (Eds.)
Copyright © 2006 WILEY-VCH Verlag GmbH & Co. KGaA, Weinheim
ISBN: 3-527-40670-0

17.2
The Beginning: Milgram's Experiments

The first real-world network which showed the small-world effect was a social network. This was the result of some experimental studies by the social psychologist, Milgram [6]. In Milgram's experiments, various recruits in remote places of Kansas and Nebraska were asked to forward letters to specific addresses in Cambridge and Boston in Massachusetts. A letter had to be hand-delivered only via persons known on a first-name basis. Surprisingly, it was found that, on average, six people were required for a successful delivery. The number six is approximately $\log(N)$, where N is the total number of people in the USA. In this particular experiment, if a person A gave the letter to B, A and B are said to have a link between them. If B next hands over the mail to C, the effective "distance" between A and C is 2; while between A and B, as well as between B and C, it is 1. Defining distance in this way, Milgram's experiments show that two persons in a society are separated by an average distance of six steps. This property of having a small average distance in the society is what is known as the small-world effect.

It took around thirty years to realise that this small-world effect is not unique to the human society but rather possessed by a variety of other real (both natural and artificial) networks. These networks include social networks, Internet and WWW networks, power grid networks, biological networks, transport networks, etc.

17.3
Topological Properties of Networks

Before analyzing a society from the network point of view, it will be useful to summarise the topological properties characterizing common small-world networks.

A network is nothing but a graph having nodes as the vertices and links between nodes as edges. A typical network is shown in Fig. 17.1 where there are ten nodes and ten links.

The shortest distance S between any two nodes A and B is the number of edges on the shortest path from A to B through connected nodes. In Fig. 17.1, the shortest path from A to D goes through three edges and $S_{AD} = 3$. The diameter D of a network is the largest of the shortest distances S and in a small-world network (SWN), both the average shortest distance $\langle S \rangle$ and D scale in the same way with N the number of nodes in the network.

Erdős and Rényi studied the random graph [7] in which any two vertices have a finite probability to get linked. In this network or graph, both D and $\langle S \rangle$ were found to vary as $\log(N)$. (This result is true when a minimum number of edges is present in the graph so that a giant structure is formed; if in the graph with N nodes, the connectivity probability of any two nodes is p, a giant structure is formed for $pN > 1$.)

The network of human population or, for that matter, many other networks, can hardly be imagined to be a random network, although the latter has the property of a small average shortest distance between nodes. The other factor which is bound to be present in a social network (like the one considered by Milgram) is the clustering tendency. This precisely means that if A is linked to both B and C there is a strong likelihood that C is also linked to B. This property is not expected to be present in a random network.

Let a_{ij} define the adjacency matrix of the network; $a_{ij} = 1$ if nodes i and j are connected and zero otherwise. Then one can quantify the clustering coefficient C_i of node i as

$$C_i = \sum_{j_1 \neq j_2} \frac{2}{k_i(k_i+1)} a_{j_1 j_2} \qquad (17.1)$$

where j_1, j_2 are nodes connected to i and $k_i = \sum_j a_{ij}$, the total number of links possessed by node i, also known as its degree (for example, in Fig. 17.1, the degree of node A is 4). Measuring the average clustering coefficient $C = \frac{1}{N} \sum_i C_i$ in several networks, it was observed that the clustering coefficient was an order of magnitude higher than that of a random network with the same number of vertices and edges. Thus it was concluded that the real-world networks were quite different from a random network as far as the clustering property is concerned. Networks with a small value of $\langle S \rangle$ or D ($O(\log(N))$) together with a clustering coefficient much larger than the corresponding random network were given the name small-world network. Mathematically, it is possi-

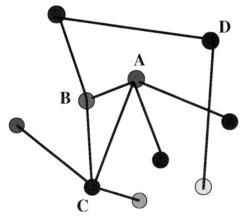

Fig. 17.1 A typical network. The shortest distance from A to C is 1 while that from A to D is 3. A,B,C form a cluster.

ble to distinguish between a small-world network and a random network by comparing their clustering coefficients.

Another type of network may be conceived in which the clustering property is high but the average shortest distance between nodes is comparable to N. Such networks are called regular networks and thus the small-world networks lie in between regular and random networks.

One can find the probability distribution of the number of neighbors of a node, commonly called the degree, of a network. An interesting feature revealed in many real-world networks is the scale-free property [8]. This means that the degree distribution $P(k)$ (i.e., the probability that a node has degree k) shows the behavior $P(k) \sim k^{-\gamma}$, implying the presence of a few nodes which are very highly connected. These highly connected nodes are called the hubs of the network.

It may be mentioned here that in the random graph, $P(k)$ has a different behavior,

$$P(k) = e^{-\langle k \rangle} \frac{\langle k \rangle^k}{k!} \qquad (17.2)$$

where $\langle k \rangle$ is the average degree of the network. $P(k)$ therefore follows a Poisson distribution, here decaying rapidly with k.

The above features constitute the main properties of networks. Apart from these, many other characteristics have been detected and analyzed as research in small-world networks increased by leaps and bounds. Some of these, e.g., closeness centrality or betweenness centrality were already quite familiar to social scientists [1].

Closeness centrality. The measure of the average shortest distance of a node to the other nodes in a network is its closeness centrality.

Betweenness centrality. The fraction of shortest paths passing through a node is its betweenness [9, 10]. The greater the betweenness of a node, the more important it is in the network since its absence will affect the small-world property to a great extent. It is not necessarily true that nodes with maximum degree will have the largest closeness or betweenness centrality.

Remaining degree distribution. If we arrive at a vertex following a random edge, the probability that it has degree k is $kP(k)$. The remaining degree distribution q_k, which is the probability that the node has k other edges, is given by

$$q_k = \frac{(k+1)P(k+1)}{\sum_j jP(j)} \qquad (17.3)$$

Assortativity. This measure is for the correlation between degrees of nodes which share a common edge. A straightforward measure will be to calculate the average degree $\langle k_{nn} \rangle$ of the neighbors of a vertex with degree k. If

$\frac{d\langle k_{nn}\rangle}{dk} > 0$, it will mean a positive correlation or assortativity in the network. A negative value of the derivative denotes disassortativity and a zero value would mean no correlation. A more rigorous method of calculating the assortativity is given in [11], where one defines a quantity r as

$$r = \frac{M^{-1}\sum_i j_i k_i - [M^{-1}\sum_i \frac{1}{2}(j_i + k_i)]^2}{M^{-1}\sum_i \frac{1}{2}(j_i^2 + k_i^2) - [M^{-1}\sum_i \frac{1}{2}(j_i + k_i)]^2} \qquad (17.4)$$

where j_i and k_i are the degrees of the vertices connected by the ith edge ($i = 1, 2, ..., M$) and M is the total number of edges in the network. Again, high assortativity means that two nodes which are both highly connected tend to be linked and $r > 0$. A negative value of r implies that nodes with dissimilar degrees are more likely to get connected. A zero value implies no correlation of node degrees and their connectivity.

Community structure. This is a property which is highly important for social networks. More often than not we find a society divided into a number of communities, e.g, based on profession, hobby, religion, etc. For the scientific collaboration network, communities may be formed with scientists belonging to different fields of research, for example, physicists, mathematicians or biologists. Within a community also, there may be different divisions in that physicists may be classified into different groups, e.g., high-energy physicists, condensed-matter physicists and so on.

It is clear that the properties of a network simply depend on the way the links are distributed among the vertices, or to be precise, on the adjacency matrix. Until now we have not specified anything about the links. Links in the networks may be both directed and undirected, e.g., in an email network [12], if A sends a mail to B, we have a directed link from A to B. Edges may also be weighted; weights may be defined in several ways depending on the type of network. In the weighted collaboration network, two authors sharing a large number of publications have a link which has more weight than that between two authors who have collaborated fewer times.

17.4
Some Prototypes of Small-world Networks

At this juncture, it is useful to describe a few important prototype small-world network models which have been considered to mimic the properties of real networks.

17.4.1
Watts and Strogatz (WS) Network

This was the first network model which was successful in reproducing the features of the small-diameter and large-clustering coefficient of a network. The small-world effect in networks of varied nature indicated a similarity in the underlying structure of the networks. Watts and Strogatz [2] conjectured that some common features in the geometry of the networks are responsible for the small-world effect. In their model, the nodes are placed on a ring. Each node has connection to k number of nearest neighbors initially. With probability p, a link is then rewired to form a random long-ranged link.

At $p = 0$, the shortest paths scale as N and the clustering coefficient of the network is quite high as it behaves as a regular network with a considerable number of nearest neighbors. The remarkable result was, even with $p \to 0$, the diameter of the network is small ($O(\log(N))$). The clustering coefficient on the other hand remains high even when $p \neq 0$ unless p approaches unity. Thus for $p \to 0$, the network has a small diameter as well as high clustering coefficient, i.e., it is a small-world network. For $p \to 1$, the network ceases to have a large clustering coefficient and behaves as a random network. This model thus displays phase transitions from a regular to a small world to a random graph by varying a single parameter p. Later it was shown to have mean-field behavior in the small-world phase [13].

The degree distribution $P(k)$ in this network, however, did not have a power-law behavior but showed an exponential decay with a peak at $\langle k \rangle$.

17.4.2
Networks with Small-world and Scale-free Properties

Although the WS model was successful in showing the small-world effect, it did not have a power-law degree distribution. The discovery of a scale-free property in many real world networks required the construction of a model which would have a small world as well as a scale-free property.

Barabási and Albert (BA) [8] proposed an evolving model in which one starts with a few nodes linked with each other. Nodes are then added one by one. An incoming node will have a probability Π_i of getting attached to the ith node already existing in the network according to the rule of preferential attachment, which means that

$$\Pi_i = k_i / \sum_i k_i \tag{17.5}$$

where k_i is the degree of the ith node. This implies that a node with higher degree will get more links as the network grows such that it has a "rich gets richer" effect. The results showed a power-law degree distribution with exponent $\gamma = 3$. While the average shortest distance grows with N slower than

$\log(N)$ in this network, the clustering coefficient vanishes in the thermodynamic limit. Several other network models have been conceived later as variants of the BA network which allow a finite value of the clustering coefficient. Also, scale-free networks have been achieved using algorithms other than the preferential attachment rule [14] or even without considering a growing network [15].

17.4.3
Euclidean and Time-dependent Networks

In many real-world networks the nodes are embedded on a Euclidean space and the link length distribution shows a strong distance dependence [16–22]. Models of Euclidean networks have been constructed for both static and growing networks [23–31]. In static models, transition between regular, random and small-world phases may be obtained by manipulating a single parameter occurring in the link length distribution [25–28]. In a growing model, a distance dependent factor is incorporated in a generalized preferential attachment scheme [29–31] giving rise to a transition between a scale-free and a nonscale-free network.

Aging is also another factor which is present in many evolving networks, e.g., the citation network [32, 33]. Here the time factor plays an important role in the linking scheme. Aging of nodes has been taken into account in a few theoretical models where the aging factor is suitably incorporated in the attachment probability. Again one can achieve a transition from a scale-free to a nonscale-free network by appropriately tuning the parameters [34–36].

17.5
Social Networks: Classification and Examples

Social networks can be broadly divided into three classes. The social networks in which links are formed directly between the nodes may be said to belong to class A. The friendship network is perhaps the most fundamental social network of class A. Other social networks like the networks of office colleagues, email senders and receivers, sexual partners etc., also belong to this class.

The second class of social networks consists of the various collaboration networks which are formed from bipartite networks. For example, in the movie actors' network, two actors are said to have a link between them if they have acted in the same movie. Similarly, in the research collaboration networks, two authors form a link if they feature in the same paper. We classify these networks as class B social networks. The difference between class A and class B networks is that in class A, it is ensured that two persons sharing a link have interacted at a personal level while it is possible that in a collaboration act, two collaborators sharing a link hardly know each other.

In some other social networks, which we classify as class C networks, the nodes are not human beings but the links connect people indirectly. Examples are citation networks and transport networks. In a citation network, links are formed between two papers when one cites the other. Transport networks consist of railways, roadways and airways networks; the nodes here are usually cities which are linked by air, road or rail routes.

Real-world data for class A networks, e.g., friendship, acquaintance or sexual network is difficult to get and usually available for small populations [3,37,38]. Communication networks like email [12] and telephone [39] networks show a small-world effect and power-law degree distribution [3]. Datasets may be constructed artificially also, e.g., as in [40], which shows a small-world effect in a friendship network.

Class B networks have been studied extensively as many databases are available. These, in general, also show the small-world effect. In the movie actors' network with over 2×10^5 nodes, the average shortest path length is 3.65. Typically in co-authorship networks, it varies between 2 and 10. The degree distribution in these networks can be of different types depending on the particular database. Some collaboration networks have a power-law degree distribution, while there exist two different power-law regimes or an exponential cutoff in the degree distribution in other databases [41,42].

In Euclidean social networks like the scientific collaboration networks, links also have strong distance dependence [17–19,21,22], usually showing a proximity bias. Recently, in a collaboration network of authors of *Physical Review Letters*, the study of time evolution of the link-length distribution was made by Sen et al. [22]. It indicated the intriguing possibility that in a few decades hence, scientists located at any distance may collaborate with equal probability, thanks to the communication revolution.

The citation network belonging to class C is also quite well studied. It is an example of a directed network. Analysis shows that the citation networks have a very interesting behavior of degree distributions and aging effects [33,43,44]. While the out-degree (number of papers referred to) shows a Poisson-like distribution, the in-degree (number of citations) has scale-free distribution.

The aging phenomenon is particularly important in citation networks; older papers are less cited, barring exceptions. Studies over a long time show [33] that the number of cited papers of age t (in years) decreases exponentially with t, while the probability that a paper gets cited after time t of its publication, decreases as a power law over an initial time period of typically 15-20 years and exponentially for large values of t. These features can be reproduced using a growing network model [36] where the preferential attachment scheme contains an age-dependent factor such that an incoming node j links with node i

with the probability

$$\Pi_{ij}(\tau_i, k_i) = \frac{k_i^\beta \exp(-\alpha \tau_i)}{\sum_i k_i^\beta \exp(-\alpha \tau_i)} \qquad (17.6)$$

where k_i and τ_i are the degree and age of node i at that moment, respectively.

Although transport networks, also belonging to class C, do not involve human beings directly as nodes, they can have great impact on social networks of both class A and B types. The idea that in a railway network, two stations are linked if at least one train stops at both, was introduced by Sen, Dasgupta et al. [45] in a study of the Indian railway network. Some details of that study is provided in the Appendix.

17.6
Distinctive Features of Social Networks

That social networks of all classes have a small diameter is demonstrated in almost all real examples. Many (although not all) social networks also show the scale-free property, and the exponent for the degree distribution generally lies between 1 and 3.

Social networks are characterized by three features: very high values of the clustering coefficient, positive assortativity and community structure.

High clustering tendency is quite understandable – naively speaking, a friend of a friend is quite often a friend also. Again, in a collaboration network of scientists, whenever a paper is written by three or more authors, there is a large contribution to the clustering coefficient.

It is customary to compare the clustering coefficients of a real network to that of its corresponding random network to show that the real network has a much larger clustering coefficient. Instead of a totally random model, one could also consider a "null" model which is a network with the same number of nodes and edges, having also the same degree distribution, but otherwise random. The clustering coefficients in nonsocial networks turn out to be comparable to that of the null model. In contrast, for social networks, this is not true.

Let us consider the random model. Suppose two neighbors of a vertex in this model have remaining degrees j and k. Then there will be a contribution to the clustering if these two nodes share an edge. With M edges in the network, the number of edges shared by these two nodes is $jk/2M$. Both j and k are distributed according to Eq. (17.3), and therefore the clustering coefficient is

$$C = \frac{1}{2M}[\sum_k k q_k]^2 = \frac{1}{N} \frac{[\langle k^2 \rangle - \langle k \rangle^2]}{\langle k \rangle^3} \qquad (17.7)$$

For networks not large enough this will still give a finite value. One can compare it with the small network of the food web of organisms in Little Rock Lake which has $N = 72$, $\langle k \rangle = 21.0$ and $\langle k^2 \rangle = 655.2$, giving $C = 0.47$ which compares well with the actual value 0.40. It can be shown that, for the null model with degree distribution $P(k) \sim k^{-\gamma}$, the clustering coefficient remains finite even for large N when $\gamma < 7/3$. The theoretical value of the clustering coefficient here is

$$C \propto N^{(7-3\gamma)/(\gamma-1)} \tag{17.8}$$

This is assuming that there can be more than one edge shared by two vertices. Ignoring multiple edges, the clustering coefficient turns out to be [46]

$$C = \sum_{jk} q_j q_k (1 - e^{-jk/2M}) \tag{17.9}$$

Comparison with nonsocial networks like the Internet, food webs or world wide webs shows that the difference between the values obtained theoretically and the actual ones are a minimum. For social networks, however, the theoretical values are at least one order of magnitude smaller than the observed ones.

Assortativity in social networks is generally positive in contrast to nonsocial networks. For nonsocial networks like the Internet, WWW, biological networks etc., r (Eq. (17.4)) lies between -0.3 and -0.1 while for social networks like the scientific collaboration networks or company directors, r is between 0.2 and 0.4 [11].

It has been argued that the large clustering coefficient and the positive assortativity in social networks arise from the community structure. Since the discovery of community structure in social networks, there has been tremendous activity in this field and we devote the next section to the details of these studies.

17.7
Community Structure in Social Networks

A society is usually divided into many groups and again the groups may regroup to form a bigger group. The community structure may reflect the self-organization of a network to optimize some task performance, for example, in searching, optimal communication pathways or even maximization of productivity in collaborations. There is no unique definition of a community but the general idea is that the members within a community have more connections within themselves and less with members belonging to other communities.

For small networks, it is possible to visually detect the community structure. But, with the availability of large-scale data in many fields in recent times, it is essential to have good algorithms to detect the community structure.

17.7.1
Detecting Communities: Basic Methods

A short review of the community detection methods is available in [47], many other methods have developed since then. Here we attempt to present an outline of some basic and some recently proposed methods.

17.7.1.1 Agglomerative and Divisive Methods

The traditional method for detecting communities in networks is agglomerative hierarchical clustering [48]. In this method, each node is assigned its own cluster so that with N nodes, one has N clusters in the beginning. Then the most similar or closest nodes are merged into a single cluster following a certain prescription so that one is left with one cluster less. This can be continued till one is left with a single cluster.

In the divisive method, the nodes belong to a single community to begin with. By some appropriate scheme, edges are removed, resulting in a splitting of the communities into subnetworks in steps and ultimately all the nodes are split into separate communities. Both the agglomerative and divisive methods give rise to what is known as a dendogram (see Fig. 17.2) which is a diagrammatic representation of the nodes and communities at different times. While in the agglomerative method one goes from bottom to top, in the divisive method it is just the opposite.

Girvan and Newman (GN) [49] proposed a divisive algorithm in which edges are removed using the measure of betweenness centrality (BC). Generalizing the idea of BC of nodes to BC of edges, edge betweenness can be defined as the number of shortest paths between pairs of vertices that run along it. An edge connecting members belonging to different communities is bound to have a large betweenness centrality measure. Hence, in this method, one calculates the edge betweenness at every step and removes the edge with the maximum measure. The process is continued till no edge is left in the network. After every removal of an edge, the betweenness centrality is to be recalculated, in this algorithm.

17.7.1.2 A Measure of the Community Structure Identification

Applying either of the above algorithms it may be possible to divide the network into communities. Even the random graph with no community structure may be separated into many classes. One needs to have a measure to see how good is the obtained structure. Also, the dendogram represents communities

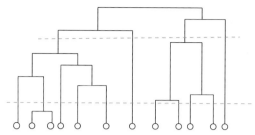

Fig. 17.2 A typical dendogram. In the divisive algorithm, one explores from top to bottom while for an agglomerative method, one goes from the bottom to the top. The number of communities is given by the number of intersection points of the horizontal dashed lines with the dendogram at different levels. The circles represent the nodes.

at all levels from a single to as many as the number of nodes and one should have an idea at which level the community division has been most sensible.

When the network has a fixed well-known community structure, a quantity to measure the performance of the algorithm for the community detection may be the fraction of correctly classified nodes.

The other measure, known as the modularity of the network, can be used when there is no *ad hoc* knowledge about the community structure of the network. This is defined in the following way [50]. Let there be k communities in a particular network. Then e_{ij} is the fraction of edges which exist between nodes belonging to communities i and j; clearly it is a $k \times k$ matrix. The trace $Tr(e) = \sum e_{ii}$ then gives the fraction of all edges in the network that connect vertices in the same community and a good division into communities should give a high value of the trace. However, the trace is simply not a good measure of the quality of the division as placing all nodes in a single community will give trivially $Tr(e) = 1$. Let us now take the case when the network does not have a community structure; in this case links will be distributed randomly. Defining the row (or column) sums $\alpha_i = \sum_j e_{ij}$, one gets the fraction of edges that connect to vertices in community i. The expected number of fraction of links within a partition is then simply the probability that a link begins at i, α_i, multiplied by the fraction of links that end at a node in i, α_i. So the expected number of intra-community links is just $\alpha_i \alpha_i$. A measure of the modularity is then

$$Q = \sum_i (e_{ii} - \alpha_i^2) \tag{17.10}$$

which calculates the fraction of edges which connect nodes of the same type minus the same quantity in a network with identical community division but random connections between the nodes. If the number of within-community edges is no better than random, $Q = 0$. The modularity plotted at every level of the bifurcation would indicate the quality of the detection procedure and at which level to look to see the best division where Q has a maximum value.

There are some typical networks to which the community detection algorithms are applied to see how efficient it is. Most of the methods are applied to an artificial computer generated graph to check its efficiency. Here 128 nodes are divided into four communities. Each node has $k = 16$ neighbors of which k_{out} connections are made to nodes belonging to other communities. The quality of separation into communities here can be measured by calculating the number of correctly classified nodes. All available detection algorithms work very well for $k_{out} \leq 6$ and the performance becomes poorer for larger values. A comparative study of the application of different algorithms to this network has been made by Danon et al. [51].

A popular network data with known community structure which is also very much in use to check or compare algorithms is the Zachary's Karate Club (ZKC) data. Over the course of two years in the early 70s, Wayne Zachary observed social interactions between the members of a karate club at an American university [52]. He constructed a network of ties between members of the club based on their social interactions both within and outside the club. As a result of a dispute between the administrator and the chief karate teacher the club eventually split into two. Interestingly, not just two, but up to five communities have been detected in some of the algorithms.

Community structure in a jazz network has also been studied. In this network, musicians are the nodes and two musicians are connected if they have performed in the same band. In a different formulation, the bands act as nodes and have a link if they have at least one common musician. This was studied mainly in [53].

The American college football network is also a standard network where community-detection algorithms have been applied. This is a network representing the schedule of games between American college football teams in a single season. The teams are divided into groups or "conferences" and intraconference groups are more frequent than inter-conference games. The conferences are the communities, so that teams belonging to the same conference would play more between themselves.

Community-detection algorithms have been applied to other familiar social networks like collaboration and email networks as well as in several nonsocial networks.

17.7.2
Some Novel Community Detection Algorithms

The method of GN to detect the communities typically involves a time of the order of $N^2 M$, where N is the number of nodes and M the number of edges. This is considerably slower. Several other methods have been proposed although none of them is really "fast". Some of these methods are just variants of the divisive method of GN (i.e., edge removal by a certain rule) and interested readers will find the details in [50,54,55]. Here we highlight those which are radically different from the GN method.

17.7.2.1 Optimization Methods

Newman [56] suggested an alternative algorithm in which one attempts to optimize the splitting for which Q is maximum. However, this is a costly programme as the number of ways in which N vertices can be divided into g nonempty groups is given by Stirling's number of the second kind $S_N^{(g)}$ and the number of distinct community division is $\sum_{g=1}^{N} S_N^{(g)}$. It increases at least exponentially in N and therefore approximate methods are required. In [56], a greedy agglomerative algorithm was used. Starting from the state where each node belongs to its own cluster, clusters are joined such that the particular joining results in the greatest increase in Q. In this way a dendogram is obtained and the corresponding Q values are calculated. Community division for which Q is maximum is then noted. The computational effort involved in this method is $O((M+N)N)$. For small networks, the GN algorithm is better but for large networks this optimization method functions more effectively. Applications of this method has been made for the ZKC and the American Football Club networks.

Duch and Arenas [57] have used another optimization method which is local using extremal optimization. In this local method, a quantity q_r for the rth node is defined as

$$q_r = \kappa_{i(r)} - k_r \alpha_{i(r)} \tag{17.11}$$

where $\kappa_{i(r)}$ is the number of links that a node r belonging to a community i has with nodes in the same community and k_r is the degree of node r, $\alpha_{i(r)}$ is the fraction of edges connected to r having either or both ends in the community i. Note that q_r is just the contribution to Q (Eq. (17.10)) from each node in the network, given a certain partition into communities, and $Q = \frac{1}{2N} \sum_r q_r$. Rescaling the local variables q_r by k_r, a proper definition for the contribution of node r to the modularity can be obtained. In this optimization method, the whole graph is initially split into two randomly. At every time step, the node with the lowest fitness $\lambda_r = q_r/k_r$ is moved from one partition to another. The process is repeated until an optimal state with a maximum value of Q is reached. After that all the inter-partition links are deleted and the optimization procedure applied to the resulting partitions. The process is continued until the modularity Q cannot be improved any further. Applications were made to the ZKC, jazz, email, cond-mat and biological networks. In each case it gives a higher modularity value compared to that of the GN method.

17.7.2.2 Spectral Methods

The topology of a network with N vertices can be expressed through a symmetric $N \times N$ matrix L, the Laplacian matrix, where $L_{ii} = k_i$ (degree of the ith node) and L_{ij} are equal to -1 if i and j are connected and zero otherwise. The column or row sum of L is trivially zero. A constant vector, with all its elements identical will thus be an eigenvector of this matrix with eigenvalue zero.

For a connected graph, the eigenvectors are nondegenerate. Since L is real-symmetric, the eigenvectors are orthogonal. This immediately implies that the sum of the components of the other eigenvectors are zero as the first one is a constant one. If the number of subgraphs is equal to two, there will be a clear-cut bifurcation in the eigenvector; the components of the second (first nontrivial) eigenvector are positive for one subgraph and negative for the other. This is the spectral bisection method which works very well when only two communities are present, e.g., in the case of the ZKC network.

For more than two subgraphs, the distinction may become fuzzy (see Fig. 17.3). In this case, a few methods have been devised recently [58–60]. The eigenvectors of L with nonzero eigenvalues are the nontrivial eigenvectors. If there is a network structure with m communities, there will be $m - 1$ nontrivial eigenvectors with eigenvalues close to zero and these will have a characteristic structure: the components corresponding to nodes within the same cluster have very similar values x_i. If the partition is sharp, the profile of each eigenvector, sorted by components, is step-like. The number of steps in

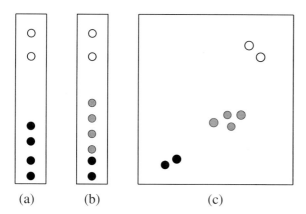

Fig. 17.3 (a) The structure of the first nontrivial eigenvector when there are only two communities. (b) An example of a case with three communities, now the first nontrivial eigenvector has a structure with fuzziness when the components corresponding to two communities are quite close (shown in gray and black). (c) When the first two nontrivial eigenvectors are projected on a two-dimensional space, the three communities are clearly shown.

the profile corresponds to the number of communities. If the partition is not so sharp, and the number of nodes is high, it is difficult to arrive at a concrete conclusion by just looking at the first nontrivial eigenvector. The method is then to combine information from the first few, say D, nontrivial eigenvectors. Consider an enlarged D-dimensional space where the components of these eigenvector are projected (i.e., the coordinates of the points in this space correspond to the values of the components of the eigenvectors for each node). The points one gets can now be treated as the "positions" of the nodes and nodes belonging to the same community are clustered in this space. Donetti and Munoz [59] used an agglomerative method to detect communities from this D-dimensional space. The choice of D, the number of eigenvectors to be used, is an important question here. One can fix a value of D, and find out the optimal value of Q – the D value corresponding to the best optimal value of Q may be the ideal choice. It was observed that D is of the order of the number of communities in the network.

Instead of the Laplacian matrix, one can also take another matrix, $X = K - A$ where K is the diagonal matrix with elements $k_{ii} = \sum_{j=1}^{N} a_{ij}$ and a_{ij} the elements of the adjacency matrix A. Here instead of zero eigenvalue, the normal matrix X has largest eigenvalue 1 associated with the trivial eigenvector. This matrix has been used successfully by Capocci et al. [58] for networks with weighted edges. It has been tested on a directed network of word association where the method has been modified slightly to take into account the directedness of the edges.

17.7.2.3 Methods Based on Dissimilarity

Two vertices are said to be similar if they have the same set of neighbors other than each other. Exact similarity is rare, so one can define some dissimilarity measure and find the communities using this measure. One of these measures is called the Euclidean distance, discussed in [47]. Recently, another measure based on the differences in their "perspective" was done. Taking any vertex "i", the distance d_{ik} to a vertex "k" can be measured easily. Taking another node "j", one can measure d_{jk}. If the nodes belong to the same community, then the difference $[d_{ik} - d_{jk}]^2$ is expected to be small. The dissimilarity index λ_{ij} was defined by Zhou [61] as

$$\lambda_{ij} = \frac{\sqrt{\sum_k [d_{ik} - d_{jk}]^2}}{N - 2} \tag{17.12}$$

The algorithm is a divisive one involving a few intricate steps and has been successfully applied to many networks.

17.7.2.4 Another Local Method

Bagrow and Bollt [62] have suggested a local method in which, starting from a particular node, one finds $k^{(l)}$, the number of its lth nearest neighbors at every stage $l = 0, 1, 2, \ldots$. By definition, $k^{(0)} = 1$. The cumulative number of degrees $K^{(l)} = \sum k^{(l)}$ is also evaluated here. The ratio $K^{(l+1)}/K^{(l)}$ is compared to a preassigned quantity α and if the ratio is less than α, the counting is terminated and all nodes upto the lth neighborhood are listed as members of the starting node's community.

17.7.3
Community-detection Methods Based on Physics

A few methods of detecting communities are based on models and methods used in Physics.

17.7.3.1 Network as an Electric Circuit

Wu and Huberman [63] have proposed a method by considering the graph to be an electric circuit. Suppose it is known that nodes A and B belong to different communities. Each edge of the graph is regarded as a resistor and all edges have identical resistance. Let a battery connect A and B such that the voltages at A and B are 1 and 0 respectively. Now the graph is simply an electric circuit with a current flowing through each resistor. By solving Kirchoff's equations one can obtain the voltage at each node. Choosing a threshold value of the voltage (which obviously lies between 0 and 1), one then decides whether a node belongs to the community containing A or that containing B. It is easy to see how this method works. First suppose that there is a node C which is a leaf or peripheral node, i.e., connected to a single node D. No current flows through CD and therefore C and D have the same voltage and belong to the same community. Next consider the case when C is connected to nodes D and E. Since the resistance is the same for all edges, $V_c = (V_D + V_E)/2$. Hence, if both D and E belong to the same community (i.e., they both have high or low voltages), C also has a comparable voltage and belongs to the same community. If, however, V_D and V_E are quite different and D and E belong to different communities then it is hard to tell which community C belongs to. This is close to reality, a node has a connection with more than one community. In general, when C is linked to n other nodes with voltages $V_1, V_2, \ldots, V_n$, $V_c = \sum V_i/n$. Hence if most of its neighbors belong to a particular community, C also belongs to it. The method works in linear time.

When it is not known initially that two nodes belong to two different communities, one can randomly choose two nodes and repeat the process of evaluation of communities a large number of times. The Chance that the detection is correct is fifty percent. However, this probability will be greater than half if these two initial nodes are chosen such that they are not connected. Majority of the results would then be regarded as correct. Applications have been made to the ZKC and the College Football Club networks.

17.7.3.2 Application of Potts and Ising Models

The q-state Potts model in Physics is a model in which the spin at site i can have a value $\sigma_i = 1, 2, \ldots, q$. If two spins interact in such a way that the energy is minimum when they have the same value, we have the ferromagnetic Potts model. The Potts model has well known applications for graph-coloring problems, where each color corresponds to a value of the spin. For community detection also, the method can be similarly applied. Reichhardt and Bornholdt [64] have used the Potts model Hamiltonian

$$H = -J \sum_{i,j} \delta_{\sigma_1, \sigma_2} + \gamma \sum_{\sigma}^{q} \frac{n_\sigma (n_\sigma - 1)}{2} \qquad (17.13)$$

where n_σ is the number of spins having spin s. Here the first term is the standard ferromagnetic Potts model and in the absence of the second term it favors a homogeneous distribution of spins which minimizes the ferromagnetic term. However, one needs to have diversity and that is incorporated by the second term. If there is only one community it will be maximum and if spin values are more or less evenly distributed over all nodes it will be minimum.

When γ is set equal to the average connection probability, then both the criteria that (a) within a community the connectivity is larger than the average connectivity, and that (b) the average connectivity is greater than the inter-community connectivity, are satisfied. With this value of γ and $J = 1$, the ground state was found by the method of simulated annealing. The communities appeared as domains of equal spin value near the ground state of the system. Applications to computer-generated graphs and biological networks were made successfully and the method can work faster than many other algorithms.

In another study by Guimerà et al. [65] in which the Potts model Hamiltonian is used again, although in a slightly different form, an interesting result was obtained. It was shown that networks embedded in low-dimensional spaces have high modularity and that even random graphs can have high modularity due to fluctuations in the establishment of links. The authors argue that like the clustering coefficient, modularity in complex networks should also be compared to that of the random graph.

Son et al. [66] have used an Ising model Hamiltonian to detect the community structure in networks. The Ising model is the case when the spins have only two states, usually taken to be $\sigma = \pm 1$. This method can be best understood in the context of the Zachary Karate Club case. Here two members started having differences and the club members eventually become polarized. In this case, the authors argue, the members will try to minimize the number of broken ties, or in other words, the breakup of ties should be in accordance with the community structure. Thus the community structure may be found by simulating the breakup caused by an enforced frustration among nodes. The breakup is simulated by studying the ferromagnetic random-field Ising model which has the Hamiltonian

$$H = -\frac{1}{2}\sum_{i,j} J_{ij}\sigma_i\sigma_j - \sum_i B_i\sigma_i \qquad (17.14)$$

where σ_i is the Ising spin variable for the ith node and the spins interact ferromagnetically with strength J_{ij} and B is the random field. J_{ij} is simply equal to the elements of the adjacency matrix, can be 1 or 0 for unweighted networks and equal to the weights for a weighted graph. The ferromagnetic interaction, as in the Potts case, represents the cost for broken bonds. The conflict in the network (e.g., as raised in the ZKC) is mimicked in the system by imposing very strong negative and positive fields for two nodes (in the ZKC case, these nodes should correspond to the instructor and administrator). The spins with the same signs in the ground state then belong to the same community. The method was applied to the Zachary Karate Club network and the co-authorship networks, with a fair amount of success. It indicated the presence of some nodes which were marginal in the sense that they did not belong to a community uniquely.

17.7.4
Overlap of Communities and a Network at a Higher Level

Division of a society into communities may be done in many ways. If the criteria is friendship, that is one way, while if it is a professional relationship it is another. A node in general belongs to more than one community. If one focusses on an individual node, it lies at the center of different communities (see Fig. 17.4). Within each of these communities also, there are subgraphs which may or may not have overlaps. Considering such a scenario, Pallà et al. [67] developed a new concept, the "network of communities", in which the communities acted as nodes and the overlap between communities constituted the links.

Here a community was assumed to consist of several complete subgraphs (k cliques [68]) which tend to share many of their nodes and a single node belongs to more than one community. Pallà et al. [67] defined certain quantities:

1. The membership number m which is the number of communities a node belongs to.

2. Size of a community n_s which is the number of nodes in it.

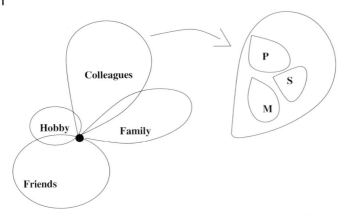

Fig. 17.4 A node (black circle) belonging to different communities. A particular community, e.g., the community of colleagues, may again contain subcommunities P,M,S, etc.

3. Overlap size $s^{\alpha,\beta}$, the number of nodes shared by communities α and β.

Two quantities related to the network of communities were also defined:

1. Community degree k_c – the number of communities a community is attached to.

2. Clustering coefficient of the community network.

A method was used in which the k cliques are first located and then the identification is carried out by a standard analysis of the clique-clique overlap matrix [68]. Identification and the calculation of the distributions for the quantities m, n_s, $s^{\alpha,\beta}$ and k_c were done for the co-authorship, word association and biological networks.

The community size distributions $P(n_s)$ for the different real networks appear to be power-law distributed ($n_s^{-\tau}$) with the value of the exponent τ between -2 and -2.6. At the node level, the degree distribution is often a power law in many networks. Whether at a higher level, i.e., at the community level it still has the same organization is the important question. The degree distribution of the communities showed an initial exponential decay followed by a power-law behavior with the same exponent τ.

The distribution of the overlap size s^{ov} and the membership number m also showed power-law variations with both the distributions having a large exponent for the collaboration and word association networks. This shows that there is no characteristic overlap size and that a node can be the member of a large number of communities.

The clustering coefficients of the community networks were found to be fairly large, e.g., 0.44 for the cond-mat collaboration network, showing that two communities overlapping with a given community have a large probability of overlapping with each other as well.

17.7.4.1 Preferential Attachment of Communities

As the community degree distribution (or the size distribution) is scale-free, the next question which arises is what happens to a new node which joins a community, i.e., what is the attachment rule chosen for the new node as regards the size and degree of the communities [69]?

To determine the attachment probability with respect to any property ρ, the cumulative distributions $P(\rho)$ at times t and $t+1$ are noted. Let the unnormalized ρ distribution of the chosen objects during the time interval t and $t+1$ be $w_{t \to t+1}(\rho)$. The value of $w_{t \to t+1}(\rho^*)$ at a given ρ^* equals the number of objects chosen which have a ρ value greater than ρ^*. If the process is uniform, then objects chosen with a given ρ are chosen at a rate given by the distribution of ρ amongst the available objects. However, if the attachment mechanism prefers high (low) ρ values, then objects with high (low) ρ are chosen with a higher rate compared to the distribution $P(\rho)$. In real systems, the application of the method indicated that, rather than a uniform attachment, there is a preferential attachment even at the higher level of networks.

17.8
Models of Social Networks

In this section we discuss some models of social networks which particularly aim at reproducing the community structure and/or positive assortativity. We also review some dynamical models which show rich phase transitions.

17.8.1
Static Models

By static models we mean a social model where the existing relations between people remain unchanged while new members can enter and form links and the size of the system may grow. Examples are collaboration networks where existing links cannot be rewired.

Newman and Park [46] have constructed a model which has a given community division. The aim is to show that it has positive assortativity. Here it is assumed that members belonging to the same community are linked with probability p similar to a bond percolation [70] problem. An individual may be attached to more than one community, and the number of members in a community is assumed to be a variable here.

The assortativity coefficient r was calculated here in terms of p and the moments of the distributions of m (the number of communities to which a member belongs) and s (the community size). The theoretical formula was applied to real systems after estimating the distributions by detecting communities using standard algorithms. The value of p was calculated by dividing the number of edges in the network by the total number of possible within-group edges. The theoretical value of r for the co-authorship network turns out to be 0.145 which is within the statistical error of the real value 0.174 ± 0.045.

Another model was proposed by Catanzaro et al. [71] to reproduce the observations of a specific database, the cond-mat archive of preprints. This network of co-authors was found to be scale-free with positive assortativity. In the model the preferential attachment scheme was used with some modification: the new node gets linked to an existing node with preferential attachment but with a probability p. Two existing nodes with degree k_1 and k_2 could also get linked stochastically with the probability being proportional to $(1-p)f(|k_1 - k_2|)$. $f(x)$ was chosen to be decaying with x either exponentially or as an inverse power law. It was shown that, by tuning the parameter p, one could obtain different values of the exponents for the degree distribution, clustering coefficient, assortativity and betweenness. The results agree fairly well for all the quantities of the actual network except the clustering coefficient.

Boguñá et al. [72] have used the concept of social distance in their model. A set of quantities h_n^i for the nth individual is used to represent the characteristic features of individuals like profession, religion, geographic locations etc. Social distance between two individuals n and m with respect to characteristic i can be quantified by the difference of h_n^i and h_m^i. It is assumed that two individuals with larger social distance will have a smaller probability of getting acquainted. If the social distance with respect to the ith feature is denoted by $d_i(h_n^i, h_m^i)$, then a connection probability is defined as

$$r_i(h_n^i, h_m^i) = \frac{1}{1 + [b_i^{-1} d_i(h_n^i, h_m^i)]^\alpha} \tag{17.15}$$

where α is a parameter measuring homophily, the tendency of people to connect to similar people and b_i^{-1} is a characteristic scale. The total probability of a link to exist between the two individuals is a weighted sum of these individual probabilities

$$r(h_n, h_m) = \sum_i w_i r_i(h_n^i, h_m^i) \tag{17.16}$$

The degree distribution, the clustering coefficient and the assortativity (in terms of $\langle k_{nn} \rangle$) were obtained analytically for the model and also compared to simulation results. The resulting degree distribution had a cutoff, the clustering coefficient showed an increase with α and the assortativity was found to be positive. The model also displayed a community structure when tested with the GN algorithm.

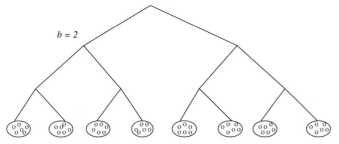

Fig. 17.5 A hierarchical structure of human population obtained using a single characteristic feature. People belonging to the same lowest group (ellipse) have social distance 1 and the distance increases as one goes up in the hierarchy. The maximum distance in this case is $l = 4$.

In the model proposed by Wong et al. [73], nodes are distributed in a Euclidean space and the connection probabilities depend on the distance separating them. Typically, a neighborhood radius is defined within which nodes are connected with a higher probability. For chosen values of the parameters in the scheme, graphs were generated which showed community structure and a small-world effect.

A hierarchical structure of the society has been assumed by Motter et al. [74] in their model in which the concept of social characteristics is used. Here a community of N people are assumed to have H relevant social characteristics. Each of these characteristic defines a nested hierarchical organization of groups, where people are split into smaller and smaller subgroups downwards in this nested structure (Fig. 17.5). Such a hierarchy is characterized by the number l of levels, the branching ratio b at each level and the average number g of people in the lowest group.

Here also distance factors corresponding to each feature can be defined and it is assumed that the distances are correlated, i.e., two individuals are expected to be different (similar) to each other in the same degree with respect to different characteristics. Connection probability decreases exponentially with the social distance and links are generated until the number of links per person reaches a preassigned average value. Networks belonging to the random and regular class are obtained for extreme values of the parameters and in between there is a wide region in which social networks fall.

A growing model of collaboration network to simulate the time-dependent feature of the link-length distribution observed in [22] has been proposed recently by Chandra et al. [75]. Here the nodes are embedded on a Euclidean two-dimensional space and an incoming node i gets attached to (a) its nearest neighbor j, (b) to the neighbors of j with a certain constant probability and (c) to other nodes with a time-dependent probability. Step (a) takes care of the proximity bias, step (b) ensures high clustering and step (c) takes care of the fact that long-distance collaborations increase in time due to rapid progress in communication. The link-length distributions obtained from the simulations are consistent with the observed results and the model also gives a positive assortativity.

17.8.2
Dynamical Models

Dynamical models allow the links to change in time whether or not the network remains constant in size. This is close to reality as human interactions are by no means static [76, 77]. These models are applicable to cases where direct human interactions are involved.

In Jin et al. [78], the model consists of a fixed number of members and links may have a finite lifetime. Here it is assumed that the probability of interaction between two individuals depends on their degrees and is enhanced by the presence of a mutual friend. A cutoff in the number of acquaintances ensures that the total number of links has an upper bound. The strength s_{ij} of a tie between individuals i and j is a function of the time t since they last met; $s_{ij} \sim \exp(-\kappa t)$. Starting with a set of individuals with no links, edges are allowed to appear with a small value of κ close to zero. When the network saturates, κ is set to a larger value and the evolution of the network studied. The resulting network shows features of large clustering and community structure for a realistic range of values of κ.

A somewhat similar model has been proposed by Ebel et al. [79] where new acquaintances are initiated by mutual friends with the additional assumption that a member may leave the network with probability p. All its links are deleted in that case. The finite lifetime of links brings the network to a stationary state which manifests large clustering coefficients, small diameter and scale-free or exponential degree distribution depending on the value of p.

Another model which aims at forming communities is due to Grönlund and Holme [80] who construct a model based on the assumption that individual people's psychology is to try to be different from the crowd. This model is based on the agent-based model known as the seceder model [81]. It involves an iterative scheme where the number of nodes remains constant and the links are generated and rewired during the evolution. Networks generated by this algorithm indeed showed community structure and a small-world effect while the degree distribution had an apparent exponential cutoff. The clustering coefficient was found to be much higher than the corresponding random network while the assortativity coefficient r showed a positive value.

Some of the dynamic models are motivated by the model proposed by Bonabeau et al. [82] in which agents are distributed randomly in an $L \times L$ lattice and assigned a "fitness" variable h. The agents perform a random walk on the lattice and during an interaction of agents i and j, the probability that i wins over j is q_i given by

$$q_i = \frac{1}{1 + \exp(\eta(h_i - h_j))} \tag{17.17}$$

If the fluctuation σ of the q_i values is small, the resulting society is egalitarian while, for large values of q, it is a society with a hierarchical structure. In the model the individuals occupy the lattice with probability p. A phase transition is observed as the value of p is changed (in a slightly modified model where η is replaced by σ). Gallos [83] has shown that when a model with random long-distant connections is considered, the phase transition remains with subtle differences in the behavior of the distribution of q.

In another variation of the Bonabeau model, Ben-Naim and Redner [84] allow the fitness variable to increase by interactions and decline by inactivity. The latter occurs with a rate r. The rate equation for the fraction of agents with a given fitness was solved to show a phase transition as r is varied; with $r \geq 1$, one has a society with a single class and for $r < 1$, a hierarchical multiple class society can exist. In the latter case, the lower class is destitute and the middle class is dynamic and has a continuous upward mobility.

Individuals are endowed with a complex set of characteristics rather than a single fitness variable. Models which allow dynamics of these social characteristics have also been proposed recently. Transitions from a perfect homogeneous society to a heterogeneous society has been found in the works of Klemm et al. [85] and Gonzalez-Avella et al. [86].

Let each trait take any of the integral values $1, 2, ..., q$. When individuals have nearest-neighbor interactions and the initial traits follow a Poisson distribution, the nonuniformity in the traits can drive the system to a heterogeneous state at $q = q_c$ [87]. Klemm et al. found that, on a small world network, the transition point q_c shifts towards higher values as the disorder (fraction of long-range bonds) is enhanced and the transition disappears in a scale-free BA network in the thermodynamic limit. In [86], the additional influence of an external field has been considered.

Shafee [88] has considered a spin-glass-like model of a social network which is described by the Hamiltonian

$$H = -\sum_{i,k,a,b} J_{ik}^{ab} s_i^a s_k^b - \sum_{i,a} h_i^a s_i^a \tag{17.18}$$

where s_i^a is the state of the ith agent with respect to trait a, J_{ik}^{ab} are the interactions between different agents and h_i^a is an external field. Zero-temperature Monte Carlo dynamics are then applied to the system. However, the simulations were restricted to systems of very small size. Results showed either punctuated equilibrium or oscillatory behavior of the trait values with time.

17.9
Is it Really a Small World? Searching: Post Milgram

Although Milgram's experiments have been responsible for inspiring research in small world networks, they have their own limitation. The chain completion percentage was too low – in the first experiment it was just five. The later experiment had a little better success rate but this time the recruit selection was by no means perfectly random. Actually, a number of subtle effects can have profound bearing on what can be termed "searching in small worlds". First, although the network may have a small-world property, searches are done mainly locally – the individual may not know the global structure of the network that would help then to find the shortest path to a target node. Secondly, many factors like the race, income, family connections, job connections, friend circle, etc., determine the dynamics of the search process. Quite a few experiments in this direction are being carried out currently. Some theoretical approaches have also been proposed for the searching mechanism.

17.9.1
Searching in Small-world Networks

The first theoretical attempt to find out how a search procedure works in a small-world network was made by Kleinberg [24] in which it was assumed that the nodes were embedded on a two-dimensional lattice. Each node was connected to its nearest neighbor and a few short cuts (long-range connections) were also present to facilitate the small-world effect. The probability that two nodes at a Euclidean distance l had a connection was taken to be $P(l) \sim l^{-\alpha}$. The algorithm used was a "greedy algorithm", i.e., each node would send the message to one of its nearest neighbors, and look for the neighbor which takes it closest to the target node. Here the source and target nodes are random. Interestingly, it was found that only for a special value of $\alpha = 2$, did the time taken (or the number of steps) scale as $\log(N)$, while for all other values it varied sublinearly with N. This result, according to [24] could be generalized to any lattice of dimension d and there always existed a unique value of $\alpha = d$ where short paths between two random nodes could be found. Navigation on the WS model [89] and a one-dimensional Euclidean network [90] agreed perfectly with the above picture.

The above results indicate that, although the network may have small average shortest paths globally, it does not necessarily mean that short chains can be realized using only local information.

17.9.2 Searching in Scale-free Graphs

In a scale-free network with degree distribution given by $P(k) \sim k^{-\tau}$, one can consider two kinds of search, one random and the other one biased. In the latter, one has the option of choosing neighboring nodes with higher degree which definitely makes the searching process more efficient.

Adamic et al. [91] have assumed that, apart from the knowledge about one's nearest neighbors, each member also has some knowledge about the contacts of the second neigbours. Using this assumption, the average search length s in the random case is given by

$$s \sim N^{3(1-2/\tau)} \tag{17.19}$$

Only for $\tau = 2$, the search length is $O(\ln(N)^2)$. The corresponding expression for the biased search, when nodes with larger degree are used, is

$$s \sim N^{2-4/\tau} \tag{17.20}$$

Simulations were done with random and biased search mechanisms where a node could scan both its first and second neighbors. The results for $\tau \neq 2$ indeed yielded the shortest paths which scale sublinearly with N, although in a slower manner compared to the theoretical predictions.

17.9.3 Search in a Social Network

A social network comprizing of people whose links are based on friendship or acquaintance can be thought of as a network where the typical degree of each node is $k = O(10^3)$. Hence the number of second neighbors should be k^2 and therefore, ideally in two steps one can access 10^6 other people and the search procedure would be complete if the order of total population is similar. Even if the degree is one order less, the number of steps is still quite small and search paths can be further shortened by taking advantage of highly connected individuals. However, there is a flaw in this argument as there is a considerable overlap of the set of one's friends and that of one's friends' friends.

It is expected that the participants in a real searching procedure would like to take advantage of certain features like geographical proximity or similarity of features such as profession and hobbies, etc. with this background the hierarchical structure of the social network is extremely significant (Fig. 17.5).

Watts et al. [92] have considered a hierarchical model in which the individuals are endowed with network ties and identities by a set of characteristics as described in Section 17.8.1. The Distance between individuals is calculated as the height of their lowest common ancestor level in the hierarchy.

The probability of acquaintance between two individuals is assumed to be proportional to $\exp(-\alpha x)$ where α is the measure of homophily and x the social distance. Unlike [74], here it is assumed that the distance between two individuals for two different features is uncorrelated. One can consider the minimum distance y_{ij}, the smallest of the x_{ij}s corresponding to all the social features, to be sufficient to denote affiliation.

It is assumed that the individual who passes the message knows only its own coordinates, its neighbors' coordinates and the coordinates of the target node. Thus the search process is based on partial information, the information about social distance and network paths are both only known locally.

In this work one important aspect of the original experiments by Milgram and his co-workers was brought under consideration, that most of the search attempts failed but when it was successful it took only a few steps. Introducing a failure probability p that a node fails to carry forward the message, it should be noted that, if the probability of a successful search of length s is r, it must satisfy $q = (1-p)^s \geq r$. This gives an upper limit for s; $s \leq \log(r)/\log(1-p)$. In the simulations, the number of traits H and the value of the parameter α were varied keeping $p = 0.25$ and $r = 0.05$ fixed (these values are in accordance with realistic values which give $s \leq 10.4$). The values of the average number of nearest neighbors and branching ratio were also kept constant. A phase diagram in the $H - \alpha$ plane showed regions where the searching procedure can be successful. It showed that almost all searchable networks display $\alpha > 0$ and $H > 1$. The best performance, over the largest range of α, is achieved for $H = 2$ or 3. In fact the model could be tuned to reproduce the experimental results of Milgram ($s \sim 6.7$).

17.9.4
Experimental Studies of Searching

A few projects which study searching in social networks experimentally have been initiated recently. Dodds et al. [93] have conducted a global, Internet-based social search by registering participants online. Eighteen target persons from thirteen countries with varied occupation were randomly selected and the participants were informed that their task was to help relay a message to their allocated target by passing the message to a social acquaintance whom they considered closest to the target. One-quarter of the participants provided their personal information and it showed that the sample was sufficiently representative of the general Internet-using population. The participants also provided data as to the basis on which they chose the contact's name and e-mail address; in maximum cases they were friends, followed by relatives and colleagues.

The links in this experimental network showed that geographical proximity of the acquaintance to the target and similarity of occupation were the two major deciding factors behind their existence. Many of the chains were terminated and not completed as in Milgram's experiment and the reason behind it was mainly lack of interest or incentive. In total, 384 chains were completed (nearly 100 000 persons registered in the beginning). It was also found that when chains did complete, they were short, the average length s being 4.05. However, this is a measure for completed chains only, and the hypothetical estimate in the limit of zero attrition comes out to be $s = 7$.

In another study, Adamic and Adar [94] derived the social network of emails at HP laboratories from the email logs by defining a social contact to be someone with whom at least six emails have been exchanged both ways over an approximate period of three months. A network of 430 individuals was generated and the degree distribution showed an exponential tail. Search experiments were simulated on this network. Three criteria for sending messages in the search strategy were tested in this simulation; degree of the node, closeness to the target in the organizational hierarchy and location with respect to the target.

In a scale-free network, seeking a high degree node was shown to be a good search strategy [91]. However, in this network, which has an exponential tail, this does not prove to be effective. This is because most of the nodes do not have a neighbor with high degree. The second strategy of utilizing the organising hierarchy worked much better, showing that the hierarchical structure in this network is quite appropriate. The probability of linking as a function of the separation in the organizational hierarchy also showed consistency with an exponential decay as in [92]. The corresponding exponent α has a value = 0.94 and is well within the searchable region identified in [92].

The relation between linkage probability and distance r turned out to be $1/r$ in contrast to $1/r^2$ where the search strategy should work best according to [24]. Using geographical proximity as a search strategy gave larger short paths compared to the search based on the organizational hierarchy, but the path lengths were still "short".

A similar experiment was conducted by Liben-Nowell et al. [95], using a real-world friendship network to see how the geographical routing alone is able to give rise to short paths. In this simulated experiment, termination of chains was allowed. Chain completion was successful in 13% of cases with an average search length a little below 6.

17.10
Endnote

The study of social networks has great practical implications with respect to many phenomena like spreading of information, epidemic dynamics, behavior under attack, etc. In these examples, the analysis of the corresponding networks can help in making proper strategies for voting, vaccination or defence programmes. In this particular review, we have mainly highlighted the theoretical aspects of social networks like structure, modeling, phase transitions and searching. Social networks have emerged truly as an interdisciplinary topic during recent times and the present review is a little biased as it is from the viewpoint of a physicist.

17.11
Appendix: The Indian Railways Network

The Indian Railways Network (IRN) is more than 150 years old and has a large number of stations and trains (running at both short and long distances). In the study of the IRN [45], a coarse graining was made by selecting a sample of trains, and the stations through which these run. The total number of trains considered was 579 which run through 587 stations. Here the stations represent the "nodes" of the graph, whereas two arbitrary stations are considered to be connected by a "link" when there is at least one train which stops at both the stations. These two stations are considered to be at unit distance of separation irrespective of the geographical distance between them. Therefore the shortest distance ℓ_{ij} between an arbitrary pair of stations s_i and s_j is the minimum number of different trains one needs to board to travel from s_i to s_j. Smaller subsets of the network were also considered to analyze the behavior of different quantities as a function of the number of nodes. The average distance between an arbitrary pair of stations was found to depend only logarithmically on the total number of stations in the country as shown in Fig. 17.6.

Like other social networks, the IRN also showed a large clustering coefficient which also depends on the number of nodes (stations). For the entire IRN, the clustering coefficient is around 0.69, the value for the corresponding random graph being 0.11. The clustering coefficient as a function of the degree k of a node showed that it is a constant for small k and decreases logarithmically with k for larger values.

The degree distribution $P(k)$ of the network, that is, the distribution of the number of stations k which are connected by direct trains to an arbitrary station was also studied. The cumulative degree distribution $F(k) = \int_k^\infty P(k)dk$ for the whole IRN approximately fits to an exponentially decaying distribution: $F(k) \sim \exp(-\alpha k)$ with $\alpha = 0.0085$ (Fig. 17.7).

The average degree $\langle k_{nn}(k) \rangle$ of the nearest neighbors of a node with degree k is plotted in Fig. 17.8 to check the assortativity behavior of the network. This data is not very indicative. The assortativity coefficient r is therefore calculated using Eq. (17.4), which gives the value $r = -0.033$. This shows that, unlike social networks of class A and B, here the assortativity is negative.

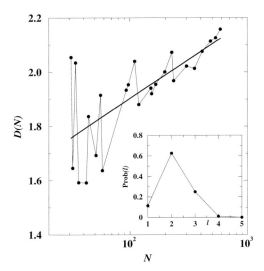

Fig. 17.6 The variation of the mean distance $\mathcal{D}(N)$ of 25 different subsets of IRN having different numbers of nodes (N). The whole range is fitted with a function like $\mathcal{D}(N) = A + B\log(N)$ where $A \approx 1.33$ and $B \approx 0.13$. The inset shows the distribution $\text{Prob}(\ell)$ of the shortest path lengths ℓ on IRN. The lengths varied to a maximum of only five link lengths and the network has a mean distance $\mathcal{D}(N) \approx 2.16$.

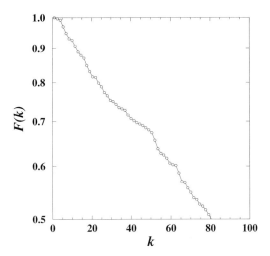

Fig. 17.7 The cumulative degree distribution $F(k)$ of the IRN with degree k is plotted on the semi-logarithmic scale.

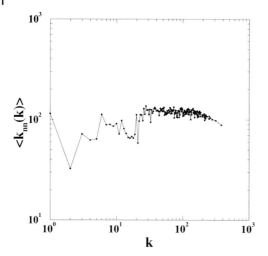

Fig. 17.8 The variation of the average degree $\langle k_{nn}(k) \rangle$ of the neighbors of a node of degree k with k in the IRN. After some initial fluctuations, $\langle k_{nn}(k) \rangle$ remains almost the same over a decade around $k = 30$–300 indicating an absence of correlations among the nodes of different degrees.

References

1. WASSERMAN, S., FAUST, K., *Social Network Analysis*, Cambridge University Press, **1994**
2. WATTS, D. J., STROGATZ, S. H., *Nature* 393 (**1998**), p. 440
3. ALBERT, R., BARABÁSI, A.-L., *Rev. Mod. Phys.* 74 (**2002**), p. 47
4. BARABÁSI, A.-L., *Linked*, Perseus, **2002**
5. DOROGOVTSEV, S. N., MENDES, J. F. F., *Evolution of Networks*, Oxford University Press, **2003**
6. MILGRAM, S., *Psychology Today* 1 (**1967**) p. 60; TRAVERS, J., MILGRAM, S., *Sociometry* 32 (**1969**), p. 425
7. ERDŐS, P., RÉNYI, A., *Publ. Math.* 6 (**1959**), p. 290
8. BARABÁSI, A.-L., ALBERT, R., *Science* 286 (**1999**), p. 509
9. FREEMAN, C., *Sociometry* 40 (**1977**), p. 35
10. GOH, K.-I., OH, E., KAHNG, B., KIM, D., *Phys. Rev. E* 67 (**2003**), p. 017101
11. NEWMAN, M. E. J., *Phys. Rev. Lett.* 89 (**2002**), p. 208701
12. EBEL, H., MIELSCH, L., BORNHOLDT, S., *Phys. Rev. E* 66 (**2002**), p. 035103R
13. BARRAT, A., WEIGT, M., *Eur. Phys. J. B* 13 (**2000**), p. 547; GITTERMAN, M., *J. Phys A* 33 (**2000**), p. 8373; KIM, B. J., HONG, H., HOLME, P., JEON, G. S., MINNHAGEN, P., CHOI, M. Y., *Phys. Rev. E* 64 (**2001**), p. 056135 HONG, H., KIM, B. J., CHOI, M. Y., *Phys. Rev. E* 66 (**2002**), p. 018101
14. HUBERMAN, B. A., ADAMIC, L. A., *Nature* 401 (**1999**), p. 131
15. MASUDA, N., MIWA, H., KONNO, N., *Phys. Rev. E* 71 (**2005**), p. 036108
16. YOOK, S.-H., JEONG, H., BARABÁSI, A.-L., *Proc. Natl. Acad. Sci. USA* 99 (**2002**), p. 13382
17. KATZ, J. S., *Scientometrics* 31 (**1994**), p. 31
18. NAGPAUL, P. S., *Scientometrics* 56 (**2003**), p. 403
19. ROSENBLAT, T. S., MOBIUS, M. M., *Quart. J. Econ* 121 (**2004**), p. 971
20. GASTNER, M. T., NEWMAN, M. E. J., cond-mat/0407680

21 OLSON, G. M., OLSON, J. S., *Hum. Comp. Inter.* 15 (**2004**), p. 139

22 SEN, P., CHANDRA, A. K., BASU HAJRA, K., DAS, P.,K., physics/0511181

23 WAXMAN, B., *IEEE J. Selec. Areas Commun.,* SAC 6 (**1988**), p. 1617

24 KLEINBERG, J. M., *Nature* 406 (**2000**), p. 845

25 JESPERSEN, S., BLUMEN, A., *Phys. Rev. E* 62 (**2000**), p. 6270

26 SEN, P., CHAKRABARTI, B. K., *J. Phys. A* 34 (**2001**), p. 7749.

27 MOUKARZEL, C. F., DE MENEZES, M. A., *Phys. Rev. E* 65 (**2002**), p. 056709

28 SEN, P., BANERJEE, K., BISWAS, T., *Phys. Rev. E* 66 (**2002**) p. 037102

29 MANNA, S. S., SEN, P., *Phys. Rev. E* 66 (**2002**), p. 066114

30 SEN P., MANNA S. S., *Phys. Rev. E* 68 (**2003**), p. 0206104

31 MANNA, S. S., MUKHERJEE, G., SEN, P., *Phys. Rev. E* 69 (**2004**), p. 017102

32 BASU HAJRA, K., SEN, P., *Physica A* 346 (**2005**), p. 44

33 REDNER, S., physics/0407137; REDNER, S., *Physics Today* 58 (**2005**), p. 49

34 DOROGOVTSEV, S. N., MENDES, J. F. F., *Phys. Rev. E* 62 (**2000**), p. 1842

35 ZHU, H., WANG, X., ZHU, J.-Y., *Phys. Rev. E* 68 (**2003**), p. 058121

36 BASU HAJRA, K., SEN, P., *Physica A* (**2006**), in press (cond-mat/0508035)

37 NEWMAN, M. E. J., *Phys. Rev. E* 67 (**2003**), p. 026126

38 SCHNEGG, M., physics/0603005

39 ABELLO, J., PARADALOS, P. M., RESENDE, M. G. C., *DIMACS Series in DIsc. Math and Theo. Comp. Sc* 50 (**1999**), p. 119

40 CSÀNYI, G., SZENDRÖI, B., *Phys. Rev. E* 69 (**2004**), p. 036131

41 NEWMAN, M. E. J., *Proc. Natl. Acad. Sci. USA* 98 (**2001**), p. 404; NEWMAN, M. E. J., *Phys. Rev. E* 64 (**2001**), p. 016131; p. 016132

42 BARABÁSI, A.-L., JEONG, H., NEDA, Z., RAVASZ, E., SCHUBERT, A., VICSEK, T., *Physica A* 311 (**2002**), p. 590

43 REDNER, S., *Eur. Phys. J. B* 4 (**1998**), p. 131

44 VAZQUEZ, A., cond-mat/0105031

45 SEN, P., DASGUPTA, S., CHATTERJEE, A., SREERAM, P. A., MUKHERJEE, G., MANNA, S. S., *Phys. Rev. E* 67 (**2003**), p. 036106

46 NEWMAN, M. E. J., PARK, J., *Phys. Rev. E* 68 (**2003**), p. 036122

47 NEWMAN, M. E. J., *Eur. Phys. J. B* 38 (**2004**), p. 321

48 JAIN, A. K., DUBES, R. C., *Algorithms for clustering data*, Prentice Hall **1988**; EVERITT, B. S., *Cluster Analysis*, Edward Arnold **1993**

49 GIRVAN, M., NEWMAN, M. E. J., *Proc. Natl. Acad. Sci. USA* 99 (**2002**), p. 7821

50 NEWMAN, M. E. J., GIRVAN, M., *Phys. Rev. E* 69 (**2004**), p. 026113

51 DANON, L., DUCH, J., DIAZ-GUILERA, A., ARENAS, A., *J. Stat. Mech.* (**2005**), p. P09008

52 ZACHARY, W. W., *J. Anthrop. Res* 33 (**1977**), p. 452

53 ARENAS, A., DANON, L., DIAZ-GUILERA, A., GLEISER, P. M., GUIMERÀ, R., *Eur. Phys. J. B* 38 (**2004**), p. 373; GLEISER, P. M., DANON, L., *Adv. in Complex Syst.* 6 (**2003**) 565

54 RADICCHI, F., CASTELLANO, C., CECCONI, F., LORETO, V., PARISI, D., *Proc Natl. Acad. Sci. USA* 101 (**2004**), p. 2658

55 FORTUNATO, S., LATORA, V., MARCHIORI, M., *Phys. Rev. E* 70 (**2004**), p. 056104

56 NEWMAN, M. E. J., *Phys. Rev. E* 69 (**2004**), p. 066133

57 DUCH, J., ARENAS, A., cond-mat/0501368

58 CAPOCCI, A., SERVEDIO, V. D. P., CALDARELLI, G., COLAIORI, F., cond-mat/0402499

59 DONETTI, L., MUNOZ, M. A., *J. Stat. Mech.* (**2004**), p. P10012

60 DONETTI, L., MUNOZ, M. A., physics/0504059

61 ZHOU, H., *Phys. Rev. E* 67 (**2003**), p. 061901

62 BAGROW, J. P., BOLLT, E. M., *Phys. Rev. E* 72 (**2005**), p. 046108

63 WU, F., HUBERMAN, B. A., *Euro Phys. J. B* 38 (**2004**), p. 331

64 REICHHARDT, J., BORNHOLDT, S., *Phys. Rev. Lett.* 93 (**2004**), p. 218701

65 GUIMERÀ, R., SALES-PARDO, M., AMARAL, L. A. N., *Phys. Rev. E* 70 (**2004**), p. 025101R

66 Son, S.-W., Jeong, H., Noh, J. D., cond-mat/0502672

67 Palla, G., Derenyi, I., Farkas, I., Vicsek, T., *Nature* 435 (**2005**), p. 814

68 Derenyi, I., Palla, G., Vicsek, T., *Phys. Rev. Lett.* 94 (**2005**), p. 160202

69 Pollner, P., Palla, G., Vicsek, T., *Europhys. Lett.* 73 (**2006**), 478

70 Stauffer, D., Aharony, A., *An Introduction to Percolation Theory*, Taylor and Francis, **1994**

71 Catanzaro, M., Caldarelli, G., Pietronero, L., *Phys. Rev. E* 70 (**2004**), p. 037101

72 Boguñá, M., Pastor-Satorras, R., Díaz-Guilera, A., Arenas, A., *Phys. Rev. E* 70 (**2004**), p. 056122

73 Wong, L. H., Pattison, P., Robins, G., physics/0505128

74 Motter, A. E., Nishikawa, T, Lai, Y.-C., *Phys. Rev. E* 68 (**2003**), p. 036105

75 Chandra, A. K., Basu Hajra, K., Das, P. K., Sen, P., to be published.

76 Holme, P., *Europhys. Lett* 64 (**2003**), p. 427

77 Roth, C., nlin.AO/0507021

78 Jin, E. M., Girvan, M., Newman, M. E. J., *Phys. Rev. E* 64 (**2001**), p. 046132

79 Ebel, H., Davidsen, J., Bornholdt, S., *Complexity* 8(2) (**2002**), p. 24

80 Grönlund, A., Holme, P., *Phys. Rev. E* 70 (**2004**), p. 036108

81 Dittrich, P., Liljeros, F., Soulier, S., Banzhaf, W., *Phys. Rev. Lett.* 84 (**2000**), p. 3205

82 Bonabeau, E., Theraulaz, G., Deneubourg, J.-L., *Physica A* 217 (**1995**), p. 373

83 Gallos, L., physics/0503004

84 Ben-Naim, E., Redner, S., *J. Stat. Mech.* (**2005**), p. L11002

85 Klemm, K., Eguíluz, V. M., Toral, R., San Miguel, M., *Phys. Rev. E* 67 (**2003**), p. 026120

86 González-Avella, J. C., Eguíluz, V. M., Cosenza, M. G., Klemm, K., Herrera, J. L., San Miguel, M., cond-mat/0601340

87 Axelrod, R., *J. of Conflict Resolution* 41 (**1997**) p. 203

88 Shafee, F., physics/0506161

89 de Moura, A. P. S., Motter, A. E., Grebogi, C., *Phys. Rev. E* 68 (**2003**), p. 036106

90 Zhu, H., Huang, Z.-X., *Phys. Rev. E* 70 (**2004**), p. 036117

91 Adamic, L. A., Lukose, R. M., Puniyani, A. R., Huberman, B. A., *Phys. Rev. E* 64 (**2001**), p. 046135

92 Watts, D. J., Dodds, P. S., Newman, M. E. J., *Science* 296 (**2002**), p. 1302

93 Dodds, P. S., Muhamad, R., Watts, D. J., *Science* 301 (**2003**), p. 827

94 Adamic, L. A., Adar, E., *Social Networks* 27 (**2005**), p. 187

95 Liben-Nowell, D., Novak, J., Kumar, R., Raghavan, P., Tomkins, A., *Proc. Natl. Acad. Sci. USA* 102 (**2005**), p. 11623

18
Emergence of Memory in Networks of Nonlinear Units: From Neurons to Plant Cells
Jun-ichi Inoue

In this chapter, we give a review of some our recent results on associative memories, which emerge as collective phenomena in a huge number of nonlinear units. We especially focus on neural networks and a model of plant intelligence. The neural network model [1,2] which we deal with in this chapter is a kind of Hopfield model in which the connections between two arbitrary neurons are dynamically diluted. The degree of the dilution is controlled by the chemical potential term in the Hamiltonian. Phase diagrams of the equilibrium state are obtained under replica symmetric calculations. On the other hand, for the plant-intelligence model, we also introduce a cellular-network model to explain the capacity of the plants as memory devices. Following earlier observations by J. C. Bose [3], we regard the plant as a network in which each of the elements (plant cells) are connected *via* negative (inhibitory) interactions. To investigate the performance of the network, we construct a model following that of Hopfield, whose energy function possesses both Hebbian spin-glass and anti-ferromagnetic terms. We discuss the critical phenomena for the case when the number of embedded patterns is extensive. The plant-intelligence model is extended to the quantum version by taking into account the microscopic tunneling process of each neuronal state from the active state to the quiescent state. By analysis of these model systems, we try to answer the question: *Do plants act as a computer?*, which was recently presented by several authors [4,5], from the statistical mechanics of disordered spin systems point of view.

18.1
Introduction

Over twenty years, associative memories based on the so-called Hopfield model have been investigated from the viewpoint of statistical mechanics. Most of these studies are restricted to neural networks with fixed structure. However, it has been reported that in our brains, a remarkable number of neurons might die every day. Moreover, in babyhood, our brain self-organizes its structure by pruning (or killing) redundant neurons to realize good performance of learning or memory. Therefore, some mathematical models are needed to make this point clear. With this requirement in mind, we propose here an associative memory in which some neurons are pruned (killed) dynamically. To realize this scenario, we use a frustrated lattice gas (FLG) [6] which is regarded as a kind of spin glass [7]. Our main goal is to draw the retrieval phase diagrams and to evaluate to what extent pruning the neurons improves the storage capacity.

As is well known, the associative memory of the Hopfield model is mathematically understood to be the collective behavior of a huge number of neurons as nonlinear units. Then, an interesting question naturally arises. Namely, *Do plants act as memory devices?* To answer this question, we next introduce a cellular-network model to explain the capacity of the plants as memory devices. Following earlier observations (Bose [3] and others [4,8,9]), we regard the plant as a network in which each of the elements (plant cells) are connected *via* negative (inhibitory) interactions. To investigate the performance of the network, we construct a model following that of Hopfield, whose energy function possesses both Hebbian spin glass and antiferromagnetic terms. We discuss the critical phenomena for the system in which extensively many patterns are embedded. In this chapter, we also deal with the quantum version of the plant intelligence model in which each plant cell changes its state stochastically from an active state to a quiescent state by microscopic tunneling effects. By analysis of these model systems, we try to answer the question: *Do plants act as a computer?*, which was recently presented by several authors [4,5], from the statistical mechanics of quantum spin systems.

18.2
Neural Networks

We first review the associative memory of neural networks, namely, the so-called Hopfield model and its recent extension by the present author [10].

In the conventional Hopfield model [2], the state of each neuron is specified by a single variable s_i, which takes $+1$ (fire) or -1 (rest). Then, each neuron updates its state according to the following nonlinear equation:

$$s_i(t+1) = \text{sgn}\left[\sum_j J_{ij}s_j(t)\right] \tag{18.1}$$

where J_{ij} means a connection between the neurons s_i and s_j. We assume here that every neuron is connected to all the other neurons. In other words, each neuron is located on the node of *complete graph* K_N. The choice of the connection J_{ij} is an essential point to realize associative memories. We usually choose them as

$$J_{ij} = \frac{1}{N} \sum_{\mu=1}^{p} \xi_i^\mu \xi_j^\mu \qquad (18.2)$$

where $\xi^\mu = (\xi_1^\mu, \cdots, \xi_N^\mu)$ is μth embedded pattern for the network and p is the total number of the stored patterns. This method of connection is referred to as the *Hebb rule*.

One can show that for these connections $\{J_{ij}\}$, the following energy function:

$$H = -\frac{1}{N} \sum_{ij} J_{ij} s_i s_j \qquad (18.3)$$

decreases by the update rule (18.1).

The performance of the retrieval of an arbitrary pattern, say, ξ^1 is measured by the overlap:

$$m = \frac{1}{N}(\xi^1 \cdot s) = \frac{1}{N} \sum_i \xi_i^1 s_i \qquad (18.4)$$

The main goal for the quantitative analysis of the associative memory is to evaluate this overlap as a function of the number of embedded patterns p or the noise level of the network. In the following, we show a simple extension of the Hopfield model.

18.2.1
The Model System

We consider the Hopfield model described by the following Hamiltonian [10]:

$$\mathcal{H} = -\sum_{ij} J_{ij} n_i n_j s_i s_j + \frac{\gamma}{N} \sum_{ij} n_i n_j - \mu \sum_i n_i \qquad (18.5)$$

with the Hebb rule (18.2). In this Hamiltonian, each label n_i $(i = 1, \cdots, N)$ is a dynamical variable and takes $n_i \in \{0, 1\}$. Thus, each neuron i is labeled so as to survive, $n_i = 1$, or so as to die, $n_i = 0$ (see Fig. 18.1). The rate of survival is determined by the chemical potential μ appearing in the third term of the Hamiltonian (18.5). If μ satisfies $\mu \gg 1$, most of the neurons are occupied, and the model described by the Hamiltonian (18.5) is reduced to the conventional Hopfield model [1, 2]. On the other hand, if we consider the case of $\mu \ll 1$, the situation is quite different from the conventional model. The second term with positive γ has a role in preventing the localization of the "killed" neuron. In this paper, we set the γ to a positive value $\gamma = 1$. This is because in a real brain, such a localization of killed neurons might result in fatal damage for the

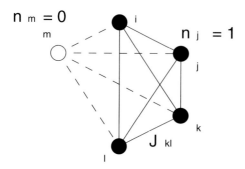

Fig. 18.1 The model system. The neuron i labeled $n_i = 0$ is killed, whereas the neuron labeled $n_i = 1$ survives.

brain functions such as learning or memory. In the field of condensed matter, the system described by the Hamiltonian Eq. (18.5) is referred to as a *Frustrated Lattice Gas* (FLG) (J_{ij} is the exchange interaction and is randomly distributed in this context) and was recently introduced by Nicodemi and Coniglio [6].

In this chapter, we investigate the retrieval phase diagrams of the model described by the above Hamiltonian Eq. (18.5). Our interests are, for example,

- how many patterns can be embedded in the network?
- how does the retrieval diagram change by dilution?

We investigate these problems by using replica-symmetric calculations [7].

18.2.2
Equations of States

In this section, we derive equations of state for the spin systems of the Hamiltonian (18.5). We consider the two cases separately, namely, the case of $p = \mathcal{O}(1)$ and the case of $p = \mathcal{O}(N)$ for the number of the embedded pattern p.

18.2.2.1 $p = 1$ Case

We first investigate the case of finite p, in particular, $p = 1$ and $N \to \infty$. For this case, the Hamiltonian $\mathcal{H}$ is rewritten as

$$\mathcal{H} = -\frac{1}{2N} \sum_{\mu} \left(\sum_i n_i \zeta_i^{\mu} s_i \right)^2 + \frac{1}{2N} \left(\sum_i n_i \right)^2 - \mu \sum_i n_i \tag{18.6}$$

Then, the partition function $Z = \text{tr}_{\{n,s\}} e^{-\beta \mathcal{H}}$ leads to

$$Z = \text{tr}_{\{n,s\}} \exp \left[\frac{\beta}{2N} \sum_{\mu} \left(\sum_i n_i \zeta_i^{\mu} s_i \right)^2 - \frac{\beta}{2N} \left(\sum_i n_i \right)^2 + \beta \mu \sum_i n_i \right] \tag{18.7}$$

This partition function is rewritten in terms of the Hubbard–Stratonovich transformation with respect to a quadratic term in the exponential as

$$Z = \text{tr}_{\{n,s\}} \int_{-\infty}^{\infty} \prod_\mu \frac{dm_\mu}{\sqrt{2\pi/N\beta}} \int_{-i\infty}^{+i\infty} \frac{idw}{\sqrt{2\pi/N\beta}} \exp\left[Nf(m,w)\right] \quad (18.8)$$

with the following free energy per neuron:

$$f(m,w) \equiv -\frac{\beta}{2}(m)^2 + \frac{\beta}{2}w^2 + \log 2\left(1 + e^{-\beta(w-\mu)}\cosh(\beta(m\cdot\xi))\right) \quad (18.9)$$

We should bear in mind that, from the saddle-point condition in the argument of the exponential appearing in the second line of equation (18.8), the meanings of the order parameters m and w are

$$m = \frac{1}{N}\sum_{i=1}^{N} n_i \xi_i s_i, \quad w = \frac{1}{N}\sum_{i=1}^{N} n_i \quad (18.10)$$

respectively.

We here consider the simplest case, namely the case of only a single pattern being embedded and we set $\xi = (1, 0, \ldots, 0)$. For this case, the free-energy density f leads to

$$f = -\frac{\beta}{2}m^2 + \frac{\beta}{2}w^2 + \log 2\left(1 + e^{-\beta(w-\mu)}\cosh(\beta m)\right) \quad (18.11)$$

The saddle-point equations with respect to m and w lead to

$$m = \frac{e^{-\beta(w-\mu)}\sinh(\beta m)}{1 + e^{-\beta(w-\mu)}\cosh(\beta m)} \quad (18.12)$$

$$w = \frac{e^{-\beta(w-\mu)}\cosh(\beta m)}{1 + e^{-\beta(w-\mu)}\cosh(\beta m)} \quad (18.13)$$

respectively. Then we consider the limit of $\mu \to \infty$. We easily obtain

$$m = \tanh(\beta m), \quad w = 1 \quad (18.14)$$

We see that, in this limit, our model corresponds to the conventional Hopfield model. In Fig. 18.2(a), we plot the overlap m as a function of T. In this figure, we choose the chemical potential μ as $\mu = -0.05, 0, 0.05, 0.2, 1, 2$ and $\mu = 3$. At the critical temperature T_c, the system changes from the ferromagnetic retrieval phase ($m \neq 0$) to the paramagnetic phase ($m = 0$).

We next evaluate the transition temperature T_c between the ferromagnetic retrieval and the paramagnetic phases. By expanding the Eq. (18.12) and (18.13) up to $\mathcal{O}(m)$, the critical temperature $T_c = \beta_c^{-1}$ is a solution of the following equations

$$\frac{1}{\beta_c} = T_c = \frac{e^{-\beta_c(w_c-\mu)}}{1 + e^{-\beta_c(w_c-\mu)}} = w_c \quad (18.15)$$

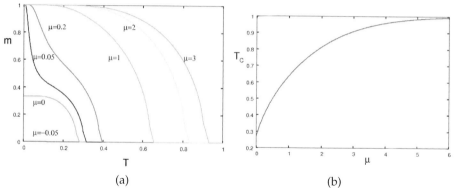

Fig. 18.2 (a) Overlap of m as a function of T. We choose the chemical potential $\mu = -0.05, 0, 0.05, 0.2, 1, 2$ and $\mu = 3$. At the critical temperature T_c, the system changes from the ferromagnetic retrieval phase to the paramagnetic phase. (b) shows the critical temperature T_c at which the system undergoes the second-order phase transition to the paramagnetic state ($m = 0$). T_c is obtained as a solution of the equation: $T_c = w_c = e^{\mu/w_c}/(e + e^{\mu/w_c})$.

namely, T_c is given by a solution of $T_c = e^{\frac{\mu}{T_c}}/e + e^{\frac{\mu}{T_c}}$. In Fig. 18.2(b), we plot the critical temperature T_c as a function of μ. Below the critical temperature T_c, the network acts as an associative memory device. As μ increases, the critical temperature increases monotonically to 1.

18.2.2.2 Entropy of the System

We next evaluate the entropy of the system. The entropy s of the system per neuron is given in terms of the derivative of the free-energy density:

$$f_s = \frac{m^2}{2} - \frac{w^2}{2} - \beta^{-1}\log 2[1 + e^{-\beta(w-\mu)}\cosh(\beta m)] \tag{18.16}$$

with respect to inverse temperature β, namely, $s = \beta^2(\partial f_s/\partial \beta)$ as

$$s = \beta w^2 - \beta m^2 + \log 2[1 + e^{-\beta(w-\mu)}\cosh(\beta m)] - \beta\mu w \tag{18.17}$$

We first check the hight-temperature limit $\beta \to 0$ of the entropy s. In this limit, we easily obtain

$$s(\beta = 0) = \log 4 = 2\log 2 \tag{18.18}$$

This result is reasonable because at high temperature, the possible state of the system is $s \otimes n \in \Re^{4N}$, and the entropy per site is naturally given by $(\log 4^N)/N = 2\log 2$. We next consider the entropy at the ground state. If we consider the case of $m > 0$, the saddle-point equations with respect to m and

w lead to

$$m = \frac{e^{-\beta(w-\mu)}\frac{e^{\beta m}}{2}}{1+e^{-\beta(w-\mu)}\frac{e^{\beta m}}{2}} = w \tag{18.19}$$

As a result, we conclude $m = w$. Substituting this result into the expression of the entropy, we obtain

$$s(T=0) = \log 2 \left[1 + e^{-\beta(w-\mu)} \cdot \frac{e^{\beta m}}{2}\right] - \beta\mu = \log e^{\beta\mu} - \beta\mu = 0. \tag{18.20}$$

On the other hand, for $m < 0$, we easily obtain $m = -w$, then the entropy density at the ground state also becomes zero.

18.2.2.3 Internal Energy Density

We next evaluate the internal energy density of the system. Using the thermodynamic relation $f_s = u_s + s/\beta$ and (18.16) and (18.17), we obtain

$$u_s = \frac{w^2}{2} - \frac{m^2}{2} - \mu w \tag{18.21}$$

where m and w are solution of the saddle-point equations (18.12) and (18.13).

18.2.2.4 Compressibility

The compressibility κ of our *neuronal gas* system is easily obtained by using the relation $\kappa = \beta^{-1}(\partial w/\partial \mu)$. Substituting

$$\frac{\partial w}{\partial \mu} = \frac{\beta e^{\beta(w-\mu)}\cosh(\beta m)}{[e^{\beta(w-\mu)} + \cosh(\beta m)]^2} \tag{18.22}$$

into the expression of κ, we obtain

$$\kappa = \frac{e^{\beta(w-\mu)}\cosh(\beta m)}{[e^{\beta(w-\mu)} + \cosh(\beta m)]^2} \tag{18.23}$$

where m and w obey the following saddle-point equations (18.12) and (18.13). In Fig. 18.3(a), we plot the compressibility κ as a function of μ for the case of $T = 0.2, 0.05$ and $T = 0.005$. We find that at the ground state, the compressibility κ diverges at $\mu = 0 \equiv \mu_c$. In (b), we also plot the corresponding w. At the ground state, w changes from $w = 1$ to $w = 0$ discontinuously at $\mu = \mu_c = 0$. At the ground state, $m = w = \Theta(\mu)$ holds and κ leads to

$$\kappa(T=0) = \beta^{-1}\left(\frac{\partial w}{\partial \mu}\right)\bigg|_{\mu=0} = \beta^{-1}\left(\frac{\partial \Theta(\mu)}{\partial \mu}\right)\bigg|_{\mu=0} = \delta(\mu) \tag{18.24}$$

Thus, the divergence of the compressibility at $\mu = 0$ is analytically observed.

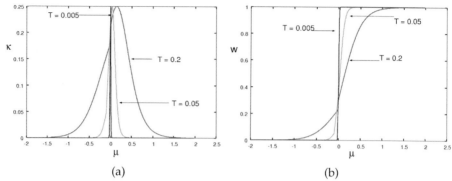

Fig. 18.3 (a) The compressibility κ as a function of μ for the case of $T = 0.2, 0.05$ and $T = 0.005$. (b) we plot the corresponding order parameter w as a function of μ. (For a color version please see color plate on page 602.)

18.2.2.5 Overlap at the Ground State for $\mu = 0$

We check the overlap m for $\mu = 0$. The saddle-point equations for $\mu = 0$ lead to

$$m = \frac{e^{-\beta w} \sinh(\beta m)}{1 + e^{-\beta w} \cosh(\beta m)} \tag{18.25}$$

$$w = \frac{e^{-\beta w} \cosh(\beta m)}{1 + e^{-\beta w} \cosh(\beta m)} \tag{18.26}$$

From these two equations, we find $(m/w) = \tanh(\beta m)$, namely,

$$w = \frac{m}{\tanh(\beta m)} \tag{18.27}$$

Substituting this relation into the Eq. (18.25), we obtain

$$\sinh(\beta m) = m \, e^{\frac{\beta m}{\tanh(\beta m)}} + m \cosh(\beta m) \tag{18.28}$$

When we restrict ourselves to the case of $m > 0$ in the limit of $\beta \to \infty$, $\tanh(\beta m) = 1, \sinh(\beta m) = \cosh(\beta m) = e^{\beta m}/2$ hold and we have $e^{\beta m}/2 = m \, e^{\beta m} + m \, e^{\beta m/2}$. This equation is easily solved with respect to m and we obtain $m = 1/3$.

18.2.3
Replica Symmetric Calculations for the Case of Extensive Patterns

We next consider the system in which extensively many patterns are embedded, namely, $p = N = \mathcal{O}(N)$ and the ratio $\alpha \equiv p/N = \mathcal{O}(1)$. In order to average the free energy over the embedded patterns ξ, that is, $\ll \log Z \gg_\xi$, we use the replica method:

$$\ll \log Z \gg_\xi = \lim_{n\to 0} \frac{\ll Z^n \gg_\xi - 1}{n} \tag{18.29}$$

Then, we introduce the replica spins s_i^α and n_i^α for each i, and write the replicated partition function Z^n as follows.

$$Z^n = \exp\left[\frac{\beta}{N}\sum_{\alpha\mu}\sum_{ij}\xi_i^\mu \xi_j^\mu s_i^\alpha s_j^\alpha n_i^\alpha n_j^\alpha - \frac{\beta}{N}\sum_{\alpha\mu}\sum_{ij} n_i^\alpha n_j^\alpha + \mu\beta\sum_\alpha\sum_i n_i^\alpha\right] \tag{18.30}$$

By standard replica calculations, the average of the replicated partition function $\ll Z^n \gg_\xi$ leads to

$$\ll Z^n \gg_\xi = \prod_{\alpha\beta\mu} \int_{-\infty}^{\infty} \frac{dm_\mu^\alpha}{\sqrt{2\pi/N\beta}} \int_{-i\infty}^{+i\infty} \frac{dt^\alpha}{i\sqrt{2\pi/N\beta}} \int_{-\infty}^{\infty} \frac{dw_\alpha}{2\pi i} \int_{-\infty}^{\infty} dq_{\alpha\beta}$$

$$\times \int_{-i\infty}^{+i\infty} \frac{dr_{\alpha\beta}}{2\pi i} \exp[-N f_{\beta\mu}(m,q,w,\hat{w},r,t)] \tag{18.31}$$

where we define the free energy density as

$$f_{\beta\mu}(m,q,w,\hat{w},r,t) = \frac{\beta}{2}\sum_{\alpha\mu}(m_\mu^\alpha)^2 - \frac{\beta}{2}\sum_\alpha (t^\alpha)^2$$
$$- \beta\sum_\alpha t^\alpha w_\alpha + \beta\mu\sum_\alpha w_\alpha$$
$$+ \sum_\alpha w_\alpha \hat{w}_\alpha + \sum_{\alpha\beta} r_{\alpha\beta} q_{\alpha\beta} + \frac{\alpha}{2}\sum_\alpha \log \hat{\lambda}_\alpha \tag{18.32}$$
$$- \ll \log \operatorname{tr}_{\{n,s\}} \exp[\sum_\alpha \hat{w}_\alpha n^\alpha + \mu\beta\sum_\alpha n^\alpha$$
$$+ \sum_{\alpha\beta} r_{\alpha\beta}(n^\alpha s^\alpha)(n^\beta s^\beta) + \beta\sum_{\alpha\mu} m_\mu^\alpha \xi^\mu s^\alpha n^\alpha]\gg_\xi$$

where $\ll \cdots \gg_\xi$ means the average over the condensed patterns for $\mu \leq s$. Namely, we assume $m_\mu^\alpha = \mathcal{O}(1)$ for $\mu \leq s$ and $m_\mu^\alpha = \mathcal{O}(1/\sqrt{N})$ for $\mu > s$. In other words, among the extensive patterns $p = \alpha N$, the finite fraction of the patterns $s = \mathcal{O}(1)$ has an overlap of order 1. We call these *condensed patterns*.

In order to calculate the spin trace explicitly, we use the next replica symmetric approximation:

$$\begin{cases} m_\mu^\alpha &= m_\mu \\ t^\alpha &= t \\ w_\alpha &= w \\ \hat{w}_\alpha &= \hat{w} \\ r_{\alpha\beta} &= r \\ q_{\alpha\beta} &= q \end{cases} \tag{18.33}$$

Among these order parameters, m^α_μ, w_α and q_α have these own meaning, namely, diluted overlap between the embedded pattern ξ and the neuronal state s, number of survived neurons, and diluted spin-glass order parameter, which are given by

$$m^\alpha_\mu = \frac{1}{N}\sum_i \xi_i S^\alpha_i n^\alpha_i, \quad w_\alpha = \frac{1}{N}\sum_i n^\alpha_i \tag{18.34}$$

$$q_{\alpha\beta} = \begin{cases} (1/N)\sum_i (n^\alpha_i s^\alpha_i)(n^\beta_i s^\beta_i) & (\alpha \neq \beta) \\ 0 & (\alpha = \beta) \end{cases} \tag{18.35}$$

Under this approximation, Eq. (18.32) is rewritten as

$$\begin{aligned}f^{(RS)}_{\beta\mu}(m,q,w,\hat{w},r,t) &= \frac{n}{2}\beta m^2 - \frac{n\beta}{2}t^2 - n\beta tw + nw\hat{w} - \frac{nrq}{2} \\ &+ \frac{\alpha}{2}\sum_\alpha \log \lambda_\alpha - \ll \log \operatorname{tr}_{\{n,s\}} \exp\left[(\hat{w}+\mu\beta)\sum_\alpha n_\alpha \right. \\ &\left. + r\sum_{\alpha\beta}(n^\alpha s^\alpha)(n^\beta s^\beta) + \beta(\boldsymbol{m}\cdot\boldsymbol{\xi})\sum_\alpha s^\alpha n^\alpha \right]\gg_\xi \end{aligned} \tag{18.36}$$

where we used the following vector representations

$$m^2 \equiv \sum_\mu (m^\alpha_\mu)^2, \quad \boldsymbol{m}\cdot\boldsymbol{\xi} \equiv \sum_\mu m^\alpha_\mu \xi^\mu. \tag{18.37}$$

Therefore, we obtain

$$\begin{aligned}\frac{1}{n}f^{(RS)}_{\mu\beta}(m,q,w,\hat{w},r,t) &= \frac{1}{2}\beta(\boldsymbol{m})^2 - \frac{\beta}{2}t^2 - \beta tw + w\hat{w} - \frac{1}{2}rq \\ &+ \frac{\alpha}{2}\log[1-\beta(w-q)] - \frac{\alpha\beta q}{2[1-\beta(w-q)]} \\ &- \ll \int_{-\infty}^{\infty} Dz \log 2\left[1 + e^{\hat{w}+\mu\beta-\frac{r}{2}}\cosh(\beta(\boldsymbol{m}\cdot\boldsymbol{\xi}) + z\sqrt{r})\right]\gg_\xi\end{aligned} \tag{18.38}$$

Let us consider the case in which only one of the patterns is retrieved, that is, $m_\mu = \delta_{\mu 1} m$. Then, we obtain

$$\begin{aligned}\frac{1}{n}f^{(RS)}_{\mu\beta}(m,q,w,\hat{w},r,t) &= \frac{1}{2}\beta m^2 - \frac{1}{2}\beta t^2 - \beta tw + w\hat{w} - \frac{1}{2}rq \\ &+ \frac{\alpha}{2}\log[1-\beta(w-q)] - \frac{\alpha\beta q}{2[1-\beta(w-q)]} \\ &- \int_{-\infty}^{\infty} Dz \log 2\left[1 + e^{\hat{w}+\mu\beta-\frac{r}{2}}\cosh(\beta m + z\sqrt{r})\right]\end{aligned} \tag{18.39}$$

In the following, we evaluate the saddle-point surface of the above function $f^{(RS)}_{\mu\beta}$.

18.2.4
Evaluation of the Saddle Point

Next we consider the saddle point of the free-energy density $f_{\mu\beta}^{(RS)}$. We first consider the saddle-point equation with respect to t. This is easily obtained as

$$t = -w \qquad (18.40)$$

The replica symmetric free-energy density is rewritten in terms of the above saddle-point equations as

$$\frac{1}{n} f_{\mu\beta}^{(RS)}(m, q, r, w, \hat{w}) = \frac{1}{2}\beta m^2 + \frac{1}{2}\beta w^2 + w\hat{w} - \frac{1}{2}rq$$
$$+ \frac{\alpha}{2}\log[1 - \beta(w - q)] - \frac{\alpha\beta q}{2[1 - \beta(w - q)]} \qquad (18.41)$$
$$- \int_{-\infty}^{\infty} Dz \log 2\left[1 + e^{\hat{w}+\mu\beta-r/2}\cosh(\beta m + z\sqrt{r})\right]$$

The saddle-point equations with respect to $m, \hat{w}, q, r$, and w lead to the following coupled equations.

$$m = \int_{-\infty}^{\infty} Dz \left[\frac{e^{\beta(\mu-w)+\frac{\alpha\beta r}{2q}[1-\beta(w-q)]}\sinh\beta(m+z\sqrt{\alpha r})}{1 + e^{\beta(\mu-w)+\frac{\alpha\beta r}{2q}[1-\beta(w-q)]}\cosh\beta(m+z\sqrt{\alpha r})}\right] \qquad (18.42)$$

$$w = \int_{-\infty}^{\infty} Dz \left[\frac{e^{\beta(\mu-w)+\frac{\alpha\beta r}{2q}[1-\beta(w-q)]}\cosh\beta(m+z\sqrt{\alpha r})}{1 + e^{\beta(\mu-w)+\frac{\alpha\beta r}{2q}[1-\beta(w-q)]}\cosh\beta(m+z\sqrt{\alpha r})}\right] \qquad (18.43)$$

$$r = \frac{q}{[1 - \beta(w - q)]^2} \qquad (18.44)$$

$$q = w - e^{\beta(\mu-w)+\frac{\alpha\beta r}{2q}[1-\beta(w-q)]}$$
$$\times \int_{-\infty}^{\infty} Dz \frac{e^{\beta(\mu-w)+\frac{\alpha\beta r}{2q}[1-\beta(w-q)]} + \cosh\beta(m+z\sqrt{\alpha r})}{[1 + e^{\beta(\mu-w)+\frac{\alpha\beta r}{2q}[1-\beta(w-q)]}\cosh\beta(m+z\sqrt{\alpha r})]^2} \qquad (18.45)$$

where we have replaced $r \to \alpha\beta^2 r$ for convenience. We should keep in mind that, in the limit of $\mu \to \infty$, our model corresponds to the conventional Hopfield model.

Here we should note that the above equations are simplified as

$$\Theta \equiv \beta(\mu - w) + \frac{\alpha\beta r}{2q}[1 - \beta(w - q)]$$
$$= \beta(\mu - w) + \frac{\alpha\beta}{2[1 - \beta(w - q)]} = \beta\theta \qquad (18.46)$$

$$\theta \equiv \mu - w + \frac{\alpha}{2(1 - C)} \qquad (18.47)$$

$$C \equiv \beta(w - q) \qquad (18.48)$$

as

$$m = \ll \langle \xi_i^1 n_i s_i \rangle \gg_\xi = \int_{-\infty}^{\infty} Dz \frac{\sinh\beta(m + z\sqrt{\alpha r})}{e^{-\Theta} + \cosh\beta(m + z\sqrt{\alpha r})} \tag{18.49}$$

$$w = \ll \langle n_i \rangle \gg_\xi = \int_{-\infty}^{\infty} Dz \frac{\cosh\beta(m + z\sqrt{\alpha r})}{e^{-\Theta} + \cosh\beta(m + z\sqrt{\alpha r})} \tag{18.50}$$

$$q = \ll \langle n_i s_i \rangle^2 \gg_\xi = \int_{-\infty}^{\infty} Dz \frac{\sinh^2\beta(m + z\sqrt{\alpha r})}{[e^{-\Theta} + \cosh\beta(m + z\sqrt{\alpha r})]^2} \tag{18.51}$$

$$r = \frac{q}{[1 - \beta(w - q)]^2} = \frac{q}{(1 - C)^2} \tag{18.52}$$

where $\langle \cdots \rangle$ means the thermal average.

We should mention here the case of $\alpha = 0$. From the saddle-point equations, $\Theta = \beta(\mu - w)$ and m, w and q lead to

$$m = \left(\frac{\sinh(\beta m)}{e^{-\Theta} + \cosh(\beta m)} \right) \int_{-\infty}^{\infty} Dz = \frac{\sinh(\beta m)}{e^{-\Theta} + \cosh(\beta m)} \tag{18.53}$$

$$w = \frac{\cosh(\beta m)}{e^{-\Theta} + \cosh(\beta m)}, \quad q = \left(\frac{\sinh(\beta m)}{e^{-\Theta} + \cosh(\beta m)} \right)^2 = m^2 \tag{18.54}$$

For the case of $\mu \neq 0$ and $\beta \to \infty$, we find

$$m = \frac{\frac{1}{2} e^{\beta(m - w + \mu)}}{1 + \frac{1}{2} e^{\beta(m - w + \mu)}} = w = \Theta(m - w + \mu) = \Theta(\mu) \tag{18.55}$$

namely,

$$m = w = \Theta(\mu), \quad q = m^2 = \Theta(\mu) \tag{18.56}$$

On the other hand, for the case of $\mu = 0$, we easily find $m = 1/3$. Thus, the order parameters at the ground state are summarized as follows.

$$m = w = \begin{cases} 1 & (\mu > 0) \\ \frac{1}{3} & (\mu = 0) \\ 0 & (\mu < 0) \end{cases}, \quad q = \begin{cases} 1 & (\mu > 0) \\ \frac{1}{9} & (\mu = 0) \\ 0 & (\mu < 0) \end{cases} \tag{18.57}$$

In Fig. 18.4(a), we plot the three order parameters $m(= w)$ and q as a function of T at $\alpha = 0$.

18.2.5
Phase Diagrams

In this subsection, we investigate the phase diagram of the equilibrium states for the Hamiltonian (18.5) by solving the saddle-point equations (18.42)–(18.45).

18.2.5.1 Para-spin-glass Phase Boundary

We first investigate the para-spin-glass phase boundary. In the paramagnetic phase, the magnetization m and the spin-glass order parameter q should hold $m = 0$ and $q = 0$, respectively. Moreover, the paramagnetic phase should appear in the temperature region $1 - \beta > 0$, namely, $T > 1$. Thus, by expanding the saddle-point equations around the $q = m = 0$, the Eq. (18.50) leads to

$$w_{SG} \simeq \frac{1}{e^{-\Theta} + 1} \tag{18.58}$$

The Eq. (18.51) is approximately rewritten as

$$q \simeq \frac{\beta^2 \alpha r}{(e^{-\Theta} + 1)^2} \int_{-\infty}^{\infty} Dz\, z^2 = \frac{\beta^2 \alpha}{(e^{-\Theta} + 1)^2} \frac{q}{(1 - \beta w)^2} \tag{18.59}$$

Thus, at the transition temperature $T = T_{SG} = \beta_{SG}^{-1}$, the following relation should be satisfied.

$$(1 - \beta_{SG} w)^2 = \frac{\beta_{SG}^2 \alpha}{(e^{-\Theta} + 1)^2} \tag{18.60}$$

By solving this equation with respect to w, we obtain

$$w_{SG} \equiv \frac{1}{\beta_{SG}(1 + \sqrt{\alpha})} = \frac{T_{SG}}{1 + \sqrt{\alpha}} \tag{18.61}$$

We should keep in mind that on the phase boundary between the para and the spin-glass phases, the order parameter w behaves as (18.61).

Thus, at the critical temperature, Θ is written as

$$\Theta_{SG} = -\frac{1}{1 + \sqrt{\alpha}} \left(1 - \frac{\mu}{T_{SG}}(1 + \sqrt{\alpha}) - \frac{\sqrt{\alpha}(1 + \sqrt{\alpha})^2}{2T_{SG}}\right) \tag{18.62}$$

From the Eqs (18.58) and (18.61), we obtain the critical temperature as a function of α and μ as follows.

$$T_{SG} = \frac{1 + \sqrt{\alpha}}{\exp\left(\frac{1}{1+\sqrt{\alpha}}\left\{1 - \frac{\mu}{T_{SG}}(1 + \sqrt{\alpha}) - \frac{\sqrt{\alpha}(1+\sqrt{\alpha})^2}{2T_{SG}}\right\}\right) + 1} \tag{18.63}$$

In the limit of $\mu \to \infty$, we easily recover the well-known result for the conventional Hopfield model as follows.

$$T_{SG} = 1 + \sqrt{\alpha} \tag{18.64}$$

For example, if we set the parameter μ as a solution

$$1 - \frac{\mu}{T_{SG}}(1 + \sqrt{\alpha}) - \frac{\sqrt{\alpha}(1 + \sqrt{\alpha})^2}{2T_{SG}} = 1 \tag{18.65}$$

namely, $\mu = -\sqrt{\alpha}/(1+\sqrt{\alpha})2$, then, the critical temperature T_{SG} is given by

$$T_{SG} = \frac{1+\sqrt{\alpha}}{e^{\frac{1}{\sqrt{\alpha}+1}}+1} \tag{18.66}$$

18.2.5.2 Critical Chemical Potential μ_c at $T = 0$

We first investigate the critical chemical potential parameter μ_c above which the order parameter w goes to 1. Obviously, the spin-glass order parameter q goes to 1 in the limit of $\beta \to \infty$. Thus, we introduce the parameter C as $C = \beta(1-q)$. Then, Θ is written in terms of this C as follows.

$$\Theta = \beta(\mu - w) + \frac{\alpha\beta}{2(1-C)} \tag{18.67}$$

Therefore,

$$\mu > \mu_c = 1 - \frac{\alpha}{2(1-C)} \tag{18.68}$$

the parameter w becomes 1.

For $\mu = \mu_c$, the saddle-point equation with respect to m, r and the new order parameter C, itself leads to

$$m = 2H\left(-\frac{m}{\sqrt{\alpha r}}\right) - 1 \tag{18.69}$$

$$r = \frac{1}{(1-C)^2} \quad C = \frac{2e^{-\frac{m^2}{2\alpha r}}}{\sqrt{2\pi\alpha r}}. \tag{18.70}$$

The critical value of the chemical potential μ_c above which the order parameter w becomes $w = 1$ is given by

$$\mu_c = 1 - \frac{\sqrt{\alpha}}{2}\left(\frac{2H(-y)-1}{y}\right) \tag{18.71}$$

where y is a solution of the equation

$$y\left(\sqrt{\alpha} + \sqrt{\frac{2}{\pi}}e^{-\frac{y^2}{2}}\right) = 2H(-y) - 1. \tag{18.72}$$

This equation is identical to the result for the conventional Hopfield model.

18.2.5.3 Saddle-point Equations for $\mu < \mu_c$ at $T = 0$

We next consider the case of $\mu < \mu_c$. In this case, w satisfies $w < 1$ and the equations obtained in the previous subsection for the limit of $\beta \to \infty$ become incorrect.

In order to derive the correct saddle-point equations which are valid for $\mu < \mu_c$ and $\beta \to \infty$, we set

$$\Theta = \beta\theta, \quad \theta \equiv \mu - w + \frac{\alpha}{2(1-C)} \tag{18.73}$$

Then, we rewrite the saddle-point equations for w, m and C as follows.

$$w = H\left(\frac{-\theta - m}{\sqrt{\alpha r}}\right) + H\left(\frac{-\theta + m}{\sqrt{\alpha r}}\right) \tag{18.74}$$

$$m = H\left(\frac{-\theta - m}{\sqrt{\alpha r}}\right) - H\left(\frac{-\theta + m}{\sqrt{\alpha r}}\right) \tag{18.75}$$

$$C = \frac{1}{\sqrt{2\pi\alpha r}} \left\{ e^{-\frac{(\theta+m)^2}{2\alpha r}} + e^{-\frac{(\theta-m)^2}{2\alpha r}} \right\} \tag{18.76}$$

$$r = \frac{1}{(1-C)^2} \quad \theta = \mu - w + \frac{\alpha}{2(1-C)} \tag{18.77}$$

In Fig. 18.4(a), we show the phase diagrams of the model system. In this fig-

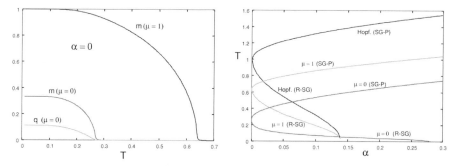

Fig. 18.4 (a) Three order parameters $m(=w)$ and q as a function of T for the case of $\alpha = 0$. (b) shows phase diagrams of the model system as associative memories. The label (R-SG) means the phase boundary between the ferromagnetic-retrieval phase and the spin-glass phase. The phase boundary between the spin-glass phase and the paramagnetic phase is labeled as (SG-P). The lines are plotted for the case of $\mu = 0, 1$ and $\mu = \infty$ (Hopf.).

ure, the label (R-SG) means the phase boundary between the ferromagnetic-retrieval phase and the spin-glass phase. The phase boundary between the spin-glass phase and the paramagnetic phase is labeled by (SG-P). The lines are plotted for the case of $\mu = 0, 1$ and $\mu = \infty$ (Hopf.).

18.2.6
Entropy of the System

In this subsection, we evaluate the entropy of the system and investigate its behavior at both high temperature and the ground state. The negative value of the entropy at low temperature indicates the replica symmetry breaking. After some algebra, we obtain the entropy of the system as

$$\begin{aligned}
s &= \beta^2 \left(\frac{\partial f_s}{\partial \beta} \right) \\
&= -\frac{\beta^2 \alpha r w^2}{2q} + \frac{\beta^2 \alpha q r}{2} - \frac{\alpha}{2} \log[1 - \beta(w-q)] \\
&\quad + \frac{\beta \alpha (q-w)(1 - \beta w + 2\beta q)}{2[1 - \beta(w-q)]^2} \\
&\quad + \int_{-\infty}^{\infty} Dz \log 2[1 + e^{\Theta} \cosh \beta(m + z\sqrt{\alpha r})] \\
&\quad + \beta w^2 - \beta m^2 - \beta \mu w - \frac{\beta \alpha r w^2}{2q}[1 - 2\beta(w-q)]
\end{aligned} \qquad (18.78)$$

with

$$\Theta = \beta(\mu - w) + \frac{\alpha \beta r}{2q}[1 - \beta(w-q)] \qquad (18.79)$$

18.2.6.1 High-temperature Limit

We fist investigate the entropy density at high temperature, that is, $\beta \to 0$. In this limit, we obtain a reasonable result as follows.

$$s(\beta = 0) = \lim_{\beta \to 0} s = \int_{-\infty}^{\infty} Dz \log 2(1+1) = 2 \log 2 \qquad (18.80)$$

This is quite reasonable as we saw in the case of $p = \mathcal{O}(1)$. This is because at high temperature the degree of the system $s \otimes n$ naturally becomes 4^N. As a result, the entropy density of the system becomes $\log 4^N / N = 2 \log 2$.

18.2.6.2 At the Ground State

In this subsection, we evaluate the entropy density s at the ground state. As is well known in the field of the mean-field spin-glass model, the replica symmetric calculations cause the negative in the low-temperature region. Therefore, the explicit form of the entropy density at the ground state $\beta \to \infty$ gives an important signal of the replica-symmetry breaking in the phase space.

At the ground state $\beta \to \infty$, the difference of two order parameters $w - q$ behaves as $w - q \simeq 0$, however, the product of β and $w - q$ remains finite. Thus, we put $C = \beta(w-q)$. Then, the entropy per neuron, that is to say, the entropy density of the system s at the ground state is written in terms of this C as follows.

$$s(T = 0) = -\frac{\alpha}{2} \log(1 - C) - \frac{\alpha C}{2(1 - C)} + (1 - w) \log 2 \qquad (18.81)$$

In Fig. 18.5(a), we plot the entropy at the ground state as a function of α for the case of $\mu = 1$. From this figure, we find that as α increases, the entropy becomes negative. This result indicates the replica symmetry breaking at low temperature.

18.2.7
Internal Energy

In this section, we derive the internal energy density u_s as a function of β. The internal energy density u_s leads to

$$u_s = f_s + \frac{s}{\beta} = \frac{w^2}{2} - \frac{m^2}{2} - \mu w - \frac{\alpha q}{2(1-C)^2} - \frac{\alpha C}{2\beta(1-C)} \tag{18.82}$$

From this expression, we first investigate the internal energy at high temperature. In the limit of $\beta \to 0$, we find

$$\lim_{\beta \to 0} C = \lim_{\beta \to 0} \beta(w - q) = 0 \tag{18.83}$$

The order parameters at high temperature behave as $m, q, r \to \infty$ and w leads to

$$w = \int_{-\infty}^{\infty} Dz \frac{\cosh \beta(m + z\sqrt{\alpha r})}{e^{-\Theta} + \cosh \beta(m + z\sqrt{\alpha r})}$$
$$= \left(\frac{1}{e^{-\Theta} + 1}\right) \int_{-\infty}^{\infty} Dz = \frac{1}{e^{-\Theta} + 1} = \frac{1}{2} \tag{18.84}$$

for finite μ, namely, $\mu < \infty$. Thus, we obtain

$$u_s(T \to \infty) = \frac{1}{2}\left(\frac{1}{2}\right)^2 - \frac{\mu}{2} = \frac{1}{2}\left(\frac{1}{4} - \mu\right) \tag{18.85}$$

Therefore, the internal energy at high temperature is independent of α and changes its sign at $\mu_* = 1/4$.

We next consider the internal energy at the ground state. For the case of $\mu > \mu_c$, $w = 1$ holds and we easily find that $u_s(T = 0)$ is given as

$$u_s(T = 0) = \frac{1}{2} - \frac{m^2}{2} - \mu - \frac{\alpha}{2(1-C)^2} \tag{18.86}$$

where m, r and C are

$$m = 2H\left(-\frac{m}{\sqrt{\alpha r}}\right) - 1 \tag{18.87}$$

$$r = \frac{1}{(1-C)^2} \qquad C = \sqrt{\frac{2}{\pi}} \frac{e^{-\frac{m^2}{2\alpha r}}}{\sqrt{\alpha r}} \tag{18.88}$$

On the other hand, for the case of $\mu < \mu_c$, $u_s(T=0)$ is written as

$$u_s(T=0) = \frac{w^2}{2} - \frac{m^2}{2} - \mu w - \frac{\alpha}{2(1-C)^2} \tag{18.89}$$

where the order parameters w, m, r and C are given as (18.74)–(18.77).

18.2.8
The Compressibility

We next calculate the compressibility κ which is defined as

$$\kappa \equiv \beta^{-1}\left(\frac{\partial w}{\partial \mu}\right) = 2\left\{H\left(\frac{-\theta - m}{\sqrt{\alpha r}}\right) + H\left(\frac{-\theta + m}{\sqrt{\alpha r}}\right) - 1\right\} \tag{18.90}$$

where the order parameters m, r and C obey the following saddle-point equations at the ground state

$$m = H\left(\frac{-\theta - m}{\sqrt{\alpha r}}\right) - H\left(\frac{-\theta + m}{\sqrt{\alpha r}}\right) \tag{18.91}$$

$$C = \frac{1}{\sqrt{2\pi\alpha r}}\left\{e^{-\frac{(\theta+m)^2}{2\alpha r}} + e^{-\frac{(\theta-m)^2}{2\alpha r}}\right\} \qquad r = \frac{1}{(1-C)^2} \tag{18.92}$$

with

$$\theta = \mu - w + \frac{\alpha}{2(1-C)} \tag{18.93}$$

In Fig. 18.5(b), we plot the compressibility κ as a function of μ for the case of $T = 0.08, 0.05$ and $T = 0.005$. We set $\alpha = 0.08$.

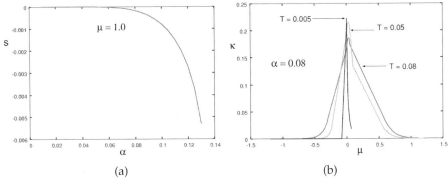

(a) (b)

Fig. 18.5 (a) The entropy at the ground state as a function of α for the case of $\mu = 1.0$. (b) shows the compressibility κ as a function of μ for the case of $T = 0.08, 0.05$ and $T = 0.005$. We set $\alpha = 0.08$.

18.3
Summary: Neural Networks

We have introduced an associative memory based on the FLG. Within the replica symmetric calculations, we found that reaping the redundant neurons effectively leads to reduction of the frustration and, as a result, the storage capacity α_c increases. However, problems of the replica symmetry breaking remain unsolved. Actually, we found that the replica symmetry might be broken from the viewpoint of the negative entropy at $T = 0$.

18.4
Plant Intelligence: Brief Introduction

Since the pioneering work by an Indian scientist J. C. Bose [3], plants have been regarded as networks which are capable of intelligent responses to environmental stimuli. For example, the dodder coil, which is a plastic plant, explores a new host tree within hours of their initial touch contact [4]. This sort of behavior might be regarded as *plant intelligence*. If that is the case, do the plants compute, learn or memorize various spacial and temporal patterns in different environments as does a computer or our brain?

Recently, Peak et al. [11] pointed out that the plants may regulate their uptake and loss of gases by distributed computation. As is well known, the function of neural networks, which are a mathematical model of the brain, is also based on parallel and distributed computation. Therefore, similarities between neural-network models of the brain and the plant network should be discussed. Although the behavior of the dodder coil which we mentioned above is due to emergence of the intelligence as a *macroscopic function*, it is important for us to investigate the *microscopic reason*.

Almost eighty years ago, J. C. Bose [3] detected electrical signaling between plants cells. Since his experiments, many examples of cross-talk, namely, the biochemical signaling pathways in plants have been found. In particular, a Boolean representation of the networks of signaling pathways is possible in terms of logical gates like AND, OR and XOR, etc. These Boolean descriptions make it possible to draw analogies between plant networks and neural-network models.

Recently, Brüggemann et al. [12,13] found that the plant vascular membrane current-voltage characteristic is established to be equivalent to that of the Zenner diode. Inspired by their work, Chakrabarti and Dutta [8] utilized such threshold behavior of the plant cell membranes to develop or model gates for performing simple logical operations. They found that the plant network connections are all positive (*excitatory* by means of neuronal states) or all negative (*inhibitory*), compared to the randomly positive-negative distributed synaptic connections in real brains. As a result, the plant network does not involve any frustration in their computational capabilities and might lack the distributed parallel computational ability like associative memories.

With this fact in mind, Inoue and Chakrabarti [14] investigated the equilibrium properties of the Hopfield model in which both ferromagnetic retrieval and antiferromagnetic terms coexist.

In their study [14], they investigated the pattern retrieval properties of the plant intelligence model. We studied the robustness of the stability of the memory state against the noise induced by thermal fluctuations. In this section, we review the results [14].

18.5
The I–V Characteristics of Cell Membranes

In this section, we briefly mention several results concerning properties of the plant units, namely, current (I) – voltage (V) characteristics of their cell membrane according to [14]. In Fig. 18.6, we show the typical nonlinear I–V characteristics of cell membranes for the logic gates. From this figure, we find

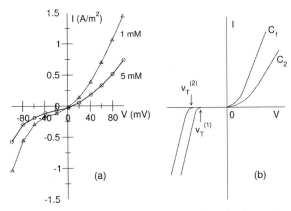

Fig. 18.6 The nonlinear I–V characteristics of cell membranes.

that the I–V characteristics are equivalent to that of a Zenner diode. From the viewpoint of input-output logical units like perceptrons for neural networks, the output of the ith unit O_i is given by

$$O_i = \Theta\left(\sum_{j=1}^{N} w_{ij} I_j\right) \quad (18.94)$$

where the strength of each connection w_{ij} is all positive or negative, while in the Hopfield model it is given as $\pm$ a randomly distributed weight matrix in terms of the Hebb rule:

$$w_{ij} = \frac{1}{N} \sum_{\mu=1}^{p} \xi_i^\mu \xi_j^\mu, \quad \xi_i^\mu \in \{-1, +1\} \tag{18.95}$$

From these experimental results and simple observations, we now have an obvious question, that is, could the plants act as memory devices in the way a real brain does? Obviously, in the above definition of a single unit of the plant, there is no frustration as in animal brains. Thus, one scope of this paper is to make this problem clear, that is to say, to what extent this kind of limitation in the sign of the weight matrix influences the ability of pattern retrieving as associative memories.

For this purpose, we introduce the simplest plant-intelligence model based on a Hopfield-like model in which ferromagnetic retrieval and antiferromagnetic ordered phases coexists. In the next section, we explain the detail.

18.6
A Solvable Plant-intelligence Model and its Replica Analysis

We start from the following Hamiltonian:

$$\mathcal{H} = \frac{1}{N} \sum_{ij} \left(\lambda - \sum_{\mu=1}^{p} \xi_i^\mu \xi_j^\mu \right) S_i S_j \equiv \mathcal{H}_{\text{AF}} + \mathcal{H}_{\text{FR}} \tag{18.96}$$

$$\mathcal{H}_{\text{AF}} \equiv \frac{\lambda}{N} \sum_{ij} S_i S_j, \quad \mathcal{H}_{\text{FR}} \equiv \frac{1}{N} \sum_{ij\mu} \xi_i^\mu \xi_j^\mu S_i S_j \tag{18.97}$$

where $\xi^\mu = (\xi_1^\mu, \cdots, \xi_N^\mu)$ is the μth embedded pattern and $S = (S_1, \cdots, S_N)$ represents the neuronal states. A single parameter λ determines the strength of the antiferromagnetic order, that is to say, in the limit of $\lambda \to \infty$, the system is completely determined by H_{AF}. On the other hand, in the limit of $\lambda \to 0$, the system becomes identical to the conventional Hopfield model. The purpose of this paper is to investigate the λ-dependence of the system, namely, to study the λ-dependence of the optimal loading rate $\alpha_c(\lambda)$ at $T = 0$ by using the technique of statistical mechanics for spin glasses.

18.6.1
Replica Symmetric Solution

In order to evaluate macroscopic properties of the system, we first evaluate the averaged free energy:

$$\ll \log Z \gg_\xi = \ll \log \text{tr}_{\{S\}} e^{-\beta H} \gg_\xi \tag{18.98}$$

where $\ll \cdots \gg$ means the quenched average over the $p = N\alpha$ patterns. To carry out this average and spin trace, we use the replica method [2, 15] by using the relation

$$\ll \log Z \gg_{\xi} = \lim_{n \to 0} \frac{\ll Z^n \gg_{\xi} - 1}{n} \tag{18.99}$$

After standard algebra [2, 15], we obtain the pattern-averaged replicated partition function as follows.

$$\ll Z^n \gg_{\xi} = \prod_{\alpha,\mu} \int_{-\infty}^{\infty} \frac{dM_\mu^\alpha}{\sqrt{2\pi/N\beta}} \int_{-i\infty}^{+i\infty} \frac{dm_\alpha}{i\sqrt{2\pi/N\beta\lambda}} \int_{-\infty}^{\infty} dq_{\alpha\beta}$$
$$\times \int_{-i\infty}^{+i\infty} \frac{dr_{\alpha\beta}}{2\pi i} \exp[-Nf(\mathbf{m}, \mathbf{q}, \mathbf{M}, \mathbf{r})] \tag{18.100}$$

By assuming the replica symmetric ansatz, namely,

$$M_\mu^\alpha = M \qquad m_\alpha = m \qquad q_{\alpha\beta} = q \qquad r_{\alpha\beta} = r \tag{18.101}$$

we obtain the free-energy density per replica number n as follows.

$$\frac{f(m, q, M, r)}{n} = \frac{\beta}{2} M^2 - \frac{\beta\lambda}{2} m^2 + \frac{\alpha\beta^2 r}{2}(1 - q)$$
$$+ \frac{\alpha}{2} \left\{ \log[1 - \beta(1 - q)] - \frac{\beta q}{1 - \beta(1 - q)} \right\} \tag{18.102}$$
$$- \log \int_{-\infty}^{\infty} Dz \log 2 \cosh \beta(\lambda m + \sqrt{\alpha r} z + M)$$

where we defined $Dz \equiv dz e^{-z^2/2}/\sqrt{2\pi}$. We should keep in mind that physical meanings of m and M are the magnetization of the system and the overlap between the neuronal state S and a specific recalling pattern ξ^1 among αN embedded patterns, respectively. q represents the spin-glass order parameters.

In the next subsection, we evaluate the saddle point of this free-energy density f and draw phase diagrams to specify the pattern-retrieval properties of the system.

18.6.2
Phase Diagrams

In this subsection, we investigate the phase diagram of the system by solving the saddle-point equations.

18.6.2.1 Saddle-point Equations

By taking the derivatives of f with respect to M, m, r and q, we obtain the saddle-point equations.

$$M = \int_{-\infty}^{\infty} Dz \tanh \beta[(1-\lambda)M + z\sqrt{\alpha r}] = -m \tag{18.103}$$

$$q = \int_{-\infty}^{\infty} Dz \tanh^2 \beta[(1-\lambda)M + z\sqrt{\alpha r}] \tag{18.104}$$

$$r = \frac{q}{[1 - \beta(1-q)]^2} \tag{18.105}$$

We solve the equations numerically to obtain the phase diagram.

18.6.2.2 $T = 0$ Noiseless Limit

We fist investigate the $T = 0$ limit. In this limit, obviously, $q \to 1$. After some algebra, we find that the optimal loading rate α_c is determined by the point at which the solution of the following equation with respect to y vanishes.

$$y \left\{ \sqrt{\alpha} + \sqrt{\frac{2}{\pi}}(1-\lambda) e^{-\frac{y^2}{2}} \right\} = (1-\lambda)\{1 - 2H(y)\} \tag{18.106}$$

where $H(x)$ is defined by $H(x) = \int_x^{\infty} Dz$. In Fig. 18.7(a), we plot the optimal loading rate α_c as a function of λ. From this figure, we see that the optimal loading rate $\alpha_c(\lambda)$ monotonically decreases. This means that the ferromagnetic retrieval order was destroyed by adding the antiferromagnetic term to the Hamiltonian. Thus we conclude that, if the weight matrix of the networks is all positive, the plant-intelligence model does not act as a memory device.

18.6.2.3 Spin-glass Para-phase Boundary

Before we solve the saddle-point equations for $T \neq 0$, it is important to determine the phase boundary between the spin-glass and para-magnetic phases. The phase transition between these two phases is of first order; by expanding the saddle-point equations around $M = q = 0$, we obtain

$$q \simeq \beta^2 \alpha r \int_{-\infty}^{\infty} z^2 Dz = \beta^2 \alpha \frac{q}{(1-\beta)^2} \tag{18.107}$$

Solving this equation by scaling $\beta \to (1-\lambda)\beta$ and $T = \beta^{-1}$, we obtain the phase boundary line:

$$T_{SG} = (1-\lambda)(1 + \sqrt{\alpha}) \tag{18.108}$$

18.6.3
Phase Diagrams for $T \neq 0$

In this subsection, we investigate the phase diagram for $T \neq 0$ by solving the saddle-point equations (18.103),(18.104) and (18.105) numerically. The result is plotted Fig. 18.7(b).

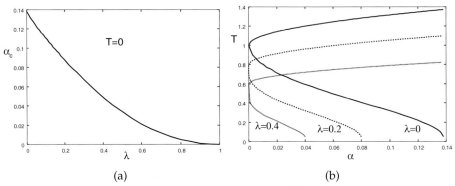

Fig. 18.7 The optimal loading rate α_c as a function of λ. $\alpha_c(\lambda)$ decreases monotonically (a). (b) shows the phase diagram of the system. The ferro-retrieval phase shrinks to zero as λ increases to 1. The paraspin glass boundary is analytically obtained as $T_c = (1-\lambda)(1+\sqrt{\alpha})$ and is independent of λ.

18.6.4
Negative λ case

In this subsection, we consider the case of negative λ. From the Hamiltonian, we find

$$\mathcal{H} = -\frac{1}{N}\sum_{ij}\sum_{\mu=1}^{p} \xi_i^\mu \xi_j^\mu S_i S_j - \frac{\lambda'}{N}\sum_{ij} S_i S_j \tag{18.109}$$

$$\lambda' = -\lambda \,(>0)$$

When λ increases, the system changes to the pure ferromagnet. Let us think about the limit of $\lambda \to -\infty$ in the saddle-point equation (18.49). Then, the term $(1-\lambda)M$ appearing in the argument of $\tanh[\beta(\cdots)]$ becomes dominant, namely, $(1-\lambda)M \gg z\sqrt{\alpha r}$ even if the loading rate α is large. Consequently, Eq. (18.49) leads to

$$M \simeq \int_{-\infty}^{\infty} Dz \tanh[\beta(1-\lambda)M] = \tanh[\beta(1-\lambda)M] \tag{18.110}$$

If the factor $1-\lambda$ is large enough, the term $\tanh[\beta(1-\lambda)M]$ becomes $\text{sgn}[\beta M]$ and the saddle-point equation (18.49) leads to

$$M = \text{sgn}[\beta M] \tag{18.111}$$

Apparently, this equation has always a positive solution even if the temperature $T = \beta^{-1}$ is large. In this sense, the factor $(1 - \lambda)$ represents the temperature rescaling. It is also possible for us to understand this result from a different point of view. In the saddle-point equation:

$$M = \int_{-\infty}^{\infty} Dz \tanh \beta[(1-\lambda)M + z\sqrt{\alpha r}] \tag{18.112}$$

the second term appearing in the argument of tanh, $z\sqrt{\alpha r}$ represents *cross-talk noise* from the other patterns ξ^μ, $(\mu = 2, \cdots, p)$ and obeys a Gaussian distribution $e^{-z^2/2}/\sqrt{2\pi}$. On the other hand, the first term $(1-\lambda)M$ represents the *signal* of the retrieval pattern ξ^1. Therefore, if the second term $z\sqrt{\alpha r}$ is dominant, the system cannot retrieve the embedded pattern ξ^1. Usually, r in the

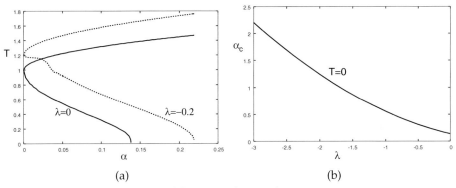

Fig. 18.8 (a) The phase diagram of the system for negative $\lambda = -0.2$. (b) The optimal loading rate α_c as a function of λ (< 0) at $T = 0$.

second term grows rapidly as T increases. And, obviously, if α increases, the noise term $z\sqrt{\alpha r}$ also increases. As a result, the signal part $(1-\lambda)M$ becomes relatively small and the system moves from the retrieval phase to the spin-glass phase. However, if λ is negative and large, the signal part is dominant and the noise part becomes vanishingly small. This is an intuitive reason why the optimal loading rate increases for negative λ. In Fig. 18.8(b), we plot the optimal loading late α_c as a function of λ (< 0) at $T = 0$. As we mention, the optimal loading rate α_c monotonically increases as λ goes to $-\infty$.

18.7
Summary and Discussion

In this chapter, we have reviewed some recent results on the associative memory of neural networks and plant networks. In particular, we introduced a simple model based on a Hopfield-like network model to explain the intelligence of the plant. According to experiments and observations [8, 12], we constructed a Hopfield model in which the ferromagnetic retrieval and the antiferromagnetic terms coexist. The strength of the disturbance of pattern retrieval by the antiferromagnetic order is controlled by a single parameter. We found that the antiferromagnetic order prevents the system from recalling a pattern. This result means that the ability of the plant as a memory device is very weak if we set all weight connections to positive values. Our analysis in this paper was done for fully-connected networks and, of course, for real plants, the cell membrane should be located in a finite-dimensional lattice [16] or in a scale-free network [17]. Moreover, it might be very interesting to extend the system to the situation in which the plant as a memory device is affected by a kind of noise induced by *quantum fluctuations*. This situation is easily realized by introducing the following Hamiltonian

$$\hat{H} = -\frac{1}{N}\sum_{ij}\left(\sum_{\mu=1}^{p}\xi_i^\mu\xi_j^\mu - J\right)\hat{S}_i^{z(A)}\hat{S}_j^{z(B)} - \Gamma_A\sum_i\hat{S}_i^{x(A)} - \Gamma_B\sum_i\hat{S}_i^{x(B)} \quad (18.113)$$

where $\xi^\mu = (\xi_1^\mu, \xi_2^\mu, \cdots, \xi_N^\mu)$ ($\mu = 1, 2, \cdots, p$) is the embedded pattern vector and $\hat{S}_i^{z(A,B)}$ and $\hat{S}_i^{x(A,B)}$ mean z, x-components of the Pauli matrix defined by

$$\hat{S}_i^{z(A,B)} = \begin{pmatrix} 1 & 0 \\ 0 & -1 \end{pmatrix} \quad \hat{S}_i^{x(A,B)} = \begin{pmatrix} 0 & 1 \\ 1 & 0 \end{pmatrix} \quad (18.114)$$

The parameters Γ_A and Γ_B are amplitudes of the transverse tunneling-field effect on the neuronal states S_A of the subgroup A and S_B of the subgroup B. Obviously, the model is reduced to the classical version [14] by setting the amplitudes of the transverse fields Γ_A and Γ_B to zero.

In order to explain how the transverse tunneling field affects the system as a noise intuitively, we consider a single unit problem as follows.

Let us define $^t(1,0)$ as the state of the *active unit* $|S^z = +1\rangle$ and $^t(0,1)$ as the state of the *quiescent unit* $|S^z = -1\rangle$. Thus, for the z-component of the Pauli matrix, the following equations should be satisfied.

$$\begin{pmatrix} 1 & 0 \\ 0 & -1 \end{pmatrix}\begin{pmatrix} 1 \\ 0 \end{pmatrix} = 1\begin{pmatrix} 1 \\ 0 \end{pmatrix}$$
$$\begin{pmatrix} 1 & 0 \\ 0 & -1 \end{pmatrix}\begin{pmatrix} 0 \\ 1 \end{pmatrix} = -1\begin{pmatrix} 0 \\ 1 \end{pmatrix} \quad (18.115)$$

Therefore, if the term like ΓS^x appears in the effective Hamiltonian, the eigenstates $^t(1,0)$ and $t(0,1)$ change as

$$\begin{pmatrix} 0 & 1 \\ 1 & 0 \end{pmatrix}\begin{pmatrix} 1 \\ 0 \end{pmatrix} = \begin{pmatrix} 0 \\ 1 \end{pmatrix}$$
$$\begin{pmatrix} 0 & 1 \\ 1 & 0 \end{pmatrix}\begin{pmatrix} 0 \\ 1 \end{pmatrix} = \begin{pmatrix} 1 \\ 0 \end{pmatrix} \quad (18.116)$$

This means that the probability of flipping states, that is, from *active* to *quiescent* or *quiescent* to *active* is controlled by Γ, which can be defined by

$$\Gamma^2 = |\langle S^z = \pm 1 | \Gamma S^x | S^z = \mp 1 \rangle|^2 \tag{18.117}$$

This fluctuation remains, even if the model system is surrounded by a zero-temperature heat bath. Analysis of this model is now ongoing.

Acknowledgment

The author thanks Professor Bikas K. Chakrabarti for fruitful discussions and useful comments.

References

1 AMIT, D. J., GUTFREUND, H., SOMPOLINSKY, H., *Phys. Rev. Lett.* 55 (**1985**), 1530

2 HERTZ, J., KROUGH, A., PALMER, R. G., *Introduction to the Theory of Neural Computation*, Addison-Wesley, **1991**

3 BOSE, J. C., *The Nervous Mechanism of Plants*, Longmans, London, **1923**

4 TREWAVAS, A., *Nature* 415 (**2002**), 841

5 BALL, P., *Do plants act like computers?*, Nature Science Update Weekly Highlights: 26 January 2004, http://info.nature.com/cgi-bin24/DM/y/eNem0CdLrY0C30Hcz0AW.

6 NICODEMI, M., CONIGLIO, A., *J. Phys. A: Math. Gen.* 27 (**1997**), L187

7 MÉZARD, M., PARISI, G., VIRASORO, M. A., *Spin Glass Theory and Beyond*, World Scientific, Singapore, **1987**

8 CHAKRABARTI, B. K., DUTTA, O. *Ind. J. Phys. A* 77 (**2003**), 549; cond-mat/0210538

9 BOSE, I., KARMAKAR, R., *Physica Scripta* T106 (**2003**), 9

10 INOUE, J., *Prog. Theor. Phys. Supplement*, 157 (**2005**), 262

11 PEAK, D. A., WEST, J. D., MESSINGER, S. M., MOTT, K. A., *Proceedings of Academy of Science USA* 101 (**2004**), 918

12 BRÜGGEMANN, L. I., POTTOSIN, I. I., SCHÖNKNECHT, G., *The Plant Journal* 16 (**1998**), 101

13 GENOUD, T., METRAUX, J. -P., *Trends in Plant Science* 4 (**1999**), 503

14 INOUE, J., CHAKRABARTI, B. K., *Physica A* 346 (**2005**), 58

15 NISHIMORI, H., *Statistical Physics of Spin Glasses and Information Processing*, Oxford Univ. Press, **2001**

16 KOYAMA, S., *Phys. Rev. E* 65 (**2002**), 016124

17 STAUFFER, D, AHARONY, A., DE FONTOURA COSTA, L., ADLER, J. *Euro. Phys. Journal B* 32 (**2003**), 395

19
Self-organization Principles in Supply Networks and Production Systems
Dirk Helbing, Thomas Seidel, Stefan Lämmer, and Karsten Peters

19.1
Introduction

In the past decade, physicists have been increasingly interested in interdisciplinary fields such as biophysics, traffic physics, econophysics and sociophysics [1–3]. However, the study of production processes has become attractive only recently [4–6], although the title of the book *Factory Physics* [7] suggests that there should be some connection. In fact, it is quite natural to study production and logistics from the point of view of material flows [8]. Therefore, many-particle approaches such as Monte Carlo simulations and fluid-dynamic models [9–12] should be applicable to production problems. As we will discuss in this chapter, this is in fact the case. Previous publications focussed on the bullwhip-effect [13–17] describing increasing variations of inventories in temporally perturbed supply chains [9, 18]. In particular, it has been suggested to interpret business cycles as the effect of decentralized adjustments of production rates in different sectors of the economy and their network interactions through product flows [19] (see Figs 19.1 and 19.2). An interesting observation is that, when coupling overdamped production dynamics through a delivery network, the overall dynamics could become oscillatory or even unstable in time [20]. That is, a network can behave very differently in time than its single elements. This generally raises serious concerns regarding the stability and efficiency of networked systems. The "slower-is-faster effect" that we discuss below shows quite clearly that obstruction effects and inefficiencies already occur in relatively simple networks such as the single intersections of two flows. In some sense, this problem can be related to the prisoner's or social dilemma in game theory, where the resulting system behavior can deviate a lot from the system optimum [21]. It may also be related to nonlinearly interacting many-particle systems, where the system may get stuck in a local optimum ("frustrated systems"). In fact, the optimization of production or logistic systems is often NP-hard, i.e., the numerical effort to find the global optimum explodes with system size.

Econophysics and Sociophysics: Trends and Perspectives.
Bikas K. Chakrabarti, Anirban Chakraborti, Arnab Chatterjee (Eds.)
Copyright © 2006 WILEY-VCH Verlag GmbH & Co. KGaA, Weinheim
ISBN: 3-527-40670-0

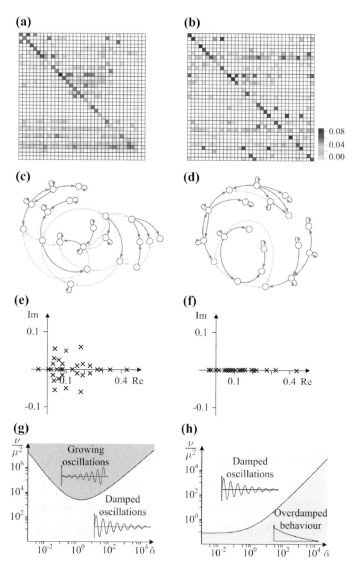

Fig. 19.1 Properties of a dynamic model of supply networks for a characteristic, empirical input matrix (left) and for a synthetic input matrix generated by random changes of input matrix entries until the number of complex eigenvalues was eventually reduced to zero (right). Subfigures (a), (b) illustrate the color-coded input matrices $\mathbf{A}$, (c), (d) the corresponding network structures, when only the strongest links (commodity flows) are shown, (e), (f) the eigenvalues $\omega_i = \text{Re}(\omega_i) + i\,\text{Im}(\omega_i)$ of the respective input matrix $\mathbf{A}$, and (g), (h) the phase diagrams indicating the stability behavior of the related model of material flows between economics sectors on a double-logarithmic scale as a function of model parameters. Surprisingly, for empirical input matrices $\mathbf{A}$, one never finds an overdamped, exponential relaxation to the stationary equilibrium state, but network-induced oscillations due to complex eigenvalues ω_i (after [19]). (For a color version please see color plate on page 603.)

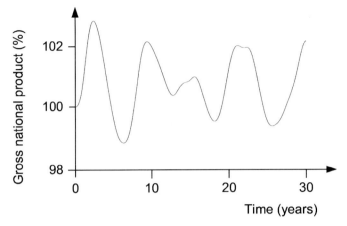

Fig. 19.2 Typical simulation result of the time-dependent gross domestic product in percent, i.e., relative to the initial value. The input matrix was chosen as in Fig. 19.1(a–d). Note that irregular oscillations with frequencies between 4 and 6 years and amplitudes of about 2.5% are qualitatively compatible with empirical business cycles. Our material flow model can explain w-shaped, nonperiodic oscillations without having to assume technological shocks or externally induced perturbations (after [19]).

Another problem of production systems can be connected to traffic dynamics, namely the instability of material flows [10,23]. For example, breakdowns in a smooth flow as in stop-and-go waves are caused by delays in adaptating to varying conditions. As a consequence, one is interested in changing the interactions among the system elements in a way that causes stable dynamics towards the system optimum. Modern driver-assistance systems are pursuing this approach [25].

Our chapter is organized as follows. In Section 19.2 we will show that production systems can easily display complex dynamics including deterministic chaos. Section 19.3 discusses the slower-is-faster effect in traffic and production systems. Then Section 19.4 presents adaptive-control approaches for production systems and their similarities with traffic systems. Section 19.5 will summarize this chapter and present an outlook.

19.2
Complex Dynamics and Chaos

Production processes are usually not constant in time. Weekly and seasonal cycles, for example, generate oscillatory behavior. However, a closer look at power spectra also indicates period-doubling phenomena (see Fig. 19.3). Even deterministic chaotic dynamics can easily appear in production systems [26–31]. Figure 19.4 illustrates a system in which a robot has to serve three different processes. Such systems are sometimes mapped to so-called switched

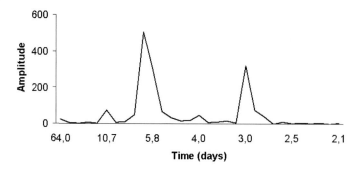

Fig. 19.3 This power spectrum of the recorded actual inventory in a supply chain shows peaks corresponding to periods of three and six days. While the six-day cycle corresponds to the weekly rhythm, the three-day cycle potentially indicates a period-doubling phenomenon.

departure systems. Let us assume we have different containers which are continuously filled with fluids at certain rates. All containers have finite storage capacity, and only one container can be emptied at a time. Moreover, switching between containers takes a setup time. Therefore, switching reduces service time and is discouraged. As a consequence, one may decide to empty the currently served container completely and then switch to the fullest container in order to avoid its capacity being eventually exceeded. Such a service algorithm leads to a strange billiards kind of dynamics (i.e., trajectories reminiscent of billard balls reflecting at some non-rectangular boundaries). It can be easily shown that this dynamics tends to be chaotic [11, 30] (see Fig. 19.4). For this reason, many statistical distributions of arrival or departure rates may actually be a result of nonlinear interactions in the system [32]. The complex dynamics has important consequences for the system, making it difficult to predict the future behavior and production times. This makes control very complicated. Moreover, nonlinear interactions often imply (phase) transitions from one dynamical behavior to another at certain critical parameter thresholds. Such transitions are often very unexpected and mix up the production schedule. We actually conjecture that the phase space (i.e., a representation of dynamic behavior as a function of parameter combination) is in many cases fractal, i.e. subdivided into small and often irregularly shaped areas. An understanding of such systems requires a solid knowledge of the theory of complex systems. For this reason, production and logistics are interesting areas for fundamental and applied research in theoretical physicists.

We should particularly underline that, due to the many parameters in production systems (e.g., production speeds, minimum or maximum treatment times in time-critical processes, etc.), the sensitivity to (even small) parameter changes is mostly large. Therefore, the performance for a new combination of

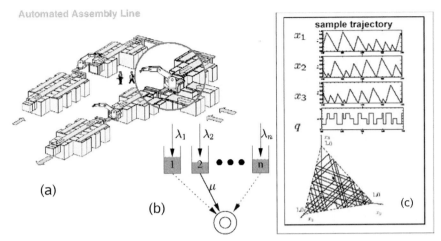

Fig. 19.4 Illustration of a robot serving different production processes (a) and its representation by a switched departure model (b). The graphics in (c) illustrates the strange billiard dynamics of the trajectories, which is often chaotic (after [11]).

parameter values is only poorly estimated by interpolation between previous experimental measurements, which are usually available for a few parameter sets only (see Fig. 19.5). That is, the predictability and robustness of production processes is often low, while they are an important aspect for efficient and reliable production. This also implies that production systems, which are normally optimally designed for given boundary conditions (order flows, product spectrum, availability of capacity, machine parameters, etc.), will perform suboptimally in most realistic situations, in which the consumption rate is varying, machines break down, workers are ill or on holiday, and machine parameters have been changed. Therefore, we favor adaptive scheduling strategies, that are made to deal with everyday and unexpected variabilities (like disasters or attacks) in a flexible way. We expect that suitable adaptive strategies can reach a better performance on average than precalculated schedules optimized for idealized conditions. This will be the subject of Section 19.4.

19.3
The Slower-is-faster Effect

One characteristic feature of multicomponent or many-particle systems operated near capacity is the possibility of mutual obstructions due to competition for scarce resources (time, space, energy, materials, etc.). From game theory it is known that the selfish, local optimization of the behavior of all elements can lead to system states far off the system optimum. That is, even if all parts of

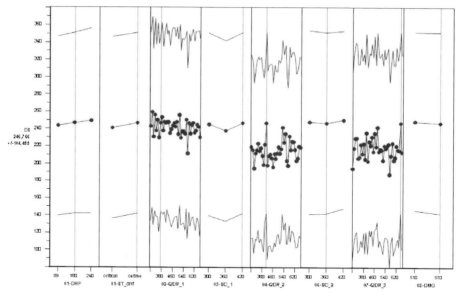

Fig. 19.5 Measurements of the performance of a wet bench in semi-conductor production for different parameter values. The large variation of the results indicates the sensitivity to parameter changes and makes an accurate prediction of performance by interpolation between previous measurements for nearby parameter combinations impossible (after [41]).

the system perform well and serve the local demands best, the overall result can still be very bad. This is obviously the case, if the different processes are not well synchronized or otherwise coordinated. Therefore, it can be better to wait for some time rather than starting activities immediately upon availability. Such action could produce overcrowding, bottlenecks, and inefficiencies in other parts of the system. We will illustrate this, sometimes rather surprising effect, for different examples of traffic and production systems. The crucial question, of course, is to determine the best moment to start activities, as delays are equally bad. Unfortunately, this question does not have a simple answer for systems with nonlinear interactions. Optimal strategies may even depend on different regimes of operation, which are hard to parametrize. Another problem is the optimal transition from one kind of operation to another one, when the boundary conditions have changed. These are questions which must be intensively addressed by future research. Some ideas will be presented in Section 19.4. Moreover, approaches developed for self-organized synchronization [42] and chaos control [43] show much promise.

19.3.1
Observations in Traffic Systems

The discovery of the "faster-is-slower" or "slower-is-faster effect" has been made in various traffic systems [3]. We will give examples for pedestrian traffic, freeway traffic, and interactions of both.

19.3.1.1 Panicking Pedestrians

It is known that the dynamics of pedestrian crowds can be well understood by means of a social-force model. This describes the accelerations and decelerations of pedestrians based on the repulsive and frictional interactions between them when they want to reach their desired velocity. It has been found that such a simple model reproduces the smooth vacation of a room under normal conditions, including a good coordination at bottlenecks. However, when pedestrians are in a rush (under conditions of "panic"), the model predicts frictional effects due to pushing, clogging, and arch formation, similar to granular materials. As a consequence, outflows may be stopped intermittently, and evacuation times may be larger [24]. The clogging has also been observed in reality, e.g., at the narrow emergency exits of Sheffield's soccer stadium during the historical crowd stampede (see Fig. 19.6). An interesting prediction of computer simulations is the reduction of the "faster-is-slower effect" by suitably designed and placed columns, although such obstacles are expected to have the opposite effect. The improvement of the flow efficiency results from an absorption of pressure in the crowd, which facilitates an easier motion, particularly at the bottleneck. Figure 19.6 shows the experimental results, which confirm this effect. While the clogging effect interrupts the flow for certain time periods, a suitable obstacle can improve the outflow by about 30–40%. In production systems, such improvements are considered to be sensational.

19.3.1.2 Freeway Traffic

Not only in pedestrian traffic, but also on freeways we find that faster is sometimes slower. This is illustrated by Fig. 19.7 for a two-lane freeway with an on-ramp at kilometer 0. Due to the on-ramp, the vehicle density downstream of the ramp is somewhat higher then upstream of it. The inflow into the freeway and the ramp have been chosen as constant, apart from a temporary reduction in the density and flow at the upstream end of the freeway. One would think that this reduction in the density would lead to higher speeds and, therefore, to a more efficient operation of the freeway. In the metastable traffic-flow regime specified, here, small perturbations would actually fade away. However, for the supercritical perturbation in our simulation, vehicles accelerate into areas of lower density and cause a local compression later on. When propagating downstream with the cars, this "localized cluster" eventually grows,

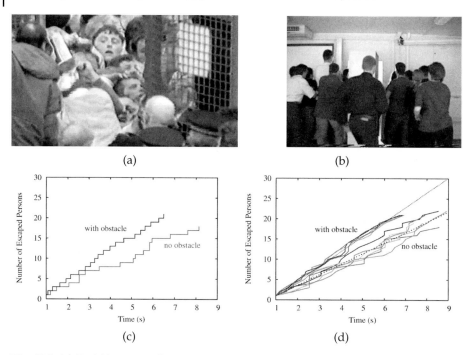

Fig. 19.6 (a) Panicking soccer fans trying to escape the football stadium in Sheffield. For some time, hardly anybody manages to pass the open door, because of the clogging effect occuring at high pressures. (b) Photograph of an experiment with an obstacle to reduce the pressure on the people at the exit. (c) Without an obstacle, the clogging effect means that nobody can pass the bottleneck for quite some time (long steps). (d) A suitably designed and placed obstacle can increase the outflow (after [36]).

and when it reaches a certain size, the propagation direction turns backwards ("boomerang effect"). This is because vehicles in traffic jams are standing. Leaving the jam shortens it at the downstream end, while new cars are joining it at the upstream end. Therefore, a traffic jam is moving upstream. Once it reaches the location of the on-ramp, traffic flow breaks down and causes a growing queue (here: oscillating congested traffic). This breakdown under constant inflow conditions indicates a reduction in the flow capacity. In fact, the outflow from traffic jams is smaller than the maximum homogeneous traffic flow. The reason is the increased time gaps when vehicles are leaving a traffic jam. As the inverse of the time gap determines the capacity, increased time gaps produce a reduction in the effective road capacity when the traffic flow breaks down. This is the reason why future traffic assistance systems, by dampening local perturbations, can successfully suppress or reduce traffic jams [25]. In conclusion, multicomponent systems with delayed adaptation can suffer from noise-induced breakdowns, which can furthermore reduce the capacity of the system.

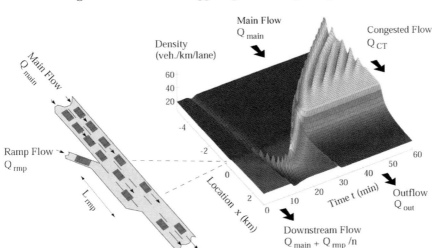

Fig. 19.7 Simulation of traffic flow on a two-lane freeway with an on-ramp entering at kilometer 0. Under certain conditions, a temporary reduction in the inflow to the freeway can finally cause a breakdown of traffic flow with a spatially extended, growing vehicle queue (here: "oscillating congested traffic").

19.3.1.3 Intersecting Vehicle and Pedestrian Streams

Another surprise occurs when pedestrians cross a street at places without traffic lights or pedestrian crossings, in particular if the vehicle speed is limited, say, to 30 kilometers per hour. The crossing of the street depends on safety considerations. If pedestrians are patient enough or careful, they will wait for large enough gaps, allowing them to cross the street without stopping vehicles. However, if they are daring or in a hurry, they may use small gaps and force vehicles to stop. In the first case, pedestrians will cross the street one by one or in small groups between moving vehicles. In the second case, however, newly arrived pedestrians will enter the street as well, as cars are standing. Given a large enough pedestrian arrival rate, vehicles cannot move for quite some time. As a consequence, a long vehicle queue may form (see Fig. 19.8).

Once there is a large enough gap between subsequently arriving pedestrians (this is just a matter of statistics), the first car will accelerate and pass the crossing point. Then, all the vehicles in the queue will pass the crossing point as well, since the gaps are too small to cross the street, even for daring pedestrians. Consequently, pedestrians have to wait until the whole vehicle queue has passed. This can take a long time, so that many pedestrians accumulate, waiting for a gap to cross. In conclusion, if pedestrians are too impatient, not only will cars suffer a long average waiting time, but so will pedestrians. Again, there is a faster-is-slower effect in this system [33]. By the way, for a simple vehicle model, the relevant quantities in this coupled queuing system can be analytically calculated [34].

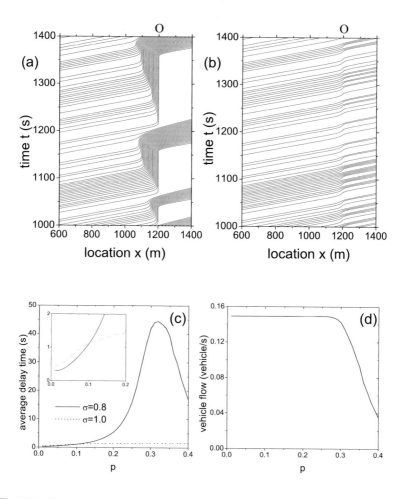

Fig. 19.8 Representative vehicle trajectories (a) if (daring) pedestrians use small gaps to cross the street, (b) if (careful) pedestrians wait for larger gaps. (c) Apart from small pedestrian-arrival rates p, the average delay time of daring pedestrians (solid line) is higher, as their stopping of cars can cause long vehicle queues. Later on, these will prevent pedestrians from passing the road for a long time (after [33]).

19.3.2
Relevance to Production and Logistics

"More haste less speed" is a well-known saying. This also applies to the organization of many logistic or production systems. We will give three examples below, all for different systems.

19.3.2.1 Semi-conductor Chip Manufacturing

We have studied a process in chip manufacturing, the so-called etching bench. Its chemical treatments of siliconwavers can be compared with the development of a photographic film. In fact, the silicon wavers are exposed to high-frequency light in a process previously called Lithiography. Then, the silicium wavers are subsequently put into several chemicals in order to remove material through an etching process, while the essential structures of the chip remain. This, however, requires certain minimum and maximum treatment times, which depend on the respective chemical. After each chemical treatment, the silicium wavers are washed in separate water basins.

It is common to process several sets of wavers at the same time, but in different chemicals. The wavers are typically moved around by a device called the "handler" or "gripper". Therefore, several sets of wavers may need to be moved at the same time. However, all but one have to wait. This may cause extended treatment times, but the maximum treatment times are not to be exceeded. Otherwise, poor quality would be produced, which is not affordable. Hence, waiting times between subsequent "runs" (i.e., treatments of different sets of wavers) are needed (see Fig. 19.9). The previous scheduling ("recipe") worked with small treatment times with the aim of reaching large throughputs. For this, however, large waiting times were required, which ruined the performance. Therefore, we have suggested extending the treatment times, specifically in a way that minimizes conflicts in requests for gripper movement, as the gripper is the bottleneck of the system. By this approach, the waiting times could be enormously reduced. The new recipe was implemented in several etching benches, and its success was quite convincing. The increase in the throughputs was between 17 and 39%.

19.3.2.2 Container Terminals

We have also studied a logistics system, namely a container terminal in a harbor. There, containers had to be moved from the ships to the container storage area and back. Nowadays, this is often done by automated guided vehicles (AGV), i.e., without any drivers. These vehicles move along virtual tracks. However, instead of moving most of the time, they often obstruct each other. This is because each vehicle is surrounded by a safety zone, which may not be entered in order to avoid accidents and damage of goods.

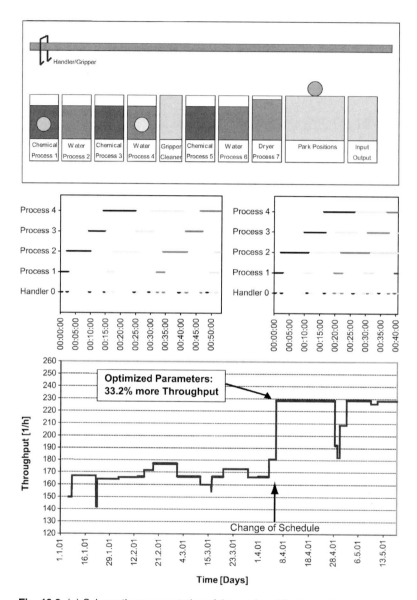

Fig. 19.9 (a) Schematic representation of the successive processes of a wet bench, i.e., a particular supply chain in semiconductor production. (b) The Gantt diagramms illustrate the treatment times of the first four of several more processes, where we have used the same colors for processes belonging to the same run, i.e., the same set of wafers. The left diagram shows the original schedule, while the right one shows an optimized schedule based on the "slower-is-faster effect" (see Section 19.3). (c) The increase in the throughput of a wet bench by switching from the original production schedule to the optimized one was found to be 33%, in some cases even higher (after [41]). (For a color version please see color plate on page 604.)

According to the "slower-is-faster effect", the logistic processes may be more efficient if the speed of the automated guided vehicles is reduced. In fact, there should be an optimal speed. Reducing the speed will, of course, increase the run times. However, it also allows one to reduce the safety zones, since these are essentially proportional to the speed. As a consequence, the vehicles will not obstruct each other so often. Thus, their waiting times will go down. This effect can overcompensate the increase in the run times. Therefore, reducing speed can sometimes make logistic processes more efficient.

19.3.2.3 Packaging and Other Industries

Such an observation can also be made in the packaging industry and other manufacturing plants. In some factories producing packaging materials (see Fig. 19.10), the most expensive machine, the corrugator, breaks down quite frequently, if it is no longer new. The corrugator produces the packaging material (corrugated paper), i.e., the basic input for all other machines. Therefore, the breakdowns are usually considered as causing the main bottleneck and limiting the profitability of the factory. As a consequence, the corrugator is often run at full speed whenever it is operational. Potentially, this causes earlier

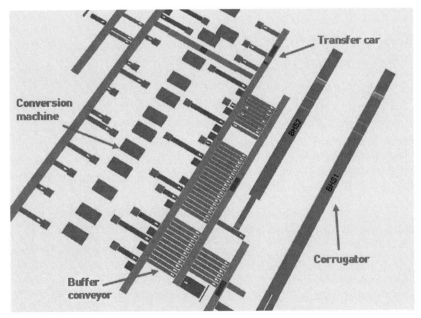

Fig. 19.10 Illustration of a plant in the packaging industry producing boxes. The corrugator produces corrugated paper board, which is processed (e.g., cut and printed) by conversion machines. Transfer cars transport the materials, and buffers allow for a temporary storage.

breakdowns, which could be avoided by a lower speed of operation. Morever, it also produces congestion in the buffer system. Whenever the system exceeds a certain utilization ("work in process"), so-called cycling procedures are needed to find the stacks which were required for further processing (see Fig. 19.11). These cycling procedures take an overproportionate amount of time, when the buffer system becomes fuller. In this way, the buffer operations become quite inefficient. As a consequence, the real bottleneck is the buffer system. In principle, there are two ways to solve this problem: either to

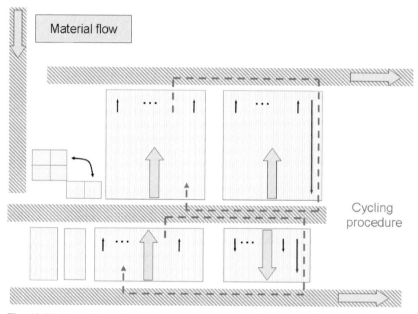

Fig. 19.11 Illustration of the cycling procedure, by which a specific stack is moved out of the buffer for further processing. This procedure can be quite time-consuming, if there is a lot of work-in-progress (WIP) in the buffer.

increase the buffer storage capacity or to stop the corrugator, when the buffer system reaches a certain utilization (see Fig. 19.12). While the corrugator is turned off, it can be cleaned and fixed. This proactive maintenance can reduce the number and duration of unplanned downtimes. Therefore, reducing the speed of the system increases its throughput and efficiency again. Our event-driven, calibrated simulations of the production system indicate that the expected increase in the production efficiency would be of the order of 14%.

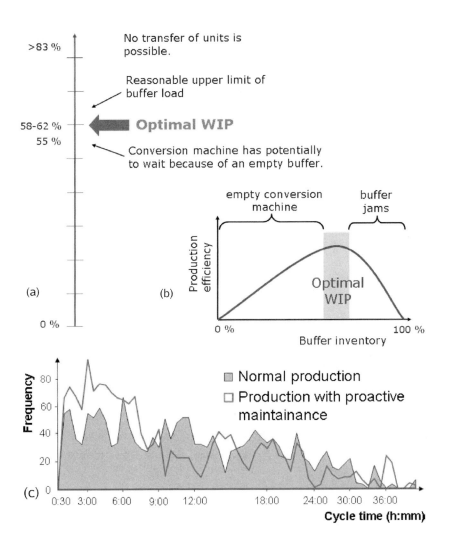

Fig. 19.12 (a) and (b) The optimum buffer utilization is neither close to zero nor close to capacity. In the first case, the buffer may be emptied, so that further production steps are delayed. In the second case, buffer operation becomes inefficient or even almost impossible. In the illustrated practical example, the optimum buffer utilization is between 58% and 62% (only!). (c) Stopping further transfer to the buffer at this level is reasonable and allows for proactive maintenance of the upstream production machine(s). According to computer simulations, this strategy is expected to reach a reduction in production times by 14%.

19.4
Adaptive Control

As we have discussed in Section 19.2, the optimization of production processes is often an NP-hard problem. For this reason, it is common to use precalculated schedules and designs determined off-line for certain assumed boundary conditions (e.g., given order flows). This is mostly done with methods from Operations Research (OR) or event-driven simulations. However, in reality the boundary conditions are varying in an unknown way, so that the outcome may be far from optimal, as the optimal solution is sensitive to parameter changes (see Section 19.2). Therefore, adaptive on-line control strategies would be desirable. Although they cannot be expected to be system-optimal, the higher degree of flexibility promises a higher average performance, if the adaptation manages only to drive production *close* to the system optimum.

As the exact system optimum can mostly not be determined on-line, one needs to find suitable heuristics. Typical and powerful heuristic methods are, for example, genetic or evolutionary algorithms. However, their speed is also not sufficient for adaptive on-line control. Therefore, we are currently seeking better approaches that make use of characteristic properties of material-flow systems, such as conservation laws. A typical example is the continuity equation for material flow. Such equations can also be formulated for merging, diverging, and intersecting flows, although this is sometimes quite demanding. In the following, we will call them "traffic equations" for production systems (see Fig. 19.13). In fact, both traffic and production networks may be viewed as coupled dynamical queuing systems. Road sections correspond to buffers, machines or service stations to traffic lights at intersections, accidents to machine breakdowns, different origin–destination flows to different production paths (products), traffic congestion to crowded buffers, etc. Despite the differences between both systems, insights in both areas are mutually stimulating and beneficial. For example, it is typical for both systems that congestion builds up quickly, but it takes a long time to resolve it again. Therefore, the primary goal must be the prevention of congestion in the system. As a consequence, one must take care that the demand (i.e., the jobs fed into the system) do not exceed capacity. Capacity, however, depends itself on operation mode. We will come back to this in Section 19.4.3.

19.4.1
Traffic Equations for Production Systems

Let us briefly give the traffic equations for the simplest case of production systems (see [22,44,45]. The nodes of the production system represent machines or service stations, while the links i correspond to the transport connections for

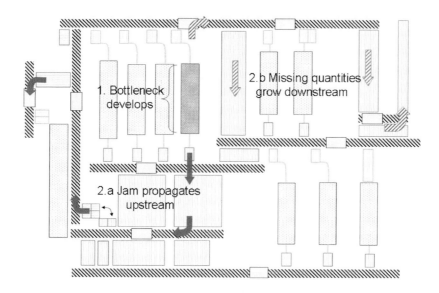
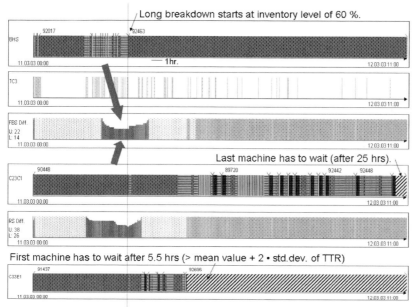

Fig. 19.13 (a) The material flows in production systems can be described by generalized traffic equations. While congestion propagates upstream, the lack of materials propagates downstream (see Section 19.4.3 for the issue of gap propagation). (b) Illustration of the state variables for different parts of the production system. Green represents progress according to time schedule, yellow critical operation, and red serious inefficiencies. (For a color version please see color plate on page 605.)

different kinds of material flow. Here, we will focus on one machine which can produce different kinds of products j from the educts i, but only subsequently and with a finite intermediate setup time. We will denote the arrival flow of products of kind j by $A_j(t)$, the maximum production rate by $\widehat{A}_j$, the (actual) departure flow of materials i by $O_i(t)$, and the potential departure flow by $\widehat{O}_i(t)$. Furthermore, we will assume that the buffer capacities are large enough to take up all incoming materials (which is not always the case).

If producing one unit of product j requires a_{ij} units of educt i, one has

$$A_j(t) = \gamma_j(t) \min\left(\widehat{A}_j, \left\{\frac{\widehat{O}_i(t)}{a_{ij}}\right\}\right) \tag{19.1}$$

and

$$O_i(t) = a_{ij} A_j = \gamma_j(t) \min\left(a_{ij}\widehat{A}_j, \widehat{O}_i(t)\right) \tag{19.2}$$

$\gamma_j(t)$ is 1, when product j is produced, otherwise 0. Moreover,

$$\widehat{O}_i(t) = \begin{cases} A_i(t - T_i) & \text{if } \Delta N_i(t) = 0 \\ \widehat{Q}_i & \text{if } \Delta N_i(t) > 0 \end{cases} \tag{19.3}$$

as the inflow of educts is a maximum when they are already available to be processed, otherwise it is determined by the arrival flow of materials i. $\widehat{Q}_i$ denotes the maximum delivery rate of educts, $\Delta N_i(t)$ the units of materials i waiting to be processed, and T_i the transport time for materials along the transport connection. $\Delta N_i(t) \geq 0$ can be determined via the equation

$$\Delta N_i(t) = \int_0^t dt' \, [A_i(t - T_i) - O_i(t)] \tag{19.4}$$

This model may be generalized in many different ways.

19.4.2
Re-routing Strategies and Machine Utilization

For simplicity, the above model has assumed a single path per product through the system. However, often materials can be processed by multiple machines with different technical and performance specifications. In the simplest case, one has I identical machines for parallel processing. The question is, which stacks should be sent to which machine? Therefore, if different alternative production paths are available, adaptive routing is an issue (see Fig. 19.14). Due to the finite setup times, it is normally not reasonable to send different stacks of the same job to different machines. Moreover, depending on capacity utilization, it may be costly to use *all* available machines.

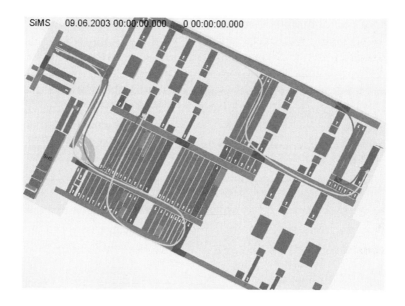

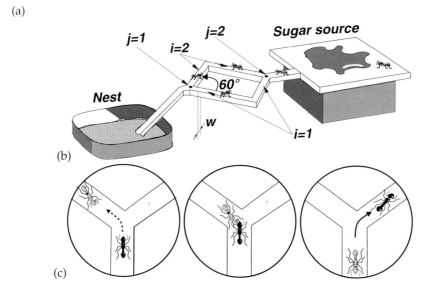

Fig. 19.14 (a) Illustration of different routing options for material flows in a packaging factory. (b) Illustration of an experiment on the route-choice behavior of ants. Neither of the two branches of the bridge from the nest to the food source provided enough transport capacity alone. (c) Although the chemical attraction is in favor of one ant trail only, repulsive pushing interactions can lead to the establishment of additional trails, if the transport capacity of one trail is too low (after [35]).

Obviously, the optimum usage of the parallel processing capacity must be load-dependent. Here, a biologically inspired approach can help. When the trails are wide enough, ants are known to establish one path between the nest and a single food source. After some time, this path corresponds approximately to the shortest path. This means a minimization of "transport costs", i.e., optimum use of resources. The underlying mechanism is a random exploration behavior in combination with a pheromone-based attraction via particular chemicals, which leads to a reinforcement of the most utilized trails. In the end, one of the trails survives, while the others fade away.

It is interesting to see what happens if the capacity of the ant trail is too small to keep up the desired level of material flow (food). This has been tested by connecting nest and food source by only two narrow bridges [35]. In such cases, ants use additional trails to keep up the desired level of throughput. The underlying mechanism is a repulsive interaction among ants. In an encounter of two ants at a bifurcation point, an ant is likely to be pushed to the alternative bridge. This can lead to the establishment of additional and stable ant trails [35, 37, 38] (see Fig. 19.14).

Therefore, optimal machine utilization could be implemented via an ant algorithm [39, 40]. When the utlization (demand) is low, just one of the machines would be operated. Otherwise, based on suitably defined repulsive interactions, jobs would be distributed over more alternative machines or routes, depending on queue lengths and capacities.

19.4.3
Self-organized Scheduling

Adaptive re-routing is only one element of adaptive scheduling. It is usually based on minimizing waiting or overall production times, see [22, 44, 45]. However, the adaptation of $\gamma_j(t)$, i.e., the determination of the optimal starting times and durations is equally important for the optimal production of products.

Obviously, one is interested in maximizing overall throughputs and minimizing cycle times by suitable outflow control. Conventionally, this is done by a centralized control approach. However, a decentralized approach promises a better adjustment to the local flow conditions (i.e., the respective, potentially time-dependent capacities and their utlization in time). On the other hand, too much sensitivity to the local situation will cause coordination problems in the system, see Section 19.3. So, is centralized control unavoidable? The answer is "no". Too much coordination (too strong coupling of all processes) will reduce the flexibility with respect to local, unexpected variations. This will finally affect performance. The challenge is to reach the right balance between a flexible adjustment to the local conditions and a coordination

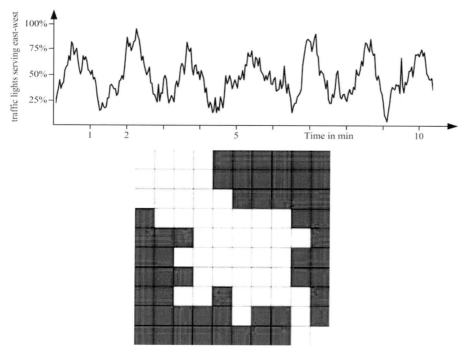

Fig. 19.15 Top: Adaptive traffic light scheduling as an example for a potentially self-organized coordination among locally coupled service stations. The fraction of green lights for the east-west direction in a Manhattan-like road network varies in an oscillatory way, as expected. However, in contrast to precalculated traffic light schedules, the oscillations are irregular due to the flexible response to the local, stochastically varying traffic situation. Bottom: The color-coded represenation shows that neighboring traffic lights tend to display a green (or red) light for the same driving directions. However, the shapes of the equally colored areas are changing irregularly in time, which shows the sensitivity to the local traffic situations (see www.trafficforum.org/trafficlights). The difficulty is to reach the right balance between flexibility and coordination. Too much flexibility will cause uncoordinated behavior among neighboring traffic lights, while too much coordination will not leave enough room to respond to the respective local traffic conditions (after [45]).

among all system elements. For example, a synchronization of machines or service stations may be favorable for the overall throughput of the system (see Fig. 19.15). It may be reached by a coupling of neighboring control elements, while a centralized control is not necessarily required. In fact, when suitable local interaction mechanisms are established, local coordination may eventually spread all over the system. This is one of the interesting features of self-organizing systems. Based on nonlinear interactions, a locally emerging pattern may have global effects. This can be viewed as a phase transition in the system. At such phase transitions, a scale-free behavior of the system is often observed, i.e., correlations can occur over very long distances.

We should, however, note that synchronization may not necessarily be a good strategy. This rather depends on the flow conditions. In fact, one needs to distinguish several operation regimes, in which different optimization strategies apply:

1. In the free-flow, low-utilization regime, demand is considerably below capacity. From the point of view of minimizing production times (not considering setup costs, here), the application of a first-in-first-out principle would perform best. Individual jobs would be served just upon their arrival. Moreover, a large spectrum of different products, i.e., a large variety can be produced. However, the overall throughput is low due to the small order flow.

2. If the order flow becomes higher, one enters the mutually obstructed, platoon-formation regime. That is, conflicts of usage are likely, and waiting times are unavoidable. Queues form in the system, but they actually help to reach a higher throughput. It is more efficient to serve educts already waiting to be processed rather than having to wait for their arrival. A minimization of waiting and production times leads to the following strategy: longer queues should be prioritized compared with shorter ones. However, interrupting a smoothly progressing production process is unfavorable. Therefore, dense moving platoons are prioritized compared with short enough standing ones. This can lead to a synchronization of successive machines or service stations (but with a transportation-time related phase shift). In traffic systems, such operation modes are called "green waves".

3. When demand exeeds capacity, one faces the congested, queue-dominated regime. This should be avoided by ensuring that jobs enter the production system with suitable delays. Otherwise, queues tend to form everywhere. A promising optimization strategy in such cases is the application of a flow-bundling principle. This basically means to reduce the variety of the products in order to minimize conflicts of usage and the heterogeneity in the system. Thereby, it is possible to increase the overall throughput of the system even under conditions of high utilization.

4. At extreme utilizations, when demand is considerably above capacity, one will end up in the heavily congested, gap-propagation regime. In that case, almost all the storage capacity is used up everywhere in the system. Now, the focus is on uninterrupted gap propagation, i.e., "green waves" for gaps. Larger gaps should be given priority compared to smaller gaps.

The above list of operation regimes is not meant to be complete. It should rather indicate that there is not a single, simple strategy to optimize the operation of a production system. Note that optimization is not only needed within each operation regime: another challenge is to manage optimal transitions between these regimes. It is also difficult to define exactly under which conditions these transitions take place. As a consequence, there are currently more questions than answers. However, based on their special methodological knowledge, physicists are expected to make make a lot of application-oriented, but fundamental contributions here.

19.5 Summary and Outlook

In this chapter, we have sketched the new research direction of econophysics, the physics of logistics and production systems [8]. This field is related to many-particle and traffic physics, but additionally the understanding of network-related properties is of primary importance [46–49, 51]. This is not only a matter of network topology, but also of material flows in complex networks. Due to conflicts of usage, optimal flows in networks are typically not stationary. They are rather organized in an oscillatory way, comparable to traffic flows at highly utilized, signalized intersections. From this point of view, it is not anymore surprising that there are many examples of oscillatory material flow in ecological and metabolic systems [52–55, 57–59]. As a consequence, the subject of queuing networks is mostly a time-dependent problem and related to issues of optimization. Optimization of production systems, however, tends to be NP-hard and cannot be performed in real-time. Therefore, new heuristics are needed to reach an adaptive, but highly performable operation. In this respect, we favor a self-organization approach. It requires a suitable specification of the interactions in the system, otherwise the system will end up in a local optimum or behave in an unstable way, so that breakdowns and slower-is-faster effects occur.

Biologically inspired methods are a promising approach, here. We have mentioned ant algorithms, but we also call for an exploration of the potentials of *biologistics*. In Ref. [60], the challenge has been stated as follows: "By 'biologistics' we mean both the production of artifacts by biological systems and the transfer of biological organization principles to production plants and logistic processes. In fact, the human body can be considered as one of the most complex logistic systems. It manages to deliver billions of different substances to billions of different locations in the body. This material transport is quite specific and very efficient. The basic body functions consume energy at the rate of only a 100 Watt light bulb. This incredible efficiency is the result of an evolutionary optimization process and competition over millions of years." We should be able to transfer this efficiency to production and logistics as well, but obviously this would need completely different organization principles in the future.

Acknowledgments

The authors would like to thank Dieter Armburster, Thomas Brenner, Audrey Dussutour, Illés Farkas, Dominique Fasold, Rui Jiang, Anders Johansson, Alexander Mikhailov, Andrew Riddell, Dick Sanders, Martin Treiber, Tamas Vicsek, Ulrich Witt and others for the great collaboration and interesting discussions. Furthermore, they are grateful for partial financial support by SCA Packaging, the EU project MMCOMNET, and the German Research Foundation (DFG).

References

1 NAGATANI, T., *Rep. Prog. Phys.* 65 (**2002**), p. 1331

2 CHOWDHURY, D., SANTEN, L., SCHADSCHNEIDER, A., *Phys. Rep.* 329 (**2000**), p. 199

3 HELBING, D., *Rev. Mod. Phys.* 73 (**2001**), p. 1067.

4 MIKHAILOV, A. S., CALENBUR, V., *From Cells to Societies: Models of Complex Coherent Action*, Springer, Berlin, **2002**

5 RADONS, G., NEUGEBAUER, R., Eds., *Nonlinear Dynamics of Production Systems*, Wiley, New York, **2004**

6 ARMBRUSTER, D., MIKHAILOV, A. S., KANEKO, K. (Eds.)*Networks of Interacting Machines: Production Organization in Complex Industrial Systems and Biological Cells* World Scientific, Singapore, **2005**

7 HOPP, W. J., SPEARMAN, M. L., *Factory Physics*, McGraw-Hill, Boston, **2000**.

8 HELBING, D., ARMBRUSTER, D., MIKHAILOV, A., LEFEBER, E., (Eds.) *Special Issue: Information and Material Flows in Complex Networks*, Physica A 363(1) (**2006**)

9 DAGANZO, C., *A Theory of Supply Chains*, Springer, New York, **2003**

10 HELBING, D., *New Journal of Physics* 5 (**2003**), p. 90

11 PETERS, K., WORBS, J., PARLITZ, U., WIENDAHL, H.-P., in *Nonlinear Dynamics of Production Systems*, Ed. G. Radons and R. Neugebauer, Wiley, New York (**2004**), p. 39

12 ARMBRUSTER, D., MARTHALER, D., RINGHOFER, C., *SIAM J. Multiscale Modeling and Simulation* 2 (1) (**2004**), p. 43

13 FORRESTER, J. W., *Industrial Dynamics*, MIT Press, Cambridge, MA, **1961**

14 CHEN, F., DREZNER, Z., RYAN, J. K., SIMCHI-LEVI, D., *Management Science* 46 (3) (**2000**), p. 436

15 BAGANHA, M., COHEN, M., *Operations Research* 46 (3) (**1998**), p. 72

16 KAHN, J., *Econom. Rev.* 77 (**1987**), p. 667.

17 LEE, H., PADMANABHAN, P., WHANG, S., *Sloan Management Rev.* 38 (**1997**), p. 93

18 STERMAN, J. D., *Business Dynamics*, McGraw-Hill, Boston, **2000**

19 HELBING, D., LÄMMER, S., BRENNER, T., WITT, U., *Physical Review E* 70 (**2004**), p. 056118

20 HELBING, D., LÄMMER, S., SEIDEL, T., ŠEBA, P., PŁATKOWSKI, T., *Physical Review E* 70 (**2004**), p. 066116.

21 HELBING, D., SCHÖNHOF, M., STARK, H.-U., HOLYST, J. A., *Advances in Complex Systems* 8 (**2005**), p. 87

22 HELBING, D., LÄMMER, S., pending patent DE 10 2005 023 742.8.

23 NAGATANI, T., HELBING, D., *Physica A* 335 (**2004**), p. 644

24 HELBING, D., FARKAS, I., VICSEK, T., *Nature* 407 (**2000**), p. 487

25 KESTING, A., TREIBER, M., SCHÖNHOF, M., KRANKE, F., HELBING, D., in: *Traffic and Granular Flow '05*, Springer, Berlin, (**2006**), to be published

26 BEAUMARIAGE, T., KEMPF, K., in *Proceedings of the 5th IEEE/SEMI Advanced Semiconductor Manufacturing Conference*, (**1994**)p. 169

27 CHASE, C. J., SERRANO, J., RAMADGE, P., *IEEE Trans. Autom. Control.* 38 (**1993**), p. 70

28 USHIO, T., UEDA, H., HIRAI, K., *Systems & Control Letters* 26 (**1995**), p. 335.

29 KATZORKE, I., PIKOVSKY, A., *Discrete Dynamics in Nature and Society* 5 (**2000**), p. 179

30 PETERS, K., PARLITZ, U., *Int. J. Bifurcation and Chaos*, 13 (9) (**2003**), p. 2575

31 DIAZ-RIVERA, I., ARMBRUSTER, D., TAYLOR, T., *Mathematics and Operations Research* 25, 2000, p. 708

32 HELBING, D., TREIBER, M., KESTING, A., *Physica A* 363 (1) (**2006**), p. 62

33 JIANG, R., HELBING, D., SHUKLA, P. K., WU, Q.-S., *Physica A*, in print (**2006**)

34 HELBING, D., JIANG, R., TREIBER, M., *Physical Review E* 72 (**2005**), p. 046130

35 DUSSUTOUR, A., FOURCASSIÉ, V., HELBING, D., DENEUBOURG, J.-L., *Nature* 428 (**2004**), p. 70

36 HELBING, D., BUZNA, L., JOHANSSON, A., WERNER, T., *Transportation Science* 39(1) (**2005**), p. 1

37 PETERS, K., JOHANSSON, A., HELBING, D., *Kuenstliche Intelligenz*, 4 (**2005**), p. 11

38 PETERS, K., JOHANSSON, A., DUSSUTOUR, A., HELBING, D., *Advances in Complex Systems* (**2006**), in print

39 BONABEAU, E., DORIGO, M., THERAULAZ, G., *Swarm Intelligence*, Oxford University Press, Oxford, **1999**

40 ARMBRUSTER, D., DE BEER, C., FREITAG, M., JAGALSKI, T., RINGHOFER, C., *Physica A* 363 (1) (**2006**), p. 104

41 FASOLD, D., Master's thesis, TU Dresden, (**2001**)

42 LÄMMER, S., KORI, H., PETERS, K., HELBING, D., *Physica A* 363 (1) (**2006**), p. 39

43 HOŁYST, J. A., HAGEL, T., HAAG, G., WEIDLICH, W., *Journal of Evolutionary Economics* 6 (**1996**), p. 31

44 HELBING, D., *Journal of Physics A: Mathematical and General* 36 (**2003**), p.L593

45 HELBING, D., LÄMMER, S., LEBACQUE, P., in: C. Deissenberg, R.F. Hartl (eds.), *Optimal Control and Dynamic Games* Springer, Dortrecht,(**2005**), p. 239

46 WATTS, D. J., STROGATZ, S. H., *Nature* 393 (**1998**), p. 440

47 ALBERT, R., BARABÁSI, A.-L., *Reviews of Modern Physics* 74 (**2002**), p. 47

48 MASLOV, S., SNEPPEN, K., *Science* 296 (**2002**), p. 910

49 BORNHOLDT, S., SCHUSTER, H. G., *Handbook of Graphs and Networks*, Wiley, Weinheim, **2003**

50 BORNHOLDT, S., ROHLF, T., *Physical Review Letters* 84 (**2000**), p. 6114

51 DOROGOVTSEV, S. N., MENDES, J. F. F., *Evolution of Networks*, Oxford University Press, Oxford, **2004**

52 ELOWITZ, M. B., LEIBLER, S., *Nature* 403 (**2000**), p. 335

53 ALMAAS, E., KOVÁCS, B., VICSEK, T., OLTVAI, Z. N., BARABÁSI, A.-L., *Nature* 427 (**2004**), p. 839

54 MIKHAILOV, A. S., HESS, B., *J. Biol. Phys.* 28 (**2002**), p. 655

55 KOYA, S., UEDA, T., *ACH Models in Chemistry* 135 (3) (1998), p. 297

56 ALBERTS, B., JOHNSON, A., LEWIS, J., RAFF, M., ROBERTS, K., WALTER, P., *Molecular Biology of the Cell*, Garland Science, New York, **2002**

57 SCHUSTER, S., MARHL, M., HÖFER, T., *Eur. J. Biochem.* 269 (**2002**), p. 1333

58 KUMMER, U., OLSEN, L. F., DIXON, C. J., GREEN, A. K., BOMBERG-BAUER, E., BAIER, G., *Biophysical Journal* 79 (**2000**), p. 1188

59 MEINHOLD, L., SCHIMANSKY-GEIER, L., *Physical Review E* 66 (**2002**), p. 050901(R)

60 HELBING, D., ARMBRUSTER, D., MIKHAILOV, A., LEFEBER, E., *Physica A* 363 (1) (**2006**), p.xi

20
Can we Recognize an Innovation?: Perspective from an Evolving Network Model
Sanjay Jain and Sandeep Krishna

"Innovations" are central to the evolution of societies and the evolution of life. But what constitutes an innovation? We can often agree after the event when its consequences and impact over a long term are known, whether or not something was an innovation, and whether it was a "big" innovation or a "minor" one. But can we recognize an innovation "on the fly" as it appears? Successful entrepreneurs often can. Is it possible to formalize that intuition? We discuss this question in the setting of a mathematical model of evolving networks. The model exhibits self-organization, growth, stasis, and collapse of a complex system with many interacting components, reminiscent of real-world phenomena. A notion of "innovation" is formulated in terms of graph-theoretic constructs and other dynamical variables of the model. A new node in the graph gives rise to an innovation provided it links up "appropriately" with existing nodes; in this view innovation necessarily depends upon the existing context. We show that innovations, as defined by us, play a major role in the birth, growth and destruction of organizational structures. Furthermore, innovations can be categorized in terms of their graph-theoretic structure as they appear. Different structural classes of innovation have potentially different qualitative consequences for the future evolution of the system, some minor and some major. Possible general lessons from this specific model are briefly discussed.

20.1
Introduction

In everyday language, the noun "innovation" stands for something new that brings about a change; it has a positive connotation. Innovations occur in all branches of human activity – in the world of ideas, in social organization, in technology. Innovations may arise by conscious and purposeful activity, or serendipitously; in either case, innovations by humans are a consequence of cognitive processes. However, the word innovation does not always refer to a product of cognitive activity. In biology, we say, for example, that photosynthesis, multicellularity, and the eye, were evolutionary innovations. These were products not of any cognitive activity, but of biological evolution. It nevertheless seems fair to regard them as innovations; these novelties certainly transformed the way organisms lived. The notion of innovation seems to presuppose a context provided by a complex evolutionary dynamics; for example, in everyday language the formation of the earth, or even the first star, is not normally referred to as an innovation.

Innovations are a crucial driving force in chemical, biological and social systems, and it is useful to have an analytical framework to describe them. This subject has a long history in the social sciences (see, e.g., [1,2]). Here we adopt a somewhat different approach. We give a mathematical example of a complex system that seems to be rich enough to exhibit what one might intuitively call innovation, and yet simple enough for the notion of innovation to be mathematically defined and its consequences analytically studied. The virtue of such a stylized example is that it might stimulate further discussion about innovation, and possibly help clarify the notion in more realistic situations.

Innovations can have "constructive" and "destructive" consequences at the same time. The advent of the automobile (widely regarded as a positive development) was certainly traumatic for the horse-drawn carriage industry and several other industries that depended upon it. When aerobic organisms appeared on the earth, their more efficient energy metabolism similarly caused a large extinction of several anaerobic species [3]. The latter example has a double irony. Over the first two billion years of life on earth, there was not much oxygen in the earth's environment. Anaerobic creatures (which did not use free oxygen for their metabolism) survived, adapted, innovated new mechanisms (e.g., photosynthesis) in this environment, and spread all over the earth. Oxygen in the earth's environment was largely a by-product of photosynthetic anaerobic life, a consequence of anaerobic life's "success". However, once oxygen was present in the environment in a substantial quantity, it set the stage for another innovation, the emergence of aerobic organisms which used this oxygen. Because of their greater metabolic efficiency the aerobic organisms out-competed and decimated the anaerobic ones. In a very real sense, therefore, anaerobic organisms were victims of their own success. Innovation has this dynamic relationship with "context": what constitutes "successful" innovation depends upon the context, and successful innovation then alters the context. Our mathematical example exhibits this dynamic and explicitly illustrates the two-faced nature of innovation. We show that the ups and downs of our evolutionary system as a whole are also crucially related to innovation.

20.2
A Framework for Modeling Innovation: Graph Theory and Dynamical Systems

Systems characterized by complex networks are often represented in terms of a graph consisting of nodes and links. The nodes represent the basic components of the system, and links between them their mutual interactions. A graph representation is quite flexible and can represent a large variety of situations [4]. For a society, nodes can represent various agents such as individuals, firms and institutions, as well as goods and processes. Links between nodes can represent various kinds of interaction, such as kinship or communication links between individuals, inclusion links (e.g., a directed link from a node representing an individual to a node representing a firm, implying that the individual is a member of the firm), production links (from a firm to an article that it produces), links that specify the technological web (for every process node, incoming links from all the goods it needs as input and outgoing links to every article it produces), etc. In an ecological setting, nodes can represent biological species, and links their predator-prey or other interactions. In cellular biology, nodes might represent molecules such as metabolites and proteins as well as genes, and links their biochemical interactions.

A graph representation is useful for describing several kinds of innovation. Often, an innovation is a new article, process, firm, or institution. This is easily represented by inserting a new node in the graph, together with its links to existing nodes. Of course, not every such insertion can be called an innovation; other conditions have to be imposed. The existing structure of the graph provides one aspect of the "context" in which a prospective innovation is to be judged, reflecting its "location" or relationship with other entities. In this formulation it is clear that innovations such as the ones mentioned above are necessarily a change in the graph structure. Thus a useful modeling framework for innovations is one where graphs are not static but change with time. In real systems graphs are always evolving: new nodes and links constantly appear, and old ones often disappear as individuals, firms and institutions die, goods lose their usefulness, species become extinct or any of these nodes lose some of their former interactions. It is in such a scenario that certain kinds of structures and events appear that earn the nomenclature "innovation". We will be interested in a model where a graph evolves by the deletion of nodes and links as well as the insertion of new ones. Insertions will occasionally give rise to innovations. We will show that innovations fall into different categories that can be distinguished from each other by analyzing the instantaneous change in the graph structure caused by the insertion, locally as well as globally. We will argue that these different "structural" categories have different "dynamical" consequences for the "well-being" of other nodes and the evolution of the system as a whole in the short as well as the long run.

In addition to an evolving graph, another ingredient seems to be required for modeling innovation in the present approach: a graph-dependent dynamics of some variables associated with nodes or links. In a society, for example, there are flows of information, goods and money between individuals that depend upon their mutual linkages, which affect node attributes such as individual wealth, power, etc. The structure of an ecological food web affects the populations of its species. Thus, a change in the underlying graph structure has a direct impact on its "node variables". Deciding whether a particular graph change constitutes an innovation must necessarily involve an evaluation of how variables such as individual wealth, populations, etc., are affected by it. Changes in these variables in turn trigger further changes in the graph itself, sometimes leading to a cascade of changes in the graph and other variables. For example, the decline in wealth of a firm (node variable) may cause it to collapse; the removal of the corresponding node from the market (graph change) may cause a cascade of collapses. The invention of a new product (a new node in the graph) which causes the wealth of the firm inventing it to rise (change in a node variable) may be emulated by other firms causing new linkages and further new products.

In order to "recognize an innovation on the fly" it thus seems reasonable to have a framework which has (a) a graph or graphs representing the network of interactions of the components of the system; (b) the possibility of graph evolution (the appearance and disappearance of nodes and links); and (c) a graph-dependent dynamics of node or link variables that in turn has a feedback upon the graph evolution. The example discussed below has these features. They are implemented in a simple framework that has only one type of node, one type of link and only one type of node variable.

20.3
Definition of the Model System

The example is a mathematical model [5] motivated by the origin of life problem, in particular, the question of how complex molecular organizations could have emerged through prebiotic chemical evolution [6–10]. There are s interacting molecular species in a "prebiotic pond", labelled by $i \in S \equiv \{1, 2, \ldots, s\}$. Their interactions are represented by the links of a directed graph, of which these species are nodes. The graph is defined by its $s \times s$ adjacency matrix $C = (c_{ij})$, with $c_{ij} = 1$ if there exists a link from node j to node i (chemically that means that species j is a catalyst for the production of species i), and $c_{ij} = 0$ otherwise. c_{ii} is assumed zero for all i: no species in the pond is self-replicating. Initially the graph is chosen randomly, each c_{ij} for $i \neq j$ is chosen to be unity with a small probability p and zero with probability

$1 - p$. p represents the "catalytic probability" that a given molecular species will catalyze the production of another randomly chosen one [11].

The pond sits by the side of a large body of water like a sea or river, and periodically experiences tides or floods which can flush out molecular species from the pond and bring in new ones, changing the network. We use a simple graph update rule in which exactly one node is removed from the graph (along with all its links) and one new node is added whose links with the remaining $s - 1$ nodes are chosen randomly with the same probability p. We adopt the rule that the species with the least relative population (or, if several species share the least relative population, one of them chosen randomly) is removed. This is where selection enters the model: species with smaller populations are punished. This is an example of "extremal" selection [10] in that the least populated species is removed; the results of the model are robust in relaxing the extremality assumption [12, 13].

In order to determine which node will have the least population, we specify a population dynamics which depends upon the network. The dynamics of the relative populations, x_i ($0 \leq x_i \leq 1$, $\sum_{i=1}^{s} x_i = 1$) is given by

$$\dot{x}_i = \sum_{j=1}^{s} c_{ij} x_j - x_i \sum_{k=1}^{s} \sum_{j=1}^{s} c_{kj} x_j \qquad (20.1)$$

This is a set of rate equations for catalyzed chemical reactions in a well-stirred chemical reactor[1]. They implement the approximate fact that under certain simplifying assumptions a catalyst causes the population of whatever it catalyzes to grow at a rate proportional to its own (i.e., the catalyst's) population [5, 14]. Between successive tides or floods the set of species and hence the graph remains unchanged, and the model assumes that each x_i reaches its attractor configuration X_i under (20.1) before the next graph update. The species with the least X_i is removed at the next graph update.

Starting from the initial random graph and random initial populations, the relative populations are evolved according to (20.1) until they reach the attractor $\mathbf{X}$, and then the graph is updated according to the above rules. The new incoming species is given a fixed relative population x_0, all x_i are perturbed about their existing values (and rescaled to restore normalization). This process is iterated several times. Note that the model has two inbuilt time scales, the population dynamics relaxes on a fast time scale, and graph evolution on a slow time scale. The above model may be regarded as an evolutionary model in nonequilibrium statistical mechanics.

---

[1] See *Derivation of Eq. (20.1)* in Appendix A.

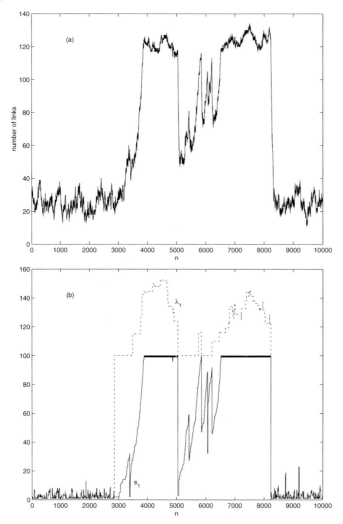

Fig. 20.1 A run with parameter values $s = 100$ and $p = 0.0025$. The x-axis shows time, n (= number of graph updates). (a) shows the number of links in the graph as a function of time. In (b), the continuous line shows s_1, the number of populated species in the attractor (= the number of nonzero components of X_i) as a function of time. The dotted line shows λ_1, the largest eigenvalue of C as a function of time. (The λ_1 values shown are 100 times the actual λ_1 value.)

20.4
Time Evolution of the System

A sample run is depicted in Figs 20.1 and 20.2. For concreteness, we will discuss this run in detail, describing the important events, processes, and the graph structures that arise, with an emphasis on the role of innovation. The same qualitative behavior is observed in hundreds of runs with the various parameter values. Quantitative estimates of average time scales, etc., as a function of the parameters s and p are discussed in [5, 14] and Appendix A. The robustness of the behavior to various changes of the model is discussed in [12, 13].

Broadly, Figs 20.1 and 20.2 exhibit the following features. Initially, the graph is sparse and random (see Figs 20.2(a-d)), and remains so until an autocatalytic set (ACS), defined below, arises by pure chance. On average the ACS arrives on a time scale $1/(p^2 s)$ in units of graph update time[2]; in the exhibited run it arrives at $n = 2854$ (Fig. 20.2(e)). In this initial regime, called the "random phase", the number of populated species, s_1, remains small. The appearance of the ACS transforms the population and network dynamics. The network self-organizes, its density of links increases (Fig. 20.1(a)), and the ACS expands (Figs 20.2(e-n)) until it spans the entire network (as evidenced by s_1 becoming equal to s, at $n = 3880$, Figs 20.1(b), 20.2(n)). The ACS grows across the graph exponentially fast, on a time scale $1/p$ [5]. This growth is punctuated by occasional drops (e.g., Fig. 20.1(b) at $n = 3387$, see also Fig. 20.2(h,i)). The period between the appearance of a small ACS and its eventual spanning of the entire graph is called the "growth phase". After spanning, a new effective dynamics arises, which can cause the previously robust ACS to become fragile, resulting in crashes (the first major one is at $n = 5041$) in which s_1 as well as the number of links, drops drastically. The system experiences repeated rounds of crashes and recoveries (Figs 20.2(o–u), see [12, 15] for a longer time scale.) The period after a growth phase and up to a major crash (more precisely, a major crash that is a "core-shift", defined below) is called the "organized phase". After a crash, the system ends up in the growth phase if an ACS still exists in the graph (as at $n = 5042$, Fig. 20.2(r)) or the random phase if it does not (as at $n = 8233$, Fig. 20.2(t)). Below, we argue that most of the crucial events in the evolution of the system, including its self-organization and collapse, are caused by "innovation".

20.5
Innovation

The word "innovation" certainly denotes something new. In the present model at each graph update a new structure enters the graph: the new node and its links with existing nodes. However, not every new thing qualifies as an innovation. In order for a novelty to bring about some change, it should confer some measure of at least temporary "success" to the new node. (A mutation must endow the organism in which it appears with some extra fitness,

2) See *Timescale for appearance and growth of the dominant ACS* in Appendix A.

568 | 20 Can we Recognize an Innovation?: Perspective from an Evolving Network Model

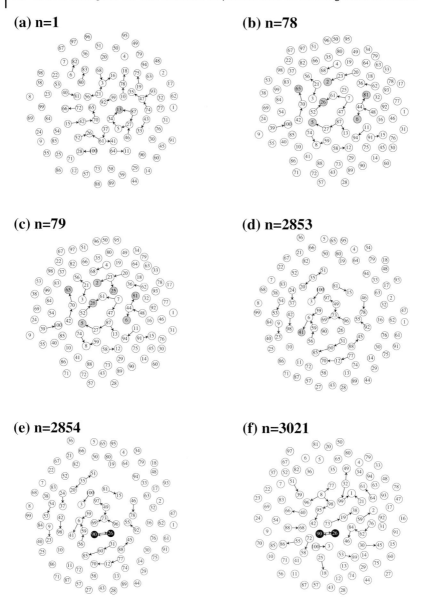

Fig. 20.2 (Caption is at the end.)

and a new product must have some sale prospects in order to qualify as an innovation.) In the present model, after a new node appears, the population dynamics takes the system to a new attractor of (20.1), which depends upon the mutual interactions of all the nodes. In the new attractor this node (denoted k) may become extinct, $X_k = 0$, or may be populated, $X_k > 0$. The

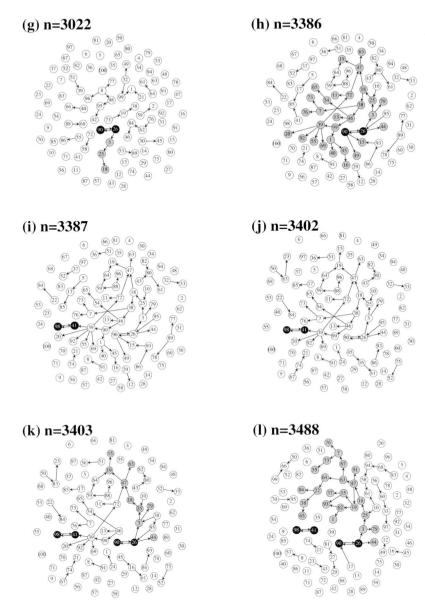

Fig. 20.2 (contd.)

only possible criterion of individual "success" in the present model is population. Thus we require the following minimal "performance criterion" for a new node k to give rise to an innovation: X_k should be greater than zero in the attractor that follows after that particular graph update. That is, the node should "survive" at least until the next graph update.

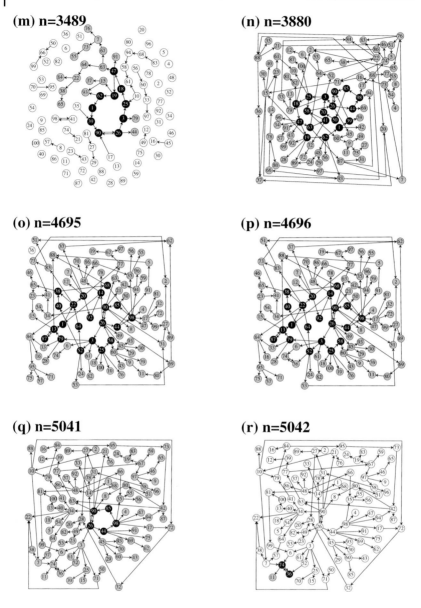

Fig. 20.2 (contd.)

This is obviously a "minimal" requirement, a necessary condition, and one can argue that we should require of an innovation more than just this "minimal performance". A new node that brings about an innovation ought to transform the system or its future evolution in a more dramatic way than merely surviving till the next graph update. Note, however, that this mini-

20.5 Innovation

(s) n=8232

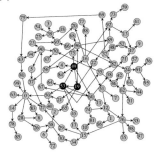

(t) n=8233

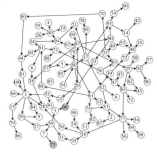

(u) n=10000

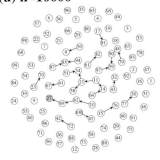

Fig. 20.2 The structure of the evolving graph at various time instants for the run depicted in Fig. 20.1. Examples of several kinds of innovation and their consequences for the evolution of the system are shown (see text for details). Nodes with $X_i = 0$ are shown in white; according to the evolution rules all white nodes in a graph are equally likely to be picked for replacement at the next graph update. Black and grey nodes have $X_i > 0$. Thus the number of black and grey nodes in a graph equals s_1, plotted in Fig. 20.1(b). Black nodes correspond to the core of the dominant ACS and grey nodes to its periphery. Only mutual links among the nodes are of significance, not their spatial location which is arranged for visual convenience. The graphs are drawn using LEDA.

mal performance criterion nevertheless eliminates from consideration a large amount of novelty that is even less consequential. Out of the 9999 new nodes that arise in the run of Fig. 20.1, as many as 8929 have $X_k = 0$ in the next population attractor; only 1070 have $X_k > 0$. Furthermore, the set of events with $X_k > 0$ can be systematically classified in the present model using a graph-theoretic description. Below we describe an exhaustive list of six categories of such events, each with a different level of impact on the system (see Fig. 20.2, discussed in detail below). One of these categories consists of nodes that disappear after a few graph updates leaving no subsequent trace on the system. Another category consists of nodes that have only an incremental impact. The remaining four categories cause (or can potentially cause) more drastic changes in the structure of the system, its population dynamics and its future evolution.

In view of this classification it is possible to exclude one or more of these categories from the definition of innovation and keep only the more "consequential" ones. However, we have chosen to be more inclusive and will regard all the above categories of events as innovations. In other words we will regard the above "minimal" performance criterion as a "sufficient" one for innovation. Thus we will call the introduction of a new node k and the graph structure so formed an *innovation* if $X_k > 0$ in the population attractor that immediately follows the event, i.e., if the node "survives" at least until the next graph update. This definition then includes both "small" and "big" innovations that can be recognized, based on their graph-theoretic structure upon appearance.

As will be seen below, it turns out that a new node generates an innovation only if it links "appropriately" to "suitable" structures in the existing graph. Thus the above definition makes the notion of innovation context dependent. It also captures the idea that an innovation rests on new linkages between structures.

20.6
Six Categories of Innovation

20.6.1
A Short-lived Innovation: Uncaring and Unviable Winners

There are situations where a node, say an agent in society or a species in an ecosystem, acquires the ability to parasite off another, without giving the system anything substantive in return. The parasite gains as long as the host survives, but often this situation does not last very long. The host dies, and eventually so does the parasite that is dependent on it. It is debatable whether the acquiring of such a parasitic ability should be termed an innovation, but from the local vantage point of the parasite, while the going is still good, it might seem like one.

Figure 20.2(b,c) shows an example of an innovation of the type that appears in the random phase of the model. Node 25 is the node which is replaced at $n = 78$ (see Fig. 20.2(b), where node 25 is colored white, implying that $X_{25} = 0$ at $n = 78$.) The new node that replaces it (also numbered 25 in Fig. 20.2(c)) receives a link from node 23, thus putting it at the end of a chain of length 2 at $n = 79$. This is an innovation according to the above definition, for, in the attractor configuration corresponding to the graph of Fig. 20.2(c), node 25 has a nonzero relative population; $X_{25} > 0$. This is because, for any graph that does not contain a closed cycle, one can show that the attractor $\mathbf{X}$ of (20.1), for generic initial conditions, has the property that only those X_i are nonzero whose nodes i are the endpoints of the longest chains in the graph[3]. The X_i for all other nodes is zero [16]. Since the longest chains in Fig. 20.2(c) are of length 2, node 25 is populated. (This explains why a node is grey or white in Figs 20.2(a-d)) Note that node 25 in Fig. 20.2(c) has become a parasite of node 23 in the sense that there is a link from 23 to 25 but none from 25 to any other node. This means that node 25 receives catalytic support for its own production from node 23, but does not give support to any other node in the system.

However, this innovation doesn't last long. Nodes 20 and 23, on whom the well-being of node 25 depends, are unprotected. Since they have the least possible value of X_i, namely, zero, they can be eliminated at subsqent graph updates, and their replacements in general do not feed into node 25. Sooner or later selection picks 23 for replacement, and then 25 also becomes depopulated. By $n = 2853$ (Fig. 20.2(d)) node 25 and all others that were populated at $n = 79$, have joined the ranks of the unpopulated. Node 25 (and others of its ilk) are doomed because they are "uncaring winners": they do not feed into (i.e., do not catalyze) the nodes upon whom their well-being depends. That is why when there are no closed cycles, all structures are transitory; the graph remains random. Of the 1070 innovations, 115 were of this type.

20.6.2
Birth of an Organization: Cooperation Begets Stability

At $n = 2853$ node 90 is an unpopulated node (Fig. 20.2(d)). It is eliminated at $n = 2854$ and the new node 90 forms a two-cycle with node 26 (Fig 20.2(e)). This is the first time (and the only time in this run) an innovation forms a closed cycle in a graph that previously had no cycles. A closed cycle between two nodes is the simplest *cooperative* graph theoretical structure possi-

---

3) See *The attractor of Eq. (20.1)* in Appendix A.

ble. Nodes 26 and 90 help each other's population grow; together they form a self-replicating system. Their populations grow much faster than other nodes in the graph; it turns out that in the attractor for this graph only nodes 26 and 90 are populated, with all other nodes having $X_i = 0$ ([16, 17]; see also Appendix A). Because node 90 is populated in the new attractor this constitutes an innovation. However, unlike the previous innovations, this one has a greater staying power, because nodes 26 and 90 do well *collectively*. At the next graph update *both* nodes 26 and 90 will be immune to removal since one of the other nodes with zero X_i will be removed. Notice that nodes 26 and 90 do not depend on nodes which are part of the least-fit set (those with the least value of X_i). The cycle has all the catalysts it needs for for the survival of each of its constituents. This property is true not just for cyclic subgraphs but for a more general cooperative structure, the *autocatalytic set* (ACS).

An ACS is a set of species which contains a catalyst for each species in the set [11, 18, 19]. In the context of the present model we can define an ACS to be a subgraph each of whose nodes has at least one incoming link from a node of the same subgraph. While ACSs need not be cycles, they must contain at least one cycle. If the graph has (one or more) ACSs, one can show that the set of populated nodes ($X_i > 0$) must be an ACS, which we call the dominant ACS[4] [5, 16]. (In Figs 20.2(e-s), the subgraph of the grey and black nodes is the dominant ACS.) Therefore none of the nodes of the dominant ACS can be hit in the next graph update as long as there is any node outside it. In other words, the collective well-being of all the constituents of the dominant ACS, ensured by cooperation inherent within its structure, is responsible for the ACSs relative robustness and hence longevity.

In societies, this kind of event is akin to the birth of an organization wherein two or more agents improve upon their performance by entering into an explicit cooperation. A booming new township or industrial district perhaps can be analyzed in terms of a closure of certain feedback loops. In prehistory, the appearance of tools that could be used to improve other tools may be regarded as events of this kind which probably unleashed a lot of artifact building. On the prebiotic earth one can speculate that the appearance of a small ACS might have triggered processes that eventually led to the emergence of life [14].

If there is no ACS in the graph then the largest eigenvalue of the adjacency matrix of the graph, λ_1 is zero. If there is an ACS then $\lambda_1 \geq 1$[5] [5, 16]. In Fig. 20.1(b), λ_1 jumped from zero to one when the first ACS was created at $n = 2854$.

4) See *Dominant ACS of a graph* in Appendix A.
5) See *Graph-theoretic properties of ACSs* in Appendix A.

20.6.3
Expansion of the Organization at its Periphery: Incremental Innovations

Consider Figs 20.2(f) and (g). Node 3, which is unpopulated at $n = 3021$, gets an incoming link from node 90 and an outgoing link to node 25 at $n = 3022$ which results in three nodes adding onto the dominant ACS. Node 3 is populated in the new attractor and hence this is an innovation. This innovation has expanded the "periphery" of the organization, defined below.

Every dominant ACS is a union of one or more "simple ACSs" each of which have a substructure consisting of a "core" and "periphery". For example, the dominant ACS in Fig. 20.2(g) has one simple ACS and in Fig. 20.2(k) it has two. For every simple ACS there exists a maximal subgraph, called the *core* of that ACS, from each of whose nodes there is a directed path to every node of that ACS. The rest of that ACS is termed its *periphery*. In Figs 20.2(e–s), the core is colored black, and the periphery grey. Thus in Fig. 20.2(g), the two-cycle of nodes 26 and 90 is the core of the dominant ACS, and the chain of nodes 3, 25 and 18 along with the incoming link to node 3 from 26 constitutes its periphery. The core of a simple ACS is necessarily an *irreducible subgraph*. An irreducible subgraph is one that contains a directed path from each of its nodes to every other of its nodes [20]. When the dominant ACS consists of more than one simple ACSs, its core is the union of their cores, and its periphery the union of their peripheries. Note that the periphery nodes by definition do not feed back into the core; in this sense they are parasites that draw sustenance from the core. The core, by virtue of its irreducible property (positive feedback loops within its structure, or cooperativity), is self-sustaining, and also supports the periphery. The λ_1 of the ACS is determined solely by the structure of its core [12, 15].

The innovation at $n = 3022$, one of 907 such innovations in this run, is an "incremental" one in the sense that it does not change the core (and hence does not change λ_1). However, such incremental innovations set the stage for major transformations later on. The ability of a core to tolerate parasites can be a boon or a bane, as we will see below.

20.6.4
Growth of the Core of the Organization: Parasites Become Symbionts

Another kind of innovation that occurs in the growth phase is illustrated in Figs 20.2(l) and (m). In Fig. 20.2(l), the dominant ACS has two disjoint components. One component, consisting of nodes 41 and 98, is just a two-cycle without any periphery. The other component has a two-cycle (nodes 26 and 90) as its core that supports a large periphery. Node 39 in Fig. 20.2(l) is eliminated at $n = 3489$. The new node 39 (Fig. 20.2(m)) gets an incoming link from the periphery of the larger component of the dominant ACS and an outgoing link to the core of the same ACS. This results in expansion of the core, with several nodes getting added to it at once and λ_1 increasing. This example illustrates two distinct processes:

1. This innovation co-opts a portion of the parasitic periphery into the core. This strengthens cooperation: 26 contributes to the well-being of 90 (and hence to its own well-being) along two paths in Fig. 20.2(m) instead of only one in Fig. 20.2(l). This is reflected in the increase of λ_1; $\lambda_1 = 1.15$ and 1 for Figs 20.2(m) and 20.2(l) respectively. The larger the periphery, the greater is the probability of such core-enhancing innovations. This innovation is an example of how tolerance and support of a parasitic periphery pays off for the ACS. Part of the parasitic periphery turns symbiont. Note that this innovation builds upon the structure generated by previous incremental innovations. In Fig. 20.1(b) each rise in λ_1 indicates an enlargement of the core [12, 15]. There are 40 such events in this run. As a result of a series of such innovations which add to the core and periphery, the dominant ACS eventually grows to span the entire graph at $n = 3880$, Fig. 20.2(n), and the system enters the "organized phase".

2. This example also highlights the competition between different ACSs. The two-cycle of nodes 41 and 98 was populated in Fig. 20.2(l), but is unpopulated in Fig. 20.2(m). Since the core of the other ACS becomes stronger than this two-cycle, the latter is driven out of business.

20.6.5
Core-shift 1: Takeover by a New Competitor

Interestingly, the same cycle of nodes 41 and 98 that is driven out of business at $n = 3489$, had earlier (when it first arose at $n = 3387$) driven the two-cycle of nodes 26 and 90 out of business. Upto $n = 3386$ (Fig. 20.2(h)), the latter two-cycle was the only cycle in the graph. At $n = 3387$ node 41 was replaced and formed a new two-cycle with node 98 (Fig. 20.2(i)). Note that at $n = 3387$ only the new two-cycle is populated; all the nodes of the ACS that was dominant at the previous time step (including its core) are unpopulated. We call such an event, where there is no overlap between the old and the new cores, a *core shift* (a precise definition is given in [15]). This innovation is an example of how a new competitor takes over.

Why does the new two-cycle drive the old one to extinction? The reason is that the new two-cycle is downstream of the old one (node 41 has also acquired an incoming link from node 39; thus there exists a directed path from the old cycle to the new one, but none from the new to the old). Both two-cycles have the same intrinsic strength, but the new two-cycle does better than the old because it draws sustenance from the latter without feeding back. In general, if the graph contains two nonoverlapping irreducible subgraphs A and B, let $\lambda_1(A)$ and $\lambda_1(B)$ be the largest eigenvalues of the submatrices corresponding to A and B. If $\lambda_1(A) > \lambda_1(B)$, then A wins (i.e., in the attractor of (20.1), nodes of A and all nodes downstream of A are populated), and nodes of B are populated if B is downstream of A and unpopulated otherwise. When $\lambda_1(A) = \lambda_1(B)$, then if A and B are disconnected, both are populated, and if one of them is downstream of the other, it wins and the other is unpopulated [12, 15]. At $n = 3387$ the latter situation applies (the λ_1 of both cycles is 1, but one is downstream of the other; the downstream cycle wins at the expense of the upstream one). Examples of new competitors taking over because their λ_1 is higher than that of the existing ACS are also seen in the model.

In the displayed run, two core shifts of this kind occurred. The first was at $n = 3387$ which has been discussed above. One more occurred at $n = 6062$ which was of an identical type with a new downstream two-cycle driving the old two-cycle to extinction. Both these events resulted in a sharp drop in s_1 (Fig. 20.1(b)). A core-shifting innovation is a traumatic event for the old core and its periphery. This is reminiscent of the demise of the horse-drawn carriage industry upon the appearance of the automobile, or the decimation of anaerobic species upon the advent of aerobic ones.

At $n = 3403$ (Fig. 20.2(k)) an interesting event (that is not an innovation) happens. Node 38 is hit and the new node 38 has no incoming link. This cuts the connection that existed earlier (see Fig. 20.2(j)) between the cycle 98-41 and the cycle 26-90. The graph now has two disjoint ACSs with the same λ_1 (see Fig. 20.2(k)). As mentioned above, in such a situation both ACSs coexist; the cycle 26-90 and all nodes dependent on it once again become populated. Thus the old core has staged a "come-back" at $n = 3402$, levelling with its competitor. As we saw in the previous subsection, at $n = 3489$ the descendant of this organization strengthens its core and in fact drives its competitor out of business (this time permanently).

It is interesting that node 38, though unpopulated, still plays an important role in deciding the structure of the dominant ACS. It is purely a matter of chance that the core of the old ACS, the cycle 26-90, did not get hit before node 38. (All nodes with $X_i = 0$ have an equal probability of being replaced in the model.) If it had been destroyed between $n = 3387$ and 3402, then nothing interesting would have happened when node 38 was removed at $n = 3403$. In that case the new competitor would have won. Examples of that are also seen in the runs. In either case an ACS survives and expands until it spans the entire graph. It is worth noting that, while overall behavior like the growth of ACSs (including their average time scale of growth) is predictable, the details are shaped by historical accidents.

20.6.6
Core-shift 2: Takeover by a Dormant Innovation

A different kind of innovation occurs at $n = 4696$. At the previous time step, node 36 is the least populated (Fig. 20.2(o)). The new node 36 forms a two-cycle with node 74 (Figs 20.2(p)). This two-cycle remains part of the periphery since it does not feed back into the core; this is an incremental innovation at this time since it does not enhance λ_1. However, because it generates a structure that is intrinsically self-sustaining (a two-cycle) this innocuous innovation is capable of having a dramatic impact in the future.

At $n = 5041$, Fig. 20.2(q), the core has shrunk to 5 nodes (the reasons for this decline are briefly discussed later). The 36-74 cycle survives in the periphery of the ACS. Now it happens that node 85 is one of those with the least X_i and gets picked for removal at $n = 5042$. Thus of the old core only the two-cycle 26-90 is left. But this is now upstream from another two-cycle 74-36 (see Fig. 20.2(r)). This is the same kind of structure as discussed above, with one cycle downstream from another. The downstream cycle and its periphery wins; the upstream cycle and all other nodes downstream from it except nodes 36, 74 and 11 are driven to extinction. This event is also a core shift and is accompanied by a huge crash in the s_1 value (see Fig. 20.1(b)). This kind of event is what we call a "takeover by a dormant innovation" [12]. The innovation 36-74 occurred at $n = 4696$. It lay dormant until $n = 5042$ when the old core had become sufficiently weakened so that this dormant innovation could take over as the new core.

In this run five of the 1070 innovations were dormant innovations. Of them only the one at $n = 4696$ later caused a core shift of the type discussed above. The others remained as incremental innovations.

At $n = 8233$ a "complete crash" occurs. The core is a simple three-cycle (Fig. 20.2(s)) at $n = 8232$ and node 50 is hit, completely destroying the ACS. λ_1 drops to zero accompanied by a large crash in s_1. Within $O(s)$ time steps most nodes are hit and replaced and the graph has become random, like the initial graph. The resemblance between the initial graph at $n = 1$ (Fig. 20.2(a)) and the graph at $n = 10\,000$ (Fig. 20.2(u)) is evident. This event is not an innovation but rather the elimination of a "keystone species" [12].

20.7
Recognizing Innovations: A Structural Classification

The six categories of innovations discussed above occur in all the runs of the model and their qualitative effects are the same as described above. The above description was broadly "chronological". We now describe these innovations structurally. Such a description allows each type of innovation to be recognized the moment it appears; one does not have to wait for its consequences to be played out. The structural recognition in fact allows us to predict qualitatively the kinds of impact it can have on the system. A mathematical classification of innovations is given in Appendix B; the description here is a plain English account of that (with some loss of precision).

As is evident from the discussion above, positive feedback loops or cooperative structures in the graph crucially affect the dynamics. The character of an innovation will also depend upon its relationship with previously existing feedback loops and the new feedback loops it creates, if any. Structurally an "irreducible subgraph" captures the notion of feedback in a directed graph. By definition, since there exists a directed path (in both directions) between every pair of nodes belonging to an irreducible subgraph, each node "exerts an influence" on the other (albeit possibly through other intermediaries).

Thus the first major classification depends on whether the new node creates a new cycle and hence a new irreducible subgraph, or not. One way of determining whether it does so is to identify the nodes "downstream" of the new node (namely those to which there is a directed path from this node) and those that are "upstream" (from which there is a directed path to this node). If the intersection of these two sets is empty the new node has not created any new irreducible subgraph, otherwise it has.

A. *Innovations in which the new node does not create any new cycles and hence no new irreducible subgraph is created.* These innovations will have a relatively minor impact on the system. There are two subclasses here which depend upon the context: whether an irreducible subgraph already exists somewhere else in the graph or not.

 A1. *Before the innovation, the graph does not contain an irreducible subgraph.* Then the innovation is a short-lived one discussed in Section 20.6.1 (Fig. 20.2(b,c)). There is no ACS before or after the innovation. The largest eigenvalue λ_1 of the adjacency matrix of the graph being zero both before and after such an innovation is a necessary and sufficient condition for it to be in this class. Such an innovation is doomed to die when the first ACS arises in the graph for the reasons discussed in the previous section.

 A2. *Before the innovation an irreducible subgraph already exists in the graph.* One can show that such an innovation simply adds to the periphery of the existing dominant ACS, leaving the core unchanged. Here the new node gets a nonzero X_k because it receives an incoming link from one of the nodes of the existing dominant ACS; it has effectively latched on to the latter like a parasite. This is an incremental innovation (section 20.6.3, Fig. 20.2(f,g)). It has a relatively minor impact on the system at the time it appears. Since it does not modify the core, the ratios of the X_i values of the core nodes remain unchanged. However, it does eat up some resources (since $X_k > 0$) and causes an overall decline in the X_i values of the core nodes. λ_1 is nonzero and does not change in such an innovation.

B. *Innovations that do create some new cycle.* Thus a new irreducible subgraph is generated. Because these innovations create new feedback loops, they have a potentially greater impact. Their classification depends upon whether or not they modify the core and the extent of the modification caused; this is directly correlated with their immediate impact.

  B1. *The new cycles do not modify the existing core.* If the new irreducible subgraph is disjoint from the existing core and its intrinsic λ_1 is less than that of the core, then the new irreducible subgraph will not modify the existing core but will become part of the periphery. Like incremental innovations, such innovations cause an overall decline in the X_i values of the core nodes but do not disturb their ratios and the value of λ_1. However, they differ from incremental innovations in that the new irreducible subgraph has self-sustaining capabilities. Thus, in the event of a later weakening of the core (through elimination of some core nodes), these innovations have the potential of causing a core-shift wherein the irreducible graph generated in the innovation becomes the new core. At that point it would typically cause a major crash in the number of populated species, as the old core and all its periphery that is not supported by the new core would become depopulated. Such innovations are the dormant innovations (Section 20.6.6, Fig. 20.2(o,p)). Note that not all dormant innovations cause core-shifts. Most, in fact, play the same role as incremental innovations.

  B2. *Innovations that modify the existing core.* If the new node is part of the new core, the core has been modified. The classification of such innovations depends on the kind of core that exists before and the nature of the modification.

    B2.1. *The existing core is nonempty,* i.e., an ACS already exists before the innovation in question arrives.

      B2.1.1. *The innovation strengthens the existing core.* In this case the new node receives an incoming link from the existing dominant ACS and has an outgoing link to the existing core. The existing core nodes get additional positive feedback, and λ_1 increases. Such an event can cause some members of the parasitic periphery to be co-opted into the core.

These are the core-enhancing innovations discussed in Section 20.6.4 (Fig. 20.2(l,m)).

B2.1.2. *The new irreducible subgraph is disjoint from the existing core and "stronger" than it.* "Stronger" means that the intrinsic λ_1 of the new irreducible graph is greater than or equal to the λ_1 of the existing core, and in the case of equality it is downstream from the existing core. Then it will destabilize the existing core and become the new core itself, causing a core-shift. The takeovers by new competitors, discussed in Section 20.6.5 (Fig. 20.2(h,i)) belong to this class.

B2.2. *The existing core is empty*, i.e., no ACS exists before the arrival of this innovation. Then the new irreducible graph is the core of the new ACS that is created at this time. This event is the beginning of a self-organizing phase of the system. This is the birth of an organization discussed in Section 20.6.2 (Fig. 20.2(d,e)). This is easily recognized graph-theoretically as λ_1 jumps from zero to a positive value.

Note that the "recognition" of the class of an innovation is contingent upon knowing graph-theoretic features like the core, periphery, λ_1, and being able to determine the irreducible graph created by the innovation.

The above rules are an analytic classification of all innovations in the model, irrespective of values of the parameters p and s. Note, however, that their relative frequencies depend upon the parameters. In particular, innovations of class A require the new node to have at least one link (an incoming one) and class B require at least two links (an incoming and an outgoing one). Thus as the connection probability p declines, for fixed s, the latter innovations (the more consequential ones) become less likely.

20.8
Some Possible General Lessons

In this model, due to the simplicity of the population dynamics, it is possible to make an analytic connection between the graph structure produced by various innovations and their subsequent effect on the short and long-term dynamics of the system. In addition, we are able to completely enumerate the different types of innovation and classify them purely on the basis of their graph structure. Identifying innovations and understanding their effects is much more complicated in real-world processes in both biological and social systems. Nevertheless, the close parallel between the qualitive categories of innovation which we find in our model and real-world examples, means that there may be some lessons to be learnt from this simple mathematical model.

One broad conclusion is that, in order to guess what might be an innovation, we need an understanding of how the patterns of connectivity influence system dynamics and vice versa. The inventor of a new product or a venture capitalist asks: what inputs will be needed, and whose needs will the product connect to? Given these potential linkages in the context of other existing nodes and links, what flows will actually be generated along the new links? How will these new flows impact the generation of other new nodes and links and the death of existing ones and how will that feed back into the flows again? The detailed rules of this dynamics are system dependent, but presumably successful entrepreneurs have an intuitive understanding of this very dynamics.

In our model, as in real processes, there are innovations which have an immediate impact on the dynamics of the system (e.g., the creation of the first ACS and core-shifting innovations) and ones which have little or no immediate impact. Innovation in real processes analogous to the former are probably easier to identify because they cause the dynamics of the system to immediately change dramatically (in this model, triggering a new round of self-organized growth around a new ACS). Of the latter, the most interesting innovations are the ones which eventually do have a large impact on the dynamics: the dormant innovations. In this model, dormant innovations sometimes lead to a dramatic change in the dynamics of the system at a later time. This suggests that in real-world processes too it might be important, when observing a sudden change in the dynamics, to examine innovations which occurred much before the change. Of course, in the model and in real processes, there are innovations which have nothing to do with any later change in the dynamics. In real processes it would be very difficult to distinguish such innovations from dormant innovations which do cause a significant impact on the dynamics. The key feature distinguishing a dormant innovation from incremental innovations in this model is that a dormant innovation creates an irreducible structure which can later become the core of the graph.

This suggests that in real-world processes it might be useful to find an analogy of the core and periphery of the system and then focus on innovations or processes which alter the core or create structures which could become the core. In the present model, it is possible to define the core in a purely graph-theoretic manner. In real systems it might be necessary to define the core in terms of the dynamics. One possible generalization is based on the observation that removal of a core node causes the population growth rate to reduce (due to the reduction of λ_1), while the removal of a periphery node leaves λ_1 unchanged. This could be used as an algorithmic way of identifying core nodes or species in more complex mathematical models, or in real systems where such testing is possible.

20.9
Discussion

As in real systems, the model involves an interplay between the force of selection that weeds out underperforming nodes, the influx of novelty that brings in new nodes and links, and an internal (population) dynamics that depends upon the mutual interactions. In an environment of nonautocatalytic structures, a small ACS is very successful and drives the other nodes to the status of "have-nots" ($X_i = 0$). The latter are eliminated one by one, and if their replacements "latch on" to the ACS, they survive, or else they suffer the same fate. The ACS "succeeds" spectacularly: eventually all the nodes join it. But this sets the stage for enhanced internal competition between the members of the ACS. Before the ACS spanned the graph, only have-nots, nodes outside the dominant ACS, were eliminated. After spanning, the eliminated node must be one of the "haves", a member of the ACS (whichever has the least X_i). This internal competition weakens the core and enhances the probability of collapse due to core-transforming innovations or elimination of keystone species. Thus the ACSs very success creates the circumstances that bring about its destruction[6]. Both its success, and a good part of its destruction, is due to innovation (see also [12]).

It is of course true that we can describe the behavior of the system in terms of attractors of the dynamics as a function of the graph without recourse to the word "innovation". The advantage of introducing the notion of innovation as defined above is that it captures a useful property of the dynamics in terms of which many features can be readily described. Further, we hope that the examples discussed above make a reasonable case that this notion of innovation is sufficiently close (as close as is possible in an idealized model such as this) to the real thing, to help in discussions of the latter.

In the present model, the links of the new node are chosen randomly from a fixed probability distribution. This might be appropriate for the prebiotic chemical scenario for which the model was constructed, but is less appropriate for biological systems and even less for social systems. While there is always some stochasticity, in these systems the generation of novelty is conditioned by the existing context, and in social systems also by the intentionality of the actors. Thus the ensemble of choices from which the novelty is drawn also evolves with the system. This feedback from the recent history of system states to the ensemble of system perturbations, though not implemented in the present version of the model, certainly deserves future investigation.

---

6) For a related discussion of discontinuous transitions in other systems, see [21–23].

20.10
Appendix A: Definitions and Proofs

In this Appendix we collect some useful facts about the model. These and other properties can be found in [5, 13, 16, 17]

20.10.1
Derivation of Eq. (20.1)

Let $i \in \{1, \ldots, s\}$ denote a chemical (or molecular) species in a well-stirred chemical reactor. Molecules can react with one another in various ways; we focus on only one aspect of their interactions: catalysis. The catalytic interactions can be described by a directed graph with s nodes. The nodes represent the s species and the existence of a link from node j to node i means that species j is a catalyst for the production of species i. In terms of the adjacency matrix, $C = (c_{ij})$ of this graph, c_{ij} is set to unity if j is a catalyst of i and is set to zero otherwise. The operational meaning of catalysis is as follows.

Each species i will have an associated non-negative population y_i in the reactor that changes with time. Let species j catalyze the ligation of reactants A and B to form the species i, $A + B \xrightarrow{j} i$. Assuming that the rate of this catalyzed reaction is given by the Michaelis–Menten theory of enzyme catalysis, $\dot{y}_i = V_{max} ab \frac{y_j}{K_M + y_j}$ [24], where a, b are the reactant concentrations, and V_{max} and K_M are constants that characterize the reaction. If the Michaelis constant K_M is very large, this can be approximated as $\dot{y}_i \propto y_j ab$. Combining the rates of the spontaneous and catalyzed reactions and also putting in a dilution flux ϕ, the rate of growth of species i is given by $\dot{y}_i = k(1 + \nu y_j) ab - \phi y_i$, where k is the rate constant for the spontaneous reaction, and ν is the catalytic efficiency. Assuming that the catalyzed reaction is much faster than the spontaneous reaction, and that the concentrations of the reactants are nonzero and fixed, the rate equation becomes $\dot{y}_i = K y_j - \phi y_i$, where K is a constant. In general, because species i can have multiple catalysts, $\dot{y}_i = \sum_{j=1}^{s} K_{ij} y_j - \phi y_i$, with $K_{ij} \sim c_{ij}$. We make the further idealization $K_{ij} = c_{ij}$ giving:

$$\dot{y}_i = \sum_{j=1}^{s} c_{ij} y_j - \phi y_i \qquad (20.2)$$

The relative population of species i is by definition $x_i \equiv y_i / \sum_{j=1}^{s} y_j$. As $0 \leq x_i \leq 1, \sum_{i=1}^{s} x_i = 1, \mathbf{x} \equiv (x_1, \ldots, x_s)^T \in J$. Taking the time derivative of x_i and using (20.2) it is easy to see that $\dot{x}_i$ is given by (20.1). Note that the ϕ term, present in (20.2), cancels out and is absent in (20.1).

20.10.2
The Attractor of Eq. (20.1)

A graph described by an adjacency matrix, C, has an eigenvalue $\lambda_1(C)$ which is a real, positive number that is greater than or equal to the modulus of all other eigenvalues. This follows from the Perron–Frobenius theorem [20] and this eigenvalue is called the Perron–Frobenius eigenvalue of C.

The attractor $\mathbf{X}$ of Eq. (20.1) is an eigenvector of C with eigenvalue $\lambda_1(C)$.

Since (20.1) does not depend on ϕ, we can set $\phi = 0$ in (20.2) without loss of generality for studying the attractors of (20.1). For fixed C the general solution of (20.2) is $\mathbf{y}(t) = e^{Ct}\mathbf{y}(0)$, where $\mathbf{y}$ denotes the s-dimensional column vector of populations. It is evident that if $\mathbf{y}^\lambda \equiv (y_1^\lambda, \ldots, y_s^\lambda)$ viewed as a column vector is a right eigenvector of C with eigenvalue λ, then $\mathbf{x}^\lambda \equiv \mathbf{y}^\lambda / \sum_i^s y_i^\lambda$ is a fixed point of (20.1). Let λ_1 denote the eigenvalue of C which has the largest real part; it is clear that $\mathbf{x}^{\lambda_1}$ is an attractor of (20.1). By the theorem of Perron–Frobenius for non-negative matrices [20] λ_1 is real and ≥ 0 and there exists an eigenvector $\mathbf{x}^{\lambda_1}$ with $x_i \geq 0$. If λ_1 is nondegenerate, $\mathbf{x}^{\lambda_1}$ is the unique asymptotically stable attractor of (20.1), $\mathbf{x}^{\lambda_1} = (X_1, \ldots, X_s)$.

20.10.3
The Attractor of Eq. (20.1) when there are no Cycles

For any graph with no cycles, in the attractor only the nodes at the ends of the longest paths have nonzero X_i. All other nodes have zero X_i.

Consider a graph consisting only of a linear chain of $r + 1$ nodes, with r links, pointing from node 1 to node 2, node 2 to 3, etc. The node 1 (to which there is no incoming link) has a constant population y_1 because the r.h.s of (20.2) vanishes for $i = 1$ (taking $\phi = 0$). For node 2, we get $\dot{y}_2 = y_1$, hence $y_2(t) = y_2(0) + y_1 t \sim t$ for large t. Similarly, it can be seen that y_k grows as t^{k-1}. In general, it is clear that for a graph with no cycles, $y_i \sim t^r$ for large t (when $\phi = 0$), where r is the length of the longest path terminating at node i. Thus, nodes with the largest r dominate for sufficiently large t. Because the dynamics (20.1) do not depend upon the choice of ϕ, $X_i = 0$ for all i except the nodes at which the longest paths in the graph terminate.

20.10.4
Graph-theoretic Properties of ACSs

1. *An ACS must contain a closed walk.*
2. *If a graph, C, has no closed walk then $\lambda_1(C) = 0$.*
3. *If a graph, C, has a closed walk then $\lambda_1(C) \geq 1$. Consequently,*

4. If a graph C has no ACS then $\lambda_1(C) = 0$.
5. If a graph C has an ACS then $\lambda_1(C) \geq 1$.

1. Let A be the adjacency matrix of a graph that is an ACS. Then, by definition, every row of A has at least one nonzero entry. Construct A' by removing, from each row of A, all nonzero entries except one that can be chosen arbitrarily. Thus A' has exactly one nonzero entry in each row. Clearly the column vector $\mathbf{x} = (1, 1, \ldots, 1)^T$ is an eigenvector of A' with eigenvalue 1 and hence $\lambda_1(A') \geq 1$. Proposition 3. therefore implies that A' contains a closed walk. Because the construction of A' from A involved only removal of some links, it follows that A must also contain a closed walk.

2. If a graph has no closed walk then all walks are of finite length. Let the length of the longest walk of the graph be denoted r. If C is the adjacency matrix of a graph then $(C^k)_{ij}$ equals the number of distinct walks of length k from node j to node i. Clearly $C^m = 0$ for $m > r$. Therefore all eigenvalues of C^m are zero. If λ_i are the eigenvalues of C then λ_i^k are the eigenvalues of C^k. Hence, all eigenvalues of C are zero, which implies $\lambda_1 = 0$. This proof was supplied by V. S. Borkar.

3. If a graph has a closed walk then there is some node i that has at least one closed walk to itself, i.e., $(C^k)_{ii} \geq 1$, for infinitely many values of k. Because the trace of a matrix equals the sum of the eigenvalues of the matrix, $\sum_{i=1}^{s}(C^k)_{ii} = \sum_{i=1}^{s} \lambda_i^k$, where λ_i are the eigenvalues of C. Thus, $\sum_{i=1}^{s} \lambda_i^k \geq 1$, for infinitely many values of k. This is only possible if one of the eigenvalues λ_i has a modulus ≥ 1. By the Perron–Frobenius theorem, λ_1 is the eigenvalue with the largest modulus, hence $\lambda_1 \geq 1$. This proof was supplied by R. Hariharan.

4. and 5. follow from the above.

20.10.5
Dominant ACS of a Graph

If a graph has (one or more) ACSs, i.e., $\lambda_1 \geq 1$, then the subgraph corresponding to the set of nodes i for which $X_i > 0$ is an ACS.

Renumber the nodes of the graph so that $x_i > 0$ only for $i = 1, \ldots, k$. Let C be the adjacency matrix of this graph. Since $\mathbf{X}$ is an eigenvector of the matrix C, with eigenvalue λ_1, we have $\sum_{j=1}^{s} c_{ij} X_j = \lambda_1 X_i \Rightarrow \sum_{j=1}^{k} c_{ij} X_j = \lambda_1 X_i$. Since $X_i > 0$ only for $i = 1, \ldots, k$ it follows that for each $i \in \{1, \ldots, k\}$ there exists a j such that $c_{ij} > 0$. Hence the $k \times k$ submatrix $C' \equiv (c_{ij})$, $i, j = 1, \ldots, k$ has at least one nonzero entry in each row. Thus each node of the subgraph corresponding to this submatrix has an incoming link from one of the other nodes in the subgraph. Hence the subgraph is an ACS. We call this subgraph the "dominant ACS" of the graph.

20.10.6
Time Scales for Appearance and Growth of the Dominant ACS

The probability for an ACS to be formed at some graph update in a graph which has no cycles, can be closely approximated by the probability of a two-cycle (the simplest ACS with one-cycles being disallowed) forming by chance, which is p^2s (= the probability that in the row and column corresponding to the replaced node in C, any matrix element and its transpose are both assigned unity). Thus, the "average time of appearance" of an ACS is $\tau_a = 1/p^2s$, and the distribution of times of appearance is $P(n_a) = p^2s(1 - p^2s)^{n_a-1}$. This approximation is better for small p.

Assuming that the possibility of a new node forming a second ACS is rare enough to neglect, and that the dominant ACS grows by adding a single node at a time, one can estimate the time required for it to span the entire graph. Let the dominant ACS consist of $s_1(n)$ nodes at time n. The probability that the new node gets an incoming link from the dominant ACS and hence joins it is ps_1. Thus in Δn graph updates, the dominant ACS will grow, on average, by $\Delta s_1 = ps_1 \Delta n$ nodes. Therefore $s_1(n) = s_1(n_a)\exp((n - n_a)/\tau_g)$, where $\tau_g = 1/p$, n_a is the time of appearance of the first ACS and $s_1(n_a)$ is the size of the first ACS. Thus s_1 is expected to grow exponentially with a characteristic time scale $\tau_g = 1/p$. The time taken from the appearance of the ACS to its spanning is $\tau_g \ln(s/s_1(n_a))$.

20.11
Appendix B: Graph-theoretic Classification of Innovations

In the main text we defined an innovation to be the new structure created by the addition of a new node, when the new node has a nonzero population in the new attractor. Here, we present a graph-theoretic hierarchical classification of innovations (see Fig. 20.3). At the bottom of this hierarchy we recover the six categories of innovation described in the main text.

Some notation follow: We need to distinguish between two graphs, one just before the new node is inserted, and one just after. We denote them by C_i and C_f respectively, and their cores by Q_i and Q_f. Note that a graph update event consists of two parts – the deletion of a node and the addition of one. C_i is the graph after the node is deleted and before the new node is inserted. The graph before the deletion will be denoted C_0; Q_0 will denote its core[7]. If a graph has no ACS, its core is the null set.

---

7) Most of the time the deleted node (being the one with the least relative population) is outside the dominant ACS of C_0 or in its periphery. Thus, in most cases the core is unchanged by the deletion: $Q_i = Q_0$. However, sometimes the deleted node belongs to Q_0. In that case $Q_i \neq Q_0$. In most such cases, Q_i is a proper subset of Q_0. In very few (but important) cases, $Q_i \cap Q_0 = \Phi$ (the null set). In these latter cases, the deleted node is a "keystone node" [15]; its removal results in a "core shift".

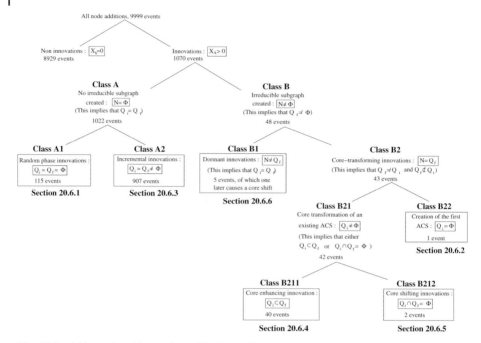

Fig. 20.3 A hierarchy of innovations. Each node in this binary tree represents a class of node-addition events. Each class has a name; the small box contains the mathematical definition of the class. All classes of events except the leaves of the tree are subdivided into two exhaustive and mutually exclusive subclasses represented by the two branches emanating downwards from the class. The number of events in each class pertain to the run of Fig. 20.1 with a total of 9999 graph updates, between $n = 1$ (the initial graph) and $n = 10\,000$. In that run, out of 9999 node addition events, most (8929 events) are not innovations. The rest (1070 events), which are innovations, are classified according to their graph-theoretic structure. The classification is general; it is valid for all runs. X_k is the relative population of the new node in the attractor of (20.1). N stands for the new irreducible subgraph, if any, created by the new node. If the new node causes a new irreducible subgraph to be created, N is the *maximal* irreducible subgraph that includes the new node. If not, $N = \Phi$ (where Φ stands for the empty set). Q_i is the core of the graph just before the addition of the node and Q_f the core just after the addition of the node. The six leaves of the innovation subtree are numbered (below the corresponding box) according to the subsection in which they are discussed in the main text. The graph-theoretic classes A, B, A1, B1, etc., are described in Section 20.7 and Appendix B.

The links of the new node may be such that new cycles arise in the graph (that were absent in C_i but are present in C_f). In this case the new node is part of a new irreducible subgraph that has arisen in the graph. N will denote the maximal irreducible subgraph which includes the new node. If the new node does not create new cycles, $N = \Phi$. If $N \neq \Phi$, then N will either be disjoint from Q_f or will include Q_f (it cannot partially overlap with Q_f because of its maximal character). The structure of N and its relationship with the core before and after the addition determines the nature of the innovation. With this notation all innovations can be grouped into two classes:

A. Innovations that do not create new cycles, $N = \Phi$. This implies $Q_f = Q_i$ because no new irreducible structure has appeared and therefore the core of the graph, if it exists, is unchanged.

B. Innovations that do create new cycles, $N \neq \Phi$. This implies $Q_f \neq \Phi$ because if a new irreducible structure is created then the new graph has at least one ACS and therefore a nonempty core.

Class A can be further decomposed into two classes:

A1. $Q_i = Q_f = \Phi$. In other words, the graph has no cycles both before and after the innovation. This corresponds to short-lived innovations discussed in Section 20.6.1 (Fig. 20.2(b,c)).

A2. $Q_i = Q_f \neq \Phi$. In other words, the graph had an ACS before the innovation, and its core was not modified by the innovation. This corresponds to incremental innovations discussed in Section 20.6.3 (Fig. 20.2(f,g)).

Class B of innovations can also be divided into two subclasses:

B1. $N \neq Q_f$. If the new irreducible structure is not the core of the new graph, then N must be disjoint from Q_f. This can only be the case if the old core has not been modified by the innovation. Therefore $N \neq Q_f$ necessarily implies that $Q_f = Q_i$. This corresponds to dormant innovations discussed in Section 20.6.6 (Fig. 20.2(o,p)).

B2. $N = Q_f$, i.e., the innovation becomes the new core after the graph update. This is the situation where the core is transformed due to the innovation.

The "core-transforming theorem" [12, 13, 17] states that an innovation of type B2 occurs whenever either of the following conditions are true:

(a) $\lambda_1(N) > \lambda_1(Q_i)$, or
(b) $\lambda_1(N) = \lambda_1(Q_i)$ and N is downstream of Q_i.

Class B2 can be subdivided as follows:

B2.1. $Q_i \neq \Phi$, i.e., the graph contained an ACS before the innovation. In this case an existing core is modified by the innovation.

B2.2. $Q_i = \Phi$, i.e., the graph had no ACS before the innovation. Thus, this kind of innovation creates an ACS in the graph. It corresponds to the birth of an organization discussed in Section 20.6.2 (Fig. 20.2(d,e)).

Finally, class B21 can be subdivided:

B2.1.1. $Q_i \subset Q_f$. When the new core contains the old core as a subset we get an innovation that causes the growth of the core, discussed in Section 20.6.4 (Fig. 20.2(l,m)).

B2.1.2. Q_i and Q_f are disjoint (Note that it is not possible for Q_i and Q_f to partially overlap, or else they would form one big irreducible set which would then be the core of the new graph and Q_i would be a subset of Q_f). This is an innovation where a core-shift is caused due to a takeover by a new competitor, discussed in Section 20.6.5 (Fig. 20.2(h,i)).

Note that each branching above is into mutually exclusive and exhaustive classes. This classification is completely general and applicable to all runs of the system. Figure 20.3 shows the hierarchy obtained using this classification.

Acknowledgment

S. J. would like to thank John Padgett for discussions.

References

1 SCHUMPETER, J. A., *The Theory of Economic Development*, Harvard University Press, Cambridge, MA, USA, **1934**

2 ROGERS, E. M., *The Diffusion of Innovations*, 4th ed., The Free Press, New York, **1995**

3 FALKOWSKI, P. G., *Science* 311 (**2006**), pp. 1724–1725

4 BORNHOLDT, S., SCHUSTER, H. G., Eds. *Handbook of Graphs and Networks: From the Genome to the Internet*, Wiley-VCH, Weinheim, **2003**

5 JAIN, S., KRISHNA, S., *Phys. Rev. Lett.* 81 (**1998**), pp. 5684–5687

6 DYSON, F., *Origins of Life*, Cambridge Univ. Press, **1985**

7 KAUFFMAN, S. A., *The Origins of Order*, Oxford Univ. Press, **1993**

8 BAGLEY, R. J., FARMER, J. D., FONTANA, W., in *Artificial Life II*, Eds. Langton, C. G., Taylor, C., Farmer, J. D., Rasmussen, S., Addison Wesley, Redwood City, **1991**, pp. 141-158

9 FONTANA, W., BUSS, L., *Bull. Math. Biol.* 56 (**1994**), pp. 1–64

10 BAK, P., SNEPPEN, K., *Phys. Rev. Lett.* 71 (**1993**), pp. 4083–4086

11 KAUFFMAN, S. A., *J. Cybernetics*, 1 (**1971**), pp. 71–96

12 JAIN, S., KRISHNA, S., *Proc. Natl. Acad. Sci. (USA)* 99 (**2002**), pp. 2055–2060

13 KRISHNA, S., Ph. D. Thesis, (**2003**); http://www.arXiv.org/abs/nlin.AO/0403050

14 JAIN, S., KRISHNA, S., *Proc. Natl. Acad. Sci. (USA)* 98 (**2001**), pp. 543–547

15 JAIN, S., KRISHNA, S., *Phys. Rev. E* 65 (**2002**), p. 026103

16 JAIN, S., KRISHNA, S., *Computer Physics Comm.* 121-122 (**1999**), pp. 116–121

17 JAIN, S., KRISHNA, S., in *Handbook of Graphs and Networks*, Eds. Bornholdt, S., Schuster, H. G., Wiley-VCH, Weinheim, **2003**, pp. 355–395

18 EIGEN, M., *Naturwissenschaften* 58 (**1971**), p. 465–523

19 ROSSLER, O. E., *Z. Naturforschung* 26b (**1971**), pp. 741–746

20 SENETA, E., *Non-Negative Matrices*, George Allen and Unwin, London, **1973**

21 PADGETT, J., in *Networks and Markets*, Eds. Rauch, J. E., Casella A., Russel Sage, New York, **2001**, pp. 211–257

22 COHEN, M. D., RIOLO, R. L., AXELROD, R., *Rationality and Society* 13 (**2001**), pp. 5–32

23 CARLSON, J. M., DOYLE, J., *Phys. Rev. E* 60 (**1999**), p. 1412

24 GUTFREUND, H., *Kinetics for the Life Sciences*, Cambridge Univ. Press, Cambridge, **1995**.

Color Plates

Chapter 3

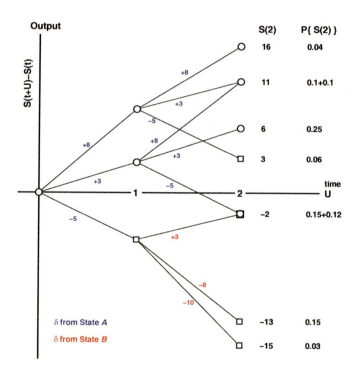

Fig. 3.5 All possible paths after two time-steps and associated probability.

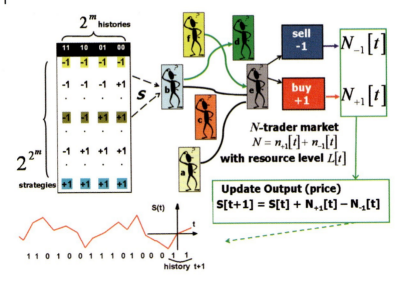

Fig. 3.14 Schematic diagram of the Binary Agent Resource (BAR) system.

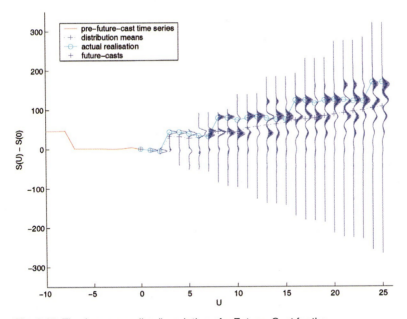

Fig. 3.15 The (non-normalized) evolution of a Future–Cast for the given $\underline{\underline{\Omega}}$, game parameters and initial state Γ. The figure shows the last 10 time-steps prior to the Future–Cast, the means of the distributions within the Future–Cast itself, and also an actual realization of the game run forward in time.

Color Plates | 595

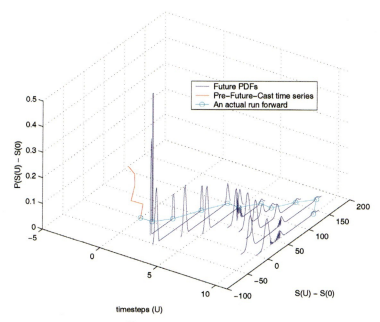

Fig. 3.17 Evolution of $\Pi^U(S)$ for a typical quenched disorder matrix $\underline{\underline{\Omega}}$.

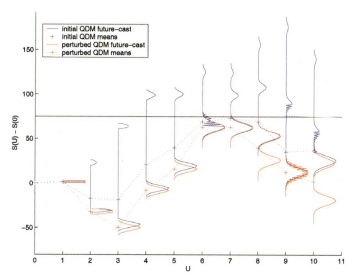

Fig. 3.20 The evolution as a result of the microscopic perturbation to the population's composition (i.e., the QDM).

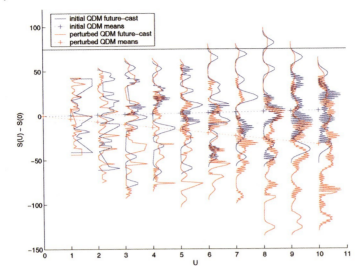

Fig. 3.21 The characteristic evolution of the initial and perturbed QDMs.

Chapter 7

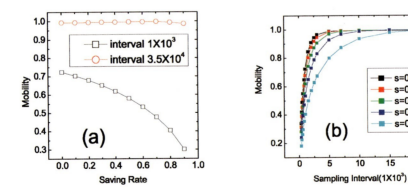

Fig. 7.5 (a) The relation between saving rate and mobility measured by employing displacement mobility index: (□) the sampling interval 1000, (○) the sampling interval 3.5×10^4; (b) The relation between sampling interval and mobility measured by employing displacement-like mobility index.

Color Plates | 597

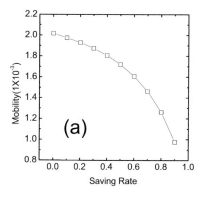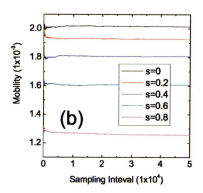

Fig. 7.6 (a) The relation between saving rate and mobility measured by employing a speed mobility index with the sampling interval 3.5×10^4; (b) The relation between sampling interval and mobility measured by employing the speed mobility index.

Chapter 10

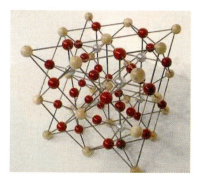

Fig. 10.8 A crystal is a three-dimensional lattice of equivalent atoms. All atoms have equivalent nearest (and second nearest) neighbor bonds.

598 Color Plates

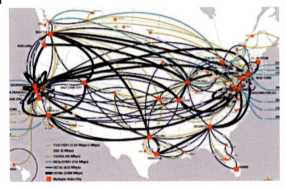

Fig. 10.11 Internet structure of UUNET, one of the world's leading Internet networks. © 2000 - UUNET, an MCI WorldCom Company.

Chapter 12

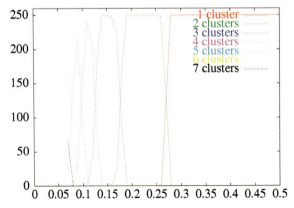

Fig. 12.6 Statistics of the number of opinion clusters as a function of d on the x axis for 250 samples ($\mu = 0.5$, $N = 1000$).

Chapter 16

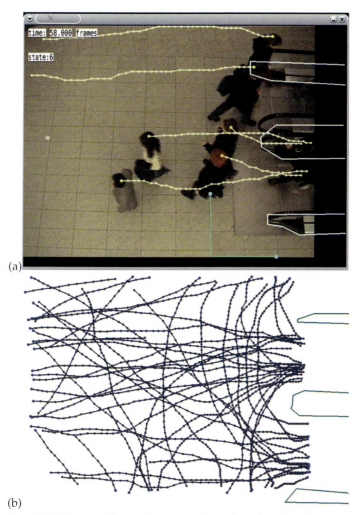

Fig. 16.3 Video tracking used to extract the trajectories of pedestrians from video recordings. (a) Illustration of the tracking of pedestrian heads; (b) Resulting trajectories after being transformed onto the two-dimensional plane.

Fig. 16.14 Simulation of pedestrian streams in the city center of Dresden, Germany, serves to assess the impact of a new theater or shopping mall.

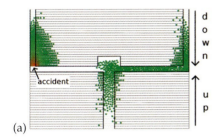
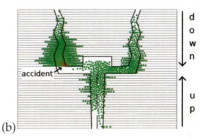
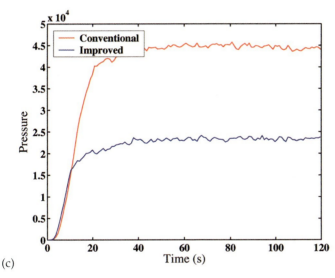

Fig. 16.15 Conventional (a) and improved (b) design of a seating sector of a sports stadium (after [5]). For safety assessment we have assumed an accident (fallen people) at the end of the left downward staircase. (c) Numerical results for the pressure level at the end of the downward staircase, in the case of a blockage of the outflows. The evacuation simulations have been performed with the social-force model.

Chapter 18

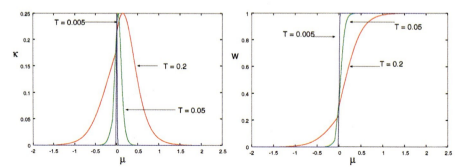

Fig. 18.3 (a) The compressibility κ as a function of μ for the case of $T = 0.2, 0.05$ and $T = 0.005$. (b) we plot the corresponding order parameter w as a function of μ.

Chapter 19

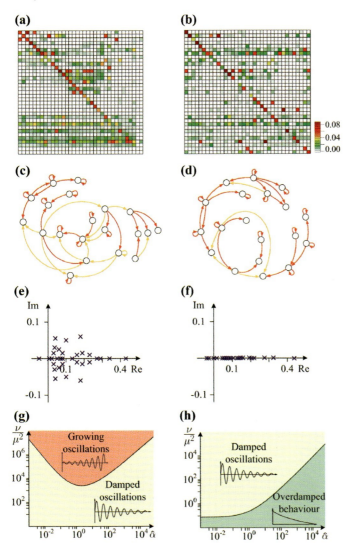

Fig. 19.1 Properties of a dynamic model of supply networks for a characteristic, empirical input matrix (left) and for a synthetic input matrix generated by random changes of input matrix entries until the number of complex eigenvalues was eventually reduced to zero (right). Subfigures (a), (b) illustrate the color-coded input matrices $\mathbf{A}$, (c), (d) the corresponding network structures, when only the strongest links (commodity flows) are shown, (e), (f) the eigenvalues $\omega_i = \mathrm{Re}(\omega_i) + i\,\mathrm{Im}(\omega_i)$ of the respective input matrix $\mathbf{A}$, and (g), (h) the phase diagrams indicating the stability behavior of the related model of material flows between economics sectors on a double-logarithmic scale as a function of model parameters. Surprisingly, for empirical input matrices $\mathbf{A}$, one never finds an overdamped, exponential relaxation to the stationary equilibrium state, but network-induced oscillations due to complex eigenvalues ω_i (after [19]).

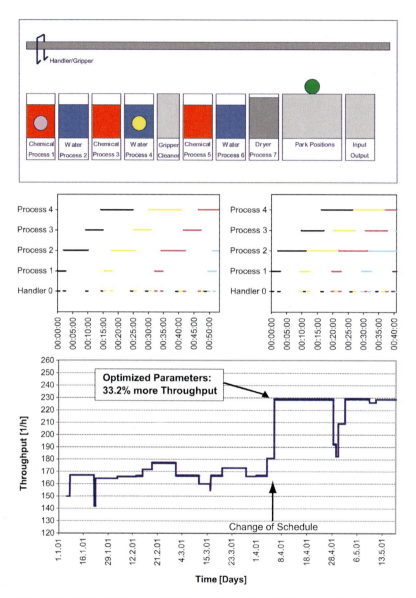

Fig. 19.9 (a) Schematic representation of the successive processes of a wet bench, i.e., a particular supply chain in semiconductor production. (b) The Gantt diagramms illustrate the treatment times of the first four of several more processes, where we have used the same colors for processes belonging to the same run, i.e., the same set of wafers. The left diagram shows the original schedule, while the right one shows an optimized schedule based on the "slower-is-faster effect" (see Section 19.3). (c) The increase in the throughput of a wet bench by switching from the original production schedule to the optimized one was found to be 33%, in some cases even higher (after [41]).

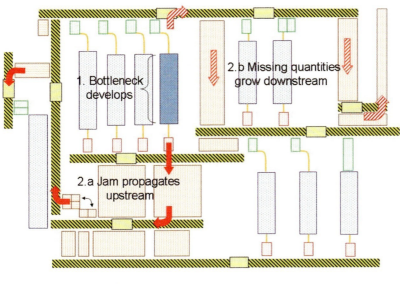

Fig. 19.13 (a) The material flows in production systems can be described by generalized traffic equations. While congestion propagates upstream, the lack of materials propagates downstream (see Section 19.4.3 for the issue of gap propagation). (b) Illustration of the state variables for different parts of the production system. Green represents progress according to time schedule, yellow critical operation, and red serious inefficiencies.

Subject Index

a
adaptive control   537, 550
ageing   320
agent-based   312, 320
agents   279, 281, 283, 284, 286, 287, 290, 291, 296, 298, 300, 303, 438–440, 442, 444, 445
agents' choice   35, 61
aggression   297, 307
Angle burglar process   147
annihilation   35, 40, 45, 46, 48, 49, 52, 54
ARCH/GARCH process   230, 259, 262
assortativity   477, 481, 482, 493, 494, 496, 502
Asterix   331
autocatalytic set   567, 574
autocorrelation function   45, 50, 51, 56, 58
automated guided vehicles (AGV)   545, 547

b
Bengal   312
betweenness   343, 476, 483, 494
bilingual   317, 320
bit-string   320, 323
blog   428–430, 447
Boltzmann distribution   12, 13, 15
books   418, 434
bottleneck   458, 459, 461, 463, 466, 471
business   361

c
calibration   452
capital   1, 5, 8, 19, 20, 24
Carnot process   1, 21, 24, 27
cellular automata   66, 344, 347, 348, 451
centrality
– betweenness   476, 483
– closeness   476, 501
Champernowne stochastic model   138
circulation
– magazine   424
– newspaper   424
citation   417, 422, 423, 429, 436

clustering   442, 475, 476, 479, 481–483, 494–496
– coefficient   475, 476, 478, 479, 481, 482, 490, 492, 494, 496, 502
clusters   345, 346, 351–354, 358, 359, 598
Cobb Douglas function   13, 14, 16, 17
collective   279, 281, 283, 289, 292, 294
– behavior   38
– choice   417, 419, 438, 440
– decision   438, 440–443
colonization   321
common beliefs   368, 369, 372, 373, 378, 390
community structure   481–485, 490, 491, 493–496
competitive population   65
complex adaptive system   65, 67
computer simulation   204, 205, 207, 208
constraints   10, 12, 15
core   571, 575–578, 580–582, 587–589
– shift   567, 576–578, 580–582, 587, 590
crash   258, 265–268, 271–273, 275
crashes   38, 50
critical
– dynamics   61
– phenomena   60, 61
– threshold   442
cross-correlation   45, 46, 54
crowd dynamics   449, 450, 470
culture   339, 350, 357, 359

d
damage   331
democracy   293, 294, 297
deposition   40, 41, 45–49, 52, 55, 56
detailed balance   105
detrended fluctuation analysis   250–254, 261
dialectometry   332
differential
– equations   312
– forms   1–3, 5
disorder   283, 287, 290, 291, 294
– annealed   173

– quenched 86, 89, 90, 96, 173, 179, 181
disparity 183, 184, 186, 188
distribution
– bimodal 170, 171, 419, 440–442
– Boltzmann 155, 163, 165, 174, 178
– cumulative 162, 163, 165
– Gamma 163, 166, 180, 181
– Gaussian 419
– Generalized Lotka–Volterra 155
– log-normal 154, 423, 425, 426, 435, 437, 445
– Pareto–Lévy 139
– polymodal 171
– price 42, 47
– stretched exponential 435
– Tsallis 141
– unimodal 439–442
– Weibull 427, 435, 437
dominance 313, 321, 324
dormant 401, 402, 404, 407–409
download 430
dynamic property 47

e
economic growth 1, 5, 26, 27, 29
economic mobility 193, 210, 211, 216
econophysics 1, 63, 98, 117, 128, 129, 158, 159, 161, 163, 189–191, 216, 247, 249, 250, 276–278
El Farol bar problem 66, 67
election 431–433
entropy 1, 8, 12, 16, 21, 165, 183, 184
evaporation 35, 40, 45–50, 52, 60
evolving graph 564, 571
exponent
– Hurst 50, 52, 54, 60, 61
– Pareto 103, 104, 109, 123, 127, 134, 135, 141, 144, 146–148, 162, 169, 182, 187, 188, 421
extremism 354–357, 367–370, 372–375, 377–382, 390

f
family 336
fat tail 222, 227, 228, 246
feedback 35, 38, 47, 50, 53, 54, 57, 59–61
– global 440, 444
financial index 250, 251
firm 99, 100, 109, 111, 113, 114, 117, 120, 122–124, 126, 127
– bankruptcy 101, 121–127
fitness 313
fixed points 372, 384, 387, 388
flag 393, 398, 405, 415
flight 324
flow optimization 463
Fokker–Planck equation 262, 263
foreign currency exchange 250, 251
Fortran 321, 324

fragmentation 313, 321, 324
frustrated lattice gas 508, 510, 525
fundamental diagram 451

g
gambling 163, 164, 166, 167, 169, 175, 186
game 66, 67, 77–79, 81–83, 87, 89, 95, 96
– minority 35, 62, 66, 67, 98
– theory 61
gap closing 55
genetic algorithm 449, 451, 452, 455, 463, 466–468, 471
geography 329, 332
Gibrat
– index 137
– law 100, 101, 103, 105, 107–109, 115, 127, 137
global terrorism 393, 409, 415
globalization 312
grammar 311
gross
– opening 426
– total 426, 427
groups 279, 283, 286, 296, 297, 308
growth 345, 346, 348

h
Hamiltonian 490, 491, 497, 507, 509, 510, 518, 527, 529, 530, 532
Hamming distance 324, 357
hierarchical structure 495, 497, 499, 501
hierarchy 291, 293, 294, 297, 298, 301
history 331, 332
Hubbard-Stratonovich transformation 511

i
Iceland 312
income 131
independent interval approximation 59, 60
individual 279–281, 283, 284, 289, 290, 292, 294
inequality 183, 184
– measure of 183, 184, 188
inflation 38, 50
initial support 369, 370, 373–381, 383, 384, 389
innovation
– constructive and destructive consequences 562
– core-transforming 583, 589
– dormant 578, 580, 582, 589
– incremental 575, 576, 578–580, 582, 589
– recognize 561, 564, 572, 579, 581
– six categories 587
– structural classification 561, 578
integration 297, 299, 305

intelligence
– plant   507, 508, 525–527, 529
– zero   35, 39, 50
intermarriage   305
irreducible subgraph   575, 577, 579–581, 588
Island ECN   41, 60

k
Keynesian multiplier   199, 201
kinetic theory   165, 180
Kolmogorov–Smirnov test   105, 107
Konya   312
Kramers–Moyal coefficient   263, 264

l
Lagrange
– function   10, 12, 13, 17
– principle   17, 279, 282, 283, 297, 304
lane formation   457, 463
language   311
– English   320
– French   320
– German   320
– Indo-European   312, 336
– Latin   311, 330
– Portuguese   311
– Romanian   311
lattice   321, 331, 343–347, 353, 354, 357, 359, 362, 441, 442, 444, 496–498, 532
learning   440–443
limit orders   45
local doubts   368, 388
log-normal   417

m
Mandelbrot
– invariance under aggregation   139
– weighted mixtures   139
market
– condition   35, 49, 50, 53, 54, 60–62
– data   35, 39, 41
– foreign exchange   219, 220, 222, 225, 236, 246, 247, 249
– limit-order   35, 37, 39–41, 51, 60
– stock   27, 103, 117, 118, 249
Markov chain   75, 79, 80, 92, 93, 96, 97
Mathew effect   422
mean field   38, 51, 53, 54, 59, 60, 171, 180, 318, 348, 441, 445
mechanism   38, 39, 47
memory   507, 508, 526, 527, 529, 532
– associative   508, 509, 512, 525, 531
minority spreading   367, 383, 389
mobility measurement   215
model   249–251, 253, 258–260, 262, 265, 267–276

– cellular automata   451
– collision   145
– Cont–Bouchaud   270–273, 275, 276
– diffusion   51
– family network   131, 143
– financial market   96, 97
– fluid dynamic   451, 472
– generalized Lotka–Volterra   131, 141
– Hopfield   507–509, 511, 517, 519, 520, 526, 527, 532
– Lévy-stable non Gaussian   228
– life-cycle   192, 198, 199
– Lux–Marchesi   274
– Maslov   268
– mass   179, 180
– minimal   38, 41, 51, 61
– minimum exchange   169
– particle   40–42, 45, 51, 60
– queueing   451
– Sato-Takayasu   220, 222, 236, 246
– social force   450, 466–468, 471, 601
– spin   220, 275
– Sznajd   220, 221
– Viviane   321
– wealth exchange   143, 163, 164
– zero intelligence   35, 39, 50, 61
modularity   484, 486, 490
monetary circulation   192, 194, 216
money
– creation   192, 193, 202–205, 208, 216
– holding time   192, 196–199, 208, 210
– transfer model   191–194, 201–204, 211, 216
– velocity of   192–199, 201, 205, 216
Monte Carlo   50, 59, 68, 74, 98, 301, 303, 312, 318, 461, 497, 535
movies   417–419, 425–427, 436, 444
multiplicative stochastic process   422, 439, 445
mutation   313, 321

n
network
– business   122
– citation   479, 480
– email   477, 480
– internet   287, 289, 474, 482, 500
– railway   480, 481
– – German   284
– – Indian   481, 502, 505
– scale-free   343, 354, 476, 478–481, 493, 494, 497, 499, 501
– shareholding   117
– small world   473–480, 495–498
– supply   535
– – stability of   535, 536, 603
New Orleans   316, 332

non-equilibrium 329
non-Gaussian 135

o

open
– path 401, 402, 404
– space 400
– – dangerous 405–411
opinion 339–344, 349–361, 363, 364, 367–370, 372, 375, 378, 379, 381, 383, 404, 409, 598
– dynamics 339, 349, 350, 355, 358–360, 367, 383, 389, 390
– formation 279, 301
order 280, 283, 286, 291, 294, 304
– as particle 35, 40–42, 45, 49, 51, 52, 54, 59, 60
– book 35–37, 39, 48
– buy 37
– clustering 42
– density profile 53, 57, 59
– diffusion 41, 47, 48
– distribution 42, 44, 45, 53
– lifetime 42, 43
– placement 35–37, 40, 52
– removal 35, 36, 40, 42, 60
– sell 37
– size distribution 42
overdiffusive behavior 42, 54, 56

p

Papua 312
Pareto law 99, 100, 103, 105, 107, 109–112, 127, 129, 132, 161, 162, 172, 177, 183, 187
– strong 132
– weak 132
partnership 297, 298
passive supporters 393, 395, 397, 398, 400, 401, 404, 409, 410, 412–415
pedestrian 449–454, 457–461, 463, 465–470, 541, 543
percolation 270, 272, 273, 393–396, 404, 408–410, 412–415
periphery 571, 575–582, 587
phase transition 38, 61, 313, 321, 324, 326, 442, 473, 478, 493, 497, 502, 512, 529
popular 417
popularity 417
positive externality 445
power law 42, 43, 50, 51, 57, 60, 61, 131
predictability 65, 66, 77, 78, 83, 84
price 36–38, 40, 41
– ask 227, 228, 235
– axis 35–37, 41, 42, 54, 60
– bid 222, 223, 227, 237–239, 244
probability 1, 11–13, 18, 27
production 1, 5–9, 13, 16, 20, 23, 27

– function 1, 9, 10, 13, 16
– system 535, 537–541, 545, 550, 556, 557
– – traffic equations 540, 550, 559
pruning neurons 508
public debate 369–371, 384, 388–390

q

quenched
– randomness 169, 173, 178, 180, 182
queue 451, 461

r

random walk 37, 51, 54–56, 60, 440
reform proposal 369, 371, 373, 384, 390
relevant variable 60, 61
replica method 515, 528
reserve ratio 193, 202, 204–207, 210
return 255–257, 259–263, 265, 270–272, 275, 276

s

sandpile 268, 273
saving 166, 167, 169, 170, 175, 176, 178, 186
– distributed 169–171, 173, 177, 178, 186
– fixed 166, 167, 176, 181
scale free 354
scheduling 539, 545, 554
searching 482, 498–500, 502
segregation 279, 291, 297, 300, 304, 306, 307
self-organization 452, 454, 457, 463, 540, 555, 557
semiconductor manufacturing 546, 604
September 11, 2001 393, 394, 410, 416
size
– city 421, 422
– company 422
– firm 101, 103–105, 107, 108, 112, 114, 121, 124, 127, 422
slower is faster effect 535, 537, 539, 541, 546, 547, 557, 604
social
– distance 494, 495, 500
– group 431
– influence 339, 343, 344, 347, 352, 361
– networks 340, 341, 343, 352, 360, 361
society 280, 283, 284, 287, 291, 296–298, 305
sociophysics 360–362
soft control 65
Spain 317
spin glass 507, 508, 516, 519–522, 527–531
spread 36, 37, 55, 60
stability 35, 60, 139
standard of living 1, 9, 16, 21, 26, 29
state
– disordered 441, 442
– ordered 441–444

– steady 53, 61
stochastic process 66, 68, 78, 83, 88
stock market 250, 254, 257, 260, 266, 268
stop-and-go waves 455
synchronization 456, 540, 555, 556

t

target 394, 397, 401–403, 405, 406, 410
terrorism 393–398, 400, 401, 404–406, 408–416
threshold 344–348, 350–355, 358, 359, 361, 362, 368–370, 390
time
– persistence 427, 437
– residence 441, 442
time series 39, 51, 60
– analysis 67, 68, 73, 76, 83
time-varying rate 60
traffic 535, 537, 540–543, 550, 556, 557
transaction 35–37, 40
– inter-market 237
– intra-market 237, 238
transfer 323
transition matrix 163, 173–180, 187
tree 331
triangular arbitrage 225–228, 230, 231, 236–238, 246
truncated Lévy
– flight 228
– process 234
turbulence 455

u

universality class 61
utility 8

v

variance 162, 184–186, 188
Verhulst 317, 324
video tracking 449
virtual impact function 45
volatility
– clustering 50, 60
voting 419, 425, 431

w

wealth 131, 161, 163–165, 167, 168, 175, 181, 183, 186
– distribution 161, 163–165, 169, 174, 176, 180, 181, 187, 188, 191, 206, 216
– exchange 163, 164, 166, 168, 173, 175, 177, 180, 184, 187
webpage 428
website 426, 428, 433, 437

y

Yule process 439

z

Zipf
– analysis 250, 254, 255
– law 99, 103, 107, 109, 110, 124, 422, 424, 428, 437, 439, 445

Author Index

a

Abbt, M.   446
Abello, J.   505
Abrams, D. M.   311, 315, 317, 324, 337, 472
Abul-Magd, A. Y.   159
Adamic, L.   447
Adamic, L. A.   428, 446, 499, 501, 504, 506
Adar, E.   501, 506
Adler, J.   533
Adler, J. L.   451, 472
Adler, M.   447
Ageon, Y.   447
Aguilar, J.-P.   277
Aharony, A.   278, 365, 416, 506, 533
Ahmed, E.   416
Ahn, Y. Y.   447
Aiba, Y.   219, 246, 247
Albert, R.   354, 365, 447, 478, 504, 559
Alberts, B.   559
Alcaraz, F. C.   63
Alesio, E.   277
Almaas, E.   559
Almeida, M. P.   447
Amaral, L. A. N.   33, 128, 129, 159, 247, 277, 278, 447, 505
Amblard, F.   364, 365, 391
Amengual, P.   391
Amit, D. J.   533
Anderson, P. W.   249, 276, 364
Anderson, S.   365
Andrade, J. S.   447
Anghel, M.   97
Angle, J.   145, 159, 168, 190
Anteneodo, C.   141, 158, 278
Aoki, M.   128, 159
Aoyama, H.   99, 128, 129, 158, 159
Arenas, A.   447, 486, 505, 506
Argyrakis, P.   337
Armbruster, D.   558, 559
Arrow, K.   249, 276
Aruka, Y.   1, 33

b

Aste, T.   159
Auerbach, F.   446
Ausloos, M.   159, 249, 271, 276–278, 337
Axelrod, R.   279, 309, 347, 357, 359, 364, 365, 447, 506, 591
Axtell, R. L.   128, 422, 446

Bachelier, L.   249, 262, 276
Badger, W. W.   128
Baganha, M.   558
Bagley, R. J.   590
Bagrow, J. P.   489, 505
Baier, G.   559
Baillie, R. T.   278
Bak, P.   62, 268, 278, 590
Ball, P.   533
Banerjee, A.   159
Banerjee, A. V.   364
Banerjee, K.   505
Banzhaf, W.   506
Barabási, A.-L.   128, 354, 365, 447, 472, 478, 504, 505, 559
Barma, M.   63, 190
Barnett, W. A.   217
Baronchelli, A.   337
Barrat, A.   504
Barro, R. J.   293, 309
Baschnagel, J.   249, 276
Bassler, K. E.   97
Basu Hajra, K.   505, 506
Beaumariage, T.   559
Becker, R.   297, 309
Bek, C.   140, 158
Ben-Naim, E.   365, 497, 506
Bernardes, A. T.   366
Bhattacharya, K.   190
Bhattacharyya, P.   445, 447
Bienenstock, E.   365
Bieniawski, S.   98
Biham, O.   158, 190
Bikhchandani, S.   447

Biswas, T.   505
Black, D.   446
Blue, V. J.   451, 472
Blumen, A.   505
Boccara, N.   97
Boguñá, M.   494, 506
Bolay, K.   450, 451, 464, 472
Bollerslev, T.   63, 247, 277, 278
Bollt, E. M.   489, 505
Boltes, M.   472
Bomberg-Bauer, E.   559
Bonabeau, E.   496, 497, 506, 559
Bonanno, G.   257, 277
Borland, L.   278
Bornholdt, S.   490, 504–506, 559, 590
Bose, I.   533
Bose, J. C.   507, 508, 525, 533
Bouchaud, J.-P.   62, 63, 128, 246, 247, 249, 265, 266, 273, 276–278
Boveroux, P.   277, 278
Boveroux, Ph.   277
Bovy, P. H. L.   472
Bowles, S.   365
Boyd, R.   365
Braun, D.   191, 217
Brechet, Y.   472
Brenner, T.   558
Breslau, L.   446
Brian Arthur, W.   66, 97, 447
Bridel, P.   217
Brighton, H.   337
Brisbois, F.   277
Briscoe, E. J.   311, 337
Broder, A.   447
Bronlet, Ph.   276, 277
Brüggemann, L. I.   525, 533
Brunner, K.   217
Buldyrev, S.   276
Buldyrev, S. V.   128, 129, 446
Bunde, A.   277
Burke, M. A.   350, 365
Buss, L.   590
Buzna, L.   472, 559

c

Cabral, L. M. B.   446
Caglioti, E.   337
Caldarelli, G.   505, 506
Calenbur, V.   558
Camerer, C.   365
Campos, P. R. A.   337
Cangelosi, A.   337
Canning, D.   33
Cao, P.   446
Capocci, A.   488, 505
Carbone, A.   277

Carlson, J. M.   591
Casella A.   591
Castellano, C.   359, 365, 505
Castelli, G.   277
Casti, J. L.   97
Catanzaro, M.   494, 506
Cauley, J.   416
Cecconi, F.   505
Chae, S.   247
Chakrabarti, B. K.   33, 128, 129, 145, 158, 159, 166, 169, 189–191, 212, 216, 247, 447, 505, 525, 526, 533
Chakraborti, A.   145, 159, 189, 190, 212, 216, 247
Challet, D.   62, 63, 97, 98
Chamley, C.   365
Champernowne, D. G.   128, 131, 138, 139, 158
Chan, D. L. C.   62
Chan, H. Y.   97
Chandra, A. K.   495, 505, 506
Chase, C. J.   559
Chatterjee, A.   33, 128, 129, 145, 158, 159, 169, 189, 190, 216, 247, 447, 505
Chatterjee, S.   350, 365
Chen, F.   558
Chessa, A.   189
Cheung, P.   97
Choe, S. C.   97
Choi, M. Y.   504
Chopard, B.   391
Chowdhury, D.   246, 275, 278, 457, 472, 558
Chung, K. H.   447
Cizeau, P.   247, 277
Clementi, F.   140, 158
Coelho, R.   131, 143, 159
Cohen, E. G. D.   158
Cohen, J. E.   350, 365
Cohen, K.   267, 278
Cohen, K. J.   63
Cohen, M.   558
Cohen, M. D.   591
Colaiori, F.   505
Cole, J. R.   446
Cole, S.   446
Coleman, J. S.   339, 364
Coniglio, A.   510, 533
Conlisk, J.   217
Cont, R.   246, 273, 277, 278
Cosenza, M. G.   506
Costello, X.   337
Cox, R. A. K.   447
Cranshaw, T.   159
Csànyi, G.   505
Culicover, P.   337

d

D'Hulst, R.   272, 276, 278
Daamen, W.   450, 459, 472
Dacorogna, M. M.   247
Daganzo, C.   558
Daniels, M. G.   62
Danon, L.   485, 505
Das, A.   190
Das, P. K.   505, 506
Dasgupta, S.   481, 505
David, B.   278
Davidsen, J.   506
Davies, J. A.   447
de Arcangelis, L.   366, 391
de Arcangelis, L. J. N.   309
de Beer, C.   559
de Fontoura Costa, L.   533
de la Rosa, A.   464, 472
de Menezes, M. A.   505
de Moura, A. P. S.   506
de Oliveira, P. M. C.   189, 337
de Oliveira, V. M.   321, 337
de Palma, A.   365
De Vany, A.   427, 446, 447
Deb, K.   464, 471, 472
Deering, B.   277
Deffuant, G.   354, 355, 364–366, 391
Deissenberg, C.   559
Delli Gatti, D.   128
Deneubourg, J.-L.   472, 506, 559
Derenyi, I.   506
Derrida, B.   63, 358, 365
Deschatres, F.   447
Di Guilmi, C.   128
Di Matteo, T.   143, 159
Díaz-Guilera, A.   447, 505, 506
Diaz-Rivera, I.   559
Ding, N.   190, 191, 217
Ding, Z.   246
Dirickx, M.   159, 337
Dittrich, P.   506
Dixon, C. J.   559
Dodds, P. S.   500, 506
Doise, W.   354, 365
Dokholyan, N. V.   276
Domany, E.   63
Domowitz, I.   63
Donetti, L.   488, 505
Dorigo, M.   559
Dorogovtsev, S. N.   504, 505, 559
Doyle, J.   591
Drăgulescu, A. A.   1, 33, 128, 158, 159, 189, 216
Drezner, D. W.   428, 447
Drezner, Z.   558
Droz, M.   63, 391

Drozd, P.   337
Drozdz, S.   278
Dubes, R. C.   505
Duch, J.   486, 505
Dufour, I.   416
Dupuis, H   278
Dussutour, A.   457, 472, 559
Dutta, O.   525, 533
Dyson, F.   590

e

Ebel, H.   496, 504, 506
Eckhardt, B.   472
Eckmann, J. P.   258, 277
Eguíluz, V. M.   272, 278, 337, 366, 506
Eigen, M.   590
Elgazzar, A. S.   416
Eliezer, D.   62, 267, 278
Elowitz, M. B.   559
Engle, R. F.   246, 247, 277
Erdős, P.   474, 504
Escobar, R.   464, 472
et Amblard, F.   365
Everitt, B. S.   505

f

Fabretti, A.   277
Falkowski, P. G.   590
Fama, E. F.   246, 249, 276
Fan, L.   446
Farkas, I.   456, 472, 506, 559
Farmer, J. D.   62, 63, 590
Farrell, H.   428, 447
Farrell, W.   366
Fasold, D.   559
Faure, T.   365
Faust, K.   504
Fehr, E.   365
Feigenbaum, J. A.   265, 266, 278
Felici, M.   337
Ferrero, J. C.   141, 158, 190
Fields, G. S.   211, 212, 214, 217
Filho, R. N. C.   447
Fischer, R.   191, 217
Fisher, D.   217
Fisher, I.   217
Flake, G. W.   447
Fogelman-Soulié, F.   365
Foley, D. K.   1, 16, 33
Föllmer, H.   365
Fontana, W.   590
Forrester, J. W.   558
Fortunato, S.   505
Fourcassié, V.   472, 559
Fowler, R.   5, 33
Francart, L.   416

Franklin, M. 391
Frappietro, V. 277
Freeman, C. 504
Freitag, M. 559
Freund, P. G. O. 265, 266, 278
Friedman, M. 195, 217, 391
Friedman, R. D. 391
Fujiwara, Y. 99, 128, 129, 158, 159

g

Gabaix, X. 128, 278, 446
Galam, S. 279, 309, 349, 354, 365, 367, 391, 393, 416
Galbi, D. A. 447
Galea, E. R. 472
Galla, T. 98
Gallegati, M. 101, 128, 140, 158
Gallos, L. 497, 506
Garfinkel, M. R. 217
Garlaschelli, D. 128
Gaskell, D. R. 217
Gastner, M. T. 504
Gavrilova, M. L. 128
Gefen, Y. 391, 416
Genoud, T. 533
Georgescu-Roegen, N. 16, 33
Gibrat, R. 131, 137, 148, 158, 422, 446
Gilbert, T. 447
Giles, C. L. 447
Giles, D. E. 447
Gillemot, L. 62, 63
Gintis, H. 365
Girvan, M. 483, 505, 506
Gitterman, M. 504
Giuliani, A. 277
Giulioni, G. 128
Glance, N. 447
Glassman, S. 446
Gleiser, P. M. 505
Glover, E. J. 447
Gode, D. 62
Goebl, H. 332, 337
Goel, T. 464, 471, 472
Goh, K.-I. 504
Goldberger, A. L. 276
Goldenfeld, N. 66, 97
Goldstein, M. L. 446
Gomes, M. A. F. 337
González, M. C. 190, 391
González-Avella, J. C. 497, 506
Gopikrishnan, P. 33, 129, 154, 159, 247, 260, 261, 271, 277, 278
Gordon, M. B. 363, 366
Gourley, S. 97
Grabher, G. 364
Granger, C. W. J. 246

Granovetter, M. 391
Grebe, R. 277
Grebogi, C. 506
Green, A. K. 559
Grimes, B. F. 329, 337
Grönlund, A. 496, 506
Gross, N. C. 339, 364
Grynberg, M. D. 63
Guardiola, X. 447
Gubernatis, J. E. 365
Guggenheim, E. A. 33
Guimerà, R. 447, 490, 505
Gutfreund, H. 533, 591
Guttal, V. 472

h

Haag, G. 559
Hackett, A. P. 434, 447
Hagel, T. 559
Hagstrom, W. O. 446
Hajnal, J. 365
Halley, J. M. 337
Hamada, K. 246
Hamlen, W. 447
Hansen, M. 309
Hart, M. 98
Hart, M. L. 98
Hart, P. E. 128
Hart, S. C. 97
Hartl, R. F. 559
Hatano, N. 219, 246, 247
Havlin, S. 128, 129, 276, 446
Hayes, B. 216
Hazari, B. R. 447
Hegazi, A. S. 416
Hegselmann, R. 350, 365
Helbing, D. 449–451, 455–459, 461, 463, 464, 467, 469, 470, 472, 535, 558, 559
Henderson, V. 446
Henkel, M. 63
Henrich, J. 350, 365
Herreiner, D. 365
Herrera, J. L. 506
Herrmann, H. J. 190, 391
Hertz, J. 533
Hess, B. 559
Higgs, P. G. 358, 365
Hindman, M. 447
Hirabayashi, T. 246
Hirai, K. 559
Hirshleifer, D. 447
Höfer, T. 559
Holme, P. 496, 504, 506
Holyst, J. A. 347, 365, 558, 559
Hong, H. 504
Hoogendoorn, S. P. 452, 459, 472

Hopp, W. J.   558
Howison, S. D.   97
Hu, B.   190
Huang, C. R.   337
Huang, Z.-X.   506
Huberman, B. A.   391, 428, 446, 489, 504–506
Hughes, B. D.   189
Hughes, R. L.   451, 472
Hui, P. M.   97, 98
Humphrey, T. M.   195, 217
Hutzler, S.   131, 146, 147, 159, 190
Hyde, T.   159
Hyman H.   346, 365

i

Ijiri, Y.   128
Inoue, J.   507, 526, 533
Iori, G.   62
Ispolatov, S.   189, 191, 216
Israeli, N.   66, 97
Ivanov, P. Ch.   189
Ivanova, K.   271, 276–278

j

Jagalski, T.   559
Jain, A. K.   505
Jain, S.   561, 590
Jan, N.   366, 391
Jarrett, T.   97
Jarvis, S.   97, 217
Jefferies, P.   97, 98
Jenkins, S. P.   217
Jeon, G. S.   504
Jeong, H.   447, 504–506
Jeong, H. C.   447
Jespersen, S.   505
Jiang, R.   559
Jin, E. M.   496, 506
Johansen, A.   265, 266, 272, 278, 447
Johansson, A.   449, 472, 559
Johnson, A.   559
Johnson, J. A.   447
Johnson, M. D.   278
Johnson, N. F.   65, 97, 98
Jonson, R.   97
Jordan, P.   63
Jung, W.-S.   247

k

Kacperski, K.   347, 365
Kadanoff, L. P.   190
Kahn, J.   558
Kahn, R. F.   217
Kahneman, D.   360, 366
Kahng, B.   504
Kaizoji, T.   128, 159

Kamphorst, S. O.   258, 277
Kaneko, K.   558
Kaplan, W.   33
Kapur, J. N.   190
Kar Gupta, A.   145, 159, 161, 190, 216
Karmakar, R.   533
Kaski, K.   145, 159, 190, 216, 247
Katori, M.   128
Katz, E.   339, 364
Katz, J. S.   504
Katzorke, I.   559
Kauffman, S. A.   590
Ke, J.   337
Keltsch, J.   472
Kempf, K.   559
Kerm, P. V.   217
Kertesz, J.   247, 366
Kesting, A.   559
Keynes, J. M.   197, 199, 217
Kichiji, N.   128
Kim, B. J.   447, 504
Kim, D.   504
Kirby, S.   337
Kirman, A.   33, 128, 129, 270, 278, 365
Kiss-Haypál, G.   276, 446
Kleinberg, J. M.   498, 505
Klemm, K.   359, 366, 497, 506
Klingsch, W.   472
Klüpfel, H.   472
Kogan, I. I.   62, 267, 278
Komarova, N. L.   315, 337
Konno, N.   504
Koponen, I.   249, 276
Kori, H.   559
Kosmidis, K.   320, 337
Kovács, B.   559
Koya, S.   559
Koyama, S.   533
Kranke, F.   559
Krapivsky, P. L.   189, 216, 365
Krause, U.   350, 365
Krishna, S.   561, 590
Krishnamurthy, S.   63, 190
Kroniss, G.   97
Kroo, I. M.   98
Kropp, J.   277
Krough, A.   533
Kulakowski, K.   391
Kullmann, L.   247
Kumar, R.   447, 506
Kumar, V.   128
Kummer, U.   559
Kurihara, S.   247
Kuznets, S. S.   217
Kwong, Y. R.   97

l

Laherrère, J. 446
Lai, Y.-C. 506
Laidler, D. 217
Laloux, L. 255, 256, 266, 277
Lambertz, M. 277
Lämmer, S. 535, 558, 559
Lamper, D. 97
Langhorst, P. 277
Langton, C. G. 590
Latané, B. 365
Latora, V. 505
Lavicka, H. 391
Lawrence, S. 447
Lazega, E. 360, 366
Le Hir, P. 391
Lebacque, P. 559
L'ecuyer, P. 128
Lee, H. 558
Lee, Y. 189, 447
Leea Y. 33
Lefeber, E. 558, 559
Legrand, G. 278
Leibler, S. 559
Lenders, W. 337
Leombruni, R. 190
Leppanen, T. 337
Leschhorn, H. 128, 129
Lévy, P. 249, 276
Levy, H. 33
Levy, M. 1, 33, 128, 158, 159, 189, 190
Levy, S. 159
Lewis, J. 559
Li, W. 446
Liben-Nowell, D. 501, 506
Liljeros, F. 506
Lima, F. W. S. 337
Limpert, E. 446
Lind, P. G. 190
Liu, Y.-H. 247
Liua, Y. 33
Lo, T. S. 97
Loreto, V. 337, 505
Lövås, G. G. 451, 472
Lukose, R. M. 506
Lundqvist, S. 62
Lux, T. 62, 147, 148, 159, 190, 274, 277, 278

m

Maass, P. 128, 129
MacDonald, G. 447
MacKay, C. 417, 446
Maghoul, F. 447
Maier, S. 278
Majumdar, S. N. 63, 190
Malamud, B. D. 277
Malcai, O. 143, 158, 190
Mallamace, F. 159
Mandelbrot, B. B. 128, 139, 145, 158, 189, 246, 247, 249, 259, 276
Manna, S. S. 159, 169, 190, 216, 505
Manrubia, S. C. 128
Mantegna, R. N. 62, 129, 189, 216, 247, 249, 256, 260, 276–278, 446
March, N. H. 62
Marchesi, M. 62, 274, 278
Marchiori, M. 505
Marhl, M. 559
Marjit, S. 189
Markandya, A. 217
Marsh, M. 391
Marsili, M. 63, 97, 98, 365
Marthaler, D. 558
Martin, X.S. 293, 309
Marumo, K. 246, 247
Maslov, S. 51, 62, 268, 278, 559
Masselot, A. 391
Masuda, N. 504
Mata, J. 446
Mauger, A. 416
Mavirides, M. 247
McElreath R. 365
McLaren, L. 391
McRobie, A. 472
Meinhold, L. 559
Meltzer, A. H. 217
Mendes, J. F. F. 129, 504, 505, 559
Menzel, H. 339, 364
Merton R. K. 446
Messinger, S. M. 533
Metraux, J. -P. 533
Meyer, M. 159, 247, 277, 278
Meyer-König, T. 451, 472
Meyer-Ortmanns, H. 324, 337, 354, 365, 391
Mézard, M. 63, 128, 533
Michael, F. 278
Mielsch, L. 504
Mikhailov, A. 558, 559
Mikhailov, A. S. 558, 559
Mikkelsen, H. O. 278
Milgram, S. 474, 475, 498, 500, 501, 504
Mills, M. 62
Mimkes, J. 1, 16, 33, 141, 146, 158, 159, 279, 306, 309
Minett, J. W. 337
Minguet, A. 277, 278
Minnhagen, P. 504
Mira, J. 337
Mitzenmacher, M. 189, 447
Miura, H. 246
Miwa, H. 504
Miyajima, H. 447

Miyazima, S.   435, 447
Mizuno, T.   128, 247
Mobilia, M.   391
Mobius, M. M.   504
Mohanty, P. K.   190
Molnár, P.   450, 472
Montroll, E. W.   128, 445, 447
Moon, H.-T.   247
Moosa, I.   247
Moreira, J. E.   447
Morris, S.   446
Morris, S. A.   446
Moscovici S.   365
Moscovici, S.   354, 365, 391
Moss de Oliveira, S.   189, 337, 365
Mott, K. A.   533
Motter, A. E.   495, 506
Moukarzel, C. F.   505
Muhamad, R.   506
Mühlenbein, H.   391
Mukamel, D.   63
Mukherjee, G.   190, 505
Muller, U. A.   247
Munoz, M. A.   488, 505

n
Nadal, J.-P.   363–366
Nagahara, Y.   128, 159
Nagamine, T.   447
Nagatani, T.   558, 559
Nagler, R. J.   247
Nagpaul, P. S.   504
Namatame, A.   129
Neau, D.   391
Neda, Z.   159, 456, 472, 505
Nelson, D. B.   278
Nettle, D.   312, 337
Neugebauer, R.   558
Newman, C. M.   365
Newman, M. E. J.   128, 189, 434, 435, 446, 483, 486, 493, 504–506
Ni Dhuinn, E.   159
Nicodemi, M.   508, 510, 533
Nielsen, J.   446
Nirei, M.   129, 147, 159, 190
Nishikawa, T.   506
Nishimori, H.   533
Nishinari, K.   472
Nitsch, V.   446
Niyogi, P.   337
Noh, J. D.   447, 506
Novak, J.   506
Novak, M. A.   246
Novotny V.   337
Nowak, A.   337, 365
Nowak, M. A.   313, 326, 337

o
Oh, E.   504
Ohira, T.   247
Ohlin, B.   217
Ok, E. A.   211, 212, 214, 217
Okazaki, M. P.   128, 159, 246
Okuyama, K.   129
Olsen, L. F.   559
Olsen, R. B.   247
Olson, G. M.   505
Olson, J. S.   505
Oltvai, Z. N.   559
Onnela, J.-P.   247
Orléan, A.   364
Osorio, R.   278
Ott, E.   472
Ottino, G.   97
Oulton, N.   128

p
Paczuski, M.   62
Padgett, J.   591
Padmanabhan, P.   558
Palestrini, A.   128
Palla, G.   491, 506
Palmer, R. G.   533
Pan, R. K.   417, 446, 447
Paradalos, P. M.   505
Paredes, A.   337
Pareto, V.   99, 131–133, 135–137, 140, 158, 216, 249, 276, 446
Parisi, D.   337, 505
Parisi, G.   533
Park, J.   493, 505
Park, S. M.   447
Parlitz, U.   558, 559
Pastor-Satorras, R.   506
Patriarca, M.   145, 159, 190, 216, 337
Pattison, P.   506
Paul, W.   249, 276
Peak, D. A.   525, 533
Peng, C. K.   276
Peng, C.-K.   247
Penna, T. J. P.   271, 320, 337
Pennock, D. M.   447
Peters, K.   457, 458, 472, 535, 558, 559
Pfeuty, P.   364
Phan, D.   366
Phillips, G.   446
Picozzi, S.   159
Pictet, O. V.   247
Pietronero, L.   506
Pikovsky, A.   559
Pinasco, J. P.   337
Pines, D.   249, 276
Płatkowski, T.   558

Plerou, V. 129, 159, 247, 255, 261, 277, 278
Podobnik, B. 189
Pollner, P. 506
Potters, M. 62, 63, 247, 249, 276, 277
Pottosin, I. I. 533
Prais, S. J. 217
Price, D. J. S. 423, 436, 439, 446
Puniyani, A. R. 506

q
Qiu, H. 217

r
Radicchi, F. 505
Radomski, J. P. 391
Radons, G. 558
Raff, M. 559
Raffaell, G. 63
Raghavan, P. 447, 506
Raghavendra, S. 446, 447
Rajagopalan, S. 447
Ramadge, P. 559
Ramasco, J. J. 159
Ramsden, J. J. 276, 446
Rasmussen, S. 590
Rauch, J. E. 591
Ravasz, E. 472, 505
Redner, S. 189, 216, 352, 365, 391, 423, 426, 439, 446, 447, 497, 505, 506
Reed, W. J. 159, 189, 446
Reichhardt, J. 490, 505
Reis, F. 63
Rényi, A. 474, 504
Repetowicz, P. 131, 146, 147, 159, 190
Resende, M. G. C. 505
Richiardi, M. 190
Richmond, P. 33, 129, 131, 141, 146, 147, 158, 159, 190
Ringhofer, C. 558, 559
Riolo, R. L. 591
Rittenberg, V. 63
Roberts, K. 559
Robins, G. 506
Rodgers, G. J. 272, 278
Rogers, E. M. 339, 364, 590
Rohlf, T. 559
Rokhsar, D. S. 364
Roling, N. 364
Romanelli, L. 337
Rosen, S. 447
Rosenblat, T. S. 504
Rosenow, B. 247, 277
Rosenthal, M. 434, 447
Rossler, O. E. 591
Roth, C. 506
Rothbard, M. N. 195, 217

Ruelle, D. 258, 277
Ruf, F. 278
Ryan B. 339, 364
Ryan, J. K. 558

s
Sa Martins, J. S. 189, 337
Sahimi, M 278
Sales-Pardo, M. 505
Salinger, M. A. 128, 129, 446
San Miguel, M. 337, 366, 391, 506
Sandler, T. 416
Santen, L. 558
Santos, M. A. 159
Sato, A.-H. 220, 222, 246
Sazuka, N. 247
Scafetta, N. 159
Scalas, E. 278
Schadschneider, A. 472, 558
Schelling, T. C. 391
Schelling, Th. 279, 297, 309
Schellnhuber, H.-J. 277
Schimansky-Geier, L. 559
Schnegg, M. 505
Schönhof, M. 558, 559
Schönknecht, G. 533
Schreckenberg, M. 472
Schubert, A. 505
Schulze, C. 311, 337
Schumpeter, J. A. 590
Schuster, H. G. 559, 590
Schuster, S. 559
Schütz, G. M. 63
Schwammle, V. 320, 337
Schwartz, A. J. 217
Schwartz, R. 278
Schweitzer, F. 365, 391, 472
Šeba, P. 558
Seidel, T. 535, 558
Sen, P. 473, 480, 481, 505, 506
Seneta, E. 350, 365, 591
Serletis, A. 217
Serrano, J. 559
Servedio, V. D. P. 505
Seyfried, A. 452, 472
Shafee, F. 497, 506
Shapir, Y. 391, 416
Sharma, S. D. 472
Shenker, S. 446
Shimizu, T. 246, 247
Shirky, C. 428, 447
Shlesinger, M. F. 128, 445, 447
Shorrocks, A. F. 217
Shubik, M. 62
Shukla, P. K. 559
Silber, J. 217

Silva, A. C.   33, 129, 155, 159
Silver, J.   148, 159, 190
Simchi-Levi, D.   558
Simon, H. A.   128, 439, 447
Sinha, S.   159, 190, 216, 417, 446, 447
Slanina   148
Slanina, F.   51, 62, 159, 216, 268, 278, 391
Slud, E.   148, 159, 190
Smith K.   337
Smith, D. M. D.   65
Smith, E.   62, 63
Sneppen, K.   559, 590
Sobol, I. M.   98
Solomon, S   1
Solomon, S.   33, 128, 129, 141, 158, 189, 190, 279, 309, 337, 362–364, 366, 391
Sommers, P. S.   217
Sompolinsky, H.   533
Son, S.-W.   490, 506
Soo, K. T.   446
Sornette, D.   129, 265, 266, 272, 273, 278, 446, 447
Soulier, S.   506
Souma, W.   99, 128, 129, 158, 159, 190
Sousa, A. O.   365, 391
Spearman, M. L.   558
Speth, J.   278
Sreeram, P. A.   505
Stahel, W.   446
Stanley, H. E.   1, 33, 62, 128, 129, 154, 159, 189, 216, 247, 249, 260, 276–278, 446
Stanley, M. H. R.   128, 129, 446
Stark, D.   364
Stark, H.-U.   558
Stata, R.   447
Stauffer, D.   161, 189, 246, 271, 273, 275, 278, 279, 309, 311, 337, 349, 352, 354, 364–366, 391, 416, 506, 533
Steels, L.   337
Steffen, B.   472
Stein, D. L.   364
Steindl, J.   129
Sterman, J. D.   558
Stinchcombe, R. B.   35, 62, 63, 159, 190, 216
Stone, M.   350, 365
Streib, D.   447
Strogatz, S. H.   97, 311, 315, 317, 324, 337, 365, 457, 472, 478, 504, 559
Sunder, S.   62
Sutherland, W. J.   337
Sutton, J.   129, 446
Szamrej, J.   365
Szendröi, B.   505
Sznajd, J.   365, 391
Sznajd-Weron, K.   220, 246, 349, 365, 391

t

Takamoto, K.   148, 159, 190
Takayasu, H.   128, 129, 159, 190, 220, 222, 246, 247, 278
Takayasu, M.   128, 129, 159, 246, 247
Tan, C. J. K.   128
Tang, C.   268, 278
Tang, L.-H.   63
Tarde, G.   339, 364
Tastan, H.   247
Taylor, C.   590
Taylor, L. R.   190
Taylor, T.   559
Tesileanu, T.   324, 337
Tessone, C. J.   391
Theraulaz, G.   506, 559
Thisse, J   365
Thornton, D. L.   217
Tian, G.-S.   63
Tomkins, A.   447, 506
Toral, R.   366, 391, 506
Toroczkai, Z.   97
Tosi, M. P.   62
Toth, B.   247
Toulouse, G.   364
Travers, J.   504
Treiber, M.   559
Trewavas, A.   533
Tsallis, C.   141, 156, 158, 278
Tsang, I. R.   337
Tsang, I. S.   337
Tschirhart, J. T.   416
Tsioutsiouliklis, K.   447
Tumer, K.   97
Turcotte, D. L.   277
Tversky, A.   360, 366

u

Ueda, H.   559
Ueda, T.   559
Ulubasoglu, M. A.   447
Ushio, T.   559

v

Vallacher, R. R.   365
van Kampen, N. G.   62
Vandenhouten, R.   277
Vandewalle, N.   276–278
Vannimenus, J.   366
Varian, H. R.   62
Vasconcelos, G. L.   337
Vazquez, A.   505
Vazquez, F.   365
Vespignani, A.   365
Vichniac, G. Y.   346, 365
Vicsek, T.   472, 505, 506, 559

Virasoro, M. A. 533
Viswanathan, G. M. 276
Voit, J. 247, 249, 276
von Mises, L. 217

w

Walls, W. D. 427, 446, 447
Walter, P. 559
Wang, B.-H. 190
Wang, J. 63
Wang, W. S. Y. 337
Wang, X. 505
Wang, Y. 190, 191, 217
Wasserman, S. 504
Watts, D. J. 129, 365, 418, 446, 478, 499, 504, 506, 559
Watts, J. M. 451, 472
Waxman, B. 505
Webber, C. L. 258, 277
Weidlich, W. 1, 33, 279, 309, 559
Weidmann, U. 452, 472
Weigt, M. 504
Weisbuch, G. 309, 339, 354, 364–366, 391
Weiss, G. H. 63
Welch, I. 447
Werner, T. 472, 559
Weron, R. 220, 246
West, B. J. 128, 159, 277
West, J. D. 533
Whang, S. 558
Wheeler, K. 97
Whitcomb, W. 278
Wichmann, S. 326, 337
Wicksell, K. 217
Wiendahl, H.-P. 558
Wiener, J. 447
Wiesenfeld, K. 268, 278
Wigner, E. 63
Wille, L. T. 277
Willis, G. 33, 141, 146, 158, 159
Willmann, R. D. 63
Wio, H. S. 391

Witt, U. 558
Wójcik, M. 278
Wolpert, D. H. 97
Wong, L. H. 494, 506
Worbs, J. 558
Wu, F. 391, 489, 505
Wu, Q.-S. 559

x

Xi, N. 190, 191, 217
Xie, Y.-B. 190
Xu, J. 190, 217

y

Yakovenko, V. M. 1, 33, 128, 129, 141, 150, 155, 158, 159, 189, 191, 216
Yang, J.-S. 247
Yarlagadda, S. 33, 128, 129, 158, 159, 189, 190, 216, 447
Yen, G. G. 446
Yeomans, J. M. 62
Yook, S.-H. 504
Young, H. P. 350, 365
Yule, G. U. 439, 447

z

Zachary, W. W. 505
Zanette, D. H. 128
Zbilut, J. P. 258, 277
Zecchina, R. 98
Zhang, L. 217
Zhang, Y.-C. 63, 97, 98
Zheng, D. 98
Zhou, H. 505
Zhou, T. 190
Zhu, H. 505, 506
Zhu, J.-Y. 505
Zimmermann, J. 391
Zimmermann, J.-B. 33, 129
Zimmermann, M. G. 272, 278
Zipf, G. K. 277, 420, 421, 446